MICROBIAL BIOSTIMULANTS

Biorational Pesticides for Management of Plant Pathogens

MICROBIAL BIOSTIMULANTS

Biorational Pesticides for Management of Plant Pathogens

Edited by

Faheem Ahmad, PhD

Rakesh Pandey, PhD

AAP APPLE ACADEMIC PRESS

First edition published 2025

Apple Academic Press Inc.
1265 Goldenrod Circle, NE,
Palm Bay, FL 32905 USA

760 Laurentian Drive, Unit 19,
Burlington, ON L7N 0A4, CANADA

CRC Press
2385 NW Executive Center Drive,
Suite 320, Boca Raton FL 33431

4 Park Square, Milton Park,
Abingdon, Oxon, OX14 4RN UK

Library and Archives Canada Cataloguing in Publication

CIP data on file with Canada Library and Archives

Library of Congress Cataloging-in-Publication Data

CIP data on file with US Library of Congress

ISBN: 978-1-77491-624-7 (hbk)
ISBN: 978-1-77491-625-4 (pbk)
ISBN: 978-1-00348-483-7 (ebk)

About the Editors

Faheem Ahmad, PhD
Assistant Professor and Principal Investigator of UGC-BSR Start-up Grant, Department of Botany, Faculty of Life Sciences, Aligarh Muslim University, India

Faheem Ahmad, PhD, is a Senior Assistant Professor in the Department of Botany at Aligarh Muslim University, Aligarh, India. He gained postdoctoral experience at Ehime University (Japan), North-West University (South Africa), Ege University (Turkey), and National Sun Yat-Sen University (Taiwan). His research is based on the study of plant-nematode interaction, plant disease, microbe-mediated biotic stress management, and agricultural biotechnology and microbiology. He is the author of more than 68 manuscripts, an associate editor of *the Journal of Plant Diseases and Protection* (Springer), and an Academic Editor of *Peer J-Life and Environment.* Dr. Ahmad guided one PhD student and several MSc students. He is a recipient of several fellowships and awards, including NRF, TSRO Young Scientist, SERB-DST Young Scientist, and TUBITAK foreign researcher awards. He is a fellow of the International Society for Development and Sustainability (ISDS, Japan) and the Bose Science Society (BBS, India).

Rakesh Pandey, PhD
Professor Emeritus and Emeritus Scientist, Crop Protection and Production Division, CSIR-Central Institute of Medicinal and Aromatic Plants, Lucknow, India

Rakesh Pandey, PhD, is a Professor Emeritus (Academy of Scientific & Innovative Research) and Emeritus Scientist, Crop Protection and Production Division, CSIR–Central Institute of Medicinal and Aromatic Plants, Lucknow, India. He has authored many research papers, book chapters, review articles, books, and bulletins. He holds four US patents and has developed eight disease-resistant plant varieties of medicinal and aromatic plants. He has 29 rhizospheric microbes accession numbers from the National Center for Biotechnology Information. He is a Fellow of the National Academy of Agricultural Sciences (NAAS), New Delhi. He is a recipient of several national and international awards including the Prof. H.M. Shah Memorial Award (2017), Dr. Manmohan Attavar Gold Medal in Floriculture (2017), M.R. Siddiqi Memorial Award (2018), and Prof. Uma Kant Sinha Memorial Award (2021), among others. He was President of the Indian Phytopathological Society (2022-23). Dr. Pandey has so far guided 13 PhD students and several MSc and MTech students. He has been granted several projects from national funding agencies. Dr. Pandey is an expert and task force member of the Department of Science and Technology (DST), Department of Biotechnology (DBT), Council of Scientific & Industrial Research (CSIR), and Indian Council of Agricultural Research (ICAR) for crop protection-related projects and selections.

Contents

Contributors

Mahfouz Mohamed Mostafa Abd-Elgawad
Department of Plant Pathology, National Research Center, El-Behooth St., Dokki, Giza, Egypt

Gholamreza Abdi
Department of Biotechnology, Persian Gulf Research Institute, Persian Gulf University, Bushehr, Iran

Isaac Boatey Akpatsu
Department of Crop Science, Faculty of Agriculture, Food and Consumer Sciences, University for Development Studies, Tamale, Ghana

Mukul Machhindra Barwant
Department of Botany, Sanjivani Arts, Commerce, and Science College, Kopargaon, Ahmednagar, Maharashtra, India

V. Bhuvaneswari
Department of Plant Pathology, Assam Agricultural University, Jorhat, Assam, India

Aissam Daaboub
Department of Plant Protection, Phytopathology Unit, Ecole Nationale d'Agriculture de Meknès, BPS 40, Meknès, Morocco

Ruchi Dube
School of Biotechnology and Bioinformatics, D. Y. Patil Deemed to be University, CBD Belapur, Navi Mumbai, Maharashtra, India

Sozan E. El-Abeid
Nanotechnology and Advanced Nano-Materials Laboratory (NANML); Mycology and Disease Survey Research Department, Plant Pathology Research Institute, Agricultural Research Center, Giza, Egypt

Marwa M. El-Deriny
Nematode Diseases Research Department and Central Lab of Biotechnology, Plant Pathology Research Institute, Agricultural Research Center (ARC), Giza, Egypt

Hajar El-Hamss
Department of Plant Protection, Phytopathology Unit, Ecole Nationale d'Agriculture de Meknès, BPS 40, Meknès, Morocco

Moussa El-Jarroudi
Environmental Sciences and Management Department, University of Liège – Arlon Campus Environment, Belgium

Khaled Elmnasri
Regional Center of Agricultural Research of Sidi Bouzid, CRRA, Tunisia

Said Ezrari
Department of Plant Protection, Phytopathology Unit, Ecole Nationale d'Agriculture de Meknès, BPS 40, Meknès, Morocco; Laboratory of Functional Ecology and Environmental Engineering, Sidi Mohamed Ben Abdellah University, Route d'Imouzzer, Fez, Morocco

Nabil S. Farag
Bacterial Diseases Research Department, Plant Pathology Research Institute, Agricultural Research Center (ARC), Giza, Egypt

Sami Fattouch
National Institute of Applied Sciences and Technology (INSAT), University of Carthage, Tunis, Tunisia

Mateti Gayithri
Department of Plant Pathology, Assam Agricultural University, Jorhat, Assam, India

Shimaa Ghareeb
Laboratory of Pest Physiology, Plant Protection Research Institute, Agricultural Research Center, Ministry of Agriculture and Land Reclamation, Giza, Egypt

Lopamudra Giri
Department of Plant Pathology, Assam Agricultural University, Jorhat, Assam, India

Khadija Goura
Department of Plant Protection, Phytopathology Unit, Ecole Nationale d'Agriculture de Meknès, BPS 40, Meknès, Morocco; Laboratory of Biotechnology and Valorization of Phyto-Resources, Faculty of Sciences, Moulay Ismail University, Avenue Zitoun, Meknes, Morocco

Arpita Gupte
School of Biotechnology and Bioinformatics, D. Y. Patil Deemed to be University, CBD Belapur, Navi Mumbai, Maharashtra, India

Lobna Hajji-Hedfi
Regional Center of Agricultural Research of Sidi Bouzid, CRRA, Tunisia

Hossam E. Harb
Biotechnology Lab, Faculty of Agriculture, Ain Shams University, Cairo, Egypt

Wassila Hlaoua
Department of Biological Sciences and Plant Protection, Higher Agronomic Institute of Chott-Meriem, Tunisia

Haitam Lahmamsi
Department of Plant Protection, Phytopathology Unit, Ecole Nationale d'Agriculture de Meknès, BPS 40, Meknès, Morocco; Laboratory of Microbial Biotechnology and Bioactive Molecules, Sidi Mohamed Ben Abdellah University, Route d'Imouzzer, Fez, Morocco

Dina S. S. Ibrahim
Nematode Diseases Research Department and Central Lab of Biotechnology, Plant Pathology Research Institute, Agricultural Research Center (ARC), Giza, Egypt

Frederick Kankam
Department of Crop Science, Faculty of Agriculture, Food and Consumer Sciences, University for Development Studies, Tamale, Ghana

Fadime Karabulut
Department of Biology, Firat University, Elazıg, Turkey

Jihane Kenfaoui
Department of Plant Protection, Phytopathology Unit, Ecole Nationale d'Agriculture de Meknès, BPS 40, Meknès, Morocco; Laboratory of Functional Ecology and Environmental Engineering, Sidi Mohamed Ben Abdellah University, Route d'Imouzzer, Fez, Morocco

Ashish Kumar
Department of Biotechnology, Sant Gahira Guru University Sarguja, Ambikapur, Chhattisgarh, India

Vikash Kumar
Department of Biochemistry, Central University of Haryana, Mahendragarh, Haryana, India

Rachid Lahlali
Department of Plant Protection, Phytopathology Unit, Ecole Nationale d'Agriculture de Meknès, BPS 40, Meknès, Morocco

Stephen Larbi-Koranteng
College of Agricultural Education, Akenten Appiah Menka University of Skills Training and Entrepreneurial Development, University in Mampong, Ghana

Ikram Legrifi
Department of Plant Protection, Phytopathology Unit, Ecole Nationale d'Agriculture de Meknès, BPS 40, Meknès, Morocco; Laboratory of Functional Ecology and Environmental Engineering, Sidi Mohamed Ben Abdellah University, Route d'Imouzzer, Fez, Morocco

Panagiotis Madesis
Department of Agriculture Crop Production and Rural Environment, University of Thessaly, Volos, Greece

Ivana Majić
Faculty of Agrobiotechnical Sciences Osijek, Josip Juraj Strossmayer University of Osijek, Osijek, Croatia

George Michail
Department of Ichthyology and Aquatic Environment, School of Agricultural Sciences, University of Thessaly, Volos, Greece

Svetlana Milijašević-Marčić
Institute of Pesticides and Environmental Protection, Belgrade-Zemun, Serbia

Mohammad Yaseen Mir
Center of Research for Development, University of Kashmir, Jammu and Kashmir, India

Yomna A. Mohamed
Biotechnology Lab, Faculty of Agriculture, Ain Shams University, Cairo, Egypt

Mohamed A. Mosa
Nanotechnology and Advanced Nano-Materials Laboratory (NANML); Mycology and Disease Survey Research Department, Plant Pathology Research Institute, Agricultural Research Center, Giza, Egypt

Soukaina Ben Othmen
Regional Center of Agricultural Research of Sidi Bouzid, CRRA, Tunisia

Jitendra Kumar Pal
Biotechnology Division, School of Science and Technology, Vanita Vishram Women's University, Athwalines, Surat, Gujarat, India

Ivana Potočnik
Institute of Pesticides and Environmental Protection, Belgrade-Zemun, Serbia

R. K. Mohana Pradeep
Department of Plant Pathology, Tamil Nadu Agricultural University, Coimbatore, Tamil Nadu, India

Nabil Radouane
Department of Plant Protection, Phytopathology Unit, Ecole Nationale d'Agriculture de Meknès, BPS 40, Meknès, Morocco; Laboratory of Functional Ecology and Environmental Engineering, Sidi Mohamed Ben Abdellah University, Route d'Imouzzer, Fez, Morocco

Abdelhak Rhouma
Regional Center of Agricultural Research of Sidi Bouzid, CRRA, Tunisia

Shaimaa N. Riad
Nematodes Diseases Department and Central Lab of Biotechnology, Plant Pathology Research Institute, Agriculture Research Center, Giza, Egypt

C. Sankaranarayanan
Division of Crop Protection, ICAR–Sugarcane Breeding Institute, Coimbatore, Tamil Nadu, India

Puppala Sri Saranya
Department of Entomology, Assam Agricultural University, Jorhat, Assam, India

Gabriella Kanižai Šarić
Faculty of Agrobiotechnical Sciences Osijek, Josip Juraj Strossmayer University of Osijek, Osijek, Croatia

Saurabh C. Saxena
Department of Biochemistry, Central University of Haryana, Mahendragarh, Haryana, India

Daisy Senapoty
Department of Plant Pathology, Assam Agricultural University, Jorhat, Assam, India

Sunita Singh
School of Biotechnology and Bioinformatics, D. Y. Patil Deemed to be University, CBD Belapur, Navi Mumbai, Maharashtra, India

Pranjali Sinha
Department of Plant Pathology, Indira Gandhi Krishi Vishwavidyalaya, Krishak Nagar, Raipur, Chhattisgarh, India

Nafissa Soudani
Agricultural Sciences Department, Mohamed Khider University, Biskra, Algeria

Khaoula Toumi
University of Jendouba, Higher School of Agriculture of Kef (ESAK), LR14AGR04: Support for the Sustainability of Agricultural Production Systems in the North West Region, Le Kef, Tunisia

Ioannis Vagelas
Department of Agriculture Crop Production and Rural Environment, University of Thessaly, Volos, Greece

Ab Waheed Wani
Department of Horticulture, Lovely Professional University, Punjab, India

Mohini Yadav
Department of Biochemistry, Central University of Haryana, Mahendragarh, Haryana, India

Foreword

Innovative and novel advances in sustainable agriculture without compromising plant and animal health are the most important global concerns. The indiscriminate use of agrochemicals has several negative effects. Concern over disturbance to microbial diversity, plant health, soil fertility, and economic constraints have prompted work for an eco-friendly approach for effective microbial biostimulants that can ensure competitive yield by protecting plants *via* mitigating pathogenic effects. Biostimulants offer a potentially novel approach for regulating and modifying physiological activities in plants to enhance growth, moderate stress-induced limitations, and increase yield. Since the last few decades, scientific evidence supporting the beneficial properties of microbes and their role in alleviating biotic stresses have also led to the popularity of microbial biostimulants as promising alternative measures in the management of plant pathogens, including a biorational approach to the practice of sustainable agriculture. These objectives have provided an impetus for research on the development of microorganisms actively associated with novel bio-products with various beneficial properties. Consequently, bioinoculants with tailored formulations of individual strains/consortia of known microorganisms that promote plant growth and development have been included in the fight against plant pathogens.

This book, *Microbial Biostimulants: Biorational Pesticides for Management of Plant Pathogens,* focuses on manipulating plant health by inoculating beneficial microbes and provides a resource for studying the impact of microbial biostimulants against plant pathogens, including plant-parasitic nematodes. The book has 16 chapters that cover various areas of the subject concern. These chapters provide information on the production of microbial and algal biostimulants as a response to modern agriculture, the application of various bio-control agents, including *Bacillus* species, arbuscular mycorrhizal fungi, rhizobacteria, bacterial and fungal endophytes, microbiota in the edible mushroom industry, and microbial consortia for better plant health by managing plant pathogens.

This book will be a valuable source of information on many aspects of plant health through microbial applications known for being environmentally friendly. The efforts made by the editors are commendable, and the book would be useful for academicians, researchers, students, farmers, and all the microbial-biostimulants industry stakeholders.

I look forward to reading the editors' effort and wish the book great success.

—***S. C. Dubey, PhD***
Assistant General Director,
Indian Council of Agricultural Research, New Delhi, India

Preface

Microbial biostimulants strongly influence plant health and growth. Beneficial plant-microbe interactions resist pest and pathogen attacks, enhance nutrient availability, improve root architecture, and modulate stress responses. Plant health depends upon soil microbiota for their growth, development, and nutrient acquisition. Plants, in turn, serve as food to fulfill human and livestock needs. Within the rhizospheric region, they produce secondary metabolites to protect plants against pathogens, including plant-parasitic nematodes, and tolerate biotic pressure. These microbial biostimulants are usually bacteria, fungi, or protists, including entomopathogenic microbes.

When inoculated to the plants or the soil, they often systemically colonize a plant's rhizosphere and interior while improving its health and yield. With the initiation of technological advancements in agricultural microbiology and biotechnology, new avenues have been established to identify and characterize beneficial microbes and assign their functions against plant pests and diseases. The bioactive compounds released during microbial colonization are readily biodegradable and have no residual impact on plants and animals. This condition attracts researchers to develop an eco-friendly approach to avoid or minimize the use of chemical pesticides. The most logical and acceptable approach in the current scenario is microbial-based biopesticides for managing pests and diseases.

In this book, researchers share their expertise in this area. This book is a true reflection of the sincerity of the authors, who agreed to contribute their chapters, and we believe this book will be helpful to readers, including young minds, who are likely to benefit. This book contains 16 chapters covering pests and diseases and the potential role of microbial biostimulants against plant pests, including plant parasitic nematodes. As such, this book looks at the application of microbial biostimulants in response to the modern agricultural need for productivity and plant health.

—***Faheem Ahmad, PhD***
Aligarh Muslim University, Aligarh, Uttar Pradesh, India

Rakesh Pandey, PhD
CSIR-CIMAP, Lucknow, Uttar Pradesh, India

CHAPTER 1

Pests and Diseases: A Global Threat to Plants

IOANNIS VAGELAS,[1] GEORGE MICHAIL,[2] and PANAGIOTIS MADESIS[1]

[1]*Department of Agriculture Crop Production and Rural Environment, University of Thessaly, Volos, Greece*

[2]*Department of Ichthyology and Aquatic Environment, School of Agricultural Sciences, University of Thessaly, Volos, Greece*

ABSTRACT

Plants are fundamental for food, feed, fibers, and even fuels, and they also play an important role in managing climate change and the environment on Earth. However, they suffer from multiple pathogens and pests, which reduce their yield. Furthermore, there is an increasing risk from new and invasive plant pests and pathogens globally, a risk that is constantly rising due to climate change and the global market and transport. Climate change poses an unparalleled challenge to Earth and society as we know it. Moreover, the pressure on plant health is unprecedented. As new diseases and pathogens emerge, combined with new ways of pathogen action, we must reinvent the way we think about plant-pathogen interaction and the way we protect crops. New research is needed to identify the mechanisms of interaction, and new chemical or non-chemical protection should be conducted to ensure food safety and security. Omics, molecular, and biotechnological tools offer new solutions in this research, providing knowledge and suggesting new ways of protection. Even metagenomics, as we discussed, can offer new knowledge based on past research. This scientific review discusses the most important diseases that pests and nematodes cause in plants worldwide, the impact of climate change on plant-pathogen interaction, and the new emerging threats

Microbial Biostimulants: Biorational Pesticides for Management of Plant Pathogens.
Faheem Ahmad and Rakesh Pandey (Eds.)

the olive trees and the landscape of the area [8]. Additionally, nematodes can also cause severe damage to crops. Sedentary endoparasitic nematodes, such as the genera – (i) *Globodera* spp. and *Heterodera* spp. of cyst nematodes; and (ii) *Meloidogyne* spp. of root-knot nematodes (RKN), are the most common and economically important nematode species in agriculture.

Bursaphelenchus xylophilus, the pine wilt nematode, is one of the most damaging invasive pests across Africa, North America, Asia, and Europe, with substantial ecological repercussions on pine trees (Pinus spp.) [9].

TABLE 1.1 Most Important Plant Pathogens Based on Scientific and Economic Importance

Plant-Pathogen		Common Disease Name(s)	Disease Highlighted	References
Fungal Plant Pathogens	*Botrytis cinerea*	Gray mold	*Botrytis cinerea* is one of the most necrotrophic fungi that infects over 200 plants.	[10, 11]
	Magnaporthe oryzae	Rice blast disease or rice rotten.	Rice blast disease, often known as rice rotting neck, is the most severe rice disease in the world.	[12, 13]
	Puccinia spp.	Wheat rusts.	Wheat crops have been particularly vulnerable to stem black rust, stripe rust, and leaf rust diseases caused by the *Puccinia graminis* f. sp. *tritici*, *P. striiformis* f.sp. *tritici* and *P. triticina*, respectively.	[14]
	Zymoseptoria tritici (syn. *Mycosphaerella graminicola* or*Septoria tritici*)	Wheat blotch diseases. *Septoria tritici* blotch.	Wheat productivity is severely hampered by *Mycosphaerella graminicola*, especially in temperate growing zones.	[14, 16]
	Blumeria graminis	Powdery mildews of wheat and barley, including grasses.	Cereal mildew infections reduce grain yield.	[16, 17]
	Fusarium oxysporum	Fusarium wilt.	*Fusarium osysporum* affects a wide range of plants, causing vascular wilt.	[19, 19]
	Ustilago maydis	Corn smut	Symptoms in corn can be extremely severe, however, infections on corn are usually localized (not associated with severe losses in corn yield).	[12, 20]
	Colletotrichum spp.	Anthracnose (sunken dead spots) and blights (tissue browning).	*Colletotrichum* spp. are responsible for significant losses in commercial crops, particularly fruits, and vegetables.	[21, 22]

TABLE 1.1 *(Continued)*

	Plant-Pathogen	Common Disease Name(s)	Disease Highlighted	References
Oomycete Plant Pathogens	*Phytophthora infestans*	Late blight. Potato late blight fungus.	Potato late blight is a disease caused by *Phytophthora infestans*. The disease has a disease with a long history. Between 1845 and 1846, an estimated 7,50,000 people died in Europe as a result of famine. Late blight continues to limit potato production, making it the world's third-largest basic crop and a constant threat to food security.	[23–27]
	P. ramorum	Ramorum disease or "sudden oak death."	*Phytophthora ramorum* caused a disease known as "sudden oak death," which is a devastating disease for oaks.	[28–32]
	Phytophthora cinnamomi	Root rot; dieback.	In the forestry and horticulture industry, *Phytophthora cinnamomi* has widespread and economically significant effects. Climate change is expected to have a considerable impact on disease severity and distribution.	[33, 34]
	P. capsici	Blight and fruit rot of pepper.	*Phytophthora capsici* causes blight and fruit rot of cucurbit, including pepper and tomato.	[35–37]
	P. parasitica	Root and stem rot.	In herbaceous and perennial plant species, *Phytophthora parasitica* causes severe foliar and fruit diseases, as well as collar root symptoms and crown rot.	[38, 39]
	Pythium ultimum	Damping-off and/or root rot.	The pathogen causes damping off and root rot diseases on a great variety of plants.	[40, 41]
	Plasmopara viticola	Downy mildew.	The pathogen that causes grape downy mildew, *Plasmopara viticola*, is native to the USA and Canada. The pathogen was accidentally introduced into Europe towards the end of the 19th century, inflicting serious damage to vines.	[42–45]
	Albugo candida (Pers.)	White rust (-of crucifers).	White rust is an important disease of Brassicaceae crops and weeds worldwide. In India, Canada, and Australia, the infection of *Brassica juncea* with *A. candida* has resulted in severe crop losses.	[46–49]

TABLE 1.1 *(Continued)*

Plant-Pathogen		Common Disease Name(s)	Disease Highlighted	References
Plant Pathogenic Bacteria	*Pseudomonas syringae*	*Pseudomonas syringae*, infect the phyllosphere and attacks different host species using many virulence factors.	*Pseudomonas syringae* is a plant pathogen that causes diseases in many plants all over the world. *P. syringae* cause bacterial cankers on perennial trees and induce significant economic losses all over the world.	[50–52]
	Ralstonia solanacearum	*Ralstonia solanacearum* caused vascular wilt diseases in many plants, mainly tomatoes.	Due to its extensive host range and pathogenic nature, *Ralstonia solanacearum* is the most destructive plant pathogenic bacteria on the planet.	[53–55]
	Agrobacterium tumefaciens	Crown gall disease.	Crown gall disease is caused by *Agrobacterium radiobacter* (also known *as Agrobacterium tumefaciens*), a plant pathogen that affects a variety of crop species around the world.	[56, 57]
	Xanthomonas oryzae pv. *oryzae*, *and Xanthomonas campestris* pathovars.	Bacterial blight of rice, and Black rot of crucifers.	*Xanthomonas oryzae* pv. *oryzae* causes bacterial blight of rice and is one of the most common diseases in rice-growing countries. Pathovars of *Xanthomonas campestris* affect all cultivated brassicas, causing the black rot of crucifers.	[58–64]
	Erwinia amylovora	Fire blight	*Erwinia amylovora* causes fireblight primarily in rosaceous hosts, mainly in apple and pear varieties.	[65, 66]
	Xylella fastidiosa	Pierce's disease	*Xylella fastidiosa* is the most harmful plant bacteria on the planet, producing a wide range of diseases and having a significant economic and environmental impact. Bacteria colonize xylem vessels and block them. *Xylella fastidiosa* is introduced into the xylem of host plants by its vector, the spittlebug species, *Philaenus spumarius*. The phytopathogen was the first bacterium to have its entire genome sequenced.	[67–69]
	Dickeya (formerly *Erwinia chrysanthemi*)	Bacterial wilt of chrysanthemum	Bacterial wilt is caused by *Erwinia chrysanthemi*, which was described as a new genus *Dickeya* responsible for wilt and soft-rot diseases in a wide range of crops.	[70, 71]
	Pectobacterium carotovorum and *P. atrosepticum*	Soft rot diseases (*P. carotovorum*) and blackleg disease of potatoes (*P. atrosepticum*).	*Pectobacterium carotovorum* and *P. atrosepticum*, which were initially classified as *Erwinia carotovora* subsp. *carotovora* and *E. carotovora* subsp. *Atroseptica*, caused soft rot and blackleg disease in potatoes, respectively.	[72–74]

TABLE 1.1 *(Continued)*

	Plant-Pathogen	Common Disease Name(s)	Disease Highlighted	References
Plant Pathogenic Viruses	*Tobacco mosaic virus* (TMV)	TMV-induced disease in tobacco	*Tobacco mosaic virus* (*Tobamovirus*, *Virgaviridae*) is a major economic importance plant pathogenic virus that causes a mottled browning of tobacco leaves. *The tobacco mosaic virus* also infects other crops (over 350 different species of plants), especially tomatoes.	[75–80]
	Tomato spotted wilt virus (TSWV)	*Tomato spotted wilt virus* is the only genus of the family *Bunyaviridae* that comprises plant-infecting viruses.	*Tomato spotted wilt virus* (*Tospovirus*, *Bunyaviridae*) infects over 1,000 plant species, including dicots and monocots, and has caused widespread crop losses in greenhouse-grown agricultural and ornamental plants.	[81–83]
	Tomato yellow leaf curl virus (TYLCV).	Tomato yellow leaf curl disease.	*Tomato yellow leaf curl virus* (*Begomovirus*, *Geminiviridae*), infection induces severe symptoms in tomato plants resulting in crop losses of up to 100%. The pathogen causes serious yield losses worldwide.	[84, 85]
	Cucumber mosaic virus (CMV)	Cucumber mosaic (foliar mosaic symptoms).	*Cucumber mosaic virus* (*Cucumovirus*, *Bromoviridae*) was first characterized (1934) in cucumbers (*Cucumis sativus*), showing mosaic symptoms. Since then, it has been found to infect a great variety of other plants, 1,200 species in over 100 plant families, worldwide including Antarctica.	[86–89]
	Potato virus Y (PVY)	Potato virus Y strains (leaf drop streak, rugose mosaic).	Potato virus Y (*Potyvirus*, *Potyviridae*) is the most damaging plant pathogenic virus causing significant losses in most Solanaceous crops around the world, such as potato, pepper, tomato, tobacco, and eggplant.	[90–92]
	Cauliflower mosaic virus (CaMV).	Cauliflower mosaic virus.	*Cauliflower mosaic virus* (*Caulimovirus*, *Caulimoviridae*) was the first plant virus – (i) identified as a DNA (circular double-stranded DNA) virus; (ii) whose genome, located on the 8,000-nucleotide-long pregenomic RNA, has been sequenced; and (iii) shown to be replicated by reverse transcription. CaMV can impact plant development, particularly in early infections, and flower output can be inhibited.	[93]
	African cassava mosaic virus (ACMV)	Cassava mosaic disease.	*African cassava mosaic virus* (*Begomovirus*, *Geminiviridae*), a geminivirus first reported in Tanzania, is the most important virus disease of cassava.	[94–97]
	Plum pox virus (PPV)	Plum pox virus disease (Sharka)	Plum pox potyvirus (*Potyvirus*, *Potyviridae*) causes Sharka, the most destructive disease of stone fruits.	[98, 99]

TABLE 1.1 *(Continued)*

	Plant-Pathogen	Common Disease Name(s)	Disease Highlighted	References
Plant-Parasitic Nematodes	*Meloidogyne* spp.	Root-knot nematodes attack the roots of various plants and typically incites galls on plant roots as a result of parasitism.	Nematodes parasitize almost every species of vascular plant. Large galls or "knots" can develop throughout the root system of affected plants as a result of nematode feeding.	[100–107]
	Heterodera and *Globodera* spp.	Cyst nematode disease (e.g., Soybean cyst nematode disease).	Large areas of soybean fields can be severely stunted and yellow due to the high population densities of the soybean cyst nematode.	[108–111]
	*Pratylenchus*spp.	Root lesion nematodes.	Stunting, fertilizer, and water deficits, and eventual dieback can result from high nematode populations and infection of younger plants. Over 400 plant species serve as hosts, including potatoes, peanuts, monocots, and fruits.	[112–117]
	Radopholus similis	Burrowing nematode disease.	The most dramatic disease symptom in banana plantations. Nematodes mainly cause the collapse of banana plants.	[118–121]
	Ditylenchus dipsaci	Swelling, deformation, and discoloration of above-ground plant parts.	*Ditylenchus dipsaci* is an endoparasite that feeds primarily in plant aerial parts (leaves and flowers), although it also affects bulbs and rhizomes.	[122, 123]
	Bursaphelenchus xylophilus	Pine wilt disease.	Pine wilt is a devastating disease that kills trees in just a few weeks to a few months.	[124–126]
	Xiphinema index	Virus vector nematode. Grapevine fanleaf virus (GFLV).	Grapevine fanleaf virus (GFLV) is the cause of an economically significant grapevine disease that causes stem and leaf deformations as well as widespread leaf yellowing in the spring. The grapevine flavivirus (GFLV) is found in various regions of the world where grapevines are grown.	[127–129]
	Aphelenchoides besseyi	Rice leaf nematode. White tip disease of rice.	Severe infection of *Aphelenchoides besseyi* is causing a shortened flag leaf, the grain is small and the infected plants mature late.	[130–132]

1.4 MOST IMPORTANT PESTS AFFECTING PLANTS WORLDWIDE

The present study (Table 1.2) is a synthesis of selected bibliographic articles from the SCOPUS database, applying three main criteria: (i) the scientific name of the pest; (ii) the keyword "PESTICIDES" within the scientific name of the pest; and (iii) the keyword "RESISTANT" within the scientific name of the pest. The first research question, "pest scientific name," was addressed in this study, following the ideas of the state of the world's plant health available at: https://stateoftheworldsplants.org/2017/plant-health.html.

Based on the key PDF available at: https://stateoftheworldsplants.org/2017/plant-health.html, Table 1.2 lists the top 20 insect pests in terms of scholarly papers in CABI abstracts during the last five years. These pests are widely distributed, have a wide host range, and are resistant to a variety of pesticides. Among the top 20 insect pests (Table 1.2), the cotton bollworm (*Helicoverpa armigera*) ranks at the top of the list. This pest is a major threat to maize, cotton, tomatoes, and sorghum, and has recently invaded Latin American countries, showing rapid spread [133]. Due to its ability to migrate and reproduce quickly, *H. armigera* is an economically significant pest. Furthermore, with the impact of climate change, pests like *H. armigera*, which cause severe damage to economically important crops, are becoming increasingly destructive, posing a growing threat to food security and the environment.

In Kritikos and co-authors' article [133], it has been shown that *H. armigera* is climatically suited to South America's agricultural regions. Therefore, it is not surprising that it is rapidly spreading across Argentina, Bolivia, Uruguay, Paraguay, Brazil, Puerto Rico, and Costa Rica. The successful establishment and expansion of this species appear to be unaffected by the presence of closely related Heliothines (Heliothine pest complex), which occupy the same or similar ecological niche. The authors [133] concluded that the invasion of *H. armigera* and its subsequent spread into Latin American countries have altered the nature of the invasion threat to North America. This is not only due to the change in the dispersal route but also because of the shift in the invasion pathway based on *H. armigera's* potential bioclimatic niche. The most significant point to note is that the total value of crop production at risk from *H. armigera* in the USA exceeds $72 billion per year. *Helicoverpa armigera* is reported as the most economically significant lepidopteran pest in the world. *H. armigera* was predicted to inflict $2 billion in annual output losses to crops in Asia, Africa, Europe, and Australia before its import and establishment in Latin America [134]. The key findings of Kritikos et al. [133] show that the impact of climate change has altered the *H. armigera* migratory routes and geographical distribution and hence cost the global economy over $722+ billion. This issue should serve as a warning on how climate change may influence the infections and spread of pests such as *H. armigera* throughout the world.

Furthermore, *H. armigera* is believed to be the target of half of all insecticides used for pest control on crops in Asian countries, primarily India and China [134]. The spread of pests not only poses challenges to the plant health community and global crop production, but also presents problems such as insecticide resistance and environmental pollution associated with pesticides.

To overcome these threats, it is crucial to preserve plant health in order to achieve Sustainable Development Goals. Plant health is a vital aspect of our efforts to enhance the efficiency, inclusiveness, resilience, and long-term sustainability of agri-food systems. Additionally, the impact of climate change on Agricultural Insect Pests could have a significant effect on crop production and impose substantial environmental costs on society.

TABLE 1.2 Search Criteria and the Number of Results Based on Scopus Documents Published from 2000*

Total Number of Documents per Main Query and Search Using Keywords Within Results				
Pest (Scientific Name)	**Rank**	**Main Query**	**Search Within Results (Using the Keyword Pesticide or Resistance)**	
		Pest	**Pesticides**	**Resistant**
Helicoverpa armigera	1	4,001	1,559	1,256
Spodoptera frugiperda	2	3,668	1,086	897
Bemisia tabaci	3	3,929	1,210	954
Tribolium castaneum	4	3,019	1,049	599
Plutella xylostella	5	2,967	1,404	1,064
Tetranychus urticae	6	2,794	1,381	745
Myzus persicae	7	2,680	830	763
Spodoptera litura	8	2,145	987	469
Spodoptera exigua	9	1,730	563	398
Nilaparvata lugens	10	1,701	698	736
Ceratitis capitate	11	1,693	294	103
Aphis gossypii	12	1,658	596	430
Acyrthosiphon pisum	13	1,493	183	254
Frankliniella occidentalis	14	1,481	429	271
Spodoptera littoralis	15	1,382	495	298
Callosobruchusmaculatus	16	1,203	401	298
Cydia pomonella	17	1,197	411	240
Diaphorina citri	18	1,155	305	192
Tuta absoluta	19	905	420	169
Thrips tabaci	20	823	247	162

*The ranking is based on Pest searches in the Scopus database.

1.5 COEVOLUTION OF PLANT PARASITES

Coevolution, by definition, is an evolutionary process that leads to adaptive genetic changes in animals that have close and direct ecological interactions due to reciprocal selection. This coevolutionary process can result in local adaptation patterns in the presence of plant–parasite interactions and a geographically diverse environment, where parasite populations are separated in different locations. Consequently, a parasite population performs better in its own environment (i.e., the host) compared to a different environment, and/or in its native habitat compared to a different habitat. Coevolution occurs due to the intimate relationship between plants and parasites, as well as the intense selection pressure that each imposes on the other.

The introduction of new crops can expand the distribution of plant-parasite interactions. The introduction of new pathogens or pests into a new ecosystem can cause significant damage since plant parasites and hosts have not coevolved, resulting in the absence of disease resistance mechanisms. Plants and their parasites have mutually benefited from this coevolution, leading to a stable balance between hosts and parasites within their specific environments. Parasites limit plant growth, disrupt development, and reduce plant productivity. Plants face selection pressure to develop resistance against parasites, while parasites face selection pressure to overcome host defenses. This ongoing conflict drives coevolution, where the fitness of genotypes in one species is determined by the fitness of genotypes in the other, resulting in a variety of host defenses and plant-parasite mechanisms [135].

Many host-parasite interactions have been studied, and meta-analyses have revealed that there is a general proclivity for local plant-parasite adaptation. This trend is more typical for plant-parasite infection than for plant-parasite proliferation. The life cycles of crops and their pests and diseases are strongly adjusted since the most important crops are annuals, such as wheat, rather than perennials, and large areas include genetically uniform crop varieties. Therefore, the host and parasite life cycles are connected in large regions of uniform crops. Native host plants in non-farming regions of rural Australia are thought to contribute to populations of *Helicoverpa armigera* (Hubner), which travel in the spring to infest cropping zones of southeast Australia and southwestern Australia. As a result, we believe that *H. armigera* has adjusted to deal with chemical defenses in new areas and that current relationships reflect an ongoing evolutionary process known as 'coevolution.' We also believe that *H. armigera* has adjusted to plant defenses that they ordinarily encounter as a result of coevolution.

1.6 PESTS AND PATHOGENS DISTRIBUTION IN RESPONSE TO CLIMATE CHANGE

1.6.1 THE CURRENT WORLD PICTURE

The Intergovernmental Panel on Climate Change (IPCC) has predicted the following: (i) an increase in global temperatures, resulting in warmer and/or fewer cold days and nights across most land areas such as Asia, Australasia, Europe, and North America; (ii) an increase in precipitation, particularly heavy precipitation events, across the majority of land regions worldwide (IPCC report, https://www.ipcc.ch/report/ar6/wg1/downloads/report/IPCC_AR6_WGI_SPM_final.pdf). The research also indicates that global anthropogenic CO2 emissions scenario is comparable to temperature changes, as historical accumulative CO2 emissions have already caused significant warming, while future emissions are likely to further contribute to the warming of the Earth.

1.6.1.1 XYLELLA FASTIDIOSA HOST-PATHOGEN ASSOCIATIONS IN EUROPE

Wells et al. [136] initially described *Xylella fastidiosa* as one of the most hazardous plant-pathogenic bacteria in the world, with a list of host plants that includes more than 563 plant species from America (USA and Latin America), the Middle East, Asia, and Europe [137]. Insect vectors that feed on the xylem of host plants naturally spread the bacterium. Mainly two taxa – of the order *Hemiptera*: *CicadellidaeCicadellinae* (known as sharpshooters) and *Cercopoidea* (known as froghoppers or spittlebugs) – are reported as the bacterial vectors [138, 139].

The spittlebug species, *Philaenus spumarius* (*Hemiptera, Aphrophoridae*), plays an important epidemiological role in the European agro-ecosystem. It appears as the predominant xylem-sap feeder and has been reported as the first confirmed European vector for the epidemic spread of the bacterium in the olive groves in the Apulian region [140].

In Mediterranean regions, the primary and currently the only epidemiologically relevant vector of the bacterium *Xylella fastidiosa* is the spittlebug species *Philaenus spumarius* (L.) [141, 142]. *Philaenus spumarius* is a widely distributed pest and is very common throughout most of Europe (from Portugal to Russia).

Adults of *P. spumarius* feed on herbaceous vegetation in Mediterranean locations with high precipitation seasonality (macroclimatic tolerances) in the spring, then change to woody hosts – including a wide range of agricultural plants – as the herbaceous flora dries out in the early summer [143]. The macroclimatic tolerances of *P. spumarius* – the only epidemiologically significant vector of *X. fastidiosa* in Europe, as mentioned above – were modeled by Godefroid et al. [143]. The models published by Godefroid et al. [143] anticipate a significant regional variance in environmental appropriateness for *P. spumarius* across the European continent. Overall, bioclimatic models predict that much of Western Europe will have good climatic suitability, whereas the driest and warmest Mediterranean areas – principally eastern Greece, the Iberian Peninsula, and Turkey – would be moderately or poorly suitable.

In contrast, some of these South European areas (e.g., the Iberian Peninsula) were predicted to be suitable for the establishment of different subspecies of *X. fastidiosa* bacterium. However, no severe *X. fastidiosa* outbreak has ever occurred in Southeast Europe (Greece), which is expected to be less suitable for the *P. spumarius* species. One possible explanation for this trend is that the climate conditions in these areas prevent the establishment of large populations of effective vectors, thus reducing the likelihood of *X. fastidiosa* transfer to plants and subsequent plant-pathogen dissemination. To adequately test this concept, we must first determine whether *X. fastidiosa* was previously introduced in these areas but did not spread due to a lack of abundant vector populations.

Apart from the insects of the Hemiptera order that transmit *X. fastidiosa*, the cotton bollworm (*Helicoverpa armigera*) was studied to investigate how variations in climate and crop planting systems affected pest population dynamics. The population size of *H. armigera* increased with global climate warming in all three generations, particularly in the third generation [144]. As specified by the study presented by Huang et al. [144], climate change has a wide range of consequences on pest phenology and population growth. Climate change has advanced the phenology of G1 to G3 moths, as well as cotton flowering dates, at varied rates, raising the population numbers of *H. armigera*, especially G3 moths [144].

1.7 META-ANALYSIS

1.7.1 META-ANALYSIS IN PLANT PATHOLOGY

A meta-analysis is a collection of statistical processes for combining research findings from several studies. Each study collects an estimation of a statistical result, such as the variance in disease severity between plants with and without a management treatment (e.g., application of pesticide or not), as well as a measurement of the difference in the effect estimate. When the results of multiple studies are combined, the statistical power increases, making it simpler to detect modest effects. Comparing the extent of the outcome as a function of additional predictor variables such as geographic location or plant-pathogen species may also be accomplished by combining results from multiple studies.

As an example, Paul et al. [145] conducted a meta-analysis to evaluate the association between triazole fungicides on wheat yield affected by *Fusarium graminearum* and its associated toxin, deoxynivalenol (DON) [145]. The objective of this study was to define: (i) the disease-toxin complex; and (ii) the mean yield and test-weight reactions of wheat to treatment with triazole fungicides, such as propiconazole. The research was based on numerous studies conducted across different environments. The information for this study [145] was gathered from USWBSI (U.S. Wheat and Barley Scab Initiative) and concluded that triazole fungicides used for managing the *Fusarium graminearum* and DON complex contribute significantly to increasing wheat yield and grain weight. The importance of these results varies depending on the active ingredients in the fungicide, wheat variety, and the stage of pathogenicity.

Overall, the meta-analysis of epidemiological data could play a crucial role in the formulation of effective government policies. Meta-analysis also plays a significant role in government policy formation for evaluating local and global yield gaps, aiming to prevent exposure to disease-causing factors in various field crops (such as soybean) and woody plants like apple, cherry, and others.

1.8 METAGENOMIC ANALYSIS

1.8.1 METAGENOMIC ANALYSIS OF SPELEOTHEMS AS A PREDICTIVE ANALYTIC TOOL ON CROP DISEASES

McDermott et al. [146] suggest that the chronology of speleothem growth provides important information about paleoclimatic conditions. These past environmental conditions have been proposed as the main drivers of past biogeographical distribution, capable of explaining not only changes in the planet but also biodiversity and the genetic structure of life, as well as climate changes over time [147]. In the past, these geological changes had a significant impact on life on the planet in general, and today we see that these changes also greatly affect plant pests and diseases.

It is well known that speleothems, such as stalactites, serve as a natural library of paleorainfall as they are shaped by groundwater, seawater, and rainfall, respectively. It has also been demonstrated that these speleothems contain, aside from the small amounts of water trapped in so-called fluid inclusions, analogous genetic information, such as DNA from microbes like bacteria [148].

Thus, it was shown that the stalactite body has preserved DNA molecules of abundant bacterial phyla such as *Proteobacteria, Actinobacteria, Firmicutes, Acidobacteria, Bacteroidetes*, and *Verrucomicrobia*. Our metagenomic analyses (Figure 1.1) demonstrate that the stalactite's core is a significant DNA source of many plant pathogenic bacteria belonging to the *Ralstonia* species, which causes bacterial wilt in a wide range of host plants [148], or other phytopathogenic bacteria in the *Pseudomonas* genus (Figure 1.1). Moreover, our data as presented in Figures 1.2 and 1.3 show clear evidence of seasonal variations in the families *Thioprofundaceae* and *Comamonadaceae*. *Thioprofundum*: *Thioprofundaceae* is a thermophilic genus, and *Polaromonas*: *Comamonadaceae* is a psychrophilic marine bacterium found on glacier surfaces in Antarctica. Our metagenomic analysis results (Figure 1.3) support the paleoclimate seasonal variations [148].

Therefore, metagenomics analyses could collect significant information from speleothems that have trapped DNA molecules from past ages. Based on that, we suggest that climate models could be developed to simulate the future impact of pests and diseases on the most important crops in correlation to climate.

1.9 CONCLUSION

Plants play a crucial role in providing food, feed, energy, a stable environment, and the oxygen we breathe [149]. In the modern world, pests, diseases, and invasive non-native species pose a growing threat to plants. The two most significant environmental issues currently affecting the planet are invasive plant pests and pathogens and climate change [150]. Furthermore, rapid temperature changes may lead to the extinction of many species while simultaneously allowing others to colonize from nearby warmer areas.

For instance, the native bacterium *Xylella fastidiosa*, a plant endophyte, has recently been found in various places outside of its original distribution, including Europe [151]. In Apulia (Italy), the bacterium *X. fastidiosa* was discovered on olive trees in 2013. Genetic tests conducted in 2016 revealed that this bacterium strain was unintentionally imported into Apulia (Italy) from Costa Rica or Honduras via infected ornamental coffee plants [152].

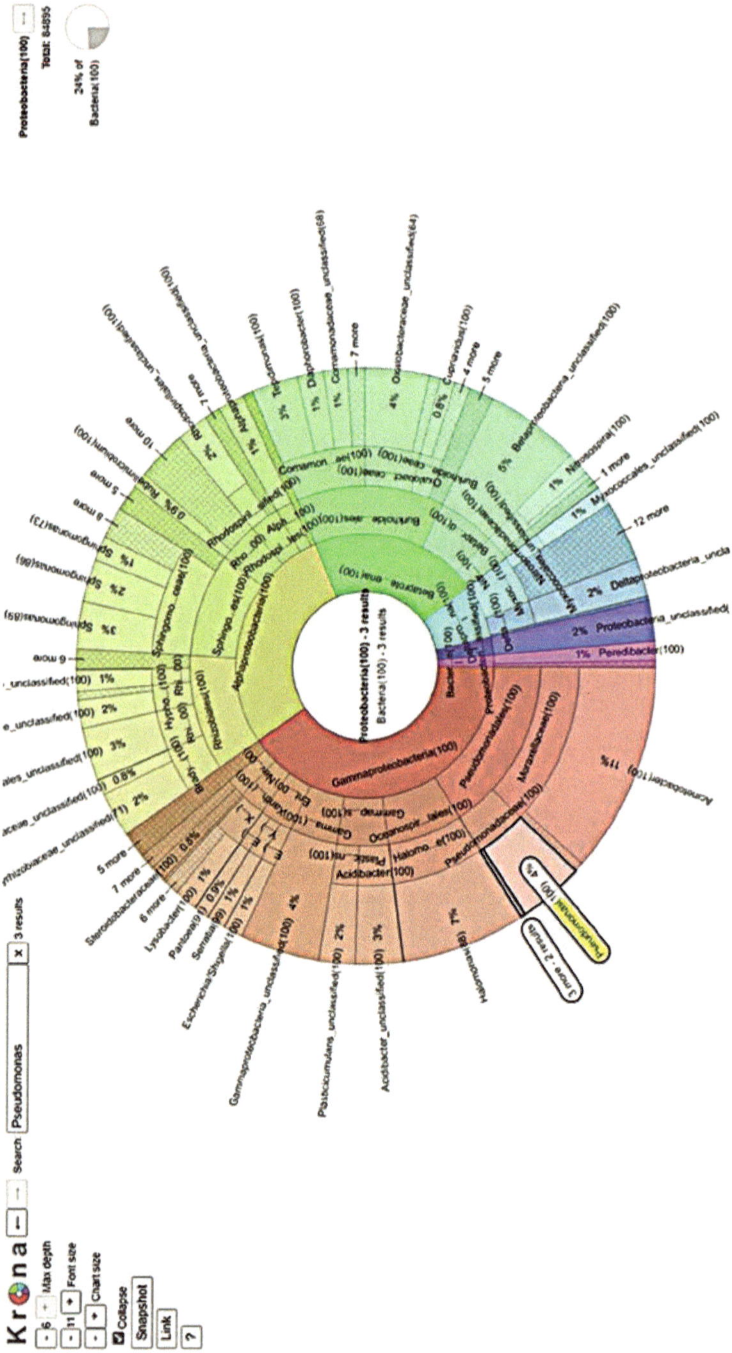

FIGURE 1.1 Stalactite metagenomic analyses show the phytopathogenic bacteria in the genera *Pseudomonas* (highlighted in yellow).

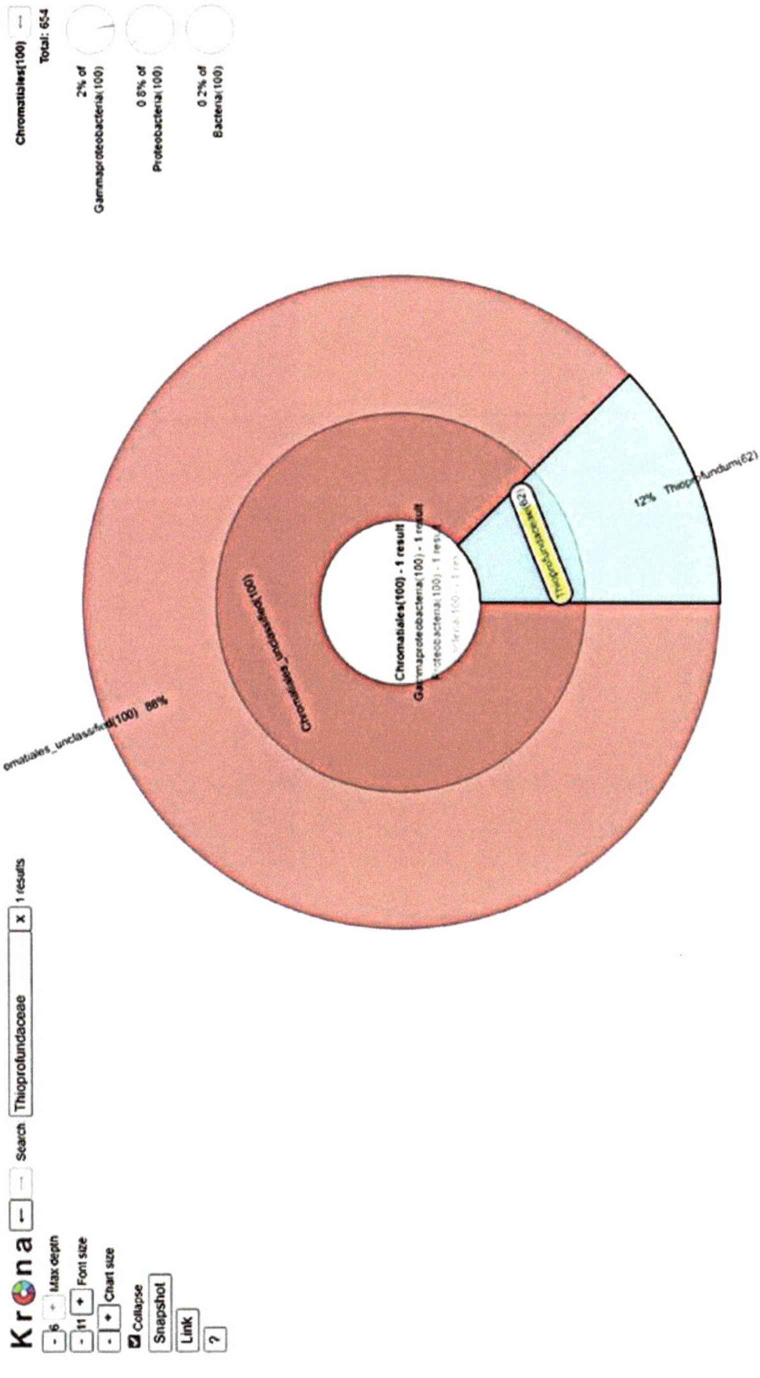

FIGURE 1.2 Stalactite metagenomic analyses show the bacteria in the family *Thioprofundaceae* (highlighted in yellow).

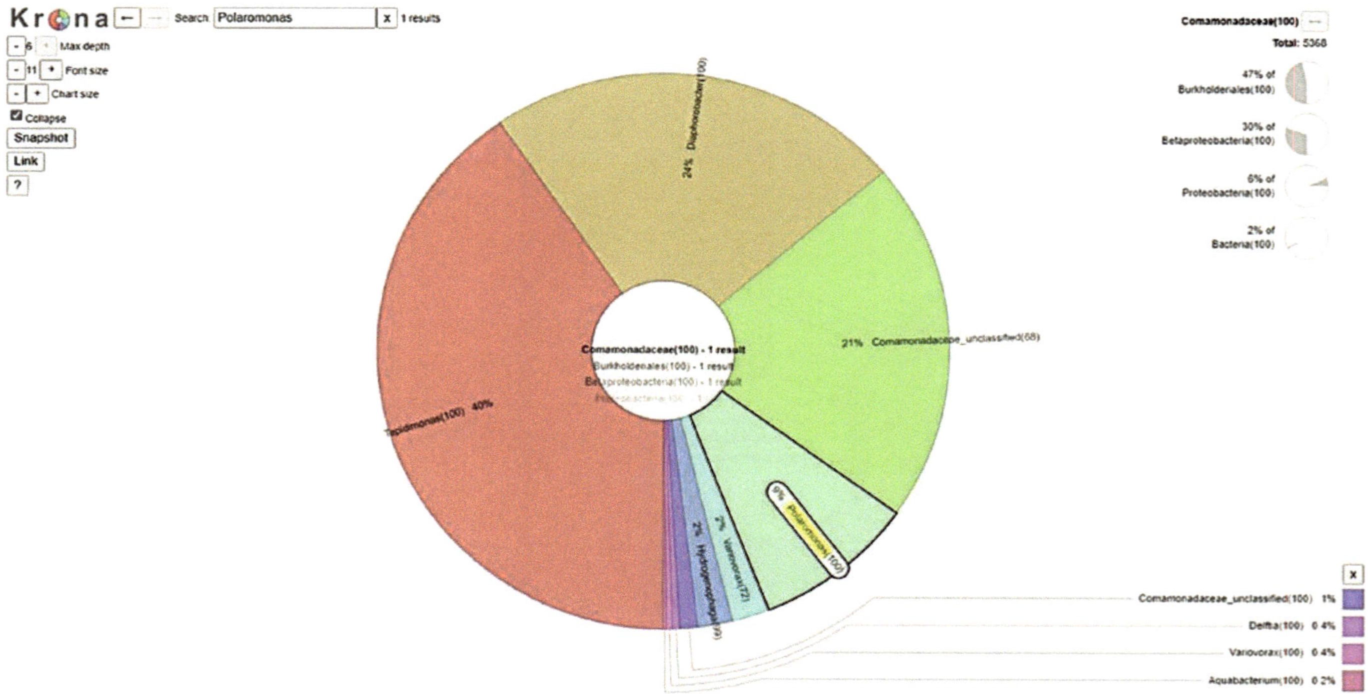

FIGURE 1.3 Stalactite metagenomic analyses show the bacteria in the family *Comamonadaceae* and the genera *Polaromonas* (highlighted in yellow).

Most of the food we eat comes from plants, and according to the Food and Agriculture Organization (FAO), between 20% and 40% of the world's crop yields are reduced annually as a result of plant pests and diseases [153]. Climate change has also created a dire situation for humanity as the world's temperatures continue to rise, which will have an immediate impact on crop production and productivity. To predict future climate, scientists use computer programs called climate models to understand how our planet is changing. There are also prediction models for plant diseases [154]. Modern big data analytic techniques and Meta-analysis [155] can be used to forecast the start or change in the severity of diseases today through the application of new information and communication technology [154].

As suggested, we can predict plant diseases caused by *X. fastidiosa* from metagenome-based data [156]. Meanwhile, Johnson et al. [156] showed that ancient pathogen genomics is a useful tool for infectious disease. Additionally, ancient plant DNA could play a big role in discussions about resilience, sustainability, and food system sovereignty [158], as well as the prognosis of new and recurring diseases [159].

This chapter suggests using new approaches to study plant-pathogen interactions in light of climate change, and using metagenomics and ancient DNA to predict future interactions and find new treatments.

KEYWORDS

- **Climate change**
- **Metagenomics**
- **Non-native species**
- **Pathogen detection**
- **Pathogen interactions**
- **Pest and diseases**
- **Plant health**

REFERENCES

1. Govindaraj, M., Pattanashetti, S. K., Patne, N., Kanatti, A. A., & Ciftci, Y. O., (2018). Breeding cultivars for heat stress tolerance in staple food crops. *Next Generation Plant Breeding* (pp. 45–74). London (UK): IntechOpen.

2. Dinham, B., & Malik, S., (2003). Pesticides, and human rights. *International Journal of Occupational and Environmental Health, 9*(1), 40–52.
3. Vagelas, I., Michail, G., & Reizopoulou, A., (2020). Bacteria isolated from cave aquatic environment exhibit significant antifungal activity against phytopathogenic fungi and oomycetes. *IRJET, 2020*, 1781–1785. https://www.irjet.net/archives/V7/i12/IRJET-V7I12328.pdf (accessed on 9 January 2023).
4. Michail, G., Reizopoulou, A., & Vagelas, I., (2020). First report of *Serratia* species isolated from subterranean cave aquatic environment. *IRJET,* 1776–1780. https://www.irjet.net/archives/V7/i12/IRJET-V7I12327.pdf (accessed on 9 January 2023).
5. Schumann, G. L., (1991). *Plant Diseases: Their Biology and Social Impact.* St Paul, USA, APS Press.
6. Saunders, D. G., Pretorius, Z. A., & Hovmøller, M. S., (2019). Tackling the re-emergence of wheat stem rust in Western Europe. *Communications Biology, 2*(1), 1–3.
7. Cooke, D. E. L., Cano, L. M., Raffaele, S., Bain, R. A., Cooke, L. R., Etherington, G. J., Deahl, K. L., et al., (2012). Genome analyses of an aggressive and invasive lineage of the Irish potato famine pathogen. *PLoS Pathogens, 8*(10), e1002940 [online]. https://doi.org/10.1371/journal.ppat.1002940.
8. Schneider, K., Van Der, W. W., Cendoya, M., Mourits, M., Navas-Cortes, J. A., Vicent, A., & Lansink, A. O., (2020). Impact of *Xylella fastidiosa* subspecies *pauca* in European olives. *PNAS, 117*, 9250–9259.
9. CABI., (2021). Bursaphelenchus xylophilus (pine wilt nematode) datasheet. In: *Invasive Species Compendium*, (2021). https://www.cabi.org/isc/datasheet/10448#todistribution (accessed on 9 January 2023).
10. Dean, R. A., Van, K. J. A., Pretorius, Z. A., Hammond-Kosack, K. E., Di Pietro, A., Spanu, P. D., Rudd, J. J., et al., (2012). The top 10 fungal pathogens in molecular plant pathology. *Molecular Plant Pathology, 13*(4), 414–430.
11. Wang, K., Zhang, H., Zhu, W., Peng, J., Li, X., Wang, Y., & Qi, Z., (2022). Preliminary study of resistance mechanism of *Botrytis cinerea* to SYAUP-CN-26. *Molecules, 27*(3), 936.
12. Ou, S. H., (1980). Pathogen variability and host resistance in rice blast disease. *Annu. Rev. Phytopathol., 18*, 167–187.
13. Ruan, H., Tian, P., Shi, N., Du, Y., Chen, F., & Chen, F., (2022). Characterization of pyraclostrobin-resistant *Magnaporthe oryzae*. *Journal of Phytopathology, 170*(4), 233–241.
14. Vagelas, I., (2021). Important foliar wheat diseases and their management: Field studies in Greece. *Mod. Concep. Dev. Agron.,8*(1). MCDA. 000680. doi: 10.31031/MCDA.2021.08.000680.
15. Orton, E. S., Deller, S., & Brown, J. K. M., (2011). *Mycosphaerella graminicola*: From genomics to disease control. *Mol. Plant Pathol., 12*, 413–424.
16. Wyand, R. A., & Brown, J. K. M., (2003). Genetic and forma specialis diversity in *Blumeria graminis* of cereals and its implications for host–pathogen coevolution. *Mol. Plant Pathol., 4*, 187–198.
17. Cieplak, M., Nucia, A., Ociepa, T., & Okoń, S., (2022). Virulence structure and genetic diversity of *Blumeria graminis* f. sp. *avenae* from different regions of Europe. *Plants, 11*(10), 1358.
18. Armstrong, G. M., & Armstrong, J. K., (1981). Formae speciales and races of *Fusarium oxysporum* causing wilt diseases. In: Cook, R., (ed.), *Fusarium: Diseases, Biology and Taxonomy* (pp. 391–399). University Park, PA: Penn State University Press.

19. Yang, X., Gong, R., Chu, Y., Liu, S., Xiang, D., & Li, C., (2022). Mechanistic insights into stereospecific antifungal activity of chiral fungicide prothioconazole against *Fusarium oxysporum* f. sp. *cubense*. *International Journal of Molecular Sciences, 23*(4), 2352.
20. Liebal, U. W., Ullmann, L., Lieven, C., Kohl, P., Wibberg, D., Zambanini, T., & Blank, L. M., (2022). *Ustilago maydis* metabolic characterization and growth quantification with a genome-scale metabolic model. *Journal of Fungi, 8*(5), 524.
21. Prusky, D., (1996). Pathogen quiescence in postharvest diseases. *Annu. Rev. Phytopathol.,34*, 413–434.
22. Guevara-Suarez, M., Cárdenas, M., Jiménez, P., Afanador-Kafuri, L., & Restrepo, S., (2022). Colletotrichum species complexes Associated with Crops in Northern South America: A review. *Agronomy, 12*(3), 548.
23. Zadoks, J. C., (2008). The potato murrain on the European continent and the revolutions of 1848. *Potato Res., 51*, 5–45.
24. Haverkort, A., Boonekamp, P., Hutten, R., Jacobsen, E., Lotz, L., Kessel, G., Visser, R., & Van, D. V. E., (2008). Societal costs of late blight in potato and prospects of durable resistance through cisgenic modification. *Potato Res., 51*, 47–57.
25. Fisher, M. C., Henk, D. A., Briggs, C. J., Brownstein, J. S., Madoff, L. C., McCraw, S. L., & Gurr, S. J., (2012). Emerging fungal threats to animal, plant, and ecosystem health. *Nature, 484*, 186–194.
26. Kamoun, S., Furzer, O. J., Jones, J. D., Judelson, H. S., Ali, G. S., Dalio, R. J., Roy, S. G., et al., (2015). The top 10 oomycete pathogens in molecular plant pathology: Top 10 oomycete plant pathogens. *Molecular Plant Pathology, 16*, 413–434.
27. Abdusattorovich, C. E., & Avazovich, S. A., (2022). Tomato plant diseases and how to control them. *International Journal of Life Science and Agriculture Research, 1*(1), 08–10.
28. Werres, S., Marwitz, R., Veld, W., De Cock, A., Bonants, P. J. M., De Weerdt, M., Themann, K., et al., (2001). *Phytophthora ramorum* sp. *Nov*., a new pathogen on *Rhododendron* and *Viburnum*. *Mycol. Res., 105*, 1155–1165.
29. Rizzo, D. M., Garbelotto, M., & Hansen, E. M., (2005). *Phytophthora ramorum*: Integrative research and management of an emerging pathogen in California and Oregon forests.*Annu. Rev. Phytopathol., 43*, 309–335.
30. Grünwald, N. J., Goss, E. M., & Press, C. M., (2008). *Phytophthora ramorum*: A pathogen with a remarkably wide host-range causing sudden oak death on oaks and ramorum blight on woody ornamentals. *Mol. Plant Pathol., 9*, 729–740.
31. Brasier, C., & Webber, J., (2010). Plant pathology: Sudden larch death. *Nature, 466*, 824–825.
32. Dale, A. L., Feau, N., Berube, J. A., Ponchart, J., Bilodeau, G. J., & Hamelin, R. C., (2022). Urban environments harbor greater oomycete and *Phytophthora* diversity, creating a bridgehead for potential new pathogens to natural ecosystems. *Environmental DNA*.
33. Thompson, S. E., Levin, S., & Rodriguez-Iturbe, I., (2014). Rainfall and temperature changes have confounding impacts on *Phytophthora cinnamomi* occurrence risk in the Southwestern USA under climate change scenarios. *Global Change Biol., 20*, 1299–1312.
34. Miyambo, T. M., Backer, R., Engelbrecht, J., Joubert, F., Van, D. M. N. A., & Van, D. B. N., (2022). The identification and characterization of endopolygalacturonases in a South African isolate of *Phytophthora cinnamomi*. *Microorganisms, 10*(5), 1061.

70. Pérombelon, A. M., & Kelman, A., (1980). Ecology of the soft rot *Erwinias*. *Annual Review of Phytopathology, 18*, 361–387.
71. Wlodarczyk, S. R., Costa-Silva, T. A., Pessoa-Jr, A., Madeira, P., & Monteiro, G., (2019). Effect of osmolytes on the activity of anti-cancer enzyme L-asparaginase II from *Erwinia chrysanthemi*. *Process Biochemistry, 81*, 123–131.
72. Hashemi, T. M., Primiceri, E., Chiriacò, M. S., Poltronieri, P., Bahar, M., & Maruccio, G., (2020). *Pectobacterium atrosepticum* biosensor for monitoring blackleg and soft rot disease of potato. *Biosensors, 10*(6), 64.
73. Czajkowski, R., Rabalski, L., Kosinski, M., De Neergaard, E., & Harding, S., (2021). High-quality complete genome resource of pathogenic bacterium *Pectobacterium atrosepticum* strain green1 isolated from potato (*Solanum tuberosum* L.) in Greenland. *Molecular Plant-Microbe Interactions, 34*(11), 1328–1333.
74. Chandrashekar, B. S., PrasannaKumar, M. K., Parivallal, P. B., Pramesh, D., Banakar, S. N., Patil, S. S., & Mahesh, H. B., (2022). Host range and virulence diversity of *Pectobacterium carotovorum* subsp. *brasiliense* strain RDKLR infecting radish in India, and development of a LAMP-based diagnostics. *Journal of Applied Microbiology, 132*(6), 4400–4412. doi: 10.1111/jam.15553.
75. Pelham, J., (1966). Resistance in tomato to tobacco mosaic virus. *Euphytica, 15*(2), 258–267.
76. McIntyre, J. L., Dodds, J. A., & Hare, J. D., (1981). Effects of localized infections of *Nicotiana tabacum* by tobacco mosaic virus on systemic resistance against diverse pathogens and an insect. *Phytopathology, 71*(3), 297–310.
77. Scholthof, K. B. G., Shaw, J. G., & Zaitlin, M., (1999). *Tobacco Mosaic Virus: One Hundred Years of Contributions to Virology*. St. Paul, MN: American Phytopathological Society Press.
78. Scholthof, K. B. G., (2004). Tobacco mosaic virus: A model system for plant biology. *Annu. Rev. Phytopathol., 42*, 13–34.
79. Scholthof, K. G., Adkins, S., Czosnek, H., Palukaitis, P., Jacquot, E., Hohn, T. D., Hohn, B., et al., (2011). Top 10 plant viruses in molecular plant pathology. *Molecular Plant Pathology, 12*(9), 938–954.
80. Abo-Zaid, G. A., Matar, S. M., & Abdelkhalek, A., (2020). Induction of plant resistance against tobacco mosaic virus using the biocontrol agent *Streptomyces cellulosae* isolate Actino 48. *Agronomy, 10*(11), 1620.
81. Best, R. J., (1968). Tomato spotted wilt virus. *Advances in Virus Research, 13*, 65–146.
82. Tsompana, M., Abad, J., Purugganan, M., & Moyer, J. W., (2005). The molecular population genetics of the Tomato spotted wilt virus (TSWV) genome. *Molecular Ecology, 14*(1), 53–66.
83. Gupta, R., Kwon, S. Y., & Kim, S. T., (2018). An insight into the tomato spotted wilt virus (TSWV), tomato and thrips interaction. *Plant Biotechnology Reports, 12*(3), 157–163.
84. Moriones, E., & Navas-Castillo, J., (2000). Tomato yellow leaf curl virus, an emerging virus complex causing epidemics worldwide. *Virus Research, 71*(1, 2), 123–134.
85. Prasad, A., Sharma, N., Hari-Gowthem, G., Muthamilarasan, M., & Prasad, M., (2020). Tomato yellow leaf curl virus: Impact, challenges, and management. *Trends in Plant Science, 25*(9), 897–911.
86. Palukaitis, P., Roossinck, M. J., Dietzgen, R. G., & Francki, R. I., (1992). Cucumber mosaic virus. *Advances in Virus Research, 41*, 281–348.

87. Polischuk, V., Budzanivska, I., Shevchenko, T., & Oliynik, S., (2007). Evidence for plant viruses in the region of Argentina islands, antarctica. *FEMS Microbiology Ecology, 59*(2), 409–417. doi: 10.1111/j.1574-6941.2006.00242.x.
88. Jacquemond, M., (2012). Cucumber mosaic virus. *Advances in Virus Research, 84*, 439–504.
89. Li, N., Yu, C., Yin, Y., Gao, S., Wang, F., Jiao, C., & Yao, M., (2020). Pepper crop improvement against cucumber mosaic virus (CMV): A review. *Frontiers in Plant Science, 11*, 598798.
90. Kerlan, C., & Moury, B., (2008). Potato virus Y. In: Granoff, A., & Webster, R., (eds.), *Encyclopedia of Virology* (3rdedn., pp. 287–296). New York: Academic Press.
91. Elwan, E. A., Rabie, M., Abdel, A. E. E., Taha, R. M., & Fattouh, F. A., (2021). Investigation into the strain typing of Potato Virus Y isolates in Egypt. *Journal of Plant Diseases and Protection, 128*(2), 369–378.
92. Mondal, S., Wintermantel, W. M., & Gray, S. M., (2021). Virus and helper component interactions favor the transmission of recombinant potato virus Y strains. *Journal of General Virology, 102*(6), 001620.
93. Bak, A., & Emerson, J. B., (2020). Cauliflower mosaic virus (CaMV) biology, management, and relevance to GM plant detection for sustainable organic agriculture. *Front. Sustain. Food Syst., 4*, 21. doi: 10.3389/fsufs.2020.00021.
94. Varma, A., & Malathi, V. G., (2003). Emerging geminivirus problems: A serious threat to crop production. *Annals of Applied Biology, 142*(2), 145–164.
95. Mansoor, S., Briddon, R. W., Zafar, Y., & Stanley, J., (2003). Geminivirus disease complexes: An emerging threat. *Trends in Plant Science, 8*(3), 128–134.
96. Fauquet, C. M., Briddon, R. W., Brown, J. K., Moriones, E., Stanley, J., Zerbini, M., & Zhou, X., (2008). Geminivirus strain demarcation and nomenclature. *Arch. Virol., 153*, 783–821.
97. Eni, A. O., Efekemo, O. P., Onile-ere, O. A., & Pita, J. S., (2021). South West and North Central Nigeria: Assessment of cassava mosaic disease and field status of African cassava mosaic virus and East African cassava mosaic virus. *Annals of Applied Biology, 178*(3), 466–479.
98. García, J. A., Glasa, M., Cambra, M., & Candresse, T., (2014). Plum pox virus and sharka: A model potyvirus and a major disease. *Molecular Plant Pathology, 15*(3), 226–241.
99. Rodamilans, B., Valli, A., & García, J. A., (2020). Molecular plant-plum pox virus interactions. *Molecular Plant-Microbe Interactions, 33*(1), 6–17.
100. Mitkowski, N. A., & Abawi, G. S., (2003). Root-knot nematodes. *The Plant Health Instructor*. doi: 10.1094/PHI-I-2003-0917-01.
101. Vagelas, I., Pembroke, B., Gowen, S., & Davies, K. G., (2007). The control of root-knot nematodes (*Meloidogyne* spp.) by *Pseudomonas oryzihabitans* and its immunological detection on tomato roots. *Nematology, 9*, 363–370.
102. Vagelas, I., Pembroke, B., & Gowen, S., (2011). Techniques for image analysis of movement of juveniles of root-knot nematodes encumbered with *Pasteuria penetrans* spores. *Biocontrol Science and Technology, 21*, 239–250.
103. Vagelas, I., Dennett, M. D., Pembroke, B., & Gowen, S., (2012). Adhering *Pasteuria penetrans* endospores affect movements of root-knot nematode juveniles. *Phytopathologia Mediterranea, 51*, 618–624.

104. Vagelas, I., Dennett, M. D., Pembroke, B., & Gowen, S., (2013). Fitting the negative binomial distribution to *Pasteuria penetrans* spore attachment on root-knot nematodes and predicting probability of spore attachments using a Markov chain model. *Biocontrol Science and Technology, 23*, 1296–1306.
105. Abd-Elgawad, M. M., & Vagelas, I. K., (2015). 12 nematophagous bacteria: Field application and commercialization. *Biocontrol Agents of Phytonematodes*, 276. doi: 10.1079/9781780643755.0276.
106. Vagelas, I. K., (2015). 13 novel bacteria species in nematode biocontrol.*Biocontrol Agents of Phytonematodes,*310. doi: 10.1079/9781780643755.0310.
107. Tapia-Vázquez, I., Montoya-Martínez, A. C., Los Santos-Villalobos, D., Ek-Ramos, M. J., Montesinos-Matías, R., & Martínez-Anaya, C., (2022). Root-knot nematodes (*Meloidogyne* spp.) a threat to agriculture in Mexico: Biology, current control strategies, and perspectives. *World Journal of Microbiology and Biotechnology, 38*(2), 1–18.
108. Davis, E. L., & Tylka, G. L., (2000). Soybean cyst nematode disease. *The Plant Health Instructor*. doi: 10.1094/PHI-I-2000-0725-01.
109. Wainer, J., & Dinh, Q., (2021). Taxonomy, morphological and molecular identification of the potato cyst nematodes, *Globodera pallida* and *G. rostochiensis*. *Plants, 10*(1), 184.
110. Davis, E. L., & MacGuidwin, A. E., (2000). Lesion nematode disease. *The Plant Health Instructor*. doi: 10.1094/PHI-I-2000-1030-02.
111. Andreoglou, F. I., Vagelas, I., Wood, M. K., Samaliev, H. Y., & Gowen, S., (2003). Influence of temperature on the motility of *Pseudomonas oryzihabitans* and control of *Globodera rostochiensis*. *Soil Biology & Biochemistry, 35*, 1095–1101.
112. Chen, D., Zebarth, B. J., Goyer, C., Comeau, L. P., Nahar, K., & Dixon, T., (2022). Effect of biofumigation on population densities of *Pratylenchus* spp. and *Verticillium* spp. and potato yield in Eastern Canada. *American Journal of Potato Research*, 1–14.
113. Brooks, F. E., (2008). Burrowing nematode. *The Plant Health Instructor.* doi: 10.1094/PHI-I-2008-1020-01.
114. Quader, M., Riley, I. T., & Walker, G. E., (2011). Spatial and temporal distribution patterns of dagger (*Xiphinema* spp.) and root lesion (*Pratylenchus* spp.) nematodes in a South Australian vineyard. *Australasian Plant Pathology, 32*, 81–86.
115. Jones, J. T., Haegeman, A., Danchin, E. G., Gaur, H. S., Helder, J., Jones, M. G., Kikuchi, T., et al., (2013). Top 10 plant-parasitic nematodes in molecular plant pathology. *Molecular Plant Pathology, 14*(9), 946–961.
116. Mokrini, F., Waeyenberge, L., Viaene, N., Andaloussi, F. A., & Moens, M., (2016). Diversity of root-lesion nematodes (*Pratylenchus* spp.) associated with wheat (*Triticum aestivum* and *T. durum*) in Morocco. *Nematology,18*, 781–801.
117. Cho, D. P., Park, K. S., Kim, Y., Koh, K., & Park, Y., (2017). Study on hot water immersion treatment for control of *Meloidogyne* spp. and *Pratylenchus* spp. in a ginger, *Zingiber officinale*. *Korean Journal of Applied Entomology*, 171–177.
118. Hölscher, D., Dhakshinamoorthy, S., Alexandrov, T., Becker, M., Bretschneider, T., Buerkert, A., Crecelius, A., et al., (2013). Phenalenone-type phytoalexins mediate resistance of banana plants (*Musa* spp.) to the burrowing nematode *Radopholus similis*. *Proceedings of the National Academy of Sciences, 111*, 105–110.
119. Seenivasan, N., (2017). Identification of burrowing nematode (*Radopholus similis*) resistance in banana (*Musa* spp) genotypes for natural and challenge inoculated populations. *Archives of Phytopathology and Plant Protection, 50*, 415–437.

120. De Rocha, A. J., Dos Ferreira, M. S., De Rocha, L. S., Oliveira, S. A., Amorim, E. P., Mizubuti, E. S.,& Haddad, F., (2020). Interaction between *Fusarium oxysporum* f. sp. *cubense* and *Radopholus similis* can lead to changes in the resistance of banana cultivars to *Fusarium* wilt. *European Journal of Plant Pathology*, 1–15.
121. Dhakshinamoorthy, S., Galon, E. J., Elsen, A., & De Waele, D., (2021). Attraction and penetration of Musa by the burrowing nematode, *Radopholus similis*, in an autotrophic bipartite culture system. *Nematology, 23*(10), 1139–1152.
122. Subbotin, S. A., Madani, M., Krall, E., Sturhan, D., & Moens, M., (2005). Molecular diagnostics, taxonomy, and phylogeny of the stem nematode *Ditylenchus dipsaci* species complex based on the sequences of the internal transcribed spacer-rDNA. *Phytopathology, 95*(11), 1308–1315.
123. Storelli, A., Kiewnick, S., Daub, M., Mahlein, A. K., Schumann, M., Beyer, W., & Keiser, A., (2021). Virulence and pathogenicity of four *Ditylenchus dipsaci* populations on sugar beet. European *Journal of Plant Pathology, 161*(1), 63–71.
124. Donald, P. A., Stamps, W. T., Linit, M. J., & Todd, T. C., (2016). Pine wilt disease. In: Todd, T. C., (ed.), *The Plant Health Instructor*. doi: 10.1094/PHI-I-2003-0130-01.
125. Filipiak, A., Malewski, T., Matczynska, E., & Tomalak, M., (2020). Molecular variation among virulent and avirulent strains of the quarantine nematode *Bursaphelenchus xylophilus*. *Molecular Genetics and Genomics, 296*, 259–269.
126. Tuomola, J., Gruffudd, H., Ruosteenoja, K., & Hannunen, S., (2021). Could pine wood nematode (*Bursaphelenchus xylophilus*) cause pine wilt disease or even establish inside healthy trees in Finland now—Or ever? *Forests, 12*(12), 1679.
127. Finetti-Sialer, M. M., & Ciancio, A., (2005). Isolate-specific detection of grapevine fanleaf virus from *Xiphinema index* through DNA-based molecular probes. *Phytopathology, 95*(3), 262–268.
128. Zyl, S. V., Vivier, M. A., & Walker, M. A., (2016). *Xiphinema index* and its relationship to grapevines: A review. *South African Journal of Enology and Viticulture, 33*, 21–32.
129. Schurig, J., Ipach, U., Hahn, M., & Winterhagen, P., (2021). Evaluating nematode resistance of grapevine rootstocks based on *Xiphinema index* reproduction rates in a fast screening assay. *EuropeanJournal of Plant Pathology, 160*(1), 233–238.
130. Todd, E. H., & Atkins, J. G., (1958). White tip disease of rice. I. Symptoms, laboratory culture of nematodes, and pathogenicity tests. *Phytopathology, 48*, 632–637.
131. Uebayashi, Y., Amano, T., & Nakanishi, I., (1976). Studies on the abnormal rice kernel "Kokutenmai." Mechanism of the symptoms development. *Japanese Journal of Nematology, 6*, 67–72.
132. Hoshino, S., & Togashi, K., (2021). Long-term population dynamics of *Aphelenchoides Besseyi* on *Oryza sativa* in a paddy field, and changes in the pathological and ecological traits of the two populations. *Nematology, 24*(4), 413–429.
133. Kriticos, D. J., Ota, N., Hutchison, W. D., Beddow, J., Walsh, T., Tay, W. T., Borchert, D. M., et al., (2015). The potential distribution of invading *Helicoverpa armigera* in North America: Is it just a matter of time? *PLoS One, 10*(3), e0119618. doi: 10.1371/journal.pone.0119618. *Erratum in: PLoS One, 10*(7), e0133224. Paula-Moreas, Silvana V [corrected to Paula-Moraes, Silvana V]. PMID: 25786260; PMCID: PMC4364701.
134. Haile, F., Nowatzki, T., & Storer, N., (2021). Overview of pest status, potential risk, and management considerations of *Helicoverpa armigera* (*Lepidoptera: Noctuidae*) for US soybean production. *Journal of Integrated Pest Management, 12*(1), 3.

135. Brown, J. K., & Tellier, A., (2011). Plant-parasite coevolution: Bridging the gap between genetics and ecology. *Annual Review of Phytopathology, 49*, 345–367.
136. Wells, J. M., Raju, B. C., Hung, H. Y., Weisburg, W. G., Mandelco-Paul, L., & Brenner, D. J., (1987). *Xylella fastidiosa* gen. *Nov*., sp. *Nov*: Gram-negative, xylem-limited, fastidious plant bacteria related to *Xanthomonas* spp. *International Journal of Systematic and Evolutionary Microbiology, 37*(2), 136–143.
137. European Food Safety Authority (EFSA), (2020). Update of the *Xylella* spp. host plant database–systematic literature search up to 30 June 2019. *EFSA Journal, 18*(4), e06114.
138. Chatterjee, S., Almeida, R. P. P., & Lindow, S., (2008). Living in two worlds: The plant and insect lifestyles of *Xylella fastidiosa. Annu. Rev. Phytopathol.*, *46*, 243–271.
139. Cornara, D., Bosco, D., & Fereres, A., (2018). *Philaenus spumarius*: When an old acquaintance becomes a new threat to European agriculture. *Journal of Pest Science, 91*(3), 957–972.
140. Saponari, M., Loconsole, G., Cornara, D., Yokomi, R. K., Stradis, A. D., Boscia, D., Bosco, D., et al., (2014). Infectivity and transmission of *Xylella fastidiosa* by *Philaenus spumarius* (*Hemiptera: Aphrophoridae*) in Apulia, Italy. *Journal of Economic Entomology, 107*(4), 1316–1319.
141. Cornara, D., Saponari, M., Zeilinger, A. R., De Stradis, A., Boscia, D., Loconsole, G., Bosco, D., et al., (2016). Spittlebugs as vectors of *Xylella fastidiosa* in olive orchards in Italy. *Journal of Pest Science, 90*, 521–530.
142. Cornara, D., Morente, M., Markheiser, A., Bodino, N., Tsai, C., Fereres, A., Redak, R. A., et al., (2019). An overview of the worldwide vectors of *Xylella fastidiosa. Entomologia Generalis, 157*–181.
143. Godefroid, M., Morente, M., Schartel, T. E., Cornara, D., Purcell, A. S., Gallego, D., Moreno, A., et al., (2021). Climate tolerances of *Philaenus spumarius* should be considered in risk assessment of disease outbreaks related to *Xylella fastidiosa. Journal of Pest Science, 95*(2), 855–868.
144. Huang, J., & Hao, H., (2020). Effects of climate change and crop planting structure on the abundance of cotton bollworm, *Helicoverpa armigera* (Hübner), (*Lepidoptera: Noctuidae*). *Ecology and Evolution, 10*(3), 1324–1338.
145. Paul, P. A., McMullen, M. P., Hershman, D. E., & Madden, L. V., (2010). Meta-analysis of the effects of triazole-based fungicides on wheat yield and test weight as influenced by *Fusarium* head blight intensity. *Phytopathology, 100*(2), 160–171.
146. McDermott, F., (2004). Palaeo-climate reconstruction from stable isotope variations in speleothems: A review. *Quaternary Science Reviews, 23*(7, 8), 901–918.
147. Chefaoui, R. M., Duarte, C. M., & Serrão, E. A., (2017). Palaeoclimatic conditions in the Mediterranean explain genetic diversity of *Posidonia oceanica* seagrass meadows. *Scientific Reports, 7*(1), 1–8.
148. Michail, G., Karapetsi, L., Madesis, P., Reizopoulou, A., & Vagelas, I., (2021). Metataxonomic analysis of bacteria entrapped in a stalactite's core and their possible environmental origins. *Microorganisms, 9*(12), 2411.
149. Spence, N., Hill, L., & Morris, J., (2019). How the global threat of pests and diseases impacts plants, people, and the planet. *Plants People Planet*. doi: 10.1002/ppp3.10088.
150. Skendžić, S., Zovko, M., Pajač Živković, I., Lešić, V., & Lemić, D., (2021). Effect of climate change on introduced and native agricultural invasive insect pests in Europe. *Insects, 12*. doi: 10.3390/insects12110985.

151. Godefroid, M., Cruaud, A., Streito, J. C., Rasplus, J., & Rossi, J., (2019). *Xylella fastidiosa*: Climate suitability of European continent. *Scientific Reports, 9*. doi: 1 0.1038/s41598-019-45365-y.
152. Martelli, G. P., Boscia, D., Porcelli, F., & Saponari, M., (2016). The olive quick decline syndrome in southeast Italy: A threatening phytosanitary emergency. *J. Phytopathol, 144*, 235–243. doi: 10.1007/s10658-015-0784-7.
153. Gautam, H. R., (2019). Impact of climate change on disease scenario in crops. *Agricultural Research Journal*. doi: 10.5958/2395-146x.2019.00001.2.
154. Fenu, G., & Malloci, F. M., (2021). Forecasting plant and crop disease: An explorative study on current algorithms. *Big Data Cogn. Comput.*, *5*, 2. doi: 10.3390/bdcc5010002.
155. Cordova, L. G., Madden, L. V., Amiri, A., Schnabel, G., & Peres, N. A., (2017). Meta-analysis of a web-based disease forecast system for control of anthracnose and botrytis fruit rots of strawberry in Southeastern United States. *Plant Disease, 101*(11), 1910–1917. doi: 10.1094/PDIS-04-17-0477-RE.
156. Johnson, M. A., Liu, H., Bush, E., Sharma, P., Yang, S., Mazloom, R., Heath, L. S., et al., (2022). Investigating plant disease outbreaks with long-read metagenomics: Sensitive detection and highly resolved phylogenetic reconstruction applied to *Xylella fastidiosa*. *Microbial Genomics, 8*. doi: 10.1099/mgen.0.000822.
157. Spyrou, M. A., Bos, K. I., Herbig, A., & Krause, J., (2019). Ancient pathogen genomics as an emerging tool for infectious disease research. *Nature Reviews Genetics,20*, 323–340. doi: 10.1038/s41576-019-0119-1.
158. Przelomska, N. A., Armstrong, C. G., & Kistler, L., (2020). Ancient plant DNA as a window into the cultural heritage and biodiversity of our food system. *Frontiers in Ecology and Evolution*. doi: 10.3389/fevo.2020.00074.
159. Gaeta, R., (2021). Ancient DNA and paleogenetics: Risks and potentiality. *Pathologica, 113*, 141–146. doi: 10.32074/1591-951X-146.

CHAPTER 2

Production of Microbial Biostimulants as a Response to the Modern Agricultural Need for Productivity and Plant Health

LOBNA HAJJI-HEDFI,[1] DINA S. S. IBRAHIM,[2] SOUKAINA BEN OTHMEN,[1] SOZAN E. EL-ABEID,[3,4] WASSILA HLAOUA,[5] MOHAMED A. MOSA,[3,4] ABDELHAK RHOUMA,[1] SHAIMAA N. RIAD,[2] SHIMAA GHAREEB,[6] MARWA M. EL-DERINY,[2] HOSSAM E. HARB,[7] YOMNA A. MOHAMED,[7] and KHALED ELMNASRI[1]

[1]*Regional Center of Agricultural Research of Sidi Bouzid, CRRA, Tunisia*

[2]*Nematodes Diseases Department and Central Lab of Biotechnology, Plant Pathology Research Institute, Agriculture Research Center, Giza, Egypt*

[3]*Nanotechnology and Advanced Nano-Materials Laboratory (NANML), Plant Pathology Research Institute, Agricultural Research Center, Giza, Egypt*

[4]*Mycology and Disease Survey Research Department, Plant Pathology Research Institute, Agricultural Research Center, Giza, Egypt*

[5]*Department of Biological Sciences and Plant Protection, Higher Agronomic Institute of Chott-Meriem, Tunisia*

[6]*Laboratory of Pest Physiology, Plant Protection Research Institute, Agricultural Research Center, Ministry of Agriculture and Land Reclamation, Giza, Egypt*

[7]*Biotechnology Lab, Faculty of Agriculture, Ain Shams University, Cairo, Egypt*

Microbial Biostimulants: Biorational Pesticides for Management of Plant Pathogens.
Faheem Ahmad and Rakesh Pandey (Eds.)

ABSTRACT

Currently, modern agriculture is increasingly in need of alternatives to synthetic chemicals, fertilizers, and pesticides. This is in response to changes in international law and regulations, as well as consumer demands for food that is free of potentially toxic residues. Microbial-based biostimulants offer an innovative biotechnological solution for sustainable agriculture, replacing conventional practices. Microbial inoculants typically consist of bacteria, fungi, and arbuscular mycorrhizal fungi (AMF). Many microorganisms have individually beneficial effects on plant health and yield, which can be further enhanced when they act together in consortia. Microbial consortia involve compatible microorganisms that have additive advantageous effects on soil and crop quality. These biological agents activate processes that improve plant development, stress tolerance, bio-control, crop production, and quality when applied to plants or the rhizosphere. Moreover, microbial products can substitute or complement chemical fertilizers and pesticides, providing these beneficial effects. This chapter expands our understanding of the numerous positive effects that microorganisms can have on crop health and yield, soil quality, biodiversity, human health, the economy, and the environment. Additionally, the potential challenges associated with the commercial use of microbial biostimulants and the development of product formulations will be discussed.

2.1 INTRODUCTION

Global ecosystems, especially agroecosystems, are being disrupted by climate change and global warming caused by climate variability [1]. Extreme climate change can greatly affect crop quality and quantity by introducing various environmental stresses to crops, impacting global food security. Rising temperatures, increased CO_2 levels, tropospheric O_3, and acid rain can create chronic stressors in plants, limiting their ability to respond to diseases [2]. Temperature and rainfall fluctuations are also important indicators of environmental stress [3]. Higher temperatures lead to increased crop respiration and evapotranspiration rates, more pest infestations, changes in weed patterns, and shorter crop duration [4]. Water scarcity is a major concern for agriculture, and climate predictions suggest it will worsen [5]. Lack of water affects a crop's evapotranspiration, causing stomata to close. This damages plants' gas exchange and photosynthesis, affecting biomass production [6]. Reduced evapotranspiration also affects nutrient absorption. Additionally, the widespread use of mineral fertilizers and pesticides has negatively impacted agro-ecosystems and human health [7, 8].

Sustainable methods are needed to support increased agricultural production with limited cultivable land. In recent decades, several projects have aimed to increase agricultural productivity by eliminating hazardous agrochemicals like pesticides and fertilizers [9, 10]. Plant biostimulants (PBs) are one such group of fertilizing products available on the market. PBs contain substances and/or microorganisms that stimulate natural processes in plants or the rhizosphere. They improve nutrient uptake and efficiency, tolerance to stress, and crop quality. Studies show that PBs do not directly act against pests, so their use does not interfere with pesticide regulations [11]. According to EU Regulation 2019/1009, PBs are products that stimulate plant nutrition processes independently of the nutrients they contain, with the sole purpose of improving characteristics such as nutrient use efficiency, abiotic stress tolerance, and nutrient availability in the soil and rhizosphere.

Many theories have been proposed to define PB arrangement. Some authors suggested classifying biostimulants based on their action mode and bioactive component origin [12]. Others proposed classifying biostimulants based on their direct action on plants and plant reactions, rather than their composition [7]. By the way, Du Jardin et al. [13] emphasized the importance of the final impact on plant productivity. They suggested that the definition of biostimulants should focus on their agricultural functions, not their constituents or modes of action.

According to EU Regulation 2019/1009, PB can contain mixtures of substances and microorganisms, so they can be categorized as microbial or non-microbial PB. Applying beneficial microbial PB is a useful approach for sustainable crop management, addressing the challenges of modern agriculture.

2.2 MICROBIAL-BASED BIOSTIMULANTS

Microorganisms have always been linked to plants, giving them many benefits through various communication mechanisms [14, 15]. Microbial inoculants are mostly bacteria, fungi, and arbuscular mycorrhizal fungi (AMF). These are found in soil, plants, plant remains, water, and composted fertilizers [16–18]. Microbial biostimulants are super interesting for plant growth and development and are found in different parts of the roots like the rhizosphere, phyllosphere, and endosphere. Plants co-evolve with these microbes and form a "microbial symbiosis," where symbionts play important roles in plant nourishment, development, and productivity, as well as increasing tolerance to stress and diseases [19, 20].

2.2.1 BACTERIA-BASED BIOSTIMULANTS

Regarding their application in agriculture, two main types of biostimulants within this group should be considered: mutualistic endosymbionts belonging to the *Rhizobium* category and mutualistic rhizobacteria, commonly called PGPRs (plant growth-promoting rhizobacteria) present in the rhizosphere. *Rhizobium* and associated taxa are commercialized as biofertilizers that help plants absorb nutrients, while PGPRs are multifunctional and influence the general plant lifecycle, including nourishment and development, reaction to aggressors and abiotic stress, as well as relations with other microorganisms in the agro-ecosystems [21–28]. *Rhizobium, Azotobacter, Azospirillum, Arthrobacter, Bacillus, Enterobacter, Clostridium, Klebsiella, Serratia,* and *Pseudomonas* are the genera that have been most extensively investigated among the numerous others, largely because of their widespread distribution [29].

The PGPR can establish contacts inside the root. A number of them can even move from the root cortex to the vascular system across the endoderm and survive as endophytes (inside different plant organs: leaves, tubers, stem…). PGPR formulations are a key biostimulant class because they boost plant development by facilitating nutrient availability and use in the root's rhizosphere. Many PGPRs are the origin of phytohormone synthesis stimulation via a variety of ways based on the biostimulant class [30].

2.2.2 FUNGI-BASED BIOSTIMULANTS

Fungi are well-known microorganisms that interact with plant roots in various ways, either as parasites or in mutualistic symbioses [31]. Fungi and plants have co-evolved since the beginning of terrestrial life, developing a wide range of interactions that can be seen in the divergence between mutualism and parasitism [32, 33]. The mutualistic associations between plant roots and fungi, known as mycorrhizae, have been recognized for their benefits and are of great interest in sustainable agriculture [34–39].

Arbuscular-forming mycorrhiza (AMF) is a common type of endomycorrhiza found in plants, where fungal hyphae of Glomeromycota species penetrate root cortical cells and form branched structures called arbuscules [31, 32]. AMF also forms tripartite connections with rhizobacteria and plants, which are important in real-world field conditions [38]. To maximize the benefits of mycorrhizal connections, crop management strategies and cultivar selection should favor positive interactions between plants and microorganisms [35, 36, 40, 41]. However, the practical challenges of disseminating AMF due to their biotrophic nature [42] and the limited information on host specificity and dynamics of mycorrhizal groups in agroecosystems are major obstacles to their use.

Other endophytic fungi, including Ascomycota and Basidiomycota (such as *Trichoderma* spp. and *Piriformospora indica* (*P. indica*) as a model organism), can exist outside of the plant in addition to their ability to colonize roots and transfer nutrients to their hosts through poorly understood mechanisms [31]. They are becoming increasingly popular as plant inoculants that are easier to replicate *in vitro*, allowing for the investigation of the mechanisms of nutrient transmission between the fungal endosymbionts and the hosts. Some of these fungi, particularly *Trichoderma* spp., have been extensively studied and utilized for their bio-pesticidal (mycoparasitic) and bio-control (promoting disease resistance) abilities, as well as serving as enzyme sources for biotechnological industries [43, 44]. Based on these effects, these fungal endophytes can be classified as biostimulants; however, their agricultural applications are currently supported by claims that they are biopesticides.

2.2.3 MICROBIAL CONSORTIA AS PLANT BIOSTIMULANTS (PBS)

Single strains of certain microorganisms are known to be effective as biostimulants, biocontrol agents, or in improving plant resilience to biotic and abiotic stresses. In order to enhance the efficiency of these microbes, complement their performance, and promote sustainable and resilient agriculture, microorganisms are used in consortia and/or in association with regenerative agricultural practices [45, 46]. Recent studies have shown synergistic or additive interactions, highlighting their potential as biostimulants and plant health protectors [47, 48]. Therefore, it is important to study these microbial mixtures *in vitro* before field application, to assess their compatibility and develop appropriate formulation recipes and commercialization technology [49].

Several successful case studies of consortia are cited, including the fungus-bacteria interaction of *Trichoderma* and *Azotobacter* [45, 46]. These microbial consortia have demonstrated high compatibility with different compounds such as polymers, seaweeds, and animal-derived products in their product formulation, resulting in significant improvements in crop yield quantity and quality, making them suitable for sustainable agriculture [45, 46] (*see* Tables 2.1 and 2.2). Recently, numerous studies have shown the promising potential of the most relevant AMF-PGPR microbial consortia, which can be utilized as agricultural probiotics. This fungi-bacteria interaction enhances the availability of minerals and nutrients, promotes plant growth and yield, and protects plants from abiotic and biotic stresses [50, 51]. However, the impact of microbial consortia inoculants on the native rhizosphere microbiota is still not well understood. Currently, metagenomics is improving our understanding of changes in microbial communities [47].

2.3 RECENT ADVANCES AND CHALLENGES IN APPLYING MICROBIAL AGENTS TO CONTROL PLANT PATHOGENS AND PESTS

Plant diseases account for 20 to 40% of global crop losses [52]. Plant-pathogen interactions are complicated by climate change because environmental factors modify plant sensitivity, parasite life cycles, pathogen life cycles, and host-pathogen physiology and interactions [53, 54]. However, the continued increase in pesticide usage (including the diversity of used chemicals, their dosages, and application frequency) [55, 56], as well as the potential loss of efficacy [54], may increase disease resistance [57]. In this situation, employing biological control based on microbial agents will become a critical choice for mitigating the increased consequences of pesticide applications. Indeed, microbial-based pest management represents a safe and eco-friendly substitute for chemical pesticides. The most significant advantage of microbial insecticides is their safety since they are restricted to a single group or species of insects [58]. They also promote the formation of good soil microbes, which promotes root and plant growth. In this way, they contribute to an increase in agricultural yield. Although most microbial pesticides are not deactivated or harmed by conventional insecticide residues, they can be employed in conjunction with chemical pesticides if necessary to guarantee more significant crop protection. Moreover, considerable benefits to agricultural science and community health programs are especially linked to bio-pesticide use [59].

TABLE 2.1 A Summary of Some Commercial Microbial Bio-Products and Their Potential Market Use

Microbial Agent	Bioproduct	Host Range	Usage
		Bacteria	
Bacillus thuringiensis var. *kurstaki*	SOKBt®, Bactur®, Bactospeine®, Worthy Attack®, Killer®, Dipel®, Futura®, Javelin® Thuricide®, Topside®, Tribactur®, Bioworm®, Caterpillar	Caterpillars (larvae of moths and butterflies).	Very effective against caterpillars that feed on plant leaves. Apply in the evening or on gloomy days, and aim some spray on lower surfaces or leaves, as it deactivates quickly in the sun.
Bacillus thuringiensis var. *tenebrinos*	Foil®, M-One®, M-Track®, Novardo®, Trident®	Larvae of Colorado potato beetle, elm leaf beetle adults.	Effective against Colorado potato beetle larvae and the elm leaf beetle. Like other *Bts*, it must be ingested. It is subject to breakdown in ultraviolet light and does not cycle extensively in the environment.
Bacillus thuringiensis var. *aizawai*	Certan®	Wax moth caterpillars.	Used only for control of moth infestations in honeybee hives.
(i)*Bacillus popilliae* (ii) *Bacillus lentimorbus*	Doom, Japidemic,® Grub Attack®	Larvae (grubs) of Japanese beetle.	The main Illinois lawn grub (the annual white grub, *Cyclocephala* sp.) is not susceptible to milky spore disease.
· *Bacillus sphaericus*	Vectolex CG®, Vectolex WDG®	Larvae of *Culex, Psorophora,* and *Culiseta* mosquitos, larvae of some *Aedes* spp.	Active only if ingested, for use against *Culex, Psorophora,* and *Culiseta* species; also effective against *Aedesvexans.* Remains effective in stagnant or turbid water.
		Fungi	
Beauveria bassiana	Botanigard®, Mycotrol®, Naturalis®	Aphids, mealy bugs, mites, thrips, whiteflies.	Effective against several pests. High moisture requirements, lack of storage longevity, and competition with other soil microorganisms are problems that remain to be solved.
Lagenidium giganteum	Laginex®	Larvae of most pest mosquito species.	Effective against larvae of most pest mosquito species; remains infective in the environment through dry periods. However, it is ineffective in high temperatures.

TABLE 2.1 *(Continued)*

Microbial Agent	Bioproduct	Host Range	Usage
		Protozoa	
Nosema locustae	NOLO Bait®, Grasshopper Attack®	European corn borer caterpillars and grasshoppers.	Useful for rangeland grasshopper and caterpillar control. It is only recommended on a large scale because the disease is slow-acting, and grasshoppers are very mobile.
		Viruses	
Gypsy moth nuclear polyhedrosis (NPV).	Gypchek® virus	Gypsy moth caterpillars.	All of the viral insecticides used for the control of forest pests are produced and used exclusively by the U.S. Forest Service.
Tussock moth NPV	TM Biocontrol-1®	Tussock moth caterpillars.	–
		Entomogenous Nematodes	
Steinernema feltiae (=*Neoaplectana carpocapsae*) *S. riobravis*, *S. carpocapsae* and other *Steinernema* species.	Biosafe®, Ecomask®, Scanmask®, and Vector®	Larvae of a wide variety of soil-dwelling and boring insects.	It is more effective against insects in damp soils or inside plant tissues due to moisture requirements. Long periods of storage or exposure to severe temperatures prior to usage may cause the nematodes to die or become incapacitated.
Heterorhabditis heliothis	Currently available on a wholesale basis for large-scale operations.	Larvae of a wide variety of solid, welling, and boring insects.	Used only in European countries and for research or large-scale commercial uses.

TABLE 2.2 Different European Registered Biocontrol Agents Commercially Used as Bio-Fungicides and Bactericides

Active Biocontrol Agent	Target Pathogen/Disease
Pseudomonas chlororaphis strain MA 342.	Seed-borne fungi.
Coniothyrium minitans strain CON/M/91-08 (DSM 9660).	*Sclerotinia* spp./*C.minitans* is a highly specialized hyperparasite. Its host range is restricted to certain sclerotia-forming species.
Gliocladium catenulatum strain J1446.	*Alternaria, Botrytis, Cladosporium, Didymella, Fusarium, Pythium, Rhizoctonia, Phytophthora, Verticillium, Helminthosporium*, and *Penicillium*
Phlebiopsis gigantea 14 strains.	Heterobasidionannosus.
Pythium oligandrum strain M1.	Soil-borne pathogenic fungi and *Alternaria* and *Sclerotium.*
Trichoderma asperellum (formerly *T. harzianum)* strains ICC012, T11, and TV1.	Soil-borne fungal plant pathogens (e.g., *Rhizoctonia* spp., *Sclerotium* spp., *Pythium* spp., *Phytophthora* spp., *Sclerotinia* spp., *Thielaviopsisbasicola, Verticillium* spp.)
Trichoderma atroviride (formerly *T. harzianum*)strains IMI 206040 and T11.	Soil-borne fungal plant pathogens.
Trichoderma polysporum strain IMI 206039.	Soil-borne fungal plant pathogens.
Trichoderma gamsii (formerly *T. viride*)strain ICC080.	Soil-borne fungal plant pathogens (e.g., *Pythium* spp., *Phytophthora* spp., *Sclerotinia* spp., *Sclerotium* spp., *Thielaviopsisbasicola, Rhizoctonia* spp., *Verticillium* spp.)
*Verticillium albo-atrum (*formerly *V. dahliae)* strain WCS850.	Dutch elm disease.
Bacillus subtilis strain QST 713.	Bacteria (fire blight) and fungal root diseases, *Alternaria, Aspergillus, Venturia*, gray mold.
Streptomyces K61(formely *S. griseoviridis*).	Root rot, gray mold, and foot decay.

Source: http://ec.europa.eu/food/plant/protection/evaluation/index en.htm.

2.3.1 MICROBIAL AGENTS IN CONTROLLING PLANT PATHOGENS

Microbial pesticides are derived from bacteria, fungi, algae, viruses, or protozoans that are either genetically modified or naturally occurring and possess unique pathogenic effects on target pests. By the end of 2001, there were 780 products and approximately 195 registered active ingredients for bio-pesticides. Biochemical pesticides, also known as "bio-pesticides," are compounds that occur naturally and utilize non-toxic methods to control pests. They consist of living microbes (natural enemies), natural products (phytochemicals, microbial enzymes), or a combination of both (semi-chemicals). Various pests can be managed using microorganisms as the active ingredient, including bacteria, fungi, viruses, and protozoa. However, each active component is specifically tailored to its intended pest. For example, certain fungi are effective against specific insects, while others target particular weeds. *Aedes aegypti* may be more susceptible to a bacterial species like *Bacillus thuringiensis* (Bt), whereas another strain of *B. sphaericus* may be more successful against a different mosquito species like *Culex quinquefasciatus* [60]. Microbial biostimulants can enhance plant resistance through various processes, such as ISR (Induction of Systemic Resistance). ISR is considered the plant's first line of defense, which, upon detecting specific elicitors, activates the plant's protective mechanisms and helps fend off potential diseases or harmful microorganisms. These elicitors are small structures known as pathogen/microbe-associated molecular patterns (PAMPs or MAMPs), which are recognized by pattern recognition receptors (PRRs) in the plant's innate immune system" [61].

ISR has been recognized as a crucial tool in the control of diseases by many biological control agents (BCA) due to its effectiveness against various diseases, nematodes, and insects [62, 63]. For instance, the PGPR *Bacillus amyloliquefaciens* stimulates the production of higher levels of jasmonic acid (JA), which triggers resistance against aphids in bean plants [64]. Additionally, insect attacks on plant roots and shoots elicit diverse responses and the release of different metabolites and hormones [65]. The synthesis of numerous extracellular compounds, acting as elicitors during PGPR-triggered ISR, aids in the development of plant resistance to various pathogens without causing significant alterations in defense-related genes [52]. Some of these metabolites include Acyl homoserine lactones (N-AHLs) [66], siderophores [67], volatile organic compounds (VOCs) [68], rhamnolipids [69], and cyclic lipopeptides. Most of these elicitors have been found in *Bacillus* and *Pseudomonas* species, while elicitors from other species remain largely unknown. However, these signaling metabolites require higher μM concentrations to induce immunological responses, suggesting that they may not be detected by high-affinity receptors [70]. The quorum-sensing compounds of PGPR (chemical signals for cell-cell communication) serve as novel elicitors that regulate plant tolerance to biotic stress. In a recent study, a PGP bacterium, *Staphylococcus equorum* strain EN21, promoted ISR against *Pseudomonas syringae* (pv. Tomato) infections in Arabidopsis and tomato plants through quorum quenching of acyl-homoserine [71]. Wheat and tomato also exhibited ISR activity of the elicitor oxo-C14-HSL against *Phytophthora infestans* and *Puccinia graminis* [72]. Cyclic lipopeptides produced by *Pseudomonas* are crucial for inducing ISR in monocots, such as rice.

Systemic acquired resistance (SAR) is another process that should be mentioned. SAR is considered an induced defense mechanism that offers enduring immunity against a wide range of microorganisms. It is a broad-spectrum-induced immunological mechanism that can be systematically expressed in all organs, is unique to plants, and is not dependent on the original infection. Through the induction of these two mentioned processes, PGPR could enhance the defense of their linked plants to overcome aggressors' attacks [73]. For instance, through the SAR-dependent signaling pathway, an active component of the fungicide Serifel® was found to elicit anti-viral action against *potato virus Y* (PVY) and *tomato spotted wilt virus* (TSWS) after plant treatment by *Bacillus amyloliquefaciens* (strain MBI600) [74]. In a different case, foliage treatment with the PGPR *Bacillus cereus* (AR156) improved resistance against *Pseudomonas syringae* (pv. tomato) in *Arabidopsis* by activating a SAR pathway [65]. Eventually, the modulation of phytohormones, signaling pathways, gene regulation and expression, secondary metabolites, the augmentation of bioactive compounds, the activation of ROS enzymes, and other processes are just a few of the ways that PGPR combats various abiotic and biotic stressors.

Microbial products come in different forms and are categorized based on their characteristics and origins. For example, entomo-pathogenic microorganisms have been used for over 150 years to reduce insect infestations. According to Tanzini et al. [75], "entomopathogens" are microorganisms that control insect populations to the point where they no longer economically harm crop plants. Some examples of commercial microbial insecticides are listed in Table 2.3.

2.3.1.1 ENTOMOPATHOGENIC BACTERIA

Entomopathogenic bacteria have developed various strategies to invade, infect, and kill their host. These interactions are believed to be ancient and to have evolved over a long period of co-evolution [76]. As a result, entomopathogenic bacteria are classified into two types: spore-forming bacteria such as *Bacillus* spp., *Clostridium* spp., and *Paenibacillus* spp., and non-spore-forming bacteria such as *Serratia, Photorhabdus, Pseudomonas, Xenorhabdus, and Yersinia genera* [77]. *B. thuringiensis* (*Bt*) is the world's best-known entomopathogenic species, accounting for about 95%

TABLE 2.3 Examples of Commercial Microbial Insecticides

Microorganism	Product Name	Target Pest
	Bacteria	
Bacillus thuringiensis subsp. *aizawai* strain GC-91.	Agree®	Lepidopteran insects (i.e., diamondback moth larvae).
Bacillus thuringiensis subsp. *kurstaki* strain EG7841.	Crymax®	Lepidopteran insects (i.e., beet armyworm, beet webworm, corn earworm).
Bacillus thuringiensis subsp. *kurstaki* strain ABTS-351.	Dipel®	Lepidopteran insects (i.e., almond moth, armyworm, bagworm, banana skipper, citrus cutworm, corn earworm).
Bacillus thuringiensis subsp. *kurstaki* strain EVB-113-19.	Bioprotec PLUS®	Lepidopteran insects (i.e., armyworm, banana skipper, azalea moth).
Bacillus thuringiensis subsp. *kurstaki* strain SA-11.	Javelin®	Lepidopteran insects (i.e., *Helicoverpa* spp., *Heliothis* spp., citrus cutworm, codling moth, alfalfa caterpillar, cotton leaf worm).
Bacillus thuringiensis subsp. *kurstaki* strain SA-12.	Thuricide® HPC-O	Lepidopteran insects (i.e., bagworm cabbage looper, elm spanworm, fall cankerworm, fall webworm, gypsy moth).
Bacillus thuringiensis subsp. *kurstaki* strain BMP 123.	BMP 123®	Lepidopteran insects (i.e., armyworm, diamondback moth, hornworms, podworms, loopers, tobacco budworm, tomato fruit worm, corn earworm).
Bacillus thuringiensis subsp. *kurstaki* strain EG2348.	Condor®	Lepidopteran insects (i.e., black-headed budworms, achemon sphinx, alfalfa caterpillar, armyworm, cabbage budworm, browntail moth).
Bacillus thuringiensis subsp. *tenebrionis* strain SA-10.	Trident®	Colorado potato beetle and elm leaf beetle.
Bacillus thuringiensis subsp. *israelensis* strain SA3 A.	Teknar®	Black flies (larvae), Chironomus midges, and mosquitoes (larvae).
Bacillus thuringiensis subsp. *israelensis* strain AM 65-52.	Bactimos®	Midges (larvae)
Bacillus thuringiensis subsp. *israelensis* strain SUM-6218.	Mosquito bits®	Fungus gnat (larvae), mosquitoes (larvae).
	Fungi	
Beauveria bassiana strain GHA	Mycotrol®	Aphids, whiteflies, borers, and mites.
Beauveria bassiana strain GHA	Botanigard®	Corn borers, aphids, mites, and other insects.
Beauveria bassiana strain ANT-03	Bioceres®	Aphids, white flies, thrips, plant bugs, beetles, and weevils.
Metarhizium anisopliae strain F52	Lalguard®	Thrips, vine weevils, ticks.

TABLE 2.3 *(Continued)*

Microorganism	Product Name	Target Pest
	Virus	
Spodoptera frugiperda Multiple Nucleopolyhedrovirus strain 3AP2.	Fawligen®	Fall armyworm and beet armyworm.
Chrysodeixis includes Nucleopolyhedrovirus isolate#460.	Chrysogen®	Soybean looper and cabbage looper.
Chrysodeixis includes nucleo-polyhedrovirus isolate #460 and *Helicoverpazea* nucleopolyhedrovirus strain ABA-NPV-U.	Surtivo®	Corn earworm, cotton bollworm, tomato fruit worm, tobacco budworm, soybean looper, and cabbage looper.
Autographa californica multiple nucleopolyhedrovirus strain R3.	Lepigen®	Corn earworm, cotton bollworm, tomato fruit worm, tobacco bollworm, soybean looper, cabbage looper, diamondback moth, velvet bean caterpillar, fall armyworm, beet army worn, southern armyworm.
Chrysodeixis includens Nucleopolyhedrovirus isolate #460, *Helicoverpazea* Nucleopolyhedrovirus strain ABA-NPV-U, *Autographa califor-nica*multiple nucleopolyhedrovirus strain R3, and *Spodoptera frugiperda* multiple nucleopolyhedrovirus strain 3AP2.	Surtivo Ultra® Surtivo Plus®	Corn earworm, cotton bollworm, tomato fruit worm, tobacco budworm, soybean looper, cabbage looper, diamondback moth, velvet bean caterpillar, fall armyworm, beet army worn, southern armyworm.

Source: U.S. Environmental Protection Agency.

of all microbes used in biological pest management [78]. This bacterium's pathogenic action is usually triggered by the ingestion of spores and crystalline inclusions, which contain entomocidal protein protoxins known as endotoxins. Research has demonstrated that *Bt* endotoxin is divided into two classes: Cyt for cytolytic and Cry for a crystal. Each type binds to epithelial cells in the midgut and causes their lysis, resulting in the death of insects. When larvae ingest inclusions, the alkaline midgut environment (pH 8-11) promotes crystal solubilization and, as a result, protoxin release. Host proteases recognize the protoxin's cleavage sites and perform the cleavage, producing an active toxin that binds to the designated receptors on the midgut epithelium [77, 79-81]. Moreover, *Bacillus thuringiensis* (Bt) is used to control lepidopteran insects (caterpillars) and coleopterans (beetles) [82]. The pathogen *Paenibacillus popilliae* causes milky disease in scarab larvae [83], while the inclusion body produced by *Bacillus sphaericus* is lethal to mosquito larvae [84]. *Bt* poison does not kill insects immediately, but insects may remain for a day or two after exposure, and during this period, insects refrain from food until they die. All parts of the plant that the insect consumes must be sprayed with biopesticides containing *Bt* to ensure that the toxic effect occurs. If this does not happen, it may reduce the effectiveness of the *Bt.* Therefore, all plant surfaces should be covered, especially the undersides of leaves [85]. However, *Brevibacillus laterosporus*, an aerobic spore-forming bacterium, is another species capable of producing protein toxins against mollusks, nematodes, and different insect orders such as Coleoptera, Lepidoptera, and Diptera [86]. The strain *Chromobacterium subtsugae*, named PRAA4-1T, which belongs to *Chromobacterium* spp., was found to exhibit strong insecticidal activity against various orders, such as Lepidoptera (*Plutella xylostella* L.), Rhynchota *(Bemisia tabaci* Gennadius; *Nezara viridula* L.), Coleoptera (*Leptinotarsa decemlineata* Say) [87, 88].

2.3.1.2 ENTOMOPATHOGENIC FUNGI

Around the world, entomopathogenic fungi play a significant role in insect bio-control. Their insect control strategy consists of spore attachment to the insect cuticle and germination, followed by penetration into the cuticle and internal dissemination [89]. The ability of entomopathogenic fungi to produce effective enzymes that break down the integument of insects and other cellular components is a key factor in determining their pathogenicity. These fungi secrete lipase, proteases, phospholipases, and chitinases [90]. Approximately 170 different microbial pesticides that rely on entomopathogenic fungi have been developed for insect control in greenhouse crops, fruits, and field vegetables. Most of these products are based on *Metarhizium anisopliae* or *Beauveria bassiana*. Imperfect fungal strains like *B. bassiana*, *Paecilomyces fumosoroseus*, and *M. anisopliae* are commonly investigated as microbiological insecticides. In the United States, effective management of target coleopteran, homopteran, and lepidopteran pests with these fungi requires large volumes of substrate. The available products are often based on *Beauveria bassiana* and are intended for use in greenhouses. In 1965, Boverin, the first fungal product, was created and utilized for the management of *Leptinotarsa decemlineata* and *Cydia pomonella*[91].

A number of the secondary metabolites generated by entomopathogenic fungi could serve as new sources of natural pesticides. Understanding which toxins are produced and why they are toxic to insects is crucial for gaining a better understanding of their mechanisms of action [92]. *Paecilomyces fumosoroseus* (Hyphomycetes) is one of the most pathogenic fungal species that affects all stages of both *Bemisia tabaci* and *Bemisia argentifolii.* In addition, *P. fumosoroseus* produces a variety of metabolites, including beauvericin, which is a cyclic hexadepsipeptide with insecticidal characteristics, beauverolides, and dipicolinic acid. Many different arthropod pests are known to be susceptible to the hypocrealean fungus *Hirsutella thompsonii, Beauveria bassiana, Isaria fumosorosea, M. brunneum,* and *Metarhizium anisopliae*. For pest control in various environments, some hypocrealean fungi products can be found commercially [93]. There is substantial proof that entomopathogenic fungi may be successfully delivered into plants as fungal endophytes utilizing a variety of artificial inoculation techniques. According to some reports, *B. bassiana* can preserve the host plant for a longer period of time when it exists in plant form as an endophyte. The fungus can survive for a very long time in the tissues of the host [94, 95].

2.3.1.3 ENTOMOPATHOGENIC VIRUS

An essential category of insect pathogens is viruses. Extensive research has been conducted on the advantages of utilizing insect baculoviruses for biological control. In contrast to chemical pesticides, the baculovirus has a prolonged death time, which can result in the death of its host within 4–14 days [96]. This issue has been addressed using molecular biology techniques. Baculoviruses are extremely small and primarily consist of double-stranded DNA, which is essential for viral replication. This genetic material allows the virus to easily disassemble if exposed to sunlight or other unfavorable conditions within the host's intestine. A protein coat, known as a polyhedron, covers the baculovirus particle to provide protection [97–100].

Around 13 viral families (out of 73 known families) have been identified as entomopathogenic viruses. Baculoviridae is a harmful family to various insect pest orders, including Diptera, Lepidoptera, Neuroptera, Hymenoptera, Isoptera, and Orthoptera. Granulovirus (GV) and nucleopolyhedrovirus (NPV) are the two genera that make up the Baculoviridae [101]. Although entomopathogenic viruses have various pathogenesis and replication strategies, infection almost always occurs through ingestion of the virus. Viral proteins bind to receptors and enter epithelial cells in the gut. In the case of baculoviruses, the infection often spreads to the hemocoel before infecting vital organs and tissues. The host dies within 5 to 14 days after infection [82]. Sandoz Inc. released Elcar™, the first viral pesticide, 47 years ago. Elcar™ was a prepared form of the relatively wide-ranging baculovirus Heliothis zea NPV, which infects numerous species from the *Helicoverpa* and *Heliothis* genera. After seven years, Sandoz decided to stop manufacturing. The HzSNPV was again of interest due to the pest resistance to several chemical pesticides, including pyrethroids, and the virus was registered as GemStar™[58].

2.3.1.4 OTHER ENTOMOPATHOGENIC MICROORGANISMS

Numerous other microorganisms are utilized as bio-control agents for insects, such as entomopathogenic nematodes. *Heterorhabditis* and *Steinernema* species of nematodes come in various commercial formulations for controlling insect pests in soil. These nematodes penetrate their hosts through the anus, spiracles, or mouth. Once inside the host, the nematodes release symbiotic bacteria that cause bacterial septicemia and ultimately lead to the host's death. Entomopathogenic nematodes can effectively manage insects such as gnats, weevils, and white grubs [77, 102–104]. Protozoa are also rarely used in biological pest control against insects such as *Nosema, Vairimorpha,* and *Malamoeba.* They kill insects when consumed. The insects' feeding behavior is disrupted, which causes them to starve to death. The only protozoan insecticide registered with the US Environmental Protection Agency (USEPA) over the past 10 years was *Nosema* spp. Protozoa are a highly effective source of microbial pesticides, particularly against some grasshopper species [100, 105, 106].

2.3.2 IMPROVING THE QUALITY AND EFFICACY OF ANTIMICROBIAL AGENTS UNDER NATURAL CONDITIONS

Aside from the unique properties and necessity of utilizing antimicrobial agents in pest disease control, there are a few strategies to improve the efficacy of antimicrobial agents. First and foremost, improper or incorrect uses/applications of such items should be avoided. Typically, failure is the result of a lack of information about how to use something or the conditions under which it should be used, resulting in incorrect use. Second, failures induced by the use of low-quality bio-inoculum should be distinguished. Ineffectiveness is frequently caused by the use of biocontrol agents in fertilizers or compost rather than approved phytosanitary products, causing a great deal of misunderstanding about bio-control agents.

Specific ways to improve the effectiveness of these products include:

- Critically apply antimicrobial agents coupled with agronomic practices in an integrated pest management strategy (IPM), such as solarization, anaerobic soil disinfestation, or bio-fumigation, rather than as a replacement for a chemical soil fumigant.
- Please verify the efficacy of the strain against a specific disease and determine the appropriate application circumstances.
- Please check the compatibility of the strain with the crop or cultivar, especially for rhizosphere colonists.
- Allow enough time for the bio-control agent to minimize the disease's inoculum.
- Apply under the appropriate conditions.
- Improve manufacturing processes to minimize the cost of bio-control agents.
- Adjust the quantity adequately to ensure the effectiveness of the products used.
- Develop specific substrates to enhance their effectiveness.
- Reduce the cost of bio-control agents by improving production systems.
- Evolve multi-year control strategies against phytopathogens, especially soil-borne ones, using bio-control agents.
- Treat only when the pathogen microbial population is low.

2.3.3 PRODUCT REGISTRATION CHALLENGES

Microbial pesticides are the largest class of pest-specific biopesticides among all currently employed biopesticides. At least 1,500 naturally produced microbial pesticides target insects [107]. Microbial biopesticides are widely used worldwide. For example, there are 53 registered products in the United States, 22 in Canada, and 21 in the European Union (EU) [109, 110]. However, reports of the use of similar products in Asia are inconsistent [108]. The number of microbial biopesticide registrations is increasing globally, thanks to the development of diverse technologies, and the trend towards creating microbial products is undoubtedly changing [111–113].

The development of microbial pesticides is a highly sophisticated process that goes through multiple stages, including discovery, manufacture, product development, efficacy evaluation, registration, and finally commercialization. Ravensberg[114] outlined the entire procedure from ideation to commercial production for insect pathogens, such as microbial pesticides. This model applies to any type of biopesticide and provides detailed descriptions of the key elements of each production phase. The biggest obstacle to the commercialization of microbial crop protection products is the registration process, which is time-consuming and costly. Additionally, many criteria do not adequately address the unique nature of these products, making registration even more challenging. To address this issue, regulatory innovation with harmonization is crucial at both the international and national levels [115]. The OECD Pesticide Program works to streamline and enhance registration processes worldwide. The registration period can range from three to five years, similar to the development phase. The costs of registration are often considered a determining factor. The main issue is that these costs are difficult to predict in advance, but they should not be ignored to avoid making significant mistakes. In general, biopesticides should be recognized by the law as low-risk active chemicals. For low-risk substances, a condensed dossier may be presented, and it must also demonstrate sufficient efficacy. Some microbial pesticides may encounter issues, as the half-life of low-risk compounds in the soil should not exceed 60 days. According to European legislation, the registration of phytosanitary products requires the approval of both the formulation and the active microorganism. The registration report should include information on the active compound (microorganism) and the product. The registration report includes details on the nature of the bacteria or fungi, the manufacturing procedures used to produce the microorganisms and the synthesized product, as well as the specific techniques used to identify the active bacteria or fungi and their preparation. Data on human toxicity and ecotoxicology, as well as environmental distribution and behavior, are taken into account when evaluating the safety of a microorganism and its final product. Communication is required between the European Food Safety Authority, the commercial claimant, the selected National recorder participant, and governmental organizations to ensure the microorganism is considered safe throughout the EU. Therefore, the scientific opinion is published by EFSA. Based on this document, the European Commission decides whether to add a bacterium or fungus to the database of secure and effective compounds. Subsequently, the product information is presented at the governmental level. Authorized institutions must follow Good Experimental Practice (GEP) during effectiveness trials in the targeted geographic zone. GEP is based on the criteria of the European Plant Protection Organization (EPPO) [116].

2.3.4 CHALLENGES WITH MICROBIAL BIOPESTICIDES EFFICACY IN FIELD APPLICATION

The appropriate application of microbial biopesticides in the field is often associated with their efficacy. Using products based on microbes offers numerous advantages, whether used alone or in rotation with chemical pesticides in Integrated Pest Management (IPM) programs. The adoption of these products requires the engagement of the farming community by educating farmers about their functionality and the benefits they provide. Regulatory processes can be overly complex, time-consuming, and costly. Challenges related to microbes include formulation, shelf life, and performance improvement. Developing an effective formulation requires a comprehensive understanding of the interaction between microbial control agents, insect pests, the environment, and other relevant organisms, with a focus on efficacy. Prior to introducing any microbial control agents, it is important to comprehend the biology of the isolated microorganisms and conduct experiments to create an appropriate and usable formulation. Subsequently, the performance of the produced formulations should be evaluated [117, 118]. It is essential to continue exploring additional microorganisms that may target different insect pests with selectivity that does not harm the environment or non-target organisms. Successful biological management relies on a thorough understanding of pest biology, natural enemies, and accurate field identification of relevant insects at various life stages. Field scouting is often necessary to monitor natural enemy activity, assess its impact on insect dynamics, and predict the need for additional control measures. While the application of microbial pesticides is generally not difficult, it does require proper training and expertise in dealing with pests and pathogens [119]. To enhance the effectiveness of microbial pesticides, it is crucial to apply them at the appropriate time. Striking a balance between the benefits of using microbial pesticides and their associated costs is a challenging task. Despite years of successful application in various agricultural and forestry systems, the operational lifespan of *Bacillus thuringiensis* (Bt) is limited and depends on several factors. Developing insecticide formulations based on microorganisms is crucial to maximize storage life, improve distribution and absorption, reduce moisture loss, and, most importantly, enhance effectiveness. During application, it is important to ensure that substrates adequately cover all parts of the plant that insects will consume, such as foliage [120].

The manufacturers, social and natural scientists, decision-makers, and regulators must work together to solve any potential new economic and regulatory issues posed by the more recent microbial pesticides [121]. There are some issues facing microbial pesticides. Due to the specific mode of action of microbial pesticides, they may only be effective against certain insect species found in a field and may not be able to manage different insect species found in the intervention area, which could result in permanent harm. The effectiveness of microbial insecticides is reduced by high temperatures, desiccation, and ultraviolet light. Microbial products are also susceptible to contamination and have a short lifespan. Markets for microbial insecticides are limited due to their specificity in targeting pests. For example, insect viruses are not widely available. Specific formulation and storage procedures are required for several microbial insecticides [113, 122].

2.4 ADVANCES AND CHALLENGES FOR APPLYING MICROBIAL BIOSTIMULANTS WITH REFERENCE TO PLANT GROWTH ENHANCEMENT AND STRESS TOLERANCE

Although a relatively recent innovation in the agricultural inputs market, biostimulants are the subject of extensive research, constituting a rich documentary base [123, 124]. They interact with plants in various ways, and their use will only be beneficial if the "Microorganism-Plant-Environment" connection is properly addressed [13]. Plant-microbe interactions can be direct through symbiotic, rhizospheric, or commensal processes, where microorganisms come into contact with the surface of roots and colonize root tissue, or indirect, where microscopic organisms are present in the soil surrounding the roots [125].

A recent study by Sangiorgio et al. [2] provided a global perspective on plant–microbe–environment interactions regarding plant responses to different stresses caused by biotic or abiotic agents. Plants interact with soil microbial effectors, which play important roles in both the ecological system and plant growth stages. Therefore, the use of bacterial biostimulants that exploit mutualism, mycorrhizal associations, and interconnections between plants represents a long-term approach to improving plant quality and profitability, even under stressful conditions. Beneficial fungi, yeasts, and cyanobacteria are examples of bacterial biostimulants that have the capacity to enhance plant nutrient uptake, development, yield, and tolerance (Figure 2.1). Plants can form symbiotic relationships with fungi known as AMFs, which enhance root surface area for nutrient uptake [127]. We can categorize microbial biostimulants into three major effects or groups, either in combination or separately:

- The first effect is the presence of microorganisms that directly affect plants.
- The second effect is the secretion of compounds that act on plant growth. At this point, we need to isolate the compound and identify the mechanisms involved in its role.
- The third facilitates plant nutrition and growth through natural assimilation or transformation to improve resource efficiency.

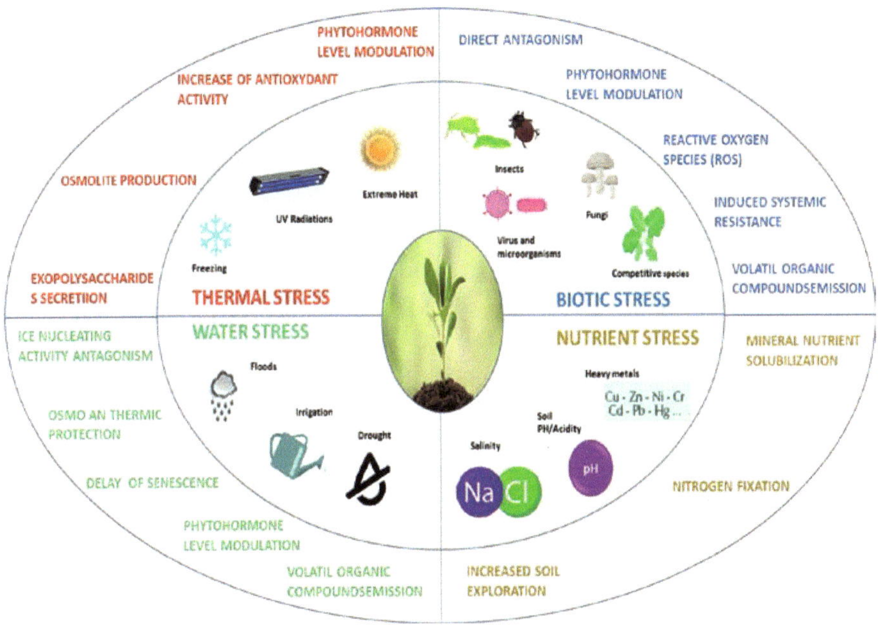

FIGURE 2.1 Protective mechanisms are used by microbial biostimulants to mitigate various stresses facing plants.

Plant microbial biostimulants enhance nutrient availability and plant productivity. They assist plants in obtaining and utilizing essential minerals by stimulating early root emission and growth. For instance, phosphorus mobilization with the help of phosphatase-releasing PGPRs can contribute to plant growth and crop yield, especially when soil inorganic phosphate is limited and the plant supports the maintenance and activity of the PGPR inoculant in the rhizosphere, such as through exudates. For instance, phosphorus mobilization via phosphatase liberation may contribute to plant development and productivity if the ionic compound in the soil is indeed restricting and the plant helps to maintain the PGPR inoculant in the rhizosphere, for example, via exudates [13] (Figure 2.2). This aids in lowering the number of nutrients wasted and the amount of energy necessary to produce fertilizers. Microbial-based biostimulants assist in increasing agricultural adaptability to climate change by reducing nutrient losses [128–131].

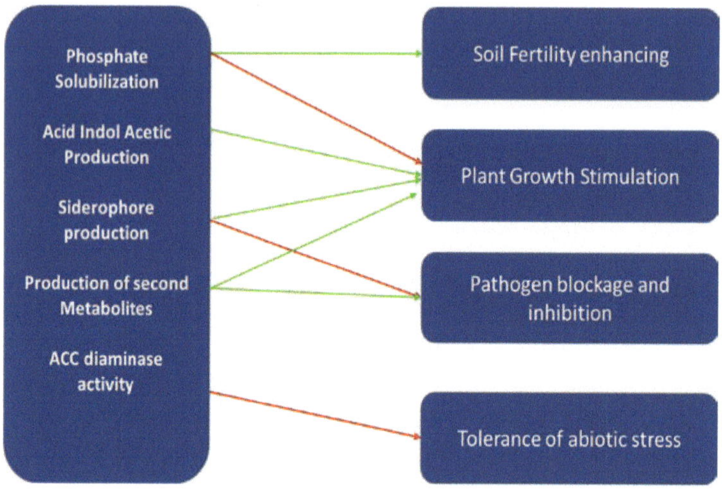

FIGURE 2.2 Mechanisms of plant-PGPR interaction (green arrows: direct effects; and red arrows: indirect effects).

2.4.1 MICROORGANISMS THAT PROMOTE PLANT GROWTH (PGPMS) AND THEIR BIOFERTILIZER ACTIVITY

PGPMs have the potential to act as biostimulants, influencing plant biological pathways and leveraging the plant's own resources to ensure its vitality and longevity [132]. Controlling bacterial behavior holds great promise for providing nourishment to plant crops. PGPR formulations, a major class of biostimulants, enhance root development, improve mineral availability, and facilitate nutrient uptake in the crop rhizosphere. PGPR bacteria stimulate phytohormone synthesis through various processes [30].

According to Kloeppe et al. [133], PGPRs are characterized by their capacity to populate the rhizosphere area, persist, proliferate, and compete with other bacteria or fungi, while stimulating plant growth. In the rhizosphere, these bacteria can be found at intra or extracellular levels [134]. At the intracellular level, the bacteria are said to be endophytic and colonize the apoplast. These bacteria are part of the Rhizobium family. Generally symbiotic, these are, in particular, the PGPRs specialized in the nodal structures of the Fabaceae. They are located on the surface or near the roots at the extracellular level, so they are rhizospheric [16]. Four main effects have been identified in these PGPRs. They can enhance nutrient uptake, regulate plant hormone production, improve tolerance to abiotic stresses, and compete with bio-aggressors [125, 135].

2.4.1.1 INCREASE IN NUTRIENT AVAILABILITY

Plant growth can be indirectly promoted by improving soil fertility in essential nutrients. Phosphorus (P) is one of the main limiting factors for plant growth, along with nitrogen and potassium. Although P is not scarce in soils, it is primarily present in insoluble forms (organic complexes, apatite), which are not readily available for plant uptake. Phosphorus solubilization offers a solution to increase the concentration of P in bioavailable forms (H_2PO^{4-}, $HPO_4{}^{2}–$) without the need for external mineral fertilizer. To achieve this, there are bacteria capable of solubilizing inorganic phosphorus, known as phosphate-solubilizing bacteria (PSB) [136]. These bacteria utilize low molecular weight molecules such as citric acid, gluconic acid, or oxalic acid [135], and the production of these secondary metabolites is controlled by the bacteria. For example, PQQ, for Pyrrolo Quinoline Quinone, is involved in gluconic acid production [137]. A second mechanism for making organic P available is carried out via phytases. These molecules are phosphatases capable of hydrolyzing organic complexes, and phytates, in which P is trapped [138]. Their hydrolysis allows the release of inorganic P in the form of phosphate groups. *Pseudomonas* and *Bacillus*, for example, have been recognized as containing phytases [139]. Kumar et al. [140] advanced that *Pseudomonas, Rhizobium, Bacillus,* and *Enterobacter* are the most potent P-solubilizing bacteria.

Due to the rhizospheric deficit of crops, Potassium is another essential element for plant development and has always been a significant production barrier. The soluble form of rhizobial potassium is also rare since it often forms insoluble structures when used as a mineral fertilizer [141]. However, PGPR can hydrolyze potassium through inorganic acid emissions, making it available to crop plants and thus improving crop wellness and yields. Potassium solubilizing agents include *Paenibacillus, Burkholderia Acido thiobacillus, Bacillus, Ferrooxidans, and Pseudomonas* [142].

Iron (Fe) is a necessary microelement for plants. It is abundantly distributed in soils in its ferric form (Fe^{3+}). However, like P, this form is hardly soluble and difficult to assimilate by plants [143]. Its absorption then requires a transporter. Some bacteria are able to release siderophores, low molecular weight molecules with a strong affinity for iron [144]. These siderophores can form siderophore-Fe^{3+} complexes to extract, chelate, and transport iron near the roots. This complex is recognized at the root level by specific protein receptors, where it can be absorbed [145]. Plants themselves produce phytosiderophores, so PGPRs help to reinforce this mechanism. Siderophores are also of interest against bio-aggressors.

Finally, the most crucial macronutrient for plant growth and development is nitrogen (N), yet most plants cannot assimilate its inert form. So-called diazotrophs are bacteria that are capable of converting tropospheric nitrogen to ammonia. The most common bacteria in symbiosis with leguminous plants is rhizobium. Additionally, there are non-symbiotic diazotrophic microorganisms belonging to the genera *Azospirillum, Azotobacter, Klebsiella, or Pseudomonas* [146, 147].

2.4.1.2 GROWTH ENHANCEMENT THROUGH PHYTOHORMONE SYNTHESIS

Phytohormones are naturally synthesized molecules by plants, which can stimulate or inhibit plant growth. Phytohormones involved in growth stimulation include molecules from three main families: auxins, cytokinins, and gibberellins. Some PGPRs can produce these hormones. Plant-derived auxins inhibit terminal bud growth but also play a role in cell growth and differentiation. They contribute to cell elongation and proliferation, as well as rhizogenesis, organogenesis, and transcription [148]. The primary and most extensively researched auxin in plants is IAA, or indole-3-acetic acid (IAA) [149]. When synthesized by bacteria, it serves two main functions. Bacterial IAA promotes root growth, allowing the plant to explore a larger volume of soil for better nutrient uptake [135]. However, excessive IAA production can inhibit root growth and normal plant development [150]. Secondly, IAA helps relax the cell walls at the rootlet level to facilitate exudation and provide energy sources for bacterial development [135]. It is also involved in germination.

The role of gibberellins is mainly expressed in cell elongation at the meristem level, inducing stem elongation and root growth. They are also involved in breaking dormancy and flower induction [151]. Bacteria can produce gibberellins as secondary metabolites, intervening in the signaling pathways of plant-bacteria interactions [152]. In this case, gibberellin helps stimulate aerial root growth and maintain plant structure (vigor) [152].

Cytokinins, in combination with auxin, are involved in tissue proliferation. This occurs through the stimulation of cell division, organogenesis, and the differentiation of specific organs [151]. Finally, these molecules slow down leaf aging. Phytohormones thus play a direct role in plant growth but can also be involved in other metabolic and signaling pathways.

2.4.1.3 TOLERANCE INCREASE TO ABIOTIC STRESSES

Stress conditions trigger signaling pathways and physiological responses in plants. Specific molecules modulate them. Ethylene (ET) is a phytohormone that plays a signaling role in many physiological processes. It intervenes particularly in response to abiotic stressors such as drought, salinity, and metal pollution [153]. When ET is produced significantly, it inhibits plant growth and induces phenomena such as senescence, chlorosis, leaf fall, and even plant death [154]. ACC, which stands for 1-aminocyclopropane carboxylic acid, is the direct precursor of ET. It can be cleaved by enzymes produced by bacteria such as *Pseudomonas putida* or *Azospirillum brasilense* [155]. These PGPRs (Plant Growth-Promoting Rhizobacteria) utilize the degradation products of ACC, namely ammonia and ketobutyrate, as an energy source and are thus able to modulate ET production [154]. As a result, plant growth is maintained under stress conditions.

In addition to PGPRs, AMFs mediate plant-abiotic stress tolerance and enhance plant growth [156]. Endophytic fungi create bioactive compounds that aid in the adaptation of the fungus and its host, as well as the utilization of nutritional elements such as nitrogen and phosphorus that are beneficial to plants [156]. They increase plant growth by secreting various hormonal substances and biochemical markers of plant protection. Some studies indicate that they have a significant impact on host plants, promoting plant growth, reducing disease severity, producing anti-herbivore/antimicrobial products, and improving defense mechanisms [157].

2.4.1.4 PATHOGEN INHIBITION

2.4.1.4.1 Pathogen Inhibition by Competition

The primary purpose of PGPRs used as biostimulants is not to compete against bio-aggressors. However, this can indirectly impact the protection and growth of the plant. Through antagonism, the activities of several microbiomes regulate the biological control of infections. PGPR generates resistance to numerous diseases through various direct or indirect pathways. Around the roots, microorganisms compete for access to nutrients such as carbon or iron, as well as space. The siderophores released by PGPRs enable the trapping of Fe in complexes that soil pathogens are unable to utilize. It also appears that the siderophores synthesized by PGPRs have a stronger affinity for Fe compared to chelators released by other organisms, giving them a competitive advantage [135, 158]. The role of siderophores is thus twofold: they serve to nourish the PGPR bacteria and the plant, while also reducing reserves for pathogenic microorganisms. Several key processes work together to limit microbial and pathogen growth in the presence of PGPR [55].

2.4.1.4.2 Pathogen Inhibition by Antibiosis

The most prominent antagonistic drugs effective against phytopathogens are antibiotics, from which PGPR is the main source for isolating these drugs. The antibiotics isolated from microbes have antibacterial properties [159]. Antibiotic synthesis typically enables better microbe competition, thereby enhancing the effectiveness of beneficial PGPR relationships. Additionally, *Bacillus* species secrete many other antibiotics, including lipopeptide biosurfactants such as bacillomycin, iturins, surfactin, etc. [160].

2.4.2 PLANT GROWTH PROMOTION (PGP) RESULTING FROM BETTER NUTRIENT UPTAKE

Microbial biostimulants increase plant metabolism, improving nutrient uptake quality and productivity. By establishing strong foundational health, plants become less susceptible to threats. Microbial-based biostimulant products can help farmers adjust their agricultural systems to a more unpredictable climate while enhancing the sustainability of food production. This contributes to the creation of a resilient and adaptable climate-smart farm model for the future [123, 130].

The nitrogenase enzyme system, which requires 16 adenosine triphosphates (ATP) to hydrolyze each fixed nitrogen, is crucial for the complex metabolic process of nitrogen fixation. Numerous studies have discussed the functions of the nif proteins involved in regulation, activation, transport, and cluster biosynthesis [161–163]. Nitrogen-fixing bacteria play a vital ecological role by contributing fixed nitrogen to the world's nitrogen cycle. However, diazotrophs must adapt to various environmental factors in each ecosystem to control the highly demanding nitrogen fixation process. All identified diazotrophs regulate nitrogen fixation at the transcriptional level.

Phosphorus (P) is essential for living organisms, as it is required for cell membranes, DNA and RNA storage, and other important organic compounds such as adenosine triphosphate (ATP) [164]. The application of phosphorus fertilizers, primarily derived from phosphate minerals, is critical for current agricultural practices and food production [165]. Based on estimates, global phosphorus supplies may be depleted as early as the next century [166, 167]. To meet the increasing demand for agricultural phosphorus, it is necessary to improve phosphorus recovery, as indicated by an analysis of documented phosphorus production and predicted production and consumption trends [168, 169]. Enhancing metabolic activities such as photosynthesis and carbohydrate accumulation, increasing the accumulation of glycine betaine-like compounds, and stimulating abscisic acid (ABA) accumulation for plant stress tolerance are important for improving plant growth, stress tolerance, and achieving sustainable agriculture. Plant Growth-Promoting Rhizobacteria (PGPRs) are considered the most environmentally friendly approach in this regard [135, 171, 172].

Mycorrhizae can solubilize and facilitate nutrient and water absorption in plant roots under drought stress [173, 174]. It has been proven that mycorrhizal fungi increase plant tolerance to salinity and provide resistance to the toxic effects of arsenic in polluted soils [175]. Mycorrhizae may also provide tolerance to excessive soil manganese and aluminum [176, 177]. Mycorrhizal fungi (MF) also contribute to increased modulation by symbiotic nitrogen-fixing bacteria. Various mechanisms, such as nitrification and phosphorus absorption, are involved in the synthesis of phytohormones (auxins, cytokinins, and gibberellins), phosphate solubility, and iron absorption. These processes have a direct and indirect impact on plant growth [24, 178, 179].

2.4.3 CHALLENGES

Production of biologically active inoculum requires the use of certain hosts, compounds, and media that stimulate the production of a large number of spores. These spores can then be added in small quantities to replenish lost biostimulants in the soil. While the efficiency of these biostimulants in facilitating plant growth and protection has been proven, their widespread use on a large scale has been limited due to various factors, including the use of pesticides and the loss of biostimulant inoculum in open fields. Additionally, many researchers utilize biological extracts or powders as biostimulant products.

2.5 BIOSTIMULANTS PRODUCTION, FORMULATION, AND COMMERCIAL APPLICATIONS

The use of microbial biostimulants at the field level is not widely applied, despite the numerous studies that underline the importance of plant growth-promoting microorganisms (PGPMs). The use of inexpensive pesticides and the loss of biostimulant inoculum in open fields are the principal reasons behind this delay. Therefore, producing a large number of biostimulants at the lowest cost and formulating them represent very important steps in developing marketable biostimulants.

2.5.1 MASS PRODUCTION OF PLANT GROWTH-PROMOTING MICROORGANISMS (PGPMS)

To achieve the required number of viable and active cells for field treatment, it is important to first produce growth promoters on a large scale. Additionally, the cost of biofertilizers, which includes expenses for raw materials, machinery, and labor, must be competitive compared to chemical fertilizers. Therefore, it is necessary to carefully select the medium used for biomass production in order to achieve significant cost savings. Conventional culture media, suitable for laboratory assays, cannot be used for this purpose as they contain expensive compounds. Several research teams have focused on identifying low-cost compounds to develop more affordable technology. For instance, corn steep liquor (CSL) and molasses have been successfully used to produce positive results in the growth of *Rhodopseudomonas palustris* strain PS3 [180]. Furthermore, a growth medium based on cheese whey (CW) has been found to be effective for the growth of *Rhizobium loti* cells [181].

Additionally, Rhizobium loti cells are adequately protected from cold and drying by a cellulose-based growth medium [181]. Three strains of plant growth-promoting rhizobacteria (PGPR) can thrive in hydrolysates made from *Laminaria* japonica processing waste (LJW) and *Enteromorphaprolifera* (EP) [182]. According to this study, algal waste hydrolysates contain a variety of mono- and polysaccharides, as well as macro- and microelements that the bacteria can utilize [182]. Crude glycerol (CG) is widely used for the production of plant-beneficial microorganisms [183]. CG, a by-product formed during the trans-esterification reaction for biodiesel production, provides excellent protection against abiotic stress such as high temperature and desiccation [183]. A wide range of low-cost substrates has been used to generate biomass for biostimulant development. These substrates typically consist of industrial effluents or breakdown products such as corn flour, soybean meal, milk sludge, etc. [182, 184–186].

Once the suitable culture medium has been selected for producing biomass, the scaling-up step becomes a crucial issue in vaccine technology. The purpose of the scaling-up process is to prevent climatic changes and their impact on bacterial activity, and to make necessary adjustments in equipment design as the bioprocess size increases. Various culture conditions were initially tested at a laboratory scale to optimize the growth parameters of microbial biostimulants [180, 186, 187]. When scaling up to large fermenters, the agitation, oxygen concentration, and pH parameters typically need to be enhanced, while maintaining a constant temperature range and growing medium [185, 188, 189].

2.5.2 FORMULATION OF PLANT GROWTH-PROMOTING MICROORGANISMS (PGPMS)

Biologically active inoculants are primarily applied directly to the soil or by coating protected seeds that are stored prior to sale or sowing [190, 191]. Several factors can influence the survival of biostimulants in their new environment, including desiccation, high temperatures, and the toxicity of grain exudates [192]. The main objective of formulation is to extend the shelf life of candidate PGPM strains and provide sufficient moisture for the survival of microbial cells during storage and/or after application. To achieve this, formulations should offer suitable support for the viability and preservation of the PGPR contained within, while enhancing its activity [193]. An ideal option for transporting microbes should possess specific qualities, such as being free from substances toxic to the environment or bacteria, having a high water-holding capacity, and exhibiting chemical and physical uniformity. Additionally, these carriers should have a pH close to neutral or easily adjustable, and be readily available at an affordable price [191, 194].

2.5.2.1 SOLID FORMULATION

Several solid materials that meet the aforementioned requirements have been tested for their carrying capacity of PGPMs. These materials include clays [195, 196], peat [197], bagasse [198, 199], cork compost [200], wheat straw, attapulgite, sepiolite [191], amorphous silica [201], bentonite, corn oil [202], mineral soils [203], vermiculite [204, 205], and perlite [206, 207]. Clays have a long history of supporting microbial inoculants applied as granules in various farming preparations, such as wettable powder (WP), suspensions (SU) powder, or dust forms. They offer exceptional preservation for dried inoculants due to their large surface area and pore diameter. In high moisture conditions, clays can retain excess water, providing a suitable moist environment for biologically active microbes [195–197]. Peat is also widely used as a carrier for fertilizing leguminous plants with rhizobacteria. It is a complex solution of organic substances that supports the growth of microorganisms and maintains the concentration of biostimulants for an extended period by protecting them against abiotic factors [197, 208].

2.5.2.2 LIQUID FORMULATION

Initially, biostimulants were introduced in liquid formulations. These conventional mixtures consist of a bacterial culture primarily suspended in water, minerals, or organic oils. Liquid formulations are easy to produce and can be applied by farmers, using inexpensive ingredients that are readily available from local suppliers. The process involves soaking the seeds in the biostimulant solution before planting or using a sprayer to evenly distribute the solution [209]. These techniques allow for the interaction between seeds and bacteria, which promote the preservation of beneficial bacteria on plant roots. However, microorganisms are not completely immune to environmental conditions and pollutants during storage, transportation, and sowing [210]. The main drawback of these formulation approaches is the susceptibility to contamination by non-target microbes, which reduces the shelf life of the product. This particular circumstance affects the stability of the final product during storage and the persistence of microbial organisms. For example, the concentration of *Azospirillum brasilense* decreased from 10^{10} to 10^{5} CFU g^{-1} within six months of storage in peat beads [211]. Similarly, *Pseudomonas* and *Bacillus subtilis* in colloidal solution showed a decrease from 10^{10} CFU/mL to 10^{7} CFU/mL within a storage period of 6 months [212].

2.5.2.3 ENCAPSULATION

Viewing the need to protect the microorganism once it has been inoculated in the ground and to guarantee its progressive and extended discharge, encapsulation technology has been widely used in biostimulant formulation. The objective of encapsulation is to create a microenvironment where biostimulants can sustain life throughout their lifespan, gradually releasing into the soil [213–215]. Sodium alginate is the most commonly used polymer for encapsulating bacterial solutions due to its ease of processing, simplicity of handling, flowability, and moisturizing properties [214]. The typical procedure for encapsulating polymeric calcium particles involves sprinkling the polymeric solution containing dispersed microbe cells into a calcium chloride solution, which acts as a "hardening agent" (Figure 2.3). The calcium ions in the polymeric solution instantly gelatinize, maintaining the structure of the polymeric drop and forming spherical beads that trap cells through diffusion in the alginate solution, thus preserving their structure [213–216]. Alternatively, the procedure can be reversed by dropping a calcium chloride solution containing dispersed microbe cells into an alginate solution. In this case, the diffusion of calcium ions into the alginate solution forms a solid membrane around the aqueous-core calcium, resulting in the formation of capsules through inverse gelation [217]. Moreover, the presence of multiple pores in the polymeric mixture allows hydrophilic molecules to diffuse between the inner and outer capsule environments [218]. Encapsulation technology provides a simple and efficient method for immobilizing biostimulants in protective capsules.

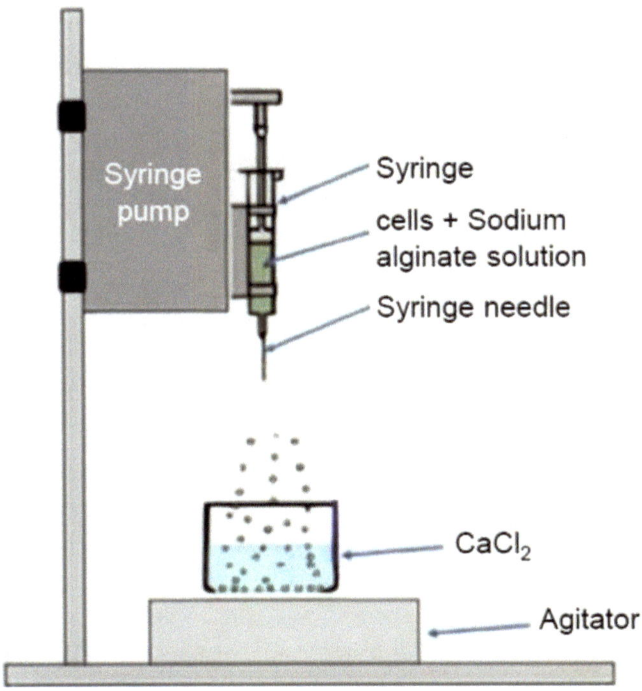

FIGURE 2.3 Immobilization of PGPMs in calcium alginate beads principle.

Nevertheless, it has notable limitations, such as cell loss that occurs during dehydration and storage. The drying process is particularly crucial for non-spore-forming and Gram-negative bacteria, which compromise the majority of PGPMs [209]. To address this issue, a filler material, such as starch or clays, can be added to the alginate matrix to enhance the survival of entrapped PGPMs. Additionally, supplementing the medium with a carbon source (such as trehalose and fructose) can protect cells during the drying process [219].

2.5.3 COMMERCIAL APPLICATION

An important part of the published literature illustrates the applicability of bacterial products as biostimulants and their agronomic supplies. Companies that create, trade, and distribute biostimulants have emerged worldwide. Table 2.4 illustrates the list of PGPR (biostimulants), some of which are currently available on the market. However, the full potential of PGPMs remains largely unexplored, and their application has no substantial impact on global food security [220]. The scarcity of marketing and utilization of biostimulants, even after years of study, is mainly due to certain limitations in their production and preservation, unlike their well-known PGP roles.

TABLE 2.4 Examples of Commercialized PGPR-based Biostimulants

Microbial Biofertilizers	Manufacturing Companies	Trade Names	Target Crops
B. amyloliquefaciens IT 45 *B. japonicum*	Lallem and plant care BASF Inc.	RHIZOCELL® GC Nodulator®	Beans, maize, carrot, rice, cotton.
B. amyloliquefaciens, Trichoderma virens	Monsanto Bio Ag	Quick Roots®	Canola, corn, field peas, lentils, small grains, and soybeans.
B. amyloliquefaciens	Bayer Agrar Schweiz	FZB25® flüssig	Potatoes and lettuce.
Bacillus mucilaginosus, B. subtilis	China Bio-Fertilizer AG	CBF	Cereals
A. brasilense, Azotobactervinelandii, B. megaterium	AGRObio	Bacto Fil A10	Maize
B. amyloliquefaciens souche IT45	Lallemand Plant Care	Rhizocell ® GC	Cereals
Azospirillum	Ajay Biotech	Ajay Azospirillum	Cereals
Pseudomonas azotoformans	Lantmannen Bioagri	Amase®	Cucumber, lettuce, tomato, pepper.

TABLE 2.4 *(Continued)*

Microbial Biofertilizers	Manufacturing Companies	Trade Names	Target Crops
Azospirillum brasilense, Azotobacter vinelandii, B. megaterium, B. polymyxa, P. fluorescens	Agro. Bio Hungary Kft.	Bacto Fil A10®	Cereals
Agrobacterium radiobacter AR 39, B. subtilis BA 51, Streptomyces spp. *SB 15*	CCS Aosta Srl	Micosat F® Uno	Fruits, vegetables, flowers.
Bacillus sp.	Organica Technologies	PGA®	Fruits, vegetables

2.6 CONCLUSION

One of the major concerns is feeding an ever-increasing population, especially when the agriculture sector deals with a wide range of complicated issues caused by global climate change. As a consequence of the variation of climatic patterns such as water resources, this worldwide phenomenon causes and amplifies already existing abiotic pressures, leading to regular storms, soil salinity, droughts, and temperature fluctuations. Therefore, these circumstances minimize the amount of beneficial microbial organisms related to plants, influence plant health, and make plants more vulnerable to biotic stress. The usage of microbial biostimulants (PB) has sparked global interest in recent years as a means of developing environmentally friendly alternatives. When it comes to crop plants, fungi, and/or bacteria have demonstrated efficacy for nutrition usage, stress reduction, and exemplary quality characteristics. They increase the availability of soil minerals such as iron, phosphorus, potassium, and zinc to plants by regulating phytohormones, synthesizing siderophores, and enhancing the production of plant stimulant and phenolic compounds, as well as their release. Moreover, Plant growth-promoting rhizobacteria combat different types of plant stresses through a variety of metabolisms or combined processes like optimization of plant hormones, transcriptional regulation and appearance, specific catabolism, phytochemical compounds optimization, and biochemical processes. In fact, in one product of a biostimulant, multiple dynamic components can exist with various ranges and mechanisms. As a result, a careful study of the different components, evaluation of the suitability of bacteria, the dedication to understanding cause and consequence in the natural agroecosystem, the invention of appropriate synthesis formulas, and the dispersion of innovation, as well as practical assistance for target consumers, are all required. However, due to their inherent diversity, biofertilizers defy regulatory classification (e.g., edition, nutrient, antifungal agent). Products may go through costly and time-consuming testing procedures before being approved, depending on the enrollment region. Therefore, the absence of a comprehensive worldwide regulatory framework makes it difficult to market products and may deter companies from creating new ones. The use of microbial biostimulants still has several drawbacks in biological and agroecological research, most of which are related to the lower effectiveness and sensitivity to environmental factors when compared to synthetic growth modulators, fertilizers, and insecticides. Furthermore, biostimulants frequently produce uneven outcomes from one crop or location to another. Thus, enhancing the efficacy of biofertilizers while ensuring the consistency of findings. Additional studies must concentrate on creating highly specific products, for example, thoroughly investigating the plant rhizosphere, defining and manipulating plant and microorganisms' interactions, either isolating microorganisms that are particularly suited to biotic and abiotic stress, or selecting a helpful microbial community rather than introducing new species. Microbial biofertilizers can be used in conjunction with farming techniques to enhance the biodiversity of agroecosystems and ensure a stable, long-lasting symbiotic interaction with crop plants. Additionally, in order to ensure that this approach can be easily implemented across different crops, regions, and climatic conditions, the testable theories of biofertilizer studies should be utilized to provide a clear example for all stakeholders, including farmers, technical assistance, and legislators. Therefore, the successful validation of scientific findings in practical contexts relies on close collaboration and ongoing information sharing between the scientific community and the stakeholders.

KEYWORDS

- **Abiotic stress**
- **Biostimulants**
- **Biotic stress**
- **Microbes**
- **Plant growth-promoting microorganisms**
- **Solid formulation**
- **Sustainable agriculture**

REFERENCES

1. Ebert, A. W., & Engels, J. M. M., (2020). Plant biodiversity and genetic resources matter. *Plants (Basel), 9*(12), 1706. doi: 10.3390/plants9121706. PMID: 33291549; PMCID: PMC7761872.
2. Sangiorgio, D., Cellini, A., Donati, I., Pastore, C., Onofrietti, C., & Spinelli, F., (2020). Facing climate change: Application of microbial biostimulants to mitigate stress in horticultural crops. *Agronomy, 10*, 794. doi: 10.3390/ agronomy10060794.
3. Raza, A., Razzaq, A., Mehmood, S. S., Zou, X., Zhang, X., Lv, Y., & Xu, J., (2019). Impact of climate change on crops adaptation and strategies to tackle its outcome: A review. *Plants, 8*(2), 34. http://dx.doi.org/10.3390/plants8020034. PMid: 30704089.
4. Malhi, G. S., Kaur, M., & Kaushik, P., (2021). Impact of climate change on agriculture and its mitigation strategies: A review. *Sustainability, 13*(3), 1–21.
5. Mancosu, N., Snyder, R. L., Kyriakakis, G., & Spano, D., (2015). Water scarcity and future challenges for food production. *Water, 7*, 975–992.
6. Bulgari, R., Franzoni, G., & Ferrante, A., (2019). Biostimulants application in horticultural crops under abiotic stress conditions. *Agronomy, 9*, 306. doi: 10.3390/ agronomy9060306.
7. Bulgari, R., Cocetta, G., Trivellini, A., Vernieri, P., & Ferrante, A., (2015). Biostimulants and crop responses: A review. *Biol. Agric. Hortic., 31*, 1–17. doi: 10.1080/01448765.2014.964649.
8. Chiaiese, P., Corrado, G., Colla, G., Kyriacou, M., & Rouphael, Y., (2018). Renewable sources of plant biostimulation: Microalgae as a sustainable means to improve crop performance. *Front Plant Sci., 9*, 1782. doi: 10.3389/fpls.2018.01782.
9. Altieri, M. A., (2018). *Agroecology, the Science of Sustainable Agriculture* (2nd edn.). CRC Press, Taylor and Francis Group, Boca Raton London New York.
10. Eyhorn, F., Müller, A., Reganold, J. P., Frison, E., Herren, H. R., Luttikholt, L., Mueller, A., et al., (2019). Sustainability in global agriculture driven by organic farming. *Nature Sustainability, 2*, 253–255. doi: 10.1038/s41893-019-0266-6.
11. European Biostimulants Industry Council EBIC, (2012). *What Are Biostimulants?* Available online at: http://www.biostimulants.eu/about/what-are-biostimulants (accessed on 9 January 2023).
12. Basak, A., (2008). Biostimulators-definitions, classification, and legislation. In: Gawronska, H., (ed.), *Monographs Series: Biostimulators in Modern Agriculture; General Aspects* (pp. 7–17). Warsaw: Wies Jutra.
13. Du Jardin, P., (2015). Plant biostimulants: Definition, concept, main categories, and regulation. *Sci. Hortic., 196*, 3–14. doi: 10.1016/j.scienta.2015.09.021.
14. Oldroyd, G. E., (2013). Speak, friend, and enter: Signaling systems that promote beneficial symbiotic associations in plants. *NatureReviewsMicrobiology, 11*, 252–263.
15. Brameyer, S., Kresovic, D., Bode, H. B., & Heermann, R., (2015). Dialkylresorcinols as bacterial signaling molecules. *Proc. Natl. Acad. Sci. USA, 112*, 572–577.
16. Vessey, J. K., (2003). Plant growth promoting rhizobacteria as biofertilizers. *Plant and Soil, 255*, 571–586. http://dx.doi.org/10.1023/A:1026037216893.
17. Berg, G., (2009). Plant-microbe interactions promoting plant growth and health: Perspectives for controlled use of microorganisms in agriculture. *Appl. Microbiol. Biotechnol., 84*, 11–18. doi: 10.1007/s00253-009-2092-7.

18. Dodd, I. C., & Ruiz-Lozano, J. M., (2012). Microbial enhancement of crop resource use efficiency. *Current Opinion in Biotechnology, 23*, 236–242. https://doi.org/10.1016/j.copbio.2011.09.005.
19. Wozniak, M., Gałazka, A., Tyskiewicz, R., & Jaroszuk-Sciseł, J., (2019). Endophytic bacteria potentially promote plant growth by synthesizing different metabolites and their phenotypic/physiological profiles in the biolog GEN III MicroPlate™ test. *Int. J. Mol. Sci., 20*, 5283.
20. Papik, J., Folkmanova, M., Polivkova, M., Suman, J., & Uhlik, O., (2020). The invisible life inside plants: Deciphering the riddles of endophytic bacterial diversity. *Biotechnol. Adv., 44*, 107614.
21. Ahmad, F., Ahmad, I., & Khan, M. S., (2008). Screening of free-living rhizospheric bacteria for their multiple plant growth promoting activities. *Microbiol. Res., 163*, 173–181. doi: 10.1016/j.micres.2006.04.001.
22. Babalola, O. O., (2010). Beneficial bacteria of agricultural importance. *Biotechnol Lett, 32*, 1559–1570.
23. Berendsen, R. L., Pieterse, C. M. J., &Bakker, P., (2012). The rhizosphere microbiome and plant health. *Trends Plant Sci, 17*, 478–486.
24. Bhattacharyya, P. N., & Jha, D. K., (2012). Plant growth-promoting rhizobacteria (PGPR): Emergence in agriculture. *World J. Microbiol. Biotechnol., 28*, 1327–1350.
25. Gaiero, J. R., McCall, C. A., Thompson, K. A., Day, N. J., Best, A. S., & Dunfield, K. E., (2013). **Inside the root microbiome: Bacterial root endophytes and plant growth promotion**. *Am. J. Bot., 100*(9), 1738–1750. 10.3732/ajb.1200572.
26. Philippot, L., Raaijmakers, J. M., Lemanceau, P., & Van, D. P. W. H., (2013). Going back to the roots: The microbial ecology of the rhizosphere. *Nat. Rev. Microbiol., 11*, 789–799.
27. Vacheron, J., Desbrosses, G., Bouffaud, M. L., et al., (2013). Plant growth-promoting rhizobacteria and root system functioning. *Front Plant Sci, 4*, 356. https://doi.org/10.3389/fpls.2013.00356.
28. Berg, G.,Zachow, C.,Müller, H.,Philipps, J., & Tilcher, R., (2013). Next-generation bio-products sowing the seeds of success for sustainable agriculture. *Agronomy, 3*(4), 648–656.
29. Pathania, P., Rajta, A., Singh, P. C., & Bhatia, R., (2020). Role of plant growth-promoting bacteria in sustainable agriculture. *Biocatal. Agric. Biotechnol., 30*, 101842.
30. Nguyen, M. L., Spaepen, S., Du Jardin, P., & Delaplace, P., (2019). Biostimulant effects of rhizobacteria on wheat growth and nutrient uptake depend on nitrogen application and plant development. *Arch Agron. Soil Sci., 65*, 58–73.
31. Behie, S. W., & Bidochka, M. J., (2014). Nutrient transfer in plant-fungal symbioses. *Trends Plant Sci., 19*, 734–740. https://doi.org/10.1016/j.tplants.2014.06.007.
32. Bonfante, P., & Genre, A., (2010). Mechanisms underlying beneficial plant-fungus interactions in mycorrhizal symbiosis. *Nature Communications, 1*, 48. doi: 10.1038/ncomms1046.
33. Johnson, N. C., & Graham, J. H., (2013). The continuum concept remains a useful framework for studying mycorrhizal functioning. *Plant Soil, 363*, 411–419. doi: 10.1007/s11104-012-1406-1.
34. Augé, R. M., (2001). Water relations, drought and vesicular-arbuscular mycorrhizal symbiosis. *Mycorrhiza, 11*, 361–365.

35. Gianinazzi, S., Gollotte, A., Binet, M. N., Van, T. D., Redecker, D., & Wipf, D., (2010). Agroecology: The key role of arbuscular mycorrhizas in ecosystem services. *Mycorrhiza, 20*, 519–530. doi: https://doi.org/10.1007/s00572-010-0333-3.
36. Hamel, C., & Plenchette, C., (2007). *Mycorrhizae in Crop Production*. The Haworth Press Inc. New York, USA.
37. Harrier, L. A., & Watson, C. A., (2004). The potential role of arbuscular mycorrhizal (AM) fungi in the bio-protection of plants against soil-borne pathogens in organic and/ or other sustainable farming systems. *Pest Manage. Sci., 60*, 149–157.
38. Siddiqui, Z. A., Akhtar, M. S., & Futai, K., (2008). *Mycorrhizae: Sustainable Agriculture and Forestry*. New Delhi: Springer.
39. Van Der, H. M. G. A., Bardgett, R. D., & Van, S. N. M., (2008). The unseen majority: Soil microbes as drivers of plant diversity and productivity in terrestrial ecosystems. *Ecol. Lett, 11*, 296–310.
40. Plenchette, C., Clermont-Dauphin, C., Meynard, J. M., & Fortin, J. A., (2005). Management of arbuscular mycorrhizal fungi in cropping systems. *Can. J. Plant Sci., 85*, 31–40.
41. Sheng, M., Tang, M., Zhang, F., & Huang, Y., (2011). Influence of arbuscular mycorrhiza on organic solutes in maize leaves under salt stress.*Mycorrhiza, 21*, 423–430.
42. Dalpé, Y., & Monreal, M., (2004). Arbuscular mycorrhiza inoculum to support sustainable cropping systems. *Crop Manage.*
43. Mukherjee, P. K., Horwitz, B. A., & Kenerley, C. M., (2012). Secondary metabolism in *Trichoderma* – a genomic perspective. *Microbiology*. https://doi.org/10.1099/ mic.0.053629-0.
44. Nicolás, C., et al., (2014). *Trichoderma* genes in plants for stress tolerance- status and prospects. *Plant Sci*. http://dx.doi.org/10.1016/j.plantsci.2014.03.005.
45. Fiorentino, N., Ventorino, V., Woo, S. L., Pepe, O., De Rosa, A., Gioia, L., et al., (2018). *Trichoderma*-based biostimulants modulate rhizosphere microbial populations and improve N uptake efficiency, yield, and nutritional quality of leafy vegetables. *Front. Plant Sci., 9*, 743. doi: 10.3389/fpls.2018.00743.
46. Manganiello, G., Sacco, A., Ercolano, M. R., Vinale, F., Lanzuise, S., Pascale, A., et al., (2018). Modulation of tomato response to *Rhizoctonia solani* by *Trichoderma harzianum* and its secondary metabolite harzianic acid. *Front. Microbiol., 9*, 1966. doi: 10.3389/fmicb.2018.01966.
47. Ventorino, V., Sannino, S., Piccolo, A., Cafaro, V., Carotenuto, R., & Pepe, O., (2018). *Methylobacterium populi* VP2: Plant growth-promoting bacterium isolated from a highly polluted environment for polycyclic aromatic hydrocarbon (PAH) biodegradation. *Sci. World J.*, 1–11. doi: 10.1155/2014/931793.
48. Aguilar-Paredes, A., Valdés, G., & Nuti, M., (2020). Ecosystem functions of microbial consortia in sustainable agriculture. *Agronomy, 10*, 1902.
49. Kong, W., Meldgin, D. R., Collins, J. J., & Lu, T., (2018). Designing microbial consortia with defined social interactions. *Nat. Chem. Biol., 14*, 821–829. doi: 10.1038/ s41589-018-0091-7.
50. Avio, L., Turrini, A., Giovannetti, M., & Sbrana, C., (2018). Designing the ideotype mycorrhizal symbionts for the production of healthy food. *Front. Plant Sci., 9*, 1–19.
51. Giovannini, L., Palla, M., Agnolucci, M., Avio, L., Sbrana, C., Turrini, A., & Giovannetti, M., (2020). Arbuscular mycorrhizal fungi and associated microbiota as

plant biostimulants: Research strategies for the selection of the best performing inocula. *Agronomy, 10*, 106.
52. Meena, M., Swapnil, P., Divyanshu, K., Kumar, S., Tripathi, Y. N., Zehra, A., Marwal, A. & Upadhyay, R. S., (2020). PGPR-mediated induction of systemic resistance and physiochemical alterations in plants against the pathogens: Current perspectives. *J. Basic. Microbiol., 60*, 828–861.
53. David, L., Kang, J., Dufresne, D., Zhu, D., & Chen, S., (2021). Multi-omics revealed molecular mechanisms underlying guard cell systemic acquired resistance. *Int. J. Mol. Sci., 22*, 191.
54. Nishad, R., Ahmed, T., Rahman, V. J., & Kareem, A., (2020). Modulation of plant defense system in response to microbial interactions. *Front. Microbiol., 11*, 1298.
55. Niu, D. D., Zheng, Y., Zheng, L., Jiang, C. H., Zhou, D. M., & Guo, J. H., (2016). Application of PSX biocontrol preparation confers root-knot nematode management and increased fruit quality in tomato under field conditions. *Biocont. Sci. Technol.*, 174–180.
56. Bukhat, S., Imran, A., Javaid, S., Shahid, M., Majeed, A., & Naqqash, T., (2020). Communication of plants with microbial world: Exploring the regulatory networks for PGPR mediated defense signaling. *Microb. Res., 23*, 126–486.
57. Zebelo, S., Song, Y., Kloepper, J. W., & Fadamiro, H., (2016). Rhizobacteria activates (+)-δ-cadinene synthase genes and induces systemic resistance in cotton against beet armyworm (*Spodoptera exigua*). *Plant Cell Environ., 39*, 935–943.
58. Usta, C., (2013). Microorganisms in biological pest control—A review (bacterial toxin application and effect of environmental factors). *Current Progress in Biological Research, 13*, 287–317.
59. Gupta, S., & Dikshit, A. K., (2010). Biopesticides: An ecofriendly approach for pest control. *Journal of Biopesticides, 3*(Special Issue), 186.
60. Lacey, L. A., Frutos, R., Kaya, H. K., & Vail, P., (2001). Insect pathogens as biological control agents: Do they have a future? *Biological Control, 21*, 230–248.
61. Abdul, M. N. A., Kumar, I. S., & Nadarajah, K., (2020). Elicitor and receptor molecules: Orchestrators of plant defense and immunity. *Int. J. Mol. Sci., 21*, 963.
62. Rashid, M. H. O., & Chung, Y. R., (2017). Induction of systemic resistance against insect herbivores in plants by beneficial soil microbes. *Front. Plant Sci., 8*, 1816.
63. Conrath, U., Beckers, G. J. M., Langenbach, C. J. G., & Jaskiewicz, M. R., (2015). Priming for enhanced defense. *Annu. Rev. Phytopathol., 53*, 97–119. doi: 10.1146/annurev-phyto-080614-120132.
64. Serteyn, L.,Quaghebeur, C.,Ongena, M.,Cabrera, N.,Barrera, A.,Molina-Montenegro, M. A., et al., (2020). Induced systemic resistance by a plant growth-promoting rhizobacterium impacts development and feeding behavior of aphids. *Insects, 11*, 234.
65. Johnson, S. N., Erb, M., & Hartley, S. E., (2016). Roots under attack: Contrasting plant responses to below- and aboveground insect herbivory. *New Phytologist., 210*, 413–418.
66. Schikora, A., Schenk, S. T., & Hartmann, A., (2016). Beneficial effects of bacteria plant communication based on quorum sensing molecules of the N-acyl homoserine lactone group. *Plant Mol. Biol., 90*, 605–612.
67. Kramer, J., Özkaya, Ö., & Kümmerli, R., (2020). Bacterial siderophores in community and host interactions. *Nat. Rev. Microbiol., 18*, 152–163.
68. Kang, S. M., Shahzad, R., Bilal, S., Khan, A. L., Park, Y. G., Lee, K. E., Asaf, S., et al., (2019). Indole-3-acetic-acid and ACC deaminase producing *Leclercia adecarboxylata*

MO1 improves *Solanum lycopersicum* L. growth and salinity stress tolerance by endogenous secondary metabolites regulation. *BMC Microbiol., 19*, 80.

69. Monnier, N., Furlan, A., Botcazon, C., Dahi, A., Mongelard, G., & Cordelier, S., (2018). Rhamnolipids from *Pseudomonas aeruginosa* are elicitors triggering *Brassica napus* protection against *Botrytis cinerea* without physiological disorders. *Front. Plant Sci., 9*, 1170.
70. Pršić, J., & Ongena, M., (2020). Elicitors of plant immunity triggered by beneficial bacteria. *Front. Plant. Sci., 11*, 1675.
71. Vega, C., Rodríguez, M., Llamas, I., Béjar, V., & Sampedro, I., (2020). Silencing of phytopathogen communication by the halotolerant PGPR *Staphylococcus equorum* strain EN21. *Microorganisms, 8*, 42.
72. Hernández-Reyes, C., Schenk, S. T., Neumann, C., Kogel, K. H., & Schikora, A., (2014). N-acyl-homoserine lactones-producing bacteria protect plants against plant and human pathogens. *Microb. Biotechnol., 7*, 580–588.
73. Maithani, D., Singh, H., & Sharma, A., (2021). Stress alleviation in plants using SAR and ISR: Current views on stress signaling network. In: *Microbes and Signaling Biomolecules Against Plant Stress* (pp. 7–36). Springer: Singapore.
74. Beris, D., Theologidis, I., Skandalis, N., & Vassilakos, N., (2018). *Bacillus amyloliquefaciens* strain MBI600 induces salicylic acid dependent resistance in tomato plants against tomato spotted wilt virus and potato virus Y. *Sci. Rep., 8*, 10320. https://doi.org/ 10.1038/s41598-018-28677-3.
75. Tanzini, M. R., Alves, S. B., Setten, A., & Augusto, N. T., (2001). Compatibilidad de agent estensoactivos com *Beauveria bassiana* y *Metarhizium anisopliae*. *Manejo Integrado De Plagas, 59*, 15–18.
76. Vilcinskas, A., (2010). Coevolution between pathogen-derived proteinases and proteinase inhibitors of host insects. *Virulence, 1*(3), 206–214.
77. Dara, S. K., (2017). *Entomopathogenic Microorganisms: Modes of Action and Role in IPM* (p. 7). Agriculture and Natural Blogs, University of California.
78. Lambert, B., Höfte, H., Annys, K., Jansens, S., Soetaert, P., & Peferoen, M., (1992). Novel *Bacillus thuringiensis* insecticidal crystal protein with a silent activity against coleopteran larvae. *Applied and Environmental Microbiology, 58*(8), 2536–2542.
79. Federici, B. A., (1999). In: Bellows, T. S., & Fisher, T. W., (eds.), *Handbook of Biological Control* (pp. 575–593). Academic Press, San Diego, CA.
80. De Maagd, R. A., Bravo, A., Berry, C., Crickmore, N., & Schnepf, H. E., (2003). Structure, diversity, and evolution of protein toxins from spore-forming entomopathogenic bacteria. *Annual Review of Genetics, 37*(1), 409–433.
81. Pigott, C. R., & Ellar, D. J., (2007). Role of receptors *in Bacillus thuringiensis* crystal toxin activity. *Microbiology and Molecular Biology Reviews, 71*(2), 255–281.
82. Deka, B., Baruah, C., & Babu, A., (2021). Entomopathogenic microorganisms: Their role in insect pest management. *Egyptian Journal of Biological Pest Control, 31*(1), 1–8.
83. Koppenhöfer, A. M., & Wu, S., (2017). Microbial control of insect pests of turfgrass. In: *Microbial Control of Insect and Mite Pests* (pp. 331–341). Academic Press.
84. Baumann, P., Clark, M. A., Baumann, L., & Broadwell, A. H., (1991). *Bacillus sphaericus* as a mosquito pathogen: Properties of the organism and its toxins. *Microbiological Reviews, 55*(3), 425–436.

85. Weinzierl, R., Henn, T., Koehler, P. G., & Tucker, C. L., (1989). *Microbial Insecticides,* 1295. Cooperative Extension Service, University of Illinois at Urbana-Champaign.
86. Ruiu, L., (2013). *Brevibacillus laterosporus*, a pathogen of invertebrates and a broad-spectrum antimicrobial species. *Insects*, *4*(3), 476–492.
87. Martin, P. A., Hirose, E., & Aldrich, J. R., (2007a). Toxicity of *Chromobacterium subtsugae* to southern green stinkbug (*Heteroptera: Pentatomidae*) and corn rootworm (*Coleoptera: Chrysomelidae*). *Journal of Economic Entomology*, *100*(3), 680–684.
88. Martin, P. A., Gundersen-Rindal, D., Blackburn, M., & Buyer, J., (2007b). *Chromobacterium subtsugae* sp. Nov., a betaproteobacterium toxic to Colorado potato beetle and other insect pests. *International Journal of Systematic and Evolutionary Microbiology*, *57*(5), 993–999.
89. Mora, M. A. E., Castilho, A. M. C., & Fraga, M. E., (2018). Classification and infection mechanism of entomopathogenic fungi. *Arquivos do Instituto Biológico, 84.*
90. Moharram, A. M., Abdel-Galil, F. A., & Hafez, W. M. M., (2021). On the enzymes' actions of entomopathogenic fungi against certain indigenous and invasive insect pests. *Egyptian Journal of Biological Pest Control*, *31*(1), 1–9.
91. Faria, M. R., & Wraight, S., (2007). Mycoinsecticides and mycoacaricides: A comprehensive list with worldwide coverage and international classification of formulation types. *Biol. Control*, *43*, 237–256.
92. Strasser, H., Vey, A., & Butt, T. M., (2000). Are there any risks in using entomopathogenic fungi for pest control, with particular reference to the bioactive metabolites of *Metarhizium*, *Tolypocladium* and *Beauveria* species? Biocontrol *Science and Technology*, *10*(6), 717–735.
93. Dara, S. K., (2019). Non-entomopathogenic roles of entomopathogenic fungi in promoting plant health and growth. *Insects*, *10*(9), 277.
94. Bamisile, B. S., Dash, C. K., Akutse, K. S., Keppanan, R., Afolabi, O. G., Hussain, M., & Wang, L., (2018). Prospects of endophytic fungal entomopathogens as biocontrol and plant growth promoting agents: An insight on how artificial inoculation methods affect endophytic colonization of host plants. *Microbiological Research*, *217*, 34–50.
95. Bamisile, B. S., Senyo, A. K., Dash, C. K., Qasim, M., Ramos, A. L. C., Ashraf, H. J., & Wang, L., (2020). Effects of seedling age on colonization patterns of *Citrus limon* plants by endophytic *Beauveria bassiana* and *Metarhizium anisopliae* and their influence on seedlings growth. *Journal of Fungi*, *6*(1), 29.
96. Groner, A., (1986). Specificity and safety of baculoviruses. In: Granados, R. R., & Federici, B. A., (eds.), *The Biology of Baculoviruses* (Vol. I, pp. 177–202). CRC Press. Boca Ratón, Florida.
97. Kamita, S. G., Kang, K. D., Inceoglu, A. B., & Hammock, B. D., (2005). Genetically modified baculoviruses for pest insect control. In: Gilbert, L. I., Iatrou, K., & Gill, S. S., (eds.), *Comprehensive Molecular Insect Science* (Vol. 6, pp. 271–322). Elsevier, Oxford.
98. Inceoglu, A. B., Kamita, S. G., & Hammock, B. D., (2006). Genetically modified baculoviruses: A historical overview and future outlook. *Advances in Virus Research*, *68*, 323–360.
99. USEPA, (2008). *United States Environmental Protection Agency Biopesticide Fact Sheets*. Washington, DC: US Environmental Protection Agency.
100. Kawalekar, J. S., (2013). Role of biofertilizers and biopesticides for sustainable agriculture. *J. Bio. Innov.*, *2*(3), 73–78.

101. Murphy, F. A., Faquet, C. M., Bishop, D. H. L., Gabrial, S. A., Jarvis, A. W., & Martelli, G. P., (1995). *Classification and Nomenclature of Viruses: Sixth Report of the International Committee on Taxonomy of Viruses*. Springer-Verlag, Berlin.
102. Kaya, H. K., & Gaugler, R., (1993). Entomopathogenic nematodes. *Annual Review of Entomology, 38*, 181–206.
103. Williams, R. N., Fickle, D. S., Grewal, P. S., & Meyer, J. R., (2002). Assessing the potential of entomopathogenic nematodes to control the grape root borer *Vitacea polistiformis* (*Lepidoptera: Sesiidae*) through laboratory and greenhouse bioassays. *Biocontrol Science and Technology, 12*(1), 35–42.
104. Klein, M. G., (2018). Efficacy against soil-inhabiting insect pests. *Entomopathogenic Nematodes in Biological Control*.
105. Parmar, B. S., (2010). Biopesticides: An Indian overview. *Pesti. Res. J., 22*(2), 93–110.
106. Kumar, V., (2015). A review on efficacy of biopesticides to control the agricultural insect's pest. *International Journal of Agricultural Science Research, 4*(9), 168–179.
107. Khachatourians, G. G., (2009). Insecticides, microbials. *Applied Microbiology: Agro/Food*, 95–109.
108. Thakore, Y., (2006). The biopesticide market for global agricultural use. *Industrial Biotechnology, 2*(3), 194–208.
109. Kiewnick, S., (2007). Practicalities of developing and registering microbial biological control agents. *CAB Rev., 2*(013), 1–11.
110. PMRA, (2008). *In Update on Reduce Risk Pesticides in Canada, RR 2008-01*. Health Canada Pest Management Regulatory Agency, Ottawa, Canada. http://www.pmra-arla.gc.ca (accessed on 9 January 2023).
111. Kristoffersen, P., Rask, A. M., Grundy, A. C., Franzen, I., Kempenaar, C., Raisio, J., & Zarina, L., (2008). A review of pesticide policies and regulations for urban amenity areas in seven European countries. *Weed Research, 48*(3), 201–214.
112. Bailey, K. L., Boyetchko, S. M., & Längle, T., (2010). Social and economic drivers shaping the future of biological control: A Canadian perspective on the factors affecting the development and use of microbial biopesticides. *Biological Control, 52*(3), 221–229.
113. Koul, O., (2011). Microbial biopesticides: Opportunities and challenges. *CAB Rev., 6*, 1–26.
114. Ravensberg, W. J., (2011). *A Roadmap to the Successful Development and Commercialization of Microbial Pest Control Products for Control of Arthropods* (Vol. 10). Springer Science & Business Media.
115. Chandler, D., Davidson, G., Grant, W. P., Greaves, J., & Tatchell, G. M., (2008). Microbial biopesticides for integrated crop management: An assessment of environmental and regulatory sustainability. *Trends in Food Science & Technology, 19*(5), 275–283.
116. Kamilova, F., Okon, Y., De Weert, S., & Hora, K., (2015). Commercialization of microbes: Manufacturing, inoculation, best practice for objective field testing, and registration. In: *Principles of Plant-Microbe Interactions* (pp. 319–327). Springer, Cham.
117. Schisler, D. A., Slininger, P. J., Behle, R. W., & Jackson, M. A., (2004). Formulation of *Bacillus spp.* for biological control of plant diseases. *Phytopathology, 94*(11), 1267–1271.
118. Timmusk, S., Behers, L., Muthoni, J., Muraya, A., & Aronsson, A. C., (2017). Perspectives and challenges of microbial application for crop improvement. *Frontiers in Plant Science, 8*, 49.

119. Charropadhyay, P., Karmakar, N., Chatterjee, S., & Sen, S. K., (2014). Field efficacy of inorganic carrier based formulations of *Serratia entomophila* AB2 in *Sesamum indicum* var. Kanak. *African Journal of Biotechnology*, *13*(34).
120. Vega, F. E., & Kaya, H. K., (2012). *Insect Pathology* (2nd edn., p. 504). Elsevier: London, UK.
121. Chattopadhyay, P., Banerjee, G., & Mukherjee, S., (2017). Recent trends of modern bacterial insecticides for pest control practice in integrated crop management system. *3 Biotech*, *7*(1), 1–11.
122. Verma, D. K., Guzmán, K. N. R., Mohapatra, B., Talukdar, D., Chávez-González, M. L., Kumar, V., & Aguilar, C. N., (2021). Recent trends in plant-and microbe-based biopesticide for sustainable crop production and environmental security. In: *Recent Developments in Microbial Technologies* (pp. 1–37). Springer, Singapore.
123. Leogrande, R., El Chami, D., Fumarola, G., Di Carolo, M., Piegari, G., Elefante, M., & Dongiovanni, C., (2022). Biostimulants for resilient agriculture: A preliminary assessment in Italy. *Sustainability*, *14*(11), 6816.
124. Carletti, P., García, A. C., Silva, C. A., & Merchant, A., (2021). Towards a functional characterization of plant biostimulants. *Frontiers in Plant Science*, *12*, 677772.
125. De Souza, R. S. C., Armanhi, J. S. L., & Arruda, P., (2020). From microbiome to traits: Designing synthetic microbial communities for improved crop resiliency. *Front. Plant Sci., 11*. doi: 10.3389/fpls.2020.01179.
126. Aranda, M., Li, Y., Liew, Y. J., Baumgarten, S., Simakov, O., Wilson, M. C., & Voolstra, C. R., (2016). Genomes of coral dinoflagellate symbionts highlight evolutionary adaptations conducive to a symbiotic lifestyle. *Scientific Reports*, *6*(1), 1–15.
127. Romano, G., Greslebin, A., & Lechner, B. E., (2017). Hongos agaricoides de los bosques de Nothofagus pumilio (Chubut, Argerntina): Clave y listado de especies. *Rev del Mus Argentino Ciencias Nat NuevaSer, 19*, 36–69. https://doi.org/10.22179/REVMACN.18.495.
128. De Vasconcelos, A. C. F., & Chaves, L. H. G., (2019). Biostimulants and their role in improving plant growth under abiotic stresses. *Biostimulants in Plant Science*, 1–14.
129. Ali, Q., Shehzad, F., Waseem, M., Shahid, S., Hussain, A. I., Haider, M. Z., & Perveen, R., (2020). Plant-based biostimulants and plant stress responses. In: *Plant Ecophysiology and Adaptation Under Climate Change: Mechanisms and Perspectives* I (pp. 625–661). Springer, Singapore.
130. Nephali, L., Piater, L. A., Dubery, I. A., Patterson, V., Huyser, J., Burgess, K., & Tugizimana, F., (2020). Biostimulants for plant growth and mitigation of abiotic stresses: A metabolomics perspective. *Metabolites*, *10*(12), 505.
131. Van, O. M. J., Pepe, O., De Pascale, S., Silletti, S., & Maggio, A., (2017). The role of biostimulants and bioeffectors as alleviators of abiotic stress in crop plants. *Chemical and Biological Technologies in Agriculture*, *4*(1), 1–12.
132. Eichmann, R., Richards, L., & Schäfer, P., (2021). Hormones as go-betweens in plant microbiome assembly. *Plant J., 105*, 518–541. doi: 10.1111/tpj.15135.
133. Kloeppe, J. W., Rodríguez-Kábana, R., Zehnder, A. W., et al., (1999). Plant root-bacterial interactions in biological control of soilborne diseases and potential extension to systemic and foliar diseases. *Australasian Plant Pathology, 28*, 21–26. https://doi.org/10.1071/AP99003.

134. Gray, E. J., & Smith, D. L., (2005). Intracellular and extracellular PGPR: Commonalities and distinctions in the plant-bacterium signaling processes. *Soil Biol. Biochem, 37*, 395–412.
135. Glick, B. R., (2012). Plant growth-promoting bacteria: Mechanisms and applications. *Scientifica,* 963401.
136. Zaidi, A., Khan, M. S., Ahemad, M., & Oves, M., (2009). Plant growth promotion by phosphate solubilizing bacteria. *Acta Microbiol. Immunol. Hung., 56*, 263–284.
137. Van, S. B. J., De Mooy, O. H., Linton, J. D., Van, D. J. P., & Kuenen, J. G., (1987). PQQ-dependent production of gluconic acid by *Acinetobacter*, *Agrobacterium* and *Rhizobium* species. *J. Gen. Microbiol., 133*, 867–875.
138. Gerke, J., (2015). The acquisition of phosphate by higher plants: Effect of carboxylate release by the roots. A critical review. *J. Plant Nutr. Soil Sci., 178*, 351–364.
139. Konietzny, U., & Greiner, R., (2004). Bacterial phytase: Potential application, *in vivo* function and regulation of its synthesis. *Brazilian Journal Microbiology, 35*, 11–18.
140. Kumar, R., Saurabh, K., Kumawat, N., Sundaram, P. K., Mishra, J. S., Singh, D. K., Hans, H., et al., (2021). Sustaining productivity through integrated use of microbes in agriculture. In: *Role of Microbial Communities for Sustainability*. Springer: Singapore.
141. Gupta, M., Kiran, S., Gulati, A., Singh, B., & Tewari, R., (2012). Isolation and identification of phosphate solubilizing bacteria able to enhance the growth and aloin-a biosynthesis of *Aloe barbadensis* miller. *Microbiol. Res.*
142. Basu, A., Prasad, P., Das, S. N., Kalam, S., Sayyed, R. Z., Reddy, M. S., &El Enshasy, H., (2021). Plant growth promoting rhizobacteria (PGPR) as green bioinoculants: Recent developments, constraints, and prospects. *Sustainability*.
143. Morrissey, J., & Guerinot, M. L., (2009). Iron uptake and transport in plants: The good, the bad, and the ionome. *Chem Rev., 109*(10), 4553–4567. doi: 10.1021/cr900112r. PMID: 19754138; PMCID: PMC2764373.
144. Neilands, J. B., (1995). Siderophores: Structure and function of microbial iron transport compounds. *J. Biol. Chem., 270*, 26723–26726. doi: 10.1074/jbc.270.45.26723.
145. Shi, D., Xu, Y., Hopkinson, B. M., & Morel, F. M. M., (2010). Effect of ocean acidification on iron availability to marine phytoplankton. *Science*, *327*, 676–679.
146. Dobbelaere, S., Croonenborghs, A., Thys, A., Vande, B. A., & Vanderleyden, J., (1999). Phytostimulatory effect of *Azospirillum brasilense* wild type and mutant strains altered in IAA production on wheat. *Plant Soil, 212*, 155–164.
147. Bucao, D. S., Yapit, R. H., & Gabriel, M. L. S., (2019). Biochemical characterization of microbials and their effects on the growth and yield of multiplier onion (*Allium ascalonicum* L.) in Northwestern Philippines. In: *Plant Growth Promoting Rhizobacteria (PGPR): Prospects for Sustainable Agriculture.* Springer: Singapore.
148. Blakeslee, J. J., Peer, W. A., &Murphy, A. S. (2005). Auxin transport. *Curr. Opin. Plant Biol., 8*, 494–500 10.1016/j.pbi.2005.07.014.
149. Bohn-Courseau, I., (2010). Auxin: A major regulator of organogenesis. *C. R. Biol., 333*, 290–296. 10.1016/j.crvi.2010.01.004.
150. Xie, L., Liu, M., Ni, B., Zhang, X., & Wang, Y., (2011). Slow-release nitrogen and boron fertilizer from a functional superabsorbent formulation based on wheat straw and attapulgite. *Chemical Engineering Journal*, *167*(1), 342–348.
151. Nambara, E., (2013). Plant hormones A2-maloy, stanley. In: Hughes, K., (éd.), *Brenner's Encyclopedia of Genetics* (2nd edn., pp. 346–348). San Diego: Academic Press.;. ISBN 978-0-08-096156-9.

152. Bottini, R., Cassán, F., & Piccoli, P., (2004). Gibberellin production by bacteria and its involvement in plant growth promotion and yield increase. *Appl. Microbiol. Biotechnol., 65*, 497–503.
153. Druege, U., (2006). Ethylene and plant responses to abiotic stresses. In: Khan, N. A., (ed.), *Ethylene Action in Plants* (pp. 81–118). Berlin; Heidelberg: Springer. 10.1007/978-3-540-32846-9_5.
154. Glick, B. R., (2014). Bacteria with ACC deaminase can promote plant growth and help to feed the world. *Microbiological Research, 169*(1), 30–39.
155. Bashan, Y., & de-Bashan, L. E., (2005). Bacteria/plant growth-promotion. In: Hillel, D., (ed.), *Encyclopedia of Soils in the Environment* (Vol. 1, pp. 103–115). Elsevier, Oxford, United Kingdom.
156. Gautam, A. K., & Avasthi, S., (2019). Fungal endophytes: Potential biocontrol agents in agriculture. In: *Role of Plant Growth Promoting Microorganisms in Sustainable Agriculture and Nanotechnology* (pp. 241–283). Woodhead Publishing.
157. Issa, A., Esmaeel, Q., Sanchez, L., Courteaux, B., Guise, J. F., Gibon, Y., & Aït Barka, E., (2018). Impacts of *Paraburkholderia phytofirmans* strain PsJN on tomato (*Lycopersicon esculentum* L.) under high temperature. *Frontiers in Plant Science, 9*, 1397.
158. Kloepper, J., Leong, J., Teintze, M., et al., (1980). Enhanced plant growth by siderophores produced by plant growth-promoting rhizobacteria. *Nature, 286*, 885, 886. https://doi.org/10.1038/286885a0.
159. Ulloa-Ogaz, A. L., Muñoz-Castellanos, L. N., & Nevárez-Moorillón, G. V., (2015). Biocontrol of phytopathogens: Antibiotic production as a mechanism of control. In: Méndez-Vilas, A., (ed.), *The Battle Against Microbial Pathogens: Basic Science, Technological Advances, and Educational Programs*. Formatex Research Center: Badajoz, Spain.
160. Wang, X., Mavrodi, D. V., Ke, L., Mavrodi, O. V., Yang, M., Thomashow, L. S., Zhang, N., et al., (2015). Biocontrol and plant growth-promoting activity of rhizobacteria from Chinese fields with contaminated soils. *Microb. Biotechnol.*
161. Kessler, P. S., Blank, C., & Leigh, J. A., (1998). The nif gene operon of the methanogenic archaeon *Methanococcus maripaludis. Journal of Bacteriology, 180*(6), 1504–1511.
162. Kessler, P. S., & Leigh, J. A., (1999). Genetics of nitrogen regulation in *Methanococcus maripaludis. Genetics, 152*(4), 1343–1351.
163. Halbleib, C. M., & Ludden, P. W., (2000). Regulation of biological nitrogen fixation. *The Journal of Nutrition, 130*(5), 1081–1084.
164. Butusov, M., & Jernelöv, A., (2013). *Phosphorus an Element That Could Have Been Called Lucifer.* New York: Springer.
165. Cordell, D., Drangert, J. O., & White, S., (2009). The story of phosphorus: Global food security and food for thought. *Glob. Environ. Change, 19*, 292–305. doi: 10.1016/j.gloenvcha.2008.10.009.
166. Cordell, D., & White, S., (2014). Life's bottleneck: Sustaining the world's phosphorus for a food secure future. *Annu. Rev. Environ. Resour., 39*, 161–188. doi: 10.1146/annurev-environ-010213-113300.
167. Cordell, D., Rosemarin, A., Schröder, J. J., & Smit, A. L., (2011). Towards global phosphorus security: A systems framework for phosphorus recovery and reuse options. *Chemosphere, 84*, 747–758. doi: 10.1016/j. chemosphere.2011.02.032.
168. Chowdhury, R. B., Moore, G. A., Weatherley, A. J., & Arora, M., (2017). Key sustainability challenges for the global phosphorus resource, their implications for

global food security, and options for mitigation. *Journal of Cleaner Production, 140,* 945–963.

169. Macintosh, K. A., Doody, D. G., Withers, P. J. A., McDowell, R. W., Smith, D. R., Johnson, L. T., et al., (2019). Transforming soil phosphorus fertility management strategies to support the delivery of multiple ecosystem services from agricultural systems. *Sci. Total Environ., 649,* 90–98. doi :10.1016/j. scitotenv.2018.08.272.
170. Lata, R., Chowdhury, S., Gond, S. K., & White, Jr. J. F. (2018). Induction of abiotic stress tolerance in plants by endophytic microbes. *Letters in Applied Microbiology, 66*(4), 268–276.
171. Chandra, D., Srivastava, R., Glick, B. R., & Sharma, A. K., (2018). Drought-tolerant *Pseudomonas* spp. Improve the growth performance of finger millet (*Eleusine coracana (L.) Gaertn.)* under non-stressed and drought-stressed conditions. *Pedosphere, 28,* 227–240. doi: 10.1016/s1002-0160(18)60013-x.
172. Nagargade, M., Tyagi, V., & Singh, M. K., (2018). Plant growth-promoting rhizobacteria: A biological approach toward the production of sustainable agriculture. In: Meena, V. S., (ed.), *Role of Rhizospheric Microbes in Soil: Stress Management and Agricultural Sustainability* (Vol. 1,pp. 205–223). Singapore: Springer Singapore. doi: 10.1007/978-981-10-8402-7_8.
173. Youpensuk, S., Lordkaew, S., & Rerkasem, B., (2006). Comparing the effect of arbuscular mycorrhizal fungi on upland rice and *Macaranga denticulata* in soil with different levels of acidity. *Science Asia, 32,* 121–126.
174. Zhang, F., Zou, Y. N., & Wu, Q. S., (2018). Quantitative estimation of water uptake by mycorrhizal extraradical hyphae in citrus under drought stress. *Scientia Horticulturae, 229,* 132–136.
175. Hirrel, M. C., & Gerdemann, J. W., (1980). Improved growth of onion and bell pepper in saline soils by two vesicular-arbuscular mycorrhizal fungi. *Soil Sci. Am. J., 44,* 654–658.
176. Smith, S. E., Christophersen, H. M., Pope, S., & Smith, F. A., (2010). Arsenic uptake and toxicity in plants: Integrating mycorrhizal influences. *Plant and Soil, 327*(1), 1–21.
177. Feng, G., Zhang, F., Li, X., Tian, C., Tang, C., & Rengel, Z., (2002). Improved tolerance of maize plants to salt stress by arbuscular mycorrhiza is related to higher accumulation of soluble sugars in roots. *Mycorrhiza, 12*(4), 185–190.
178. Egamberdieva, D., & Lugtenberg, B., (2014). Use of plant growth-promoting rhizobacteria to alleviate salinity stress in plants. In: Miransari, M., (ed.), *Use of Microbes for the Alleviation of Soil Stresses* (pp. 73–96). New York, NY: Springer. doi: 10.1007/978-1-4614-9466-9_4/.
179. Shameer, S., & Prasad, T., (2018). Plant growth promoting rhizobacteria for sustainable agricultural practices with special reference to biotic and abiotic stresses. *Plant Growth Regul., 84,* 603–615. doi: 10.1007/s10725-017-0365-1.
180. Lo, K. J., Lee, S. K., & Liu, C. T., (2020). Development of a low-cost culture medium for the rapid production of plant growth-promoting *Rhodopseudomonas palustris* strain PS3. *Plos One, 15*(7), e0236739.
181. Estrella, M. J., Pieckenstain, F. L., Marina, M., Diaz, L. E., & Ruiz, O. A., (2004). Cheese whey: An alternative growth and protective medium for *Rhizobium loti* cells. *Journal oflndustrial Microbiology and Biotechnology, 31*(3), 122–126.
182. Yuan, Y., Chu, D., Fan, J., Zou, P., Qin, Y., Geng, Y., & Wang, X., (2021). Ecofriendly conversion of algal waste into valuable plant growth-promoting rhizobacteria (PGPR) biomass. *Waste Management, 120,* 576–584.

183. Vassilev, N., Malusa, E., Requena, A. R., Martos, V., López, A., Maksimovic, I., & Vassileva, M., (2017). Potential application of glycerol in the production of plant beneficial microorganisms. *Journal of Industrial Microbiology and Biotechnology, 44*(4, 5), 735–743.
184. Pandey, P., & Maheshwari, D. K., (2007). Bioformulation of *Burkholderia sp.* MSSP with a multispecies consortium for growth promotion of *Cajanus cajan. Canadian Journal ofMicrobiology, 53*(2), 213–222.
185. Lobo, C. B., Tomás, M. S. J., Viruel, E., Ferrero, M. A., & Lucca, M. E., (2019). Development of low-cost formulations of plant growth-promoting bacteria to be used as inoculants in beneficial agricultural technologies. *Microbiological Research, 219*, 12–25.
186. Zhang, H., Liu, Y., & Wang, G., (2019). Integrated use of maize bran residue for one-step phosphate bio-fertilizer production. *Applied Biochemistry and Biotechnology, 187*(4), 1475–1487.
187. Cantabella, D., Dolcet-Sanjuan, R., Solsona, C., Vilanova, L., Torres, R., & Teixidó, N., (2021). Optimization of a food industry-waste-based medium for the production of the plant growth promoting microorganism *Pseudomonas oryzihabitans* PGP01 based on agro-food industries by-products. *Biotechnology Reports, 32*, e00675.
188. Ganeshan, S., Kim, S. H., & Vujanovic, V., (2021). Scaling-up production of plant endophytes in bioreactors: Concepts, challenges, and perspectives. *Bioresources and Bioprocessing, 8*(1), 1–16.
189. Petersen, I., Paungfoo-Lonhienne, C., Marcellin, E., Nielsen, L. K., & Gonzalez, A., (2021). Towards sustainable bioinoculants: A fermentation strategy for high cell density cultivation of *Paraburkholderia sp.* SOS3, a plant growth-promoting bacterium isolated in Queensland, Australia. *Fermentation, 7*(2), 58.
190. Smith, D. E., Leffler, M., & Mackiernan, G. B., (1992). *Oxygen Dynamics in the Chesapeake Bay: A Synthesis of Recent Results.* Maryland Sea Grant College, College Park, Maryland.
191. Albareda, M., Rodríguez-Navarro, D. N., Camacho, M., & Temprano, F. J., (2008). Alternatives to peat as a carrier for rhizobia inoculants: Solid and liquid formulations. *Soil Biology and Biochemistry, 40*(11), 2771–2779.
192. Deaker, R., Roughley, R. J., & Kennedy, I. R., (2004). Legume seed inoculation technology: A review. *Soil Biol Biochem., 36*, 1275–1288.
193. Burges, H. D., & Jones, K. A., (1998). Trends in formulation of microorganisms and future research requirements. In: Burges, H. D., (eds.), *Formulation of Microbial Biopesticides*. Springer, Dordrecht. https://doi.org/10.1007/978-94-011-4926-6_10.
194. Ferreira, E. M., & Castro, I. V., (2005). Residues of the cork industry as carriers for the production of legumes inoculants. *Silva Lusitana, 13*, 159–167.
195. Goss, G. R., Baldwin, H. M., & Riepl, R. G., (2003). Clays as biological carriers. In: *Pesticide Formulations and Delivery Systems: Meeting the Challenges of the Current Crop Protection Industry* (pp. 24–34). American Society for Testing and Materials, Dallas.
196. Goss, G., (2018). Clays in biological formulations. In: *Pesticide Formulation and Delivery Systems: 37th Volume, 2018; Formulations with Ingredients on the EPA's List of Minimal Concern.* ASTM International.
197. Agbodjato, N. A., Adoko, M. Y., Babalola, O. O., Amogou, O., Badé, F. T., Noumavo, P. A., & Baba-Moussa, L., (2021). Efficacy of biostimulants formulated with *Pseudomonas*

putida and clay, peat, clay-peat binders on maize productivity in a farming environment in Southern Benin. *Frontiers in Sustainable Food Systems, 5*, 666718.
198. Doni, F., Isahak, A., Che Mohd, Z. C. R., Mohd, A. S., Mohamad, W., Nur'ashiqin, W., & Wan, Y. W. M., (2014). Formulation of *Trichoderma sp.* SL2 inoculants using different carriers for soil treatment in rice seedling growth. *Springerplus, 3*(1), 1–5.
199. Dong, M., Li, Q., Xu, F., Wang, S., Chen, J., & Li, W., (2020). Effects of microbial inoculants on the fermentation characteristics and microbial communities of sweet sorghum bagasse silage. *Scientific Reports, 10*(1), 1–9.
200. Trillas, M. I., Casanova, E., Cotxarrera, L., Ordovás, J., Borrero, C., & Avilés, M., (2006). Composts from agricultural waste and the *Trichoderma asperellum* strain T-34 suppress *Rhizoctonia solani* in cucumber seedlings. *Biological Control, 39*(1), 32–38.
201. Djaya, L., Istifadah, N., Hartati, S., & Joni, I. M., (2019). *In vitro* study of plant growth promoting rhizobacteria (PGPR) and endophytic bacteria antagonistic to *Ralstonia solanacearum* formulated with graphite and silica nanoparticles as a biocontrol delivery system (BDS). *Biocatalysis and Agricultural Biotechnology, 19*, 101153.
202. Kremer, R. J., & Peterson, H. L., (1983). Effects of carrier and temperature on survival of *Rhizobium* spp. in legume inocula: Development of an improved type of inoculant. *Appl. Environ. Microbiol., 45*, 1790–1794.
203. Chao, W. L., & Alexander, M., (1984). Mineral soils as carriers for *Rhizobium* inoculants. *Appl. Environ. Microbiol., 47*, 94–97.
204. Graham-Weis, L., Bennet, M. L., & Paau, A. S., (1987). Production of bacterial inoculants by direct fermentation on nutrient-supplemented vermiculite. *Appl. Environ. Microbiol., 53*, 2138–2140.
205. Sparow, S. D., & Ham, G. E., (1983). Survival of *Rhizobium phaseoli* in six carrier materials. *Agron. J., 75*, 181–184.
206. Daza, A., Santamaria, C., Rodriguez-Navarro, D. N., Camacho, M., Orive, R., & Temprano, F., (2000). Perlite as a carrier for bacterial inoculants. *Soil Biol. Biochem., 32*, 567–572.
207. Ronchi, A. L., Grassano, A., & Balatti, A. P., (1997). Perlite as a carrier for legume inoculants. *Agrochimica, 41*, 186–195.
208. Brar, S. K., Sarma, S. J., & Chaabouni, E., (2012). Shelf-life of biofertilizers: An accord between formulations and genetics. *J. Biofertil. Biopestici., 3*, e109. doi: 10.4172/2155-62021000e109.
209. Bashan, Y., De-Bashan, L. E., Prabhu, S. R., & Hernandez, J. P., (2014). Advances in plant growth-promoting bacterial inoculant technology: Formulations and practical perspectives (1998–2013). *Plant and Soil, 378*(1), 1–33.
210. Bashan, Y., Hernandez, J. P., Leyva, L. A., & Bacilio, M., (2002). Alginate microbeads as inoculant carrier for plant growth-promoting bacteria. *Biol Fertil Soils, 35*, 359–368.
211. Fallik, E., & Okon, Y., (1996). Inoculants of *Azospirillum brasilense*: Biomass production, survival, and growth promotion of *Setaria italica* and *Zea mays*. *Soil. Biol. Biochem., 28*, 123–126.
212. Trivedi, P., Pandey, A., & Palni, L. M. S., (2005). Carrier based formulations of plant growth promoting bacteria suitable for use in the cooler regions. *World J. Microbiol. Biotechnol., 26*(6, 7), 941–945.
213. Vadivel, M., Azees, R. A., Paramabathan, R., Norrudeen, A. M., Karthik, N., Packiaraja, L., & Mohaideen, J. A., (2020). Sodium alginate gel beads with slow-release fertilizer

from seaweeds and biostimulants of paddy growth. In: *International Conference on Frontiers in Chemical Technology–2020* (p. 48). Institute of Chemistry Ceylon.
214. Pajkic, K., (2021). *Development of Encapsulation Methods for Biostimulants*. Doctoral dissertation, Wien.
215. Saberi-Riseh, R., Moradi-Pour, M., Mohammadinejad, R., & Thakur, V. K., (2021). Biopolymers for biological control of plant pathogens: Advances in microencapsulation of beneficial microorganisms. *Polymers, 13*(12), 1938.
216. Trivedi, P., & Pandey, A., (2008). Plant growth promotion abilities and formulation of *Bacillus megaterium* strain B 388 (MTCC6521) isolated from a temperate Himalayan location. *Indian J. Microbiol, 48*, 342–347. https://doi.org/10.1007/s12088-008-0042-1.
217. Schoebitz, M., & López, B. M. D., (2016). Encapsulation techniques for plant growth-promoting rhizobacteria. In: *Bioformulations: For Sustainable Agriculture* (pp. 251–265). Springer, New Delhi.
218. Cuomo, F., Lopez, F., Piludu, M., Miguel, M. G., Lindman, B., & Ceglie, A., (2015). Release of small hydrophilic molecules from polyelectrolyte capsules: Effect of the wall thickness. *J. Colloid Interface Sci., 447*, 211–216. doi: 10.1016/j.jcis.2014.10.060.
219. Schoebitz, M., Lopez, M. D., & Roldan, A., (2013). Bioencapsulation of microbial inoculants for better soil– plant fertilization. A review. *Agron. Sustain. Dev., 4*, 751–765.
220. Lesueur, D., Deaker, R., Herrmann, L., Bräu, L., & Jansa, J., (2016). The production and potential of biofertilizers to improve crop yields. In: Arora, N. K., Menhaz, S., & Balestrini, R., (eds.), *Bioformulations for Sustainable Agriculture* (pp.71–92). Springer: New Delhi.

CHAPTER 3

Bacillus Species: Prospects and Applications for the Root-Knot Disease Management

MAHFOUZ MOHAMED MOSTAFA ABD-ELGAWAD

Department of Plant Pathology, National Research Center, El-Behooth St., Dokki, Giza, Egypt

ABSTRACT

Although plant-parasitic nematodes (PPN) cause significant agricultural losses worldwide, the majority of these losses are attributed to root-knot nematodes (RKN; *Meloidogyne* spp.). RKN are sedentary endoparasites with a wide range of hosts and a global distribution, making current management measures insufficient for complete control. Additionally, the use of chemical nematicides is being restricted or banned due to environmental and health concerns. Therefore, extensive research is being conducted on biocontrol agents as safe and effective alternatives to these chemicals. Among these agents, a comprehensive understanding of *Bacillus* spp. bacteria may help minimize these losses and contribute to the development of synthetic/natural microbial communities for RKN control. The global diversity and characteristics of these bacteria make them ideal and safe tools for managing RKN, as well as potential biocontrol agents for other pests and pathogens. When introduced to arable soil, their spores can quickly establish season-long activity, even under challenging ecological conditions and variable soil temperature, moisture, and pH. This article discusses the various stages of exploring suitable *Bacillus* species/strains, mass culturing, formulating, and applying them to optimize their effectiveness in controlling RKN. It is

Microbial Biostimulants: Biorational Pesticides for Management of Plant Pathogens.
Faheem Ahmad and Rakesh Pandey (Eds.)

recommended to integrate these strategies with other agricultural practices to maximize their control efficacy.

3.1 INTRODUCTION

The genus *Bacillus* is aptly named as it comprises numerous species of rod-shaped bacteria belonging to the class *Bacilli* and phylum *Bacillota*. *Bacillus* species may be divided into obligate aerobes and facultative anaerobes. Another notable characteristic of this genus is the ability of its species to survive under stressful conditions in a dormant state for prolonged periods, sometimes years, in the form of endospores. As a result, these bacteria can tolerate harsh conditions such as high/low temperatures, desiccation, disinfectants, absence of nutrients, and radiation [1]. Moreover, *Bacillus* spp. are found almost everywhere, including both symbiotic (known as endophytes) and independent species. The benefits of these characteristics are evident in their applications in various aspects of our lives. For instance, some *Bacillus spp.* can produce a wide range of enzymes that are used in different activities, such as generating alpha-amylase for starch hydrolysis and producing protease subtilisin for use in detergents. *Bacillus amyloliquefaciens* is a natural source of barnase, an antibiotic protein. Additionally, several *Bacillus* spp. can produce and release lipopeptides such as mycosubtilins and surfactins [2–4]. The significance of many *Bacillus* species in the food chain is reflected in their ability to produce various compounds, including antimicrobial substances, proteins, carotenoids, and vitamins.

Additionally, *Bacillus subtilis* is currently best understood as a molecular model of prokaryotes. Therefore, it is being studied as a model for gene/protein systems and for differentiating cycle scores in bacterial cells [5]. As a result, many of these *Bacillus* species and their beneficial compounds are receiving significant attention from both humans and domestic animals. Several *Bacillus* species can be considered probiotics, which are microorganisms that, when administered in proper amounts, confer health benefits to the recipient/host [6]. On the other hand, the genus *Bacillus* contains two species that are harmful to humans. These are *B. anthracis* and *B. cereus*, which produce anthrax and food poisoning, respectively. Additionally, caution should be exercised with others, such as *B. subtilis* and *B. coagulans*, as they produce antibiotics and may result in food spoilage [7].

Of particular interest among all the previous bacterial attributes, the significance of *Bacillus* spp. in the management of plant-parasitic nematodes

(PPN) ranks high. Various species and strains of these bacteria can offer environmentally friendly bio-nematicides against numerous PPN that damage many economically important crops.

3.2 THE ROOT-KNOT DISEASE: THE CAUSE AND THE LOSSES

Staggering but real estimates are reported concerning considerable losses annually imposed by PPN on qualitative and quantitative plant yields [8, 9]. The lion's share of these losses is globally inflicted by the root-knot nematodes (RKN; *Meloidogyne* spp.) that cause the root-knot disease [10]. Admittedly, many nematologists have confirmed that RKN can attack and inflict damage on almost all cultivated plants, including feed, food, and industrial crops [11, 12]. These include vegetables, field crops, as well as ornamental and medicinal plants. The broad host range of *Meloidogyne* spp. limits the availability of resistant/immune cultivars, especially in the context of crop sequences or rotations. Similar to most PPN, the *Meloidogyne* eggs hatch into the second stage juvenile (J2) and then molt to J3 and J4. Finally, J4 develops and molts into adult males and females. Thus, the first molt of J1 occurs within the egg. Upon hatching, J2 moves in the soil to come into contact with plant roots for infection. The nematodes penetrate the roots to migrate, feed, and finally develop there into females known as sedentary endoparasites [12]. Although there are over 90 species, some have several races; four species (*Meloidogyne incognita, M. hapla, M. arenaria,* and *M. javanica*) represent global and chief pests. Indeed, more species are important in local or regional ranks [12].

Meloidogyne spp. are at the top of the list of plant-nematode pathogens based on scientific and economic importance [13]. The genus comprises pathogens that are the most destructive group among the phytonematodes. This is because, in addition to the aforementioned damage caused by RKN, species of *Meloidogyne* possess a high multiplication rate and short reproduction time. Apart from their ability to cause the aptly named root-knot disease, RKN are also capable of cooperating with other soil-borne pathogens to establish complex diseases. Such diseases can severely aggravate the health and yield of many important crops such as tomato [14], pepper [15], potato [16], and eggplant [17]. Thus, economic analysts prefer to categorize these losses into different planes: the community, consumer, and grower/producer [9]. The grower/producer is the one who initially faces the qualitative and/or quantitative losses in crop production. The second level of losses is incurred by the consumer in terms of the buyer of the production, household, or/and industry. The consumer endures the quality degradation, the raised price of the product, or both together. Clearly, health hazards could occasionally be faced by applying banned/restricted nematicides or consuming/using chemical nematicides-treated produce [18]. The widest and third level is the societal/community losses. The previous levels can surely result in wasted resources with unfavorable impacts on the society/country. Such crop losses constitute a major global challenge and can worsen the standard of living, especially in economically weak countries suffering from pandemics and wars.

RKN feeding within plant roots induces unique knot-like swellings (or the so-called galls) on the roots (Figure 3.1). These galls are the result of giant cell formation induced by the nematode females in the plant roots. Since the subterranean RKN infect the underground parts of the plants, the roots are their primary targets. As a result, the roots are significantly hindered in their essential functions of absorbing and transporting water and nutrients. As RKN feed and develop, the growth of roots slows down and the formation of secondary roots decreases. In parallel, shoot development is also suppressed. If plant roots are heavily infected by RKN, the main root system is reduced to a small number of severely RKN-galled fibrous roots with a severely disordered vascular system.

Consequently, the RKN-infected plants do not flower perfectly and produce poor-quality fruits. These plants are susceptible to drought damage. Severe infections can lead to plant death, especially in the early stages of growth when rootlets are almost entirely absent. A high level of ethylene (ET) production is believed to be responsible for expressing such symptoms in highly susceptible crops like tomatoes [19]. The symptoms and damage caused by RKN depend on the host's suitability, RKN population densities, and prevailing biotic and ecological conditions. Compared to direct seeding, older transplants may sustain relatively high levels of RKN populations. These RKN galls can range from simple globe-shaped swellings to extensive areas of outspread swellings (Figure 3.1). The latter is the result of diverse and frequent RKN infections. The shape, formation, and location of the gall are always used as positive diagnostic confirmation, indicating the presence of RKN and the potential for crop damage. Nematode infections often facilitate infection by bacterial and fungal pathogens, leading to deteriorating plant health. For many highly susceptible crops or plant species, RKN control is required if any individual is detected per 100 cm^3 [9].

FIGURE 3.1 Root-knot nematodes (*Meloidogyne* spp.) induce different gall sizes and shapes on eggplant roots.

Source: Reprinted from Ref. [17].

All *Meloidogyne* species exhibit sexual dimorphism, with females being pear-shaped and males being filiform. One important characteristic of RKN, like other PPN, is their tendency to have clumped or uneven distribution in the field [9, 20, 21]. As a result, symptoms of RKN infection often appear in patchy areas of the field (Figure 3.2). In fields infested with RKN, untreated patches of these areas will continue to spread the nematodes throughout the entire field. The typical symptoms of RKN damage on susceptible plants include stunted growth, reduced plant parameters, and yield loss. Plants infected by RKN are usually more susceptible to weed damage and generally struggle to compete with other stresses [12].

FIGURE 3.2 An eggplant field showing patchy area due to root-knot nematode (*Meloidogyne* spp.) infestation.

Source: Reprinted from Ref. [17].

Synthetic chemicals, such as *carbofuran*, cadusafos, and oxamyl, have been widely used as nematicides to effectively control RKN [12, 22]. However, their high toxicity to humans, animals, and the environment can lead to various environmental and health issues. In order to maintain ecological balance, it is almost mandatory to search for other viable alternatives, such as biocontrol approaches, as one of the main objectives for safe and reliable RKN management.

3.3 PROSPECTS FOR MANAGING THE ROOT-KNOT DISEASE

Currently, RKN are managed using diverse practices such as chemicals, soil amendments, resistant plants, crop rotation, bionematicides, flooding, following, tillage, solarization, and certified transplants. As RKN spend their life cycle within the soil/plant roots, applying chemicals to the close surroundings of RKN is generally not easy [23]. Nonetheless, these nematicides are conventional means of reliable RKN control, as they can reduce *Meloidogyne* species below threshold levels [24]. Unfortunately, the potential harm of synthetic nematicides to human health and the environment, as well as the emergence of resistance-breaking RKN pathotypes, have diminished their strategic usefulness and called for safe alternatives. Bionematicides have emerged as mostly safe alternatives. However, they are often less reliable, slower acting, and more volatile/inconsistent compared to nematicides. A proper understanding of the modes of action, biology, and ecology of biocontrol agents (BCA) is necessary for their effective utilization [25]. Crop rotation is a reliable and benign technique for RKN control, if RKN resistant/immune plant varieties/cultivars are available for the crop sequence. Soil amendments are frequently used to enhance plant growth. However, careful consideration should be given to avoid promoting the reproduction of harmful PPN and beneficial BCA [26]. Therefore, amendments must be carefully managed to establish proper frameworks for the conservation/sound biocontrol of RKN. Other additions such as plant extracts, botanical matrices, and secondary metabolites have shown efficacy against RKN. However, their registration for commercial use is hindered by slow processing [27]. These materials must be validated for their lack of side effects, effectiveness against RKN, and cost-effective usage. Notably, volatile and toxic isothiocyanates (ITCs) used as active ingredients in some chemical nematicides share the same biochemical mechanisms as natural ITCs applied as biofumigants to control nematodes.

Regrettably, the unfavorable effects of ITCs, such as those found in mustard and other bio-fumigants, have resulted in the vulnerability and instability of food webs in treated soil, as well as the suppression of BCA [28]. Fallowing and flooding are utilized as measures to control RKN. However, these methods are not often cost-effective for RKN-control schemes. Following these practices can lead to the misuse of arable areas. Flooding is only successful in warm climates and requires a significant amount of water and time. Additionally, solarization is time-consuming without cultivation and its effectiveness depends on elevated temperature and a favorable cultivation scheme. Contrary to the various colors and types of polyethylene sheets used for solarization, transparent sheets have been reported [29, 30] to be the most effective in reducing RKN populations compared to colored sheets such as black, red, green, and blue. Tillage is a good agricultural practice for controlling weeds, pests, and pathogens. However, it has side effects as it directly deactivates populations of RKN-antagonistic organisms, leading to nematode damage [26]. The use of certified seeds and transplants is a prime and outstanding approach.

Nevertheless, complementary action is needed; the plants should not be cultivated in soils infested with RKN. Ultimately, the RKN-control approaches mentioned above are not widely accepted and require careful revisions for efficacy and safety [31]. Additionally, their limitations or defects often pose challenges in accurately sampling the subterranean and within plant life stages of RKN. The broad host range and diverse and clumped distribution of RKN can complicate control measures [12–14].

It is evident from the previous review that the current methods and trends used in RKN management/control have several flaws and shortcomings.

Therefore, the application and tactics for developing and using *Bacillus* spp. in this context may also have limitations, as seen in some studies. However, it is clear that the manipulation or application of RKN-antagonistic microbes, such as *Bacillus* spp., holds promise [32–38]. Therefore, there is a dire need to expand RKN management tactics and strategies using *Bacillus* spp. and related materials to address challenges posed by PPN, the increased banning/limitation of chemical nematicides, and the recurrent emergence of resistance-breaking RKN pathotypes [18]. It is crucial to adopt effective approaches to better understand the biological and ecological conditions that affect *Bacillus* spp. [25]. Like other bionematicides, *Bacillus* bacteria can effectively replace hazardous nematicides for sustainable agriculture, as depicted in Figure 3.3.

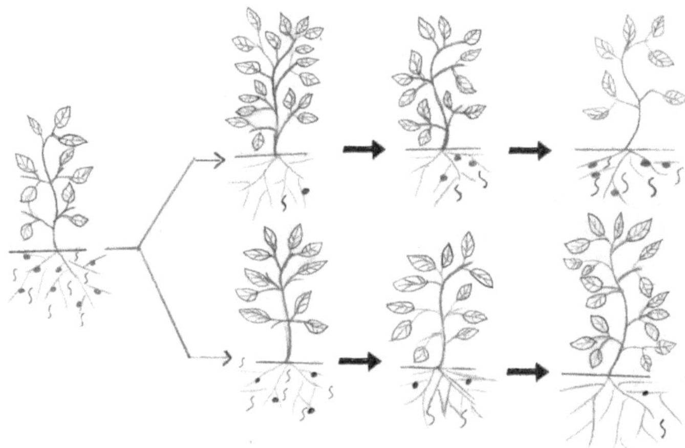

FIGURE 3.3 Effect of a chemical nematicide (upper plants) and a *Bacillus* species (lower plants) on root-knot nematodes (RKN) on susceptible plants. The chemical has a rapid and significant effect on decreasing the RKN population, but a few nematodes can escape its effect and multiply to attain damaging levels, while *Bacillus* species can work continuously to keep the RKN below the economic threshold level.

Source: Modified from: Ref. [18].

To achieve such a goal and optimize the bacterial efficacy, further investigations are still needed on *Bacillus* spp. as BCA to address the need for novel, reliable, and environmentally friendly techniques and to enhance current management schemes for root-knot nematodes (RKN). While many microbes have shown potential against PPN, *Bacillus* spp. stands out as a viable option for conservation biological control of PPN [26]. They can be mass-cultured for commercial production to combat RKN, as well as other plant pathogens and pests such as plant-pathogenic fungi and nematodes [22, 33]. Of particular interest are the synergistic or additive effects of *Bacillus* species when combined with other beneficial bacteria or fungi, which can effectively manage a wide range of plant diseases. Such combinations have the potential to enhance biocontrol efficacy across various ecological settings [18, 38].

Hence, effective approaches, including using *Bacillus* spp. as biocontrol agents on commercial/large and limited scales, alone and in combinations with other beneficial bacteria/fungi, are addressed herein. Developing these approaches for using *Bacillus* spp. will strengthen the prospects of reconciliation with the environment and human beings. They may replace pollution and health hazards imposed by chemical nematicides. Otherwise, *Bacillus* spp.-based bionematicides can play a significant role as additions to the current management programs for plant-parasitic nematodes (PPN). With the voluminous research and experimental results on *Bacillus* spp., only some recent examples of *Bacillus* applications at microcosms/controlled conditions or on definite RKN-infected crops and settings are presented in this chapter.

Nevertheless, these examples vividly express their counterparts on the other RKN-susceptible crops. In addition, they show how to successfully combat one of the most serious nematode genera by utilizing safe yet effective bionematicidal applications on *Melodogyne* spp. The role of *Bacillus* in such commercial and experimental applications has proven to be increasingly prominent [32–34, 36, 37, 39]. This is mainly because *Bacillus*, found in the plant rhizosphere, possesses favorable attributes and can readily colonize the roots, acting as a protectant against (or antagonist to) phytopathogens and promoting plant growth [40]. Ultimately, these attributes and activities of *Bacillus* species can lead to valuable returns and significant findings that contribute to maximizing farmers' income.

3.3.1 BACILLUS SPECIES: TAXA, LIFESTYLE, AND DIVERSITY

It is evident from the previous introduction and the prospects sections that *Bacillus* comprises a diverse group of bacteria. The relationships between different *Bacillus* species have been somewhat ambiguous thus far. Although recent molecular research and sequencing projects have complicated our understanding of *Bacillus* taxonomy and diversity, its fundamental scientific classification remains robust [7]. This classification includes its Domain

(Bacteria), Phylum (Bacillota), Class (Bacilli), Order (Bacillales), Family (Bacillaceae), and Genus (*Bacillus*).

Recent molecular-based taxonomic approaches have demonstrated that *Bacillus sensu lato* (in a broad sense) includes at least three bacterial families that encompass numerous genera. *Bacillus subtilis*, particularly the laboratory strain *B. subtilis* 168, has emerged as a significant model organism. This species has given rise to thousands of valuable genetic mutants and strain libraries, all of which have been developed over the past seven decades of research on *B. subtilis*, primarily using strain 168 [41, 42]. The *Bacillus* Genetic Stock Center (BGSC) has taken proper responsibility for preserving this legacy. The Center collects, maintains, distributes, and compiles information related to these mutants and genetic devices. It ensures that the mutants are always available to members of the research community with a relevant interest [42].

Species of *Bacillus* have embraced a common lifestyle. The cells of these species can propagate rapidly under favorable conditions for their growth. Conversely, in the absence of nutrient sources, various cell types may form distinct subpopulations of the original *Bacillus* population [41]. However, this formation often exhibits a high level of multicellularity. Of particular importance is the endospore type, as it is a metabolically inert resting cell with increased resistance to most of the aforementioned environmental stresses. As conditions improve for growth, these resting cells germinate and develop, establishing new populations. This process of growth is known as the *Bacillus* lifestyle [41, 42]. Due to its high degree of successful adaptation, this lifestyle has potential benefits for various aspects, such as the biological control of PPN under numerous environmental insults. The *Bacillus* lifestyle, involving species/strains of different forms, has also contributed to its diverse state.

3.4 BACILLUS SPECIES: SPECIFICITY AND ANTAGONISTIC MODES OF ACTION FOR PLANT-PARASITIC NEMATODES

Bacillus spp. represent a prominent group of rhizobacteria that can be used for the biological control of PPN [43]. Not all species are pathogens of RKN [44–46]. However, many *Bacillus* strains can not only suppress pathogens and pests of plants but also enhance plant growth. Promoting plant growth may enhance a plant's ability to withstand RKN attack. While certain species, such as *B. subtilis*, have been extensively studied for controlling RKN in the past few decades, others, such as *B. amyloliquefaciens* (syn. *B. velezensis*) and *B. cereus*, have more recently been developed as safe and effective bionematicides [44, 45]. Several studies have reported direct antagonism by various *Bacillus* spp. against numerous PPN species. These susceptible species belong to economically important genera such as *Meloidogyne, Rotylenchulus, Heterodera, Radopholus, and Ditylenchus* [25, 38, 43, 46].

The mode of action exerted by some *Bacillus* species/strains to decrease the PPN populations has been extensively investigated, primarily against *Meloidogyne* spp. Their mechanisms against the nematodes may vary from one species to another [18, 38, 46], but they generally involve antibiotic production, antagonism, and/or induced resistance. Specifically, their activities for RKN control depend on cry proteins - toxic particles of *Bacillus thuringiensis* (Bt), toxic antibiotics of *B. cereus* and *B. subtilis*, enzymatic performance of *B. firmus*, and expelling second-stage juveniles of RKN by *B. cereus* [18]. These bacteria can generally antagonize the nematodes by colonizing the plant roots. For instance, Wepuhkhulu et al. [47] reported that lytic enzymes produced by *Bacillus subtilis* have been able to suppress *Meloidogyne javanica* populations.

Specificity may be more related to the efficacy of the strain. The *B. cereus* strain BCM2 was able to inhibit *Meloidogyne incognita* infection in tomato plants by 67.1%. This strain influenced the root exudates during colonization, enhancing the secretion of specific repellent materials for *Meloidogyne incognita* J2 [48]. Lee & Kim [49] pointed out that protease and chitinase, generated by *B. pumilus* strain L1, played a role in the antagonistic effects against *Meloidogyne arenaria*. These enzymes resulted in approximately 90% mortality of J2 and 88% suppression of egg hatch. *In-vitro* activity of *Bacillus firmus* induced various mortality rates in economically important PPN species such as *M. incognita, Radopholus similis,* and *Ditylenchus dipsaci* [50]. The bacterium induced significant inhibition in *M. incognita* egg hatching when the eggs were incubated with a low concentration of *B. firmus* pure culture filtrates. The bacterial mechanism for the observed PPN paralysis before mortality was attributed to its secondary metabolites [50].

Induced systemic resistance (ISR) is another mode of action exerted by some *Bacillus* strains. For instance, *B. subtilis* can stimulate resistance to *M. incognita* on cotton [51]. This type of resistance is displayed by certain strains of *B. mycoides, B. velezensis, B. sphaericus, B. cereus, B. pasteurii, B. subtilis,* and *B. pumilus* [52]. They can induce significant decreases in the incidence/severity of numerous diseases on various plant hosts. ISR by *B. firmus* GB-126 was confirmed on *Heterodera glycines* but not on *M. incognita* in split-root design experimentation [53]. However, the bacterial strains *Rhizobium etli* G12 and *B. sphaericus* B43 were found to exert

ISR towards *Meloidogyne incognita* on tomato, as evidenced by decreased juvenile penetration in the reacting roots [54–56]. *Bacillus velezensis* strain Bve12 and *B. mojavensis* strain Bmo3, which are known to exert ISR and enhance the yield of Chinese cabbage infested by black rot pathogens [57], could also decrease *M. incognita* population levels and boost cotton yield, possibly via ISR by the strains Bve2, Bmo3, and Mixture 2 (Bve2 + abamectin + Bal13) [45].

Further avenues for reaching RKN-management schemes are displayed by various *Bacillus* spp. For instance, *B. megaterium* strain YMF 3.25 could significantly inhibit the hatching of RKN eggs and decrease infection of *M. incognita* by generating nematicidal volatiles [58]. These volatiles generated by the bacterium were chiefly 2-nonanone, benzene acetaldehyde, decanal, dimethyl disulfide, and 2-undecanone, which could control *M. incognita* juveniles and eggs at a concentration of 0.5 mmol. Three *B. thuringiensis* crystal proteins, Cry5Ba, Cry6Aa, and Cry55Aa, proved to have nematicidal efficacy against *M. incognita.* According to Peng et al. [59], the combination of Cry55Aa and Cry6Aa showed a significant synergistic effect on *M. incognita.* Their records confirmed that the mechanism of the *Bacillus* strains with such favorable activity is mostly related to the generation of secondary metabolites and bioactive components. Further research is needed to address the modes of action of the Plant growth-promoting rhizobacteria (PGPR) strains with nematicidal activity on RKN. Eventually, *Bacillus* spp. are positive candidates for managing PPN diseases via diverse mechanisms.

3.5 COMMERCIAL CULTIVATION AND PRODUCTION OF *BACILLUS* SPECIES

This process aims to produce high-quality *Bacillus* biomass with a desirable cell density, improved sporulation efficiency, and easy incorporation into an efficient formulation. To reduce costs, small-scale experiments are necessary to optimize fermentation conditions, often using Erlenmeyer flasks or similar materials, as a prerequisite for commercial production development. Promising results from small-scale experiments will lead to scaling up using larger volume bioreactors with mechanical stirring. Optimizing the cultivation process involves adjusting various factors related to the *Bacillus* culture, such as the specific species/strains used, growth medium/conditions, and fermentation type [33].

3.5.1 BACILLUS SPECIES/STRAINS

Research and screening efforts are currently underway to identify new and promising bacterial species/strains against *Meloidogyne* spp. Recently, the nematicidal activity of *Bacillus velezensis* strain Bv-25, extracted from cucumber rhizosphere soil, has shown potential in suppressing *M. incognita*-egg hatching and achieving a 100% mortality rate of *M. incognita* J2s within 12 hours of exposure to the strain fermentation broth. The RKN genes mpk-1, ord-1, and flp-18 were inhibited by the fumigation of strain Bv-25 after 48 hours. Furthermore, strain Bv-25 demonstrated superior colonizing capability on cucumber roots, with 5.94×10^7 colony-forming units/g attached to the roots within 24 hours. As a result, the root infection rate with *M. incognita* J2s decreased by 98.6% and root knots by 73.8%. The defense of cucumber plants against the nematode was enhanced by up-regulating the relevant cucumber genes pr3, pr1, and lox1 through the bacteria. In field experimentation, the root-knot disease index on cucumber roots was reduced by 61.6% [60]. The authors concluded that strain Bv-25 has great potential in controlling RKN through the use of its volatile compounds and effective colonization of plant roots. Other recent examples of novel *Bacillus* strains have been reported to significantly enhance cucumber plant height by 7.4–43.1%, leaf area by 2.7–17.8%, and root mass by 3.2–16.1% in greenhouse conditions [39]. Two of these strains, *B. velezensis* BZR 86 and *B. velezensis* BZR 277, have shown considerable potential in improving crop productivity and plant health. They are strong candidates for the development of commercial products that protect cucumber plants from *M. incognita* infection [39].

3.5.2 THE GROWTH MEDIUM

Invariably, the growth medium should be optimized to initiate the revitalization of the bacteria before fermentation. This procedure is conducted by forming an inoculum in a small volume of standard media such as Luria Bertani (LB) medium or nutrient broth. Then, a fixed amount of inoculum (usually yielding 1–10%) is taken to the proper growth medium, which is adequately set for each microorganism, and its components may vary to comply with the aim of fermentation, e.g., produce spores, multiply microorganisms, or gain certain metabolites [34]. To achieve the maximal *B. thuringiensis* spore concentration, for example, 20 g/L of glucose (carbon source) and 2.5 g/L of ammonium (NH3) sulfate (nitrogen source) are added to the medium [35]. Technically, as *Bacillus* spp. sporulate under harsh conditions, reducing glucose concentration after realizing the stationary phase leads to starvation and the initiation of sporulation [36]. Otherwise, optimizing the growth medium composition is achieved via multifactorial analysis. Other relevant examples for optimizing the growth medium to fit the aim of fermentation were reviewed [61].

3.5.3 THE GROWTH CONDITIONS

Cultivation or growth should be encouraged, considering that *Bacillus* spp. are mesophilic microorganisms (they grow best at moderate temperatures, 20–40°C). While the optimal growth temperature is suggested to be between 30–37°C, temperatures above 85°C are not only lethal for vegetative cells but also have a negative impact on spore germination [61]. Similarly, factors such as high aeration rate and moisture content, which can help reduce energy costs in cultivation, should be carefully considered [37]. Spore resistance can be enhanced through a high heat shock and cold treatment. Cold shock pretreatment can increase spore-heat resistance up to 100°C. This increase is attributed to the formation of proteins that protect the spore DNA [62]. Other factors such as pH value, fermentation duration, and concentration of available oxygen should also be taken into account. Screening tests should be conducted to determine the compounds with the highest statistical significance before optimizing the substrate composition using the response surface methodology [61]. Plackett-Burman statistical designs were used to identify the best factors for *Bacillus subtilis* biomass yield. Among the 10 factors tested, including peptone, lactose, NH3 glucose, beef extract, citrate, sodium acetate, phosphate, potassium dihydrogen sodium chloride, pH, and sodium sulfate, only pH and peptone were found to be the most significant factors. Therefore, pH and peptone were selected for further improvement using a central composite layout [63]. Additionally, in order to adequately account for the correlations among the variables, the stirring rate and the specific interphase area should be considered in a standard Erlenmeyer flask [64]. Various aspects of optimizing cultivation conditions for *Bacillus* spp. were recently reviewed [33].

3.5.4 FERMENTATION

It is an important factor for optimizing the *Bacillus* spp. cultivation process. Fermentation is often operated as batch or fed-batch. Batch cultivation is more common as it is simpler and easier than fed-batch. The latter is costly and requires continuous monitoring but offers better results [33, 61]. When cultivating *Bacillus subtilis* spore using both fed-batch and batch methods, the former method provided protection against cell lysis, resulting in higher spore concentrations. However, both methods yielded elevated vegetative cell concentrations [65]. Generally, key factors of the fermentation types, especially the continuous monitoring of the substrate volume in the medium throughout the entire process, should be followed. Alternatively, other variables such as CO_2, pH, and organic acids can be measured to determine whether supplementary substrate is required [66].

As the optimization of cultivation is perfected, the system is scaled up to initiate industrial production. It is important to ensure that the conditions and results obtained with large-scale equipment are similar to those of small-scale production [67]. Therefore, the scaling-up process should begin at the laboratory scale, then progress to a pilot plant level to optimize the process and ensure consistent kinetic parameters and process characteristics. Finally, the system is expanded to a large scale for commercial production. It is also necessary to consider the economic aspects of optimization [68].

3.6 FORMULATIONS AND APPLICATIONS OF *BACILLUS* SPECIES FOR THE MANAGEMENT OF ROOT-KNOT DISEASE

While the formulations aim to incorporate a *Bacillus* product with its corresponding inactive and active ingredients (i.e., bacterial species/their bioactive metabolites), the applications refer to their practical use. Specifically, the formulation is achieved by adding certain non-active (inert), active, and/or (functional) substances. The formulation aims to enhance the activity, ease-of-use, absorption, storage stability, or delivery of the active ingredient used (the bacterial species). Common additions to pesticide formulation ingredients may include antimicrobial agents, absorbents, anticaking agents, antioxidants (AOX), adsorbents, carriers, binders, dispersants, preservatives, humectants, solvents, surfactants, ultraviolet absorbers, and thickeners [32–34].

Bacillus species/strains could be formulated for use against numerous plant pathogens in commercial products [32, 34–37]. Multiple patterns of their formulations comprise both dry and liquid products. Granules, wettable powders (WPs), and dust are commonly dry and available products of *Bacillus*. Additionally, oils, cell suspensions (SU) in water, and emulsions serve as vivid examples of applicable liquid products [32]. Microencapsulation of these bacterial cells is also an option to protect small volumes of the active *Bacillus* ingredient with an inert layer. Successful formulations should carefully consider the microbial biomass in terms of maintaining biomass viability through rehydration, drying, and stabilization. Effective formulations should optimize target coverage, biomass delivery, and target adhesion to increase *Bacillus* biomass activity and survival after application. Therefore, many amendments are used in *Bacillus* spp. experimental/commercial formulations. These amendments can be grouped as carriers (extenders, fillers) or amendments that enhance the physical, chemical, and/or nutritional attributes of the formulated biomass. Detailed examples of types and forms of these amendments for formulating *Bacillus* biomass, as well as *Bacillus*-based biocontrol commercial products, are provided elsewhere [32–34]. Promising experimentation of *Bacillus* formulations has proven the effectiveness of the following materials in reducing plant diseases: alginate manucol, peat and chitin, clays, cellulose, methylcellulose, polyvinyl pyrrolidone, caalginate, carboxymethyl vegetable oil, nutrient medium, and peptone.

Formulations of Equity® (multiple *Bacilli* strains), BioYield® (two *Bacillus* strains), and RhizoVital® (strain FZB42) showed high biocontrol efficacy against RKN [69]. They minimized *M. incognita* eggs (gram root)$^{-1}$, juveniles (gram of soil)$^{-1}$, and galls plant^{-1} on tomato. In another study, the suppressive effects of cell-free supernatant, culture broth, and cell pellet of various B. subtilis isolates on *M. incognita* J2 activity and egg hatching were proven [70]. The authors also demonstrated that proper batch fermentation could enhance both *B. subtilis* biomass production and suppression effects on egg masses and galls of tomato roots. They found that the culture broth of two isolates, B8 and B10, at a concentration of 50%, reduced the percentages of hatched eggs to 46.3% and 44.7%, respectively. The culture broth of isolate B10 showed promising biocontrol results, significantly decreasing the number of egg masses and galls by 89.5% and 81.1%, respectively, compared to the control. When this isolate was formulated as a bionematicide at a concentration of 0.1 g/pot, it reduced egg masses and galls by 71.2% and 69.7%, respectively [70].

Recently, WP commercial formulations with 2 × 108 CFU/g of *Bacillus amyloliquifaciens* or *B. megaterium* were used against RKN at three different concentrations under laboratory and field conditions. The 4% concentration, i.e., 4 g of the powder in 100 ml distilled water, for each of the two bacterial species significantly decreased *Meloidogyne incognita* egg hatching and increased juvenile mortality [71]. The lowest mean of egg hatching was recorded by *B. amyloliquifaciens* (23.7%), followed by *B. megaterium* (24.7%), 120 hours after application. The RKN juvenile mortality was 54.46% and 53.13% for *B. megaterium* and *B. amyloliquifaciens*, respectively, at the 4% concentration, compared to 46% in the control.

Under field conditions, when seeds of bottle gourd were treated with either *B. megaterium* or *B. amyloliquifaciens* at 10 g/kg seed in a field treated with neem cake (1 ton/ha), a significant reduction in the number *of M. incognita* juveniles in soil and roots, as well as the nematode gall index, was observed compared to the non-treated check. Interestingly, separate application of neem cake, *Bacillus megaterium or B. amyloliquifaciens* revealed that the bacterial efficacy on *M. incognita* J2 populations was superior to Neem cake (1 ton/ha) in bottle gourd-infested field [71]. They concluded that *B. amyloliquifaciens* and *B. megaterium* demonstrated the potential to control *Meloidogyne incognita* whether in the laboratory or in the bottle gourd-cultivated field.

Examples of commercial products for applying *Bacillus* spp. are provided in Table 3.1, along with relevant information about the active ingredient (bacterial species), product name, and producer. Additional information about the formulation type, targeted RKN species, crop, and other *Bacillus*-based products can be obtained from the producers' websites. The labels attached to these products contain detailed information about their exact ingredients, proper application instructions, and recommended storage conditions. For instance, the commercial product of *Bacillus firmus* for controlling phytonematodes is available in three different forms: (i) Dextrose-based soluble powder (10^9 CFU/gm), (ii) Talc/Kaolin formulation at a concentration of 10^6 CFU/gm, and (iii) Glycerol-based liquid formulation (10^9 CFU/gm). The shelf life of these formulations varies from half to one year (http://www.agrinaturals.com/NEMOEND-BF.htm). Additionally, Bayer Crop Science distributes two forms of *B. firmus* products: seed treatment products (VOTiVOTM and PONCHO®/VOTiVOTM mix) and a drench product (Nortica®) [72]. The latter product is used for controlling the main groups of nematodes that infect turfgrass (e.g., *Belonolaimus* spp., *Meloidogyne* spp., and *Hoplolaimus* spp.). The product instructions recommend its application 2–7 days before planting, with post-application irrigation (minimum of 7.5 cm). The seed addressing method saves a lot of bacterial biomass per unit area. The product VOTiVO (containing *B. firmus*) was first formulated as a standalone product, but now *B. firmus* is mostly sold combined with the insecticide PONCHO (Clothianidin) as a seed treatment [72]. Therefore, PONCHO/VOTiVO mix is both insecticide and nematicide for biological seeds addressing cotton, corn, sugar beet, sorghum, and soybean. While the PONCHO ingredient shares are 40.3% in the product, the *B. firmus* component is 8.1% and 51.6% for other ingredients (2×10^9 *B. firmus* cfu ml^{-1} and 500 g l^{-1} clothianidin). As *Bacillus* spp. generate long survival-stage endospores, these products have a two-year shelf life in cool, dry conditions. Other *B. firmus* commercial formulations, such as Chancellor, BioSafe, and BioNem, are sold as WP [43].

TABLE 3.1 Examples of Commercial *Bacillus*-based Products for Managing Plant Parasitic Nematodes

Bacterium	Product name	Company/institution	Country
Bacillus subtilis	Stanes Sting	Introduced into Egypt (Gaara company) from India (T. Stanes and Company Limited).	Egypt and India
Bacillus subtilis strain CM-5 strain), *B. cereus* CM-1c.	BioStart™Defensor	Bio-Cat Microbials	USA
Bacillus thuringiensis	Avid 0.15EC (or abamectin)	Syngenta	Multinational
B. licheniformis, *B. subtilis*	Quartzo	FMC Química do Brasil Ltda.	Brazil
B. licheniformis, *B. subtilis*	Nemix C	FMC Química do Brasil Ltda.	Brazil
B. licheniformis strain FMCH001 (DSM32154) *B. subtilis* strain FMCH002 (DSM32155)	Presense	FMC Química do Brasil Ltda.	Brazil
B. firmus	(i) Bionem-WP (ii) BioSafe-WP (iii) Chancellor	Agro Green	Israel
B. firmus strain GB-126	VOTiVO®WP	Bayer Crop Science	Germany
B. methilotrophicus	Onix	Laboratorio de Bio Controle Farroupilha S.A.	Brazil
B. subtilis	Pathway Consortia®	Pathway Holdings	USA
B. chitinosporus, *B. laterosporus*, *B. licheniformis* (mixture)	BioStart® BioStart™	Bio-Cat Rincon-Vitova	USA
B. amyloliquefaciens	Nemacontrol	Simbiose Indústria e Comércio de Fertilizantes e Insumos	Brazil
B. cereus	Xian Mie	XinYi Zhong Kai Agro-Chemical Industry CO., Ltd.	China
Bacillus spp.	Nemix	Chr. Hansen	Brazil
Bacillus sp., *Pseudomonas* sp., *Rhizobacterium* sp., *Rhizobium* sp.	Micronema	Agricultural Research Center	Egypt

3.7 OPTIMIZING THE USE OF *BACILLUS* SPECIES FOR EFFECTIVE RKN CONTROL

Initially, the new biocontrol concept that uses bacterial species is generally valuable, given the unfavorable side effects of chemical nematicides [73]. Admittedly, *Bacillus* species have outstanding merits in various facets of their utilization: application, shelf-life, transportation, and commercial mass culture [32–37]. Xiang et al. [44] found that 34% out of 662 PGPR strains could induce high ($P \leq 0.05$) *M. incognita* J2 mortality comparable to that brought about by Aldicarb (common nematicide). Moreover, they reported *Bacillus* spp. as the major genus inducing the best mortality percentage of *M. incognita* J2. Their trials underlined that certain *Bacillus* strains inhibited the *Meloidogyne incognita* population level in the microplot, greenhouse, and field rating systems and enhanced seed cotton yield. Those authors reviewed the likely nematicidal efficacy of various *Bacillus* species/strains on different PPN/RKN species. They detected definite strains of the species *B. firmus*, *B. thuringiensis*, *B. velezensis*, *B. subtilis*, and *B. pumilus* with nematicidal efficacy on *M. incognita* in their tests; 17 different *Bacillus* species/subspecies were antagonistic to *Meloidogyne incognita*.

Notwithstanding these favorable attributes and effects of *Bacillus* on RKN, BCA have general drawbacks, as they are frequently more inconsistent, slower-acting, and/or less effective than chemical nematicides [22]. Therefore, it is crucial to make earnest efforts in optimizing the benefits of these bacterial species for better management of RKN. This is particularly important because RKN are underground pests that have long-lasting effects on plant residues and infested soil. In contrast, it is necessary to enable these bacteria to effectively compete, persist, and function in arable soils and/or on plant roots. There are several directions for improving the usage of *Bacillus* biocontrol. These directions should be considered at various research stages, including exploring effective species/strains, cost-effective mass-reproduction, proper formulation, integrated pest management (IPM) of multiple pests/pathogens, and successful/timely application. Recently, promising *Bacillus* species/strains for effective control of RKN have been reported [39, 44, 45, 60] and should be commercially utilized. The antagonistic potential of various *B. subtilis* isolates against *M. incognita* J2 activity and egg hatching could be enhanced through proper batch fermentation/biomass production [33, 70].

As aforementioned with the product PONCHO/VOTiVO mix, another trend is to control more than one pest, two, or a set of pests.

Other courses may include *Bacillus* products, which are also used as plant growth promoters or plant strengtheners to enhance plant tolerance against pest attacks. These *Bacillus*-based products can enhance plants' defense against RKN infection [72]. Currently, other candidate *Bacillus* products are still undergoing laboratory/pilot screening worldwide, as it is crucial to conduct laboratory, semi-field, and field tests to validate their effectiveness. Furthermore, the successful implementation of *Bacillus* applications should be achieved by integrating them into emerging or existing organic farming systems and IPM tactics/strategies, particularly through novel and effective approaches [18, 25, 26]. Therefore, *Bacillus* applications could be optimized through co-application or sequential application with compatible production inputs. Abd-Elgawad & Askary [38] have reviewed effective tactics that combine BCA with other agricultural inputs to act synergistically or additively. Promising procedures for *Bacillus* applications are still being explored, providing optimism for future advancements. Clearly, the development of multidimensional, effective, and safe IPM programs remains a current challenge. An effective way to avoid failure to fulfill any of these elements of such programs is to simultaneously use closely related or compatible species of BCA. In this respect, combining *Bacillus subtilis* with *B. pumilus* raised the average of growth parameters for *M. incognita*-infected cowpea plants relative to either bacterial species alone [74]. This trend should be adopted to provide further domination of *Bacillus* in the soil microbiome, optimum biocontrol efficacy, and/or better root protection. Thus, Abd-Elgawad [18] speculated that different *Bacillus* species/strains might possess dissimilar anti-RKN mechanisms. Another trend is the sequential usage. *Meloidogyne incognita* galls were considerably reduced because *Bacillus firmus* predated the nematicidal application of a synthetic chemical in tomato-planted cycles. Thus, the severity of galling expressed by the galling index (scale of 0–10) decreased from 7.7 in the untreated roots to 5.9, 4.3, and 4.0 for *Bacillus firmus*, oxamyl + *B. firmus*, and fosthiazate + *B. firmus*, respectively [75]. Such advantageous outputs should be grasped and extended since *Bacillus* may be compatible with many other chemicals. Moreover, an explicit strategy to circumvent the lack of appropriate indigenous *Bacillus spp.*, as antagonists to RKN in the same habitat, is to introduce the bacteria in combination with conservation agricultural practices to establish and boost the efficacy of the introduced *Bacillus*. Such practices may include proper organic amendments and well-studied crop rotations. When their spores are properly introduced to the arable soil, *Bacillus* species may swiftly initiate season-long activity, and work under relatively tough ecological conditions and variable soil temperature, moisture, and pH. Ultimately, they are recommended for integration with other production practices to optimize their control efficacy [18, 26].

Additionally, optimism in expanding the effective and safe usage of *Bacillus* spp. and other BCA by increasing growers' awareness via appropriate media should be realized. The growers must carefully follow the *Bacillus*-related instructions found on the product label. The stakeholders should increasingly be able to consult the right source, e.g., the proper internet for challenging techniques and relevant knowledge/information.

3.8 CONCLUSION

Many species of *Bacillus* are promising biological control agents (BCA). Therefore, they should be further assessed for large-scale utilization against *Meloidogyne* spp. These biologicals could potentially serve as safe and effective alternatives to synthetic nematicides, or be combined with chemicals for managing root-knot nematodes (RKN). With their eco-friendly attitude and excellent biocontrol attributes, certain *Bacillus* spp. can be included in integrated nematode management (INM) programs. Approaches that enhance and maximize their nematicidal activities, either as complementary or even surpassing chemical RKN management tactics, should be pursued. This is particularly significant because some *Bacillus* spp. have demonstrated additive or synergistic effects with other production inputs in integrated pest management (IPM) programs of agricultural systems. The combined use of *Bacillus* spp. and other pesticides should be encouraged on a large scale for safe and reliable pest management. As numerous bionematicides are already available or expected to be widely accessible soon, *Bacillus* spp. should strive to compete with them in the commercial market. To achieve this goal, it is important to identify research priorities related to *Bacillus* spp. for their application in sustainable agriculture, understand their interactions with other agricultural inputs, and explore their mode of action. These bacteria hold the promise of providing safe and effective ingredients for managing RKN.

KEYWORDS

- ***Bacillus* spp.**
- **Biological control**
- **Commercial cultivation**
- **Growth conditions**
- **Integrated pest management**
- **Nematode pests**
- **Root-knot disease**

REFERENCES

1. Christie, G., & Setlow, P., (2020). *Bacillus* spore germination: Knowns, unknowns, and what we need to learn. *Cell Signal.*, *74*, 109729. doi: 10.1016/j.cellsig.2020.109729.
2. Schallmey, M., Singh, A., & Ward, O. P., (2004). Developments in the use of *Bacillus* species for industrial production. *Canadian J. Microbiol.*, *50*(1), 1–17. doi: 10.1139/w03-076.
3. Favaro, G., Bogialli, S., Di Gangi, I. M., Nigris, S., Baldan, E., Squartini, A., et al., (2016). Characterization of lipopeptides produced by *Bacillus licheniformis* using liquid chromatography with accurate tandem mass spectrometry. *Rapid Commun. Mass. Spect.*,*30*(20), 2237–2252.doi: 10.1002/rcm.7705.
4. Nigris, S., Baldan, E., Tondello, A., Zanella, F., Vitulo, N., Favaro, G., et al., (2018). Biocontrol traits of *Bacillus licheniformis* GL174, a culturable endophyte of *Vitis vinifera* cv. Glera. *BMC Microbiol.*, *18*(1), 133. doi: 10.1186/s12866-018-1306-5.
5. Graumann, P., (2012). *Bacillus: Cellular and Molecular Biology* (2nd edn., p. 398). Caister Academic Press, Poole, UK.
6. FAO/WHO, (2002). *Guidelines for the Evaluation of Probiotics in Food.* Food and Agriculture Organization of the United Nations and World Health Organization Working Group Report. Rome: Food and Agriculture Organization.
7. Elshaghabee, F. M. F., Rokana, N., Gulhane, R. D., Sharma, C., & Panwar, H., (2017). *Bacillus* as potential probiotics: Status, concerns, and future perspectives. *Front. Microbiol.*, *8*, 1490. doi: 10.3389/fmicb.2017.01490.
8. Abad, P., Gouzy, J., Aury, J. M., Castagnone-Sereno, P., Danchin, E. G., Deleury, E., et al., (2008). Genome sequence of the metazoan plant-parasitic nematode *Meloidogyne incognita. Nat. Biotechnol.*, *26*, 909–915. doi: 10.1038/nbt.1482.
9. Abd-Elgawad, M. M. M., & Askary, T. H., (2015). Impact of phytonematodes on agriculture economy. In: Askary, T. H., & Martinelli, P. R. P., (eds.), *Biocontrol Agents of Phytonematodes* (pp. 3–49). CAB International, Wallingford, UK.

10. Kim, T. Y., Jang, J. Y., Jeon, S. J., Lee, H. W., Bae, C. H., Yeo, J. H., et al., (2016). Nematicidal activity of kojic acid produced by *Aspergillus oryzae* against *Meloidogyne incognita*. *J. Microbiol. Biotechnol.*, *26*, 1383–1391. doi: 10.4014/jmb.1603.03040.
11. Trudgill, D. L., & Blok, V. C., (2001). Apomictic, polyphagous root-knot nematodes: Exceptionally successful and damaging biotrophic root pathogens. *Annu. Rev. Phytopathol.*, *39*, 53–77. doi: 10.1146/annurev.phyto.39.1.53.
12. Perry, R. N., Moens, M., & Starr, J. L., (2009). *Root Knot Nematodes*. CAB International, Wallingford, UK.
13. Jones, J. T., Haegeman, A., Danchin, E. G. J., Gaur, H. S., Helder, J., Jones, M. G. K., Kikuchi, T., et al., (2013). Review: Top 10 plant-parasitic nematodes in molecular plant pathology. *Molecular Plant Pathology*, 12057. doi: 10.1111/mpp.12057.
14. Abd-Elgawad, M. M. M., (2020). Optimizing biological control agents for controlling nematodes of tomato in Egypt.*Egypt. J. Biol. Pest Cont.*, *30*, 58.https://doi.org/10.1186/s41938-020-00252-x.
15. Abd-Elgawad, M. M. M., (2020). Biological control agents in the integrated nematode management of pepper in Egypt. *Egypt. J. Biol. Pest Cont.*, *30*, 70. https://doi.org/10.1186/s41938-020-00273-6.
16. Abd-Elgawad, M. M. M., (2020). Biological control agents in the integrated nematode management of potato in Egypt. *Egypt. J. Biol. Pest Cont.*, *30*, 121. https://doi.org/10.1186/s41938-020-00325-x.
17. Abd-Elgawad, M. M. M., (2021). Biological control of nematodes infecting eggplant in Egypt. *Bulletin of the National Research Centre*, *45*, 6. https://doi.org/10.1186/s42269-020-00463-0.
18. Abd-Elgawad, M. M. M., (2021). Optimizing safe approaches to manage plant-parasitic nematodes. *Plants*, *10*, 1911. https://doi.org/10.3390/plants10091911.
19. Noling, J. W., (2019). *Nematode Management in Tomatoes, Peppers, and Eggplant* (pp. 1–16). University of Florida publication Series no. ENY-032, USA.
20. Abd-Elgawad, M. M. M., & Hasabo, S. A., (1995). Spatial distribution of the phytonematode community in Egyptian berseem clover fields. *Fund. Appl. Nematol.*, *18*(4), 329–334.
21. Duncan, L. W., & Phillips, M. S., (2009). Sampling root-knot nematodes. In: Perry, R. N., Moens, M., & Starr, J. L., (eds.), *Root-knot Nematodes* (pp. 275–300). St. Albans, UK.
22. Abd-Elgawad, M. M. M., (2020). Plant-parasitic nematodes and their biocontrol agents: Current status and future vistas. In: Ansari, R. A., Rizvi, R., & Mahmood, I., (eds.), *Management of Phytonematodes: Recent Advances and Future Challenges* (pp. 171–204). Springer Nature, Singapore Pte Ltd., Singapore. https://www.springer.com/gp/book/9789811540868 (accessed on 9 January 2023).
23. Chitwood, D. J., (2003). Nematicides. In: Plimmer, J. R., (ed.), *Encyclopedia of Agrochemicals* (Vol. 3, pp. 1104–1115). John Wiley & Sons, New York, USA.
24. Abd-Elgawad, M. M. M., Kour, F. F. H., Montasser, S. A., & Hammam, M. M. A., (2016). Distribution and losses of *Tylenchulus semipenetrans* in citrus orchards on reclaimed land in Egypt. *Nematology*, *18*, 1141–1150. https://doi.org/10.1163/15685411-00003020.
25. Abd-Elgawad, M. M. M., & Askary, T. H., (2020). Factors affecting success of biological agents used in controlling plant-parasitic nematodes. *Egypt. J. Biol. Pest. Cont.*, *30*, 17. https://doi.org/10.1186/s41938-020-00215-2.

26. Timper, P., (2014). Conserving and enhancing biological control of nematodes. *J. Nematol.*, *46*, 75–89.
27. Ntalli, N., Adamski, Z., Doula, M., & Monokrousos, N., (2020). Nematicidal amendments and soil remediation. *Plants*, *9*, 429. https://doi.org/10.3390/plants9040429.
28. Dutta, T. K., Khan, M. R., & Phani, V., (2019). Plant-parasitic nematode management via biofumigation using brassica and non-brassica plants: Current status and future prospects. *Curr. Plant Biol.*, *17*, 17–32. https://doi.org/10.1016/j.cpb.2019.02.001.
29. Bakr, R. A., Mahdy, M. E., & Mousa, E. M., (2013). Efficacy of soil solarization and postplanting mulch on control of root-knot nematodes. *Pak. J. Nematol.*, *31*(1), 71–76.
30. Abd-Elgawad, M. M. M., Elshahawy, I. E., & Abd-El-Kareem, F., (2019). Efficacy of soil solarization on black root rot disease and speculation on its leverage on nematodes and weeds of strawberry in Egypt. *Bull. NRC*, *43*, 175.https://doi.org/10.1186/s42269-019-0236-1.
31. Hada, A., Singh, D., Papolu, P. K., Banakar, P., Raj, A., & Rao, U., (2021). Host-mediated RNAi for simultaneous silencing of different functional groups of genes in *Meloidogyne incognita* using fusion cassettes in *Nicotiana tabacum*. *Plant Cell Rep.*, *1007*, 1–16. https://doi.org/10.1007/s00299-021-02767-5.
32. Schisler, D. A., Slininger, P. J., Behle, R. W., & Jackson, M. A., (2004). Formulation of *Bacillus* spp. for biological control of plant diseases. *Phytopathology*, *94*, 1267–1271.
33. Stojanović, S. S., Karabegović, I., Beškoski, V., Nikolić, N., & Lazić, M., (2019). *Bacillus* based microbial formulations: Optimization of the production process. *Hem. Ind.*, *73*, 169–182.
34. Shaikh, S. S., & Sayyed, R. Z., (2015). Role of plant growth-promoting rhizobacteria and their formulation in biocontrol of plant diseases. In: Arora, N. K., (ed.), *Plant Microbes Symbiosis: Applied Facets* (pp. 337–351). Springer India, New Delhi.
35. Elsayed, A. E., Othman, N. Z., Malek, R. A., Awad, H. M., Wu, K., Azizi, R., Wadaan, H. M., & Enshasy, H., (2014). Bioprocess development for high cell mass and endospore production by *Bacillus thuringiensis* var *israelensis* in semi-industrial scale. *J. Pure Appl. Microbiol.*, *8*(4), 2773–2783.
36. Castillo, H. F. D., Reyes, C. F., Morales, G. G., Herrera, R. R., & Aguilar, C., (2013). Biological control of root pathogens by plant-growth promoting *Bacillus* spp. In: Saloneski, S., & Larramnedy, M., (eds.), *Weed and Pest Control – Conventional and New Challenges* (pp. 80–103). InTech, London, UK.
37. Pryor, S. W., Gibson, D. M., Hay, A. G., Gossett, J. M., & Walker, L. P., (2007). Optimization of spore and antifungal lipopeptide production during the solid-state fermentation of *Bacillus subtilis*. *Appl. Biochem. Biotechnol.*, *143*(1), 63–79.
38. Abd-Elgawad, M. M. M., & Askary, T. H., (2018). Fungal and bacterial nematicides in integrated nematode management strategies. *Egypt. J. Biol. Pest. Cont.*, *28*, 74. https://doi.org/10.1186/s41938-018-0080-x.
39. Asaturova, A. M., Bugaeva, L. N., Homyak, A. I., Slobodyanyuk, G. A., Kashutina, E. V., Yasyuk, L. V., Sidorov, N. M., et al., (2022). *Bacillus velezensis* strains for protecting cucumber plants from root-knot nematode *Meloidogyne incognita* in a Greenhouse. *Plants*, *11*, 275. https://doi.org/10.3390/plants11030275.
40. Sun, X., Xu, Z., Xie, J., Hesselberg-Thomsen, V., Tan, T., Zheng, D., et al., (2021). *Bacillus velezensis* stimulates resident rhizosphere *Pseudomonas stutzeri* for plant health through metabolic interactions. *ISME J.*, *16*, 774–787. doi: 10.1038/s41396-021-01125-3.

41. Zeigler, D. R.,& Nicholson, W. L., (2017). Experimental evolution of *Bacillus subtilis*. *Environ. Microbiol., 19*, 3415–3422.
42. Zeigler, D. R., (2019). *The Bacillus Genetic Stock Center/ Bacillus Subtilis. From: TheBiological Resources of Model Organisms*. CRC Press.https://www.routledgehandbooks.com/doi/10.1201/9781315100999-3 (accessed on 9 January 2023).
43. Abd-Elgawad, M. M. M., & Vagelas, I. K., (2015). Nematophagous bacteria: Field application and commercialization. In: Askary, T. H., & Martinelli, P. R. P., (eds.), *Biocontrol Agents of Phytonematodes* (pp. 276–309). CAB International, Wallingford, UK.
44. Xiang, N., Lawrence, K. S., Kloepper, J. W., Donald, P. A., McInroy, J. A., & Lawrence, G. W., (2017). Biological control of *Meloidogyne incognita* by spore-forming plant growth-promoting rhizobacteria on cotton. *Plant Disease, 101*, 774–784.
45. Xiang, N., Lawrence, K. S., & Donald, P. A., (2018). Biological control potential of plant growth-promoting rhizobacteria suppression of *Meloidogyne incognita* on cotton and*Heterodera glycines* on soybean: A review. *J. Phytopathol., 66*, 449–458. doi: 10.1111/jph.12712.
46. Tian, B., Yang, J., & Zhang, K. Q., (2007). Bacteria used in the biological control of plant-parasitic nematodes: Populations, mechanisms of action, and future prospects. *FEMS Microbiology Ecology, 61*, 197–213.
47. Wepuhkhulu, M., Kimenju, J., Anyango, B., Wachira, P., & Kyallo, G., (2011). Effect of soil fertility management practices and *Bacillus subtilis* on plant parasitic nematodes associated with common bean, *Phaseolus vulgaris*. *Trop. Subtrop. Agro. Ecosyst., 13*, 27–34.
48. Li, X., Hu, H. J., Li, J. Y., Wang, C., Chen, S. L., & Yan, S. Z., (2019). Effects of the endophytic bacteria *Bacillus cereus* BCM2 on tomato root exudates and *Meloidogyne incognita* infection. *Pl. Dis., 103*, 1551–1558.
49. Lee, Y. S., & Kim, K. Y., (2016). Antagonistic potential of *Bacillus pumilus* L1 against root-knot nematode, *Meloidogyne arenaria*. *J. Phytopathol., 164*, 29–39.
50. Mendoza, A. R., Kiewnick, S., & Sikora, R. A., (2008). *In vitro* activity of *Bacillus firmus* against the burrowing nematode *Radopholus similis*, the root-knot nematode *Meloidogyne incognita* and the stem nematode *Ditylenchus dipsaci*. *Biocont. Sci. Technol., 18*, 377–389.
51. Sikora, R. A., (1988). Interrelationship between plant health promoting rhizobacteria, plant parasitic nematodes and soil microorganisms. *Commun. Agric. Appl. Biol. Sci., 53*, 867–878.
52. Kloepper, J. W., Ryu, C. M., & Zhang, S. A., (2004). Induced systemic resistance and promotion of plant growth by *Bacillus* spp. *Phytopathology, 94*, 1259–1266.
53. Schrimsher, D. W., (2013). *The Studies of Plant Host Resistance to the Reniform Nematode in Upland Cotton and the Effects of Bacillus Firmus GB-126 on Plant-Parasitic Nematode*. M.S. Thesis. Auburn University, Auburn, AL. https://etd.auburn.edu/handle/10415/3562 (accessed on 9 January 2023).
54. Hauschild, R., Hallmann, J., & Sikora, R. A., (2000). *Fusarium oxysporum* and *Meloidogyne incognita* on tomato can be controlled by antagonistic rhizobacteria. *Commun. Agric. Appl. Biol. Sci., 65*, 527–528.
55. Sikora, R. A., Schäfer, K., & Dababat, A. A., (2007). Modes of action associated with microbially induced in planta suppression of plant-parasitic nematodes. *Australas. Plant Pathol., 36*, 124–134.

56. Schäfer, K., Silva, F. C., & Sikora, R. A., (2006). Molecular investigations of rhizobacteria-induced systemic resistance towards the root-knot nematode *Meloidogyne incognita* in tomato. Multitrophic interactions in soil. *IOBCWPRS Bull.,29*, 135–140.
57. Liu, K., Garrett, C., Fadamiro, H., & Kloepper, J. W., (2016). Induction of systemic resistance in Chinese cabbage against black rot by plant growth-promoting rhizobacteria. *Biol. Cont.*, *99*, 8–13.
58. Huang, Y., Xu, C. K., Ma, L., Zhang, K. Q., Duan, C. Q., & Mo, M. H., (2010). Characterization of volatiles produced from *Bacillus megaterium* YFM3.25 and their nematicidal activity against *Meloidogyne incognita*. *Eur. J. Plant Pathol.*, *126*, 417–422.
59. Peng, D. H., Chai, L. J., Wang, F. S., Zhang, F. J., Ruan, L. F., & Sun, M., (2011). Synergistic activity between *Bacillus thuringiensis* Cry6Aa and Cry55Aa toxins against *Meloidogyne incognita*. *Microb. Biotechnol.*, *4*, 794–798.
60. Tian, X. L., Zhao, X. M., Zhao, S. Y., Zhao, J. L., & Mao, Z. C., (2022). The biocontrol functions of *Bacillus velezensis* strain Bv-25 against *Meloidogyne incognita*. *Front. Microbiol.*, *13*, 843041. doi: 10.3389/fmicb.2022.843041.
61. Stamenković, S., Beškoski, V., Karabegović, I., Lazić, M., & Nikolić, N., (2018). Microbial fertilizers: A comprehensive review of current findings and future perspectives. *Spanish J. Agric. Res.*, *16*(1), e09R01.
62. Movahedi, S., & Waites, W., (2002). Cold shock response in sporulating *Bacillus subtilis* and its effect on spore heat resistance. *J. Bacteriol.*, *184*(19), 5275–5281.
63. Sreekumar, G., & Krishnan, S., (2010). Enhanced biomass production study on probiotic *Bacillus subtilis* SK09 by medium optimization using response surface methodology. *African J. Biotechnol.*, *9*(45), 8078–8084.
64. Veljković, V. B., Nikolić, S., Lazić, M. L., & Engler, C. R., (1995). Oxygen transfer in flasks shaken on orbital shakers. *Hem Ind.*, *49*, 265–272.
65. Monteiro, S. M. S., Clemente, J. J., Carrondo, M. J. T., & Cunha, A. E., (2014). Enhanced spore production of *Bacillus subtilis* grown in a chemically defined medium. *Adv. Microbiol.*, *4*(8), 444–454.
66. Glick, B. R., (2015). *Beneficial Plant-Bacterial Interactions*. Springer International Publishing, Switzerland.
67. Wang, D. I. C., Cooney, C. L., Demain, A. L., Dunnill, P., Humphrey, A. E., & Lilly, M. D., (1997). Translation of laboratory, pilot, and plant scale data. *Ferment. Enzyme Technol.* (Vol. 73, pp. 194–211). Wiley, New York.
68. Seletzky, J. M., Noak, U., Fricke, J., Welk, E., Eberhard, W., Knocke, C., & Buchs, J., (2007). Scale-up from shake flasks to fermenters in batch and continuous mode with *Corynebacterium glutamicum* on lactic acid based on oxygen transfer and pH. *Biotechnol. Bioeng.*, *98*(4), 800–811.
69. Burkett-Cadena, M., Kokalis-Burelle, N., Lawrence, K. S., Santen, E. V., & Kloepper, J. W., (2008). Suppressiveness of root-knot nematodes mediated by rhizobacteria. *Biol. Cont.*, *47*, 55–59.
70. Basyony, A. G., & Abo-Zaid, G. A., (2018). Biocontrol of the root-knot nematode, *Meloidogyne incognita*, using an eco-friendly formulation from *Bacillus subtilis*, lab. and greenhouse studies. *Egypt. J. Biol. Pest Cont.*, *28*, 87. https://doi.org/10.1186/s41938-018-0094-4.
71. Rani, P., Singh, M., Prashad, H., & Sharma, M., (2022). Evaluation of bacterial formulations as potential biocontrol agents against the southern root-knot nematode,

Meloidogyne incognita. Egypt. J. Biol. Pest Cont., 32, 29. https://doi.org/10.1186/s41938-022-00529-3.
72. Wilson, M. J., & Jackson, T. A., (2013). Progress in the commercialization of bionematicides. *BioControl, 58*, 715–722.
73. Abd-Elgawad, M. M. M., (2021). *Photorhabdus* spp.: An overview of the beneficial aspects of mutualistic bacteria of insecticidal nematodes. *Plants, 10*, 1660. https://doi.org/10.3390/plants10081660.
74. El-Nagdi, W. M. A., Youssefi, M. M. A., Abd-El-Khair, H., Abd-Elgawad, M. M. M., & Dawood, M. G., (2019). Effectiveness of *Bacillus subtilis, B. pumilus, Pseudomonas fluorescens* on *Meloidogyne incognita* infecting cowpea. *Pak. J. Nematol., 37*, 35–43.
75. D'Errico, G., Marra, R., Crescenzi, A., Davino, S. W., Fanigliulo, A., Woo, S. L., & Lorito, M., (2019). Integrated management strategies of *Meloidogyne incognita* and *Pseudopyrenochaeta lycopersici* on tomato using a *Bacillus firmus*-based product and two synthetic nematicides in two consecutive crop cycles in greenhouse. *Crop Prot., 122*, 159–164.

CHAPTER 4

Arbuscular Mycorrhizal Fungi Protect Vegetable Crops from Pathogenic Root-Knot Nematodes

C. SANKARANARAYANAN

Division of Crop Protection, ICAR–Sugarcane Breeding Institute, Coimbatore, Tamil Nadu, India:

ABSTRACT

In vegetables, sedentary endoparasites, such as root-knot nematodes *Meloidogyne* spp., cause severe diseases affecting vegetable production. Arbuscular mycorrhizal fungi (AMF) form a symbiotic relationship with roots and provide phosphorus to the plant, and the plant, in turn, provides carbohydrates to the fungus. AMF colonizes all the vegetables and may reduce the adverse effect of *Meloidogyne*. The effect of AMF on the growth of plants infected with *Meloidogyne* is highly variable, but it generally has a positive impact by reducing nematode development. AMF, acting as biostimulators and bio-protectors, can be highly beneficial to sustainable agriculture, maintaining plant production and reducing nematode damage, ultimately helping the environment. Given the importance of sustainable crop management, this chapter aims to review in detail what AMF is, the AMF-nematode interaction, management of *Meloidogyne* spp., combined efficacy of AMF-biocontrol agents and organic amendments against root-knot nematodes (RKN) in vegetable crops, mode of action of AMF against nematodes, and conclusions and future areas of research.

Microbial Biostimulants: Biorational Pesticides for Management of Plant Pathogens.
Faheem Ahmad and Rakesh Pandey (Eds.)

4.1 INTRODUCTION

The term "*Mycorrhizae*" refers to any symbiotic associations between the roots of plants and fungi. These symbioses are characterized by carbon flows to the fungus, and inorganic nutrients move to the plant. Mycorrhizal fungi (MF) can improve the growth and development of plants in low-fertile land. As a result, mycorrhizal plants are often more competitive and better able to tolerate environmental stresses than non-mycorrhizal plants. The vast majority of plants form symbiotic associations with fungi, and an estimated 95% of all plant species belong to diverse genera that characteristically form mycorrhizae. So far, seven types of mycorrhizae have come into general use over the years based on morphology and anatomy. These are ectomycorrhiza, endomycorrhiza or arbuscular mycorrhiza, ericoid mycorrhiza, arbutoid mycorrhiza, monotropoid mycorrhiza, ect-endomycorrhiza, and orchidaceous mycorrhiza.

Taxonomically, arbuscular mycorrhizal fungi (AMF) belong to the phylum Glomeromycota, class Glomeromycetes, and four orders: Archaeosporales, Paraglomerales, Diversisporales, and Glomerales. There are eight genera, including *Glomus, Paraglomus, Sclerocystis, Scutellospora, Gigaspora, Acaulospora, Archaeospora, and Entrophospora,* which comprise approximately 214 species [1]. Fossil evidence suggests that AM fungi date back to 400 to 600 million years ago [2]. They reproduce asexually and can colonize roots through spores, hyphae, or infected root fragments [3]. Mycorrhizal root systems increase the absorptive area of roots by 10 to 1,000 times, greatly enhancing the plants' ability to utilize soil resources.

Vegetable crops are important host plants for AMF, and the fungi can induce environmental stress tolerance, diseases, and nematode resistance, as well as increase nutrient and water supply. Therefore, inoculating vegetables with AMF can be profitable, and commercial products are available in the markets [4]. The effects of agricultural practices on AM fungi have recently been comprehensively reviewed [5, 6]. Factors such as crop rotation, application of fungicides, and fertilizers can play a major role in AMF colonization. The application of soluble phosphorus has been found to decrease root colonization [7], while occasional reports suggest an increase [8]. Similarly, with nitrogen fertilizers [9, 10]. Organic farming results in higher root colonization of AMF in pasture and annual crops compared to conventional agricultural practices [11, 12].

4.2 NEMATODE AND AMF INTERACTION

Arbuscular mycorrhizal fungi (AMF) are reported to occur profusely in agricultural crops. The nematodes and AMF commonly occur together in the rhizosphere of the host plant. AMF has a positive effect, but nematodes have a negative effect on plant growth. AMF fungi can be exploited for nematode management [13–18]. Fox & Sapsoff [19] were the first to describe the interaction between these two groups of organisms. Baltruschat et al. [20] reported that AMF plants were resistant to root-knot nematode infection. Since then, numerous research papers have appeared on the beneficial role of AMF against nematodes with a variety of crop plants as test hosts in the past two decades. AMF has controlled root-knot nematodes (RKN) development in tomatoes [21], bananas [22], and grapevines [23]. The interaction of AMF with Phyto nematodes has been reviewed at regular intervals [13, 24–30]. As numerous reviews are focusing on AMF interaction effects, more emphasis will be given in this chapter on the biological control potential of AMF against Phyto nematodes with special reference to root-knot nematodes *Meloidogyne* spp on vegetables.

4.3 AM FUNGI PROTECT VEGETABLE CROPS FROM ROOT-KNOT NEMATODES (RKN)

Root-knot nematode suppression with AMF on various vegetables has been listed in Table 4.1. Around 75% reduction of eggs of *M. hapla* in AMF plants [31]. Sikora [32] observed that AMF-applied tomato plants had decreased *Meloidogyne* spp. AMF could also damage root-knot nematode galls [33, 34] and reduce nematode reproduction [35]. Sankaranarayanan [36] studied the histopathological changes in blackgram infested with *G. mosseae* and *M. incognita*. The mechanism of AMF resistance against nematodes has been proven with *M. incognita* and AMF *G. mosseae* [37].

4.3.1 COMPATIBILITY OF NEMATODE BIOAGENTS, ORGANIC AMENDMENTS, AND AMFUNGI FOR CONTROL OF ROOT KNOT NEMATODES

Generally, a single biocontrol agent is recommended for disease management [68]. The combined application of biological control organisms with different types of control mechanisms is better for achieving control

TABLE 4.1 Biological Suppression of Root Knot Nematodes by Arbuscular Mycorrhizal Fungi in Vegetable Crops

Crop	Root-Knot Nematode	AM Fungi	Effect on Nematode	References
Tomato (*Solanum lycopersicum*) and pepper (*Capsicum annuum*)	*M. incognita*	*Rhizophagus irregularis*	Decreased *M. incognita* development.	[38]
Tomato	*M. incognita*	*Glomus fasciculatum*	Reduced reproductive growth of *M. incognita* and galls.	[39]
Brinjal	*M. incognita*	*G. fasciculatum*	Root knot gall suppression up to 87%.	[40]
Chili	*M. incognita*	*G. fasciculatum*, *G. mosseae*, *G. deserticola*, and *Sclerocystis*	*G. fasciculatum* and *G. mosseae* controlled *M. incognita* better.	[41]
Cucumber	*M. incognita*	*G. fasciculatum*	Significant suppression of nematode.	[42]
Tomato	*M. incognita*	*G. mosseae* (BEG 12)	Mycorrhiza-induced resistance genes identified against root-knot nematodes.	[37]
Tomato	*M. javanica*	*G. mosseae* *G. intraradices*	Suppressed nematode development.	[43]
Solanum nigrum	*M. incognita*	*G. mosseae*	Improved plant growth in pre and simultaneous application of AMF.	[44]
Tomato	*M. incognita*	*G. mosseae*	Mycorrhiza-induced resistance against root-knot nematodes.	[15]
Cucumber	*M. incognita*	*G. versiforme*	The cucumber seedlings were protected by AM fungi, and phosphorus contents of leaves increased due to AMF.	[45]
Tomato	*M. incognita*	*G. mosseae*	Nematode penetration was affected.	[46]
Tomato	*M. incognita*	*G. mosseae*	AMF suppressed nematode infection.	[47]
Tomato and carrot	*Meloidogyne* spp.	Native AMF and commercial AMF.	Field application of AMF suppressed nematode damage.	[48]
Tomato	*M. incognita*	*G. fasciculatum*	Plant growth improvement.	[49]

TABLE 4.1 *(Continued)*

Crop	Root-Knot Nematode	AM Fungi	Effect on Nematode	References
Tobacco	*M. incognita*	*G. intraradices*	Suppression of root-knot galls.	[50]
Cowpea	*M. incognita*	*G. mosseae*	Nematode reduction by AMF both in the screen house and field experiments.	[51]
Tomato	*M. incognita*	*G. fasciculatum* (SBI-G.f)	AMF increased plant growth and yield.	[21]
Cucumber	*M. incognita*	*G. intraradices*	AMF plants have a tolerance effect to nematode infection.	[52]
Lab Lab	*M. incognita*	*Glomus* sp.	Nematodes reduction.	[53]
Cucumber	*M. incognita*	*G. intraradices*, *G. mosseae*, *Glomus versiforme*	AMF suppressed nematode galling and egg mass.	[54]
Tomato	*M. incognita*	*G. fasciculatum*	Nematode suppression.	[55]
Tomato	*M. incognita*	*G. fasciculatum*	Reduced the root penetration by nematode.	[56]
Brinjal	*M. incognita*	*G. fasciculatum*	AMF decreased the nematode development.	[57]
Tomato	*M. incognita*	*G. fasciculatum*, *G. mosseae*	*G. fasciculatum* was more efficient than *G. mosseae* in controlling root knot nematode infection.	[58]
Okra	*M. incognita*	*G. fasciculatum*	AMF and FYM could be integrated to manage *M. incognita*.	[59]
Tomato	*M. incognita*	*G. fasciculatum*	Suppressed nematode population.	[60]
Tomato	*M. incognita*	*G. fasciculatum*	AMF increased plant growth and yield.	[61]
Tomato	*M. incognita*	*G. fasciculatum*	AMF suppressed gall development by 81.2% after 6 weeks, when the mycorrhizal colonization percentage was 71%.	[62]
Brinjal	*M. incognita*	*G. mosseae*	AMF suppressed gall index.	[63]
Blackgram	*M. incognita*	*G. fasciculatum*	AMF seedling inoculation reduced nematode development.	[64]
Tomato	*M. incognita*	*G. mosseae*	33% reduction gall index.	[65]

TABLE 4.1 *(Continued)*

Crop	Root-Knot Nematode	AM Fungi	Effect on Nematode	References
Brinjal	*M. incognita*	*G. mosseae*, *G. fasciculatum*, *G. intraradices*, *G. fulvum*	AMF affected nematode fecundity.	[66]
Menthol mint	*M. incognita*	*G. mosseae*	Maximum suppression of nematode reproduction was achieved with prior inoculation of VAM fungi.	[67]

compared to individual application of any bioagent [69, 70]. Integrated application of AMF and other nematode biocontrol agents such as *Paecilomyces lilacinus* [71] and *Pochonia chlamydosporia* [72] resulted in maximum control of nematodes compared to single inoculation [73–75]. The synergistic role resulted in increased plant growth [69]. Organic farming could improve AMF development [76]. Collins and McGraw [77] observed better AMF development in organic amendments. Verma [59] observed that the integrated application of AMF with other nematode bioagents such as *Pseudomonas fluorescens, G. fasciculatum,* and farmyard manure (FYM) suppressed nematode damage in okra. Integration of AMF *G. fasciculatum*, different organic amendments (neem cake, mustard oilcake, decaffeinated tea waste, and sawdust), and carbofuran 3G controlled nematode development in green gram[78]. A combination of AMF with other bioagents and organic amendments in controlling root-knot nematodes is presented (Table 4.2).

4.4 MODE OF ACTION OF AMF AGAINST PHYTONEMATODES

AMF and plant parasitic nematode interaction involve many modes of action on a particular host. Whipps [103] categorizes possible actions of AMF against diseases, including nematodes, such as competition for space and improved growth of AMF plants with altered biochemical and higher defense mechanisms. Schouteden [29] reviewed the interaction between AMF and nematodes in detail and proposed that AMF's action against nematodes might include competition for space, food, and induced systemic resistance (ISR).

TABLE 4.2 Combination of AM Fungi with Other Nematode Bioagents and Organic Amendments in Vegetable Crops

Crop	Root-Knot Nematode	AM Fungi	Bioagents	Effect on Nematode	References
Tomato	*M. incognita*	*G. etunicatum*, *G. mosseae*, *G. deserticola*, *Gigaspora gigantea*	*Paecilomyces lilacinus* (PL GoldTM)	Suppression of nematodes.	[79]
French bean	*M. incognita*	*G. fasciculatum*	*Trichoderma harzianum*	Decreased nematode development.	[80]
Pisum sativum	*M. incognita*	*G. mosseae*	*Aspergillus awamoriPseudomonas putida*, *P. alcaligenes Paenibacillus polymyxa*	Improved plant growth.	[81]
Tomato, pepper	*M. incognita*	*G. intraradices*, *G. mosseae*	*Bacillus megaterium*	Synergistic effects from the combined use of AMFs with *B. megaterium* in pepper.	[82]
Tomato	*M. incognita*	*G. intraradices*	Abamectin Fosthiazate	Abamectin alone and fosthiazate in the half dose mixed with AMF suppressed gall formation by 60.7 and 59.7%, respectively.	[83]
Tomato	*Meloidogyne* spp.	*G. mosseae*	Organic fertilizers (manure simple, simple enriched manure ash).	Pre-inoculation with AMF reduced the galls and nematode population.	[84]
Tomato	*M. incognita*	*G. versiforme*, *G. mosseae*	*Bacillus polymyxa*	*G. mosseae* and *Bacillus* sp. Combinations gave maximum control of nematodes.	[10]
Tomato	*M. incognita*	*G. intraradices*	*Heterorhabditis indica*	Application of EPN along with AMF reduced the galls, egg masses, eggs/eggmass, and hatch rate of eggs.	[85]

TABLE 4.2 *(Continued)*

Crop	Root-Knot Nematode	AM Fungi	Bioagents	Effect on Nematode	References
Tomato	*M. incognita*	*G. intraradices*	*Steinernema abbasi*, *S. carpocapsae*, *S. feltiae*	Application of *S. carpocapsae* along with AMF reduced the galls, egg masses, eggs/eggmass, and hatch rate of eggs.	[86]
Tomato	*M. incognita*	*G. aggregatum*	*Bacillus coagulans*,vermicompost	Plant growth increased.	[87]
Cowpea	*M. incognita*	*G. geosporum*	*Rhizobium*	AMF + rhizobium produced higher roots and suppressed nematode galling.	[88]
Tomato	*M. incognita*	*G. intraradices*	*Pseudomonas putida*, organic wastes (biosolids, horse manure, sawdust, and neem leaf litter.	Reducing galling and nematode multiplication.	[75]
Tomato	*M. incognita*	*G. intraradices*	*Rhizobium etli*	Additive control of nematodes.	[89]
Coleus	*M. incognita*	*G. mosseae*	*Pseudomonas fluorescens*, *Trichoderma viride*, *Paecilomyces lilacinus*,*Verticillium lecanii*	Suppression of nematode population.	[90]
Tomato	*Meloidogyne* spp.	*G. intraradices*	EPN *Heterorhabditis* sp.	Suppression of nematode population.	[91]
Okra	*M. incognita*	*G. fasciculatum*	*Pseudomonas fluorescens* FYM	Nematode suppression.	[92]
Tomato	*M. incognita*	*G. intraradices*	*Paecilomyces lilacinus*	Increased nematode control.	[93]
Brinjal	*M. incognita*	*G. fasciculatum*	Neem cake, carbofuran	Decreased nematode index and population.	[94]
Brinjal	*M. incognita*	*G. fasciculatum*	*Verticillium chlamydosporium*	Increased plant growth.	[95]
Tomato	*M. incognita*	*Glomus* sp.	*Pasteuria penetrans*	Increased nematode control.	[96]
Tomato	*M. incognita*	*G. mosseae*	*Paecilomyces lilacinus*	Combined treatment suppressed the nematode population.	[71]

TABLE 4.2 *(Continued)*

Crop	Root-Knot Nematode	AM Fungi	Bioagents	Effect on Nematode	References
Tomato	*M. incognita*	*G. mosseae*	*Pasteuria penetrans*	Nematode suppression.	[72]
Tomato	*M. incognita*	*G. fasciculatum*	*Trichoderma harzianum*,neem cake.	AMF suppressed nematode development.	[97]
Brinjal	*M. incognita*	*G. mosseae*	*Paecilomyces lilacinus*	Combined treatment reduced nematode damage.	[98]
Tomato	*M. incognita*	*G. deserticola*	*Pasteuria penetrans*	AMF and P. penetrans combination suppressed nematodes.	[99]
Tomato	*M. incognita*	*G. deserticola*	*Verticillium chlamydosporium*	Integration of AMF with V. chlamydosporium increased the parasitization of nematode eggs by the fungus.	[100]
Tomato	*M. incognita*	*G. fasciculatum*	*Calotropis procera*	Reduction of root-knot galls and nematode eggs.	[101]
Tomato	*M. javanica*	*G. mosseae*	*Paecilomyces lilacinus*	The combined treatment completely inhibited root infection by *M. javanica*.	[102]

Root exudates of *G. mosseae* affected the motility of *M. incognita* in tomato roots [46]. The AM fungi *G. fasciculatum* can activate the defense mechanism of soybean and prompt the roots of soybean to respond quickly to *Heterodera glycine* infection. PAL and chitinase play a critical role in the resistance induced by AM fungi against the nematode. Lower β-1,3-Glucanase activities were observed in the roots of soybean plants treated with AMF and nematodes [103].

Plants treated with AMF exhibited systemic induced resistance against several nematodes and crops [15, 104, 105], indicating that AMF indirectly influences nematode parasitism by modulating plant metabolic activities. The interaction between AMF and nematodes resulted in the expression of defense-related genes [106, 107]. Vos [15] investigated the mode of action of AMF against *M. incognita* and *Pratylenchus penetrans* in tomatoes and found that AMF systemically reduced the infection by these nematodes. Vos [37] revealed the molecular mechanisms involved in the biocontrol of nematodes using AMF through the suppression subtractive hybridization (SSH) method. "Mycorrhiza-induced resistance" is also a response of AMF plants against pathogens [108].

4.5 CONCLUSION AND FUTURE AREA OF RESEARCH

Arbuscular mycorrhizal fungi (AMF) and their exploitation in the biological control of root-knot nematodes on vegetables are environmentally feasible methods, considering the health hazards associated with nematicides. Interest in AMF fungi has grown significantly in the last four decades, with numerous research papers focusing on the utilization of AMF species in both laboratory and field studies. AMF fungi can be applied to plants to achieve the dual purpose of promoting plant growth and health, as these fungi are common and natural components of our agroecosystem. Prior establishment of AMF in the field should be emphasized by cultivating maize or sorghum plants to increase AMF inoculum, which can ultimately result in a high AMF load for future crops.

Combinations of biocontrol organisms such as AMF, *Rhizobium, Pseudomonas, Bacillus, Paecilomyces lilacinus, Pochonia chlamydosporia,* and *Arthrobotrys* spp. may increase plant growth and resistance against nematodes. Mass production of AMF by *in vitro* media is difficult, but root organ culture methods [109] are feasible for larger multiplication. Molecular-level mechanisms are needed to understand the interaction between AMF and nematodes in detail and, ultimately, to develop a biological control program for nematodes.

KEYWORDS

- **Agroecosystem**
- **arbuscular mycorrhizal fungi**
- **biological control**
- ***Meloidogyne* spp.**
- **nematodes**
- **organic amendments**
- **root-knot nematodes**
- **vegetables**

REFERENCES

1. Schussler, A., Schwarzott, D., & Walker, D., (2001). A new fungal phylum, the *Glomeromycota*: Phylogeny and evolution. *Mycological Research, 105*, 1413–1421.
2. Remy, W., Taylor, T. N., Hass, H., & Kerp, H., (1994). Four hundred-million-year-old vesicular arbuscular mycorrhizae. *Proceedings of Natural Academy of Sciences, 91*, 11841–11843.
3. Klironomos, J. N. M., & Hart, M. H., (2002). Colonization of roots by arbuscular mycorrhizal fungi using different sources of inoculum. *Mycorrhiza, 12*, 181–184.
4. Baum, C., El-Tohamy, W., & Gruda, N., (2015). Increasing the productivity and product quality of vegetable crops using arbuscular mycorrhizal fungi: A review. *Scientia Horticulturae, 187*, 131–141.
5. Jansa, J., Weimken, A., & Frossard, E., (2006). The effects of agricultural practices on arbuscular mycorrhizal fungi. In: Frossard, E., Blum, W., & Warkentin, B., (eds.), *Function of Soils for Human Societies and the Environment*. Geological Society, London, UK.
6. Larsen, J., Ravnskov, S., & Sorensen, J., (2007). Capturing the benefits of arbuscular mycorrhiza in horticulture. In: Hamel, C., & Plenchette, C., (eds.), *Mycorrhizae and Crop Production* (pp. 123–150). The Haworth Press, Inc., Binghamton, New York, USA.
7. Abbott, L. K., & Robson, A. D., (1984). The effect of VA mycorrhizae on plant growth. In: Powell, C. L., & Bagyaraj, D. J., (eds.), *VA Mycorrhiza* (pp. 113–130). CRC, Boca Raton, FL.

8. Gryndler, M., Lestina, J., Moravec, V., Prikyl, Z., & Lipavsky, J., (1990). Colonization of maize roots by VAM under conditions of long-term fertilization of varying intensities. *Agriculture Ecosystem and Environment, 29*, 183–186.
9. Baltruschat, H., & Dehne, H. W., (1988). The occurrence of vesicular-arbuscular mycorrhiza in agro-ecosystems. I. Influence of nitrogen fertilizer and green manure in continuous monoculture and in crop rotation on the inoculum potential of winter wheat. *Plant and Soil107*, 279–284.
10. Liu, A., Hamel, C., Hamilton, R. I., & Smith, D. L., (2000). Mycorrhizae formation and nutrient uptake of new corn (*Zea mays* L.) hybrids with extreme canopy and leaf architecture as influenced by soil N and P levels. *Plant and Soil, 221*, 157–166.
11. Ryan, M., & Ash, J., (1999). Effects of phosphorus and nitrogen on growth of pasture plants and VAM fungi in SE Australian soils with contrasting fertilizer histories conventional and biodynamic. *Agriculture Ecosystems and Environment, 73*, 51–62.
12. Ryan, M. H., Small, D. R., & Ash, J. E., (2000). Phosphorus controls the level of colonization by arbuscular mycorrhizal fungi in conventional and biodynamic irrigated dairy pastures. *Australian Journal of Experimental Agriculture, 40*, 663–670.
13. Hallman, J.,& Sikora, R. A., (2012). Endophytic fungi. In: Davies, K., & Spiegel, Y., (eds.), *Biological Control of Plant-Parasitic Nematodes: Building Coherence Between Microbial Ecology and Molecular Mechanisms.* Springer Science Business Media B.V.
14. Harrier, L. A., & Watson, C. A., (2004). The potential role of arbuscular mycorrhizal (AM) fungi in the bioprotection of plants against soil-borne pathogens in organic and/or other sustainable farming systems. *Pest Management Science, 60*, 149–157.
15. Vos, C. M., Tesfahun, A. N., Panis, B., Waele, D. De., & Elsen, A., (2012). Arbuscular mycorrhizal fungi induce systemic resistance in tomato against the sedentary nematode *M. incognita* and the migratory nematode *Pratylenchus penetrans*. *Applied Soil Ecology, 61*, 1–6.
16. Sankaranarayanan, C., & Rajeswari, S., (2009). Reciprocal influence of arbuscular mycorrhizal fungus and root knot nematode and interaction effects on blackgram. *Nematologia Mediterranea, 37*, 197–202.
17. Sankaranarayanan, C., & Rajeswari, S., (2010). Influence of application methods of arbuscular mycorrhiza *G. mosseae* in the bio-management of root knot nematode, *M. incognita* on black gram (*Vigna mungo* L.) hepper. *Journal of Biological Control, 24*, 51–57.
18. Van Der, H. M. G., Martin, F. M., Selosse, M. A., & Sanders, I. R., (2015). Mycorrhizal ecology and evolution: The past, the present, and the future. *New Phytol., 205*, 1406–1423.
19. Fox, J. A., & Spasoff, L., (1972). Interaction of *Heterodera solanacearum* and *Endogonegigantea* on tobacco. *Journal of Nematology, 4*, 224–225.
20. Baltruschat, H., Sikora, R. A., & Schönbeck, F., (1973). Effect of VA-mycorrhiza (*Endogonemossae*) on the establishment of *Thielaviopsis basicolaand M. incognita* in tobacco. In: *Second International Congress of Plant Pathology* (p. 661). Minneapolis.
21. Kantharaju, V., Krishnappa, K., & Hemavathi, M. R., (2010). Biological control of root-knot nematode, *M. incognita*, by using indigenous isolates of *G. fasciculatum*. *Environment and Ecology, 28*, 2156–2158.
22. Jaizme-Vega, M. C., Tenoury, P., Pinochet, J., & Jaumot, M., (1997). Interactions between the root-knot nematode *M. incognita* and *G. mosseae* in banana. *Plant and Soil, 196*, 27–35.

23. Li Hai, Y., Yang, G. D., Shu, H. R., Yang, Y. T., Ye Bao, X., Nishida, I., & Zheng, C. C., (2006). Colonization by the arbuscular mycorrhizal fungus *Glomus versiforme* induces a defense response against the root-knot nematode *M. incognita* in the grapevine (*Vitis amurensis* Rupr.), which includes transcriptional activation of the class III chitinase gene VCH3. *Plant and Cell Physiology, 47*, 154–163.
24. Borowicz, V. A., (2001). Do arbuscular mycorrhizal fungi alter plant–pathogen relations? *Ecology, 82*, 3057–3068.
25. Hussey, R. S., & Roncadori, R. W., (1982). Vesicular arbuscular mycorrhizae may limit nematode activity and improve plant growth. *Plant Disease, 66*, 9–14.
26. Ingham, R. E., (1988). Interactions between nematodes and vesicular-arbuscular mycorrhizae. *Agriculture Ecosystems and Environment, 24,* 169–182.
27. Smith, G. S., (1987). Interactions of nematodes with mycorrhizal fungi. In: Veech, J. A., & Dickson, D. W., (ed.), *Vistas on Nematology* (pp. 292–300). Society of Nematologists, Hyattsville.
28. Sikora, R. A., (1995). Vesicular-arbuscular mycorrhizae: Their significance for biological control of plant-parasitic nematodes. *Biocontrol, 1*, 29–33.
29. Schouteden, N., Waele, D., Panis, B., & Vos, C. M., (2015). Arbuscular mycorrhizal fungi for the biocontrol of plant-parasitic nematodes: A review of the mechanisms involved. *Front. Microbiology, 6*, 1–12.
30. Albuquerque Da, S. C. M., (2020). Bioprotection by arbuscular mycorrhizal fungi in plants infected with *Meloidogyne* nematodes: A sustainable alternative. *Crop Protection*. doi: https://doi.org/10.1016/j.cropro.2020.105203.
31. Waceke, J. W., Waudo, S. W., & Sikora, R. A., (2001). Suppression of *Meloidogyne hapla* by arbuscular mycorrhiza fungi (AMF) on pyrethrum in Kenya. *International Journal of Pest Management, 47*, 135–140.
33. Bagyaraj, D. J., Manjunath, A., & Reddy, D. D. R., (1979). Interaction of vesicular arbuscular mycorrhia with root knot nematodes in tomato. *Plant and Soil, 51*, 397–403.
34. Kellam, M. K., & Schenck, N. C., (1980). Interactions between a vesicular-arbuscular mycorrhizal fungus and root-knot nematode on soybean. *Phytopathology, 70*, 293–296.
35. Sikora, R. A., (1979). Predisposition*to Meloidogyne* infection by the endotrophic mycorrhizal fungus *G. mosseae*. In: Lamberti, F., & Taylor, C. E., (eds.), *Root-Knot Nematode (Meloidogyne Species) Systematics, Biology, and Control* (pp. 399–404). Academic, New York.
36. Sankaranarayanan, C., (1995). *Effect of Phosphorus, Rhizobium, and Phosphobacteria on the Interaction of Vesicular Arbuscular Mycorrhizae with Root Knot Nematode Meloidogyner Incognita on Blackgram (Vigna mungo).* Ph.D. Thesis, Tamilnadu Agricultural University, Coimbatore, India.
37. Vos, C., Schouteden, N., Tuinen, D., Chatagnier, O., Elsen, A., Waele, D. De., Panis, B., & Gianinazzi-Pearson, V., (2013). Mycorrhiza-induced resistance against the root-knot nematode *M. incognita* involves priming of defense gene responses in tomato. *Soil Biology and Biochemistry, 60*, 45–54.
38. Rodriguez-Heredia, M., Djian-Caporalino, M., Ponchet, L., Lapeyre, R., Canaguier, A., Fazari, N., Marteu, B., & Industri, M., (2020). Offroy-Chave, Protective effects of mycorrhizal association in tomato and pepper against *Meloidogyne incognita* infection, and mycorrhizal networks for early mycorrhization of low mycotrophic plants. *Phytopathologia Mediterranea, 59*, 377–384.

39. Samal, B. K., Mahalik, J. K., Das, P. K., & Jitendra, P., (2018). Effect of interaction between arbuscular mycorrhizal fungi (*Glomus fasciculatum*) and root knot nematode *Meloidogyne incognita* on tomato. *Ann. Plant Protec. Science, 126*, 355–359.
40. Horta, L., (2015). *Interaction of Root Knot Nematode (Meloidogyne Incognita) & AM Fungus (Glomus Fasciculatum) on Brinjal* (p. 68). MSc Thesis, Orissa University of Agriculture and Technology, Bhubaneshwar, India.
41. Abhiniti, M., Tripti, A., & Trivedi, P. C., (2013). Evaluation of AM fungi in management of *Meloidogyne-Rhizoctonia* complex infecting *Capsicum annuum* L. *Annals of Plant ProtectionSciences, 21*, 396–399.
42. Nikhita, K., & Mahanta, B., (2013). Efficacy of *G. fasciculatum*, *Trichoderma harzianum*, neem cake and carbofuran 3G in management of *M. incognita* on cucumber. *Annals of Plant Protection Sciences, 21*, 405–407.
43. Sohrabi, F., Fadaei-Tehrani, A. A., Danesh, Y. R., & Jamali-Zavareh, A., (2012). Interaction between two arbuscular mycorrhizal fungi (*G. mosseae* and *G. intraradices*) and root knot nematode (*Meloidogyne javanica*) in tomato. *Iranian Journal of Plant Pathology, 48*, 131–134.
44. Robab, M. I., Hisamuddin, S., & Tanweer, A., (2012). Antagonistic effect of *G. mosseae* on the pathogenicity of root-knot nematode infected *Solanum nigrum*. *Crop Protection, 42*, 351–355.
45. Chen, S., Jiang, Y., Liu, H., & Cheng, Z., (2012). The effects of interaction between fungus Arbuscular mycorrhiza and root-knot nematode *M. incognita* on the growth and physiological characteristics of cucumber. *Acta Phytophylacica Sinica, 39*, 253–259.
46. Vos, C., Claerhout, S., Mkandawire, R., Panis, B., Waele, D. De., & Elsen, A., (2012). Arbuscular mycorrhizal fungi reduce root-knot nematode penetration through altered root exudation of their host. *Plant and Soil, 354*, 335–345.
47. Vos, C., Geerinckx, K., Mkandawire, R., Panis, B., Waele, D. De., & Elsen, A., (2012). Arbuscular mycorrhizal fungi affect both penetration and further life stage development of root-knot nematodes in tomato. *Mycorrhiza, 22*, 157–163.
48. Affokpon, A., Coyne, D. L., Lawouin, L., Tossou, C., Agbèdè, R. D., & Coosemans, J., (2011). Effectiveness of native West African arbuscular mycorrhizal fungi in protecting vegetable crops against root-knot nematodes. *Biology and Fertility of Soils, 47*, 207–217.
49. Okon, I. E., & Imuk, E. A., (2011). Effect of *G. fasciculatum* on the growth and yield of tomato (*Solanum lycopersicum* L.) in *M. incognita* infested soil. *Journal of Applied Horticulture, 13*, 79–81.
50. Subhashini, D. V., & Ramakrishnan, S., (2011). Interaction of VA mycorrhiza and root-knot nematode on tobacco plants: Effects of nematode inoculum density, soil texture and soil sterilization. *Indian Journal of Nematology, 41*, 56–61.
51. Odeyemi, I. S., Afolami, S. O., & Adekoyejo, A. B., (2013). Integration of *G. mosseae* with *Chromolaena odorata* powder for suppression of *M. incognita* on maize (*Zea mays* L.). *Archives of Phytopathology and Plant Protection,46*, 1589–1597.
52. Zhang, L., Zhang, J., Christie, P., & Xiaolin, L., (2009). Effect of inoculation with the arbuscular mycorrhizal fungus *G. intraradices* on the root-knot nematode *M. incognita* in cucumber. *Journal of Plant Nutrition, 32*, 967–979.
53. Ahmed, S. H., Abdelgani, M. E., & Yassin, A. M., (2009). Effects of interaction between vesicular-arbuscular mycorrhizal (VAM) fungi and root-knot nematodes on *Dolichos* bean (*Lablab Niger* Medik.) plants. *American Eurasian Journal of Sustainable Agriculture, 3*, 678–683.

54. Zhang, L., Zhang, J. L., Christie, P., & Li, X., (2008). Pre-inoculation with arbuscular mycorrhizal fungi suppresses root knot nematode (*M. incognita*) on cucumber (*Cucumis sativus*).; *Biology and Fertility of Soils, 45*, 205–211.
55. Shreenivasa, K. R., Krishnappa, K., & Ravichandra, N. G., (2007). Interaction effects of arbuscular mycorrhizal fungus *G. fasciculatum* and root-knot nematode, *M. incognita* on growth and phosphorous uptake of tomato. *Karnataka Journal of Agricultural Sciences, 20*, 57–61.
56. Shreenivasa, K. R., Krishnappa, K., & Ravichandra, N. G., (2007). Survival and penetration of *M. incognita* larvae in tomato roots in presence of arbuscular mycorrhizal fungus, *G. fasciculatum*. *Karnataka Journal of Agricultural Sciences, 20*, 166–167.
57. Aparajita, B., & Phukan, P. N., (2006). Effect of *G. fasciculatum* on nutritional and biochemical changes in brinjal infested by *M. incognita*. *Indian Journal of Nematology, 36*, 189–191.
58. Verma, K. K., & Nandal, S. N., (2006). Comparative efficacy of VAM, *G. fasciculatum* and *G. mosseae* for the management of *M. incognita* in tomato at different phosphorus levels. *National Journal of Plant Improvement, 8*, 174–176.
59. Verma, K. K., Sewak, R., & Nandal, S. N., (2006). Effect of different farm yard manure levels on managing potential of VAM, *G. fasciculatum* against *M. incognita* in okra. *Haryana Journal of Agronomy, 22*, 37–39.
60. Kumar, K. C. K., Krishnappa, K., Ravichandra, N. G., Shreenivasa, K. R., Karuna, K., & Kumar, B. R., (2006). Role of *G. fasciculatum* on growth, development and yield of tomato infested by *M. incognita*. *Environment and Ecology, 24*, 903–906.
61. Kantharaju, V., Krishnappa, K., Ravichandra, N. G., & Karuna, K., (2005). Management of root-knot nematode, *M. incognita*, on tomato by using indigenous isolates of AM fungus, *G. fasciculatum*. *Indian Journal of Nematology, 35*, 32–36.
62. Aruna, P., Ganguly, A. K., & Singh, C. S., (2003). Influence of *G. fasciculatum* on *M. incognita* infected tomato. *Annals of Plant Protection Sciences, 11*, 346.
63. Jothi, G., & Sundarababu, R., (2002). Nursery management of *M. incognita* by *G. mosseae* in eggplant. *Nematologia mediterranea, 30*, 153–154.
64. Bornali, M., & Phukan, P. N., (2004). Effect of *G. fasciculatum* on penetration, development, and survival of *M. incognita* on black gram. *Annals of Plant ProtectionSciences, 12*, 363–365.
65. Talavera, M., Itou, K., & Mizukubo, T., (2001). Reduction of nematode damage by root colonization with arbuscular mycorrhiza (*Glomus* spp.) in tomato *M. incognita* (*Tylenchida*: *Meloidogynidae*) and carrot- *Pratylenchus penetrans* (*Tylenchida: Pratylenchidae*) pathosystems. *Applied Entomology and Zoology, 36*, 387–392.
66. Jothi, G., & Rajeswari, S., (2000). Interaction of four *Glomus* spp. with *M. incognita* on brinjal (*Solanum melongena* L.). *International Journal of Tropical Plant Diseases, 18*, 147–156.
67. Ratti, N., Abdul-Khaliq, S. K., Akhtar, H., & Janardhanan, K. K., (2000). Effect of *G. mosseae* (Nicol. and Gerd.) Gerdemann and Trappe on root-knot disease of menthol mint (*Mentha arvensis* sub sp. haplocalyx Briquet) caused by *M. incognita* (Kofoid and white) chitwood. *Journal of Spices and Aromatic Crops, 9*, 129–132.
68. Wilson, M., & Backman, P. A., (1999). Biological control of plant pathogens. In: Ruberson, J. R., (ed.), *Handbook of Pest Management* (pp. 309–335). New York, Marcel Dekker, Inc.

69. Guetsky, R., Shtienberg, D., Elad, Y., Fischer, E., & Dinoor, A., (2002). Improving biological control by combining biocontrol agents each with several mechanisms of disease suppression. *Phytopathology, 92*, 976–985.
70. Sankaranarayanan, C., & Rajeswari, S., (2001). Influence of *Rhizobium* and phosphobacteria on the interaction of VA-mycorrhiza, *G. mosseae* and *M. incognita* on blackgram. *International Journal of Tropical Plant Diseases, 19*, 133–139.
71. Bhat, M. S., & Irshad, M., (2000). Role of *G. mosseae* and *Paecilomyces lilacinus* in the management of root knot nematode on tomato. *Archives of Phytopathology and Plant Protection, 33*, 131–140.
72. Rao, M. S., Reddy, P. P., & Nagesh, M., (2000). Management of *M. incognita* on tomato by integrating *G. mosseae* with *Pasteuria penetrans* under field conditions. *Pest Management in Horticultural Ecosystems, 6*, 130–134.
74. Siddiqui, Z. A., & Akhtar, M. S., (2009). Effect of plant growth promoting rhizobacteria, nematode parasitic fungi and root-nodule bacterium on root-knot nematodes *Meloidogyne javanica* and growth of chickpea. *Biocontrol Science and Technology, 19*, 511–521.
75. Siddiqui, Z. A., & Akhtar, M. S., (2008). Synergistic effects of antagonistic fungi and a plant growth promoting rhizobacterium, an arbuscular mycorrhizal fungus, or composted cow manure on populations of *M. incognita* and growth of tomato. *Biocontrol Science and Technology, 18*, 279–290.
76. Giovanetti, M., & Avio, L., (1985). VAM infection and reproduction as influenced by different organic and inorganic substances. In: Molina, R., (ed.), *Proc. 6th N. Amer. Conf. Mycorrhizae* (p. 400). Corvallis, Ore.
77. Collins, J. N., & McGraw, A. C., (1988). Vesicular-arbuscular mycorrhizae in taconite tailings. II. Effects of reclamation practices. *Agriculture Ecosystem and Environment, 21*, 143–152.
78. Neog, P. P., & Islam, M., (2008). Management of *M. incognita* on green gram using vesicular arbuscular mycorrhizal fungus, *G. fasciculatum*, organic amendments and carbofuran. *Crop Research (Hisar), 36*, 313–315.
79. Udo, I. A., Uguru, M. I., & Ogbuji, R. O., (2013). Comparative efficacy of arbuscular mycorrhizal fungi in combination with bioformulated *Paecilomyces lilacinus* against *M. incognita* on tomato in two ultisols of South-Eastern Nigeria. *Biocontrol Science and Technology, 23*, 1083–1097.
80. Gogoi, D., & Mahanta, B., (2013). Comparative efficacy of *G. fasciculatum, Trichodermaharzianum*, carbofuran and carbendazim in management of *M. incognita* and *Rhizoctonia solani* disease complex on French bean. *Annals of Plant Protection Sciences, 21*, 172–175.
81. Akhtar, M. S., & Jitendra, P., (2013). Efficacy of root-associated fungi and PGPR on the growth of *Pisum sativum* (cv. Arkil) and reproduction of the root-knot nematode *M. incognita. Journal of Basic Microbiology, 53*, 318–326.
82. Flor-Peregrin, E., Azcon, R., Salmerón, T., & Talavera, M., (2012). Biological protection conferred by *Glomus* spp. and *Bacillus megaterium* against *M. incognita* in tomato and pepper. *IOBC/WPRS Bulletin, 83*, 215–218.
83. Khalil, M. S., (2012). A comparison study with alternative biorational agents to suppress the root knot nematode populations and gall-formation in tomato plants. *International Journal of Nematology, 22*, 112–116.

84. Dodzi, B. K., Atti, T., Kodjo, T. A., Affoh, A., & Mawuena, G., (2012). Impact of organic manure applied alone and in combination with a strain of indigenous arbuscular mycorrhizal fungus *G. mosseae* on *Meloidogyne* spp. nematode parasite of tomato fruit. *Journal of Applied Biosciences, 55*, 3973–3986.
85. Hussaini, S. S., & Kumar, K. C. K., (2009). Interaction effect of *Heterorhabditis indica* and *Arbusular mycorhizal* fungus, *G. intraradices* on the development and reproduction of *M. incognita* (Kofoid and white) in tomato. *Journal of Eco-friendly Agriculture, 4*, 80–84.
86. Hussaini, S. S., Kumar, K. C. K., Alok, A., & Shakeela, V., (2009). Interaction effect of *Steinernema* spp. and arbuscular mycorrhizal fungus, *G. intraradices* on the development and reproduction of *M. incognita* (Kofoid and White) in tomato. *Trends in Biosciences, 2*, 33–36.
87. Serfoji, P., Rajeshkumar, S., & Selvaraj, T., (2010). Management of root-knot nematode, *M. incognita* on tomato cv Pusa ruby. by using vermicompost, AM fungus, *Glomusaggregatum* and mycorrhiza helper bacterium, *Bacillus coagulans*. *International Journal of Agricultural Technology, 6*, 37–45.
88. Ugwuoke, K. I., & Eze, S. C., (2010). The effect of mycorrhiza (*Glomus geosporum*), rhizobium and *M. incognita* on growth and development of cowpea (*Vigna unguiculata* L. Walp). *Research Journal of Agriculture and Biological Sciences, 6*, 937–941.
89. Reimann, S., Hauschild, R., Hildebrandt, U., & Sikora, R. A., (2008). Interrelationships between *Rhizobium etli* G12 and *G. intraradices* and multitrophic effects in the biological control of the root-knot nematode *M. incognita* on tomato. *Journal of Plant Diseases and Protection, 115*, 108–113.
90. Seenivasan, N., & Devrajan, K., (2008). Management of *M. incognita* on medicinal coleus by commercial biocontrol formulations. *Nematologia Mediterranea, 36*, 61–67.
91. Channashettar, V. A., Reena, S., & Alok, A., (2007). Effect of entomopathogenic nematode on arbuscular mycorrhizal fungi inoculated tomato plant's growth, mycorrhizal colonization, and enzyme activity of hyphae. *Mycorrhiza News, 19*, 24–26.
92. Verma, K. K., (2006). Integrated methods for controlling root-knot nematode in okra. *Haryana Journal of Horticultural Sciences, 35*, 288–289.
93. Rumbos, C., Reimann, S., Kiewnick, S., & Sikora, R. A., (2006). Interactions of *Paecilomyceslilacinus* strain 251 with the mycorrhizal fungus *G. intraradices*: Implications for *M. incognita* control on tomato. *Biocontrol Science and Technology, 16*, 981–986.
94. Aparajita, B., & Phukan, P. N., (2004). Comparative efficacy of *G. fasciculatum* with neem cake and carbofuran for the management of *M. incognita* on brinjal. *Indian Journal of Nematology, 34*, 129–132.
95. Rao, M. S., Dhananjay, N., Shylaja, M., & Reddy, P. P., (2003). Management of *M. incognita* on eggplant by integrating endomycorrhiza, *G. fasciculatum* with bio-agent *Verticilliumchlamydosporium* under field conditions. *Indian Journal of Nematology, 33*, 29–32.
96. Talavera, M., Itou, K., & Mizukubo, T., (2002). Combined application of *Glomus* sp. and *Pasteuria penetrans* for reducing *M. incognita* (*Tylenchida: Meloidogynidae*) populations and improving tomato growth. *Applied Entomology and Zoology, 37*, 61–67.

97. Reddy, P. P., Nagesh, M., & Devappa, V., (1998). Management of *M. incognita* on tomato by integration of *Trichoderma harzianum*, *G. fasciculatum* and neem cake. *Pest Management in Horticultural Ecosystems, 4*, 349–352.
98. Rao, M. S., Reddy, P. P., & Mohandas, M. S., (1998). Bio-intensive management of *M. incognita* on egg plant, by integrating *Paecilomyces lilacinus* and *G. mosseae*. *Nematologia Mediterranea, 26*, 213–216.
99. Rao, M. S., & Gowen, S. R., (1998). Bio-management of *M. incognita* on tomato by integrating *Glomus deserticola* and *Pasteuria penetrans*. *Zeitschrift-Fur-Pflanzenkrankheiten-Und-Pflanzenschutz,105*, 49–52.
100. Rao, M. S., Kerry, B. R., Gowen, S. R., Bourne, J. M., & Reddy, P. P., (1997). Management of *M. incognita* in tomato nurseries by integration of *Glomus deserticola* with *Verticillium chlamydosporium*. *Zeitschrift-fur-Pflanzenkrankheiten-und-Pflanzenschutz, 104*, 419–422.
101. Rao, M. S., Reddy, P. P., & Das, S. M., (1996). Effect of integration of *Calotropis procera* leaf and *G. fasciculatum* on the management of *M. incognita* infesting tomato. *Nematologia Mediterranea, 24*, 59–61.
102. Al-Raddad, A. M., (1995). Interaction of *G. mosseae* and *Paecilomyces lilacinus* on *Meloidogyne javanica* of tomato. *Mycorrhiza, 5*, 233–236.
103. Whipps, J. M., (2004). Prospects and limitations for mycorrhizals in biocontrol of root pathogens. *Canadian Journal of Botony, 82*, 1198–1227.
104. Hao, Z., Fayolle, L., Van, T. D., Chatagnier, O., Xiaolin, L., Gianinazzi, S., & Gianinazzi-Pearson, V., (2012). Local and systemic mycorrhiza-induced protection against the ectoparasitic nematode *Xiphinema index* involves priming of defense gene responses in grapevine. *Journal of Experimental Botany, 63*, 3657–3672.
105. Elsen, A., Baimey, H., Swennen, R., & Waele, D. De., (2003). Relative mycorrhizal dependency and mycorrhiza-nematode interaction in banana cultivars (*Musa* spp.) differing in nematode susceptibility. *Plant and Soil, 256*, 303–313.
106. Alguacil, M. Del. M., Torrecillas, E., Lozano, Z., & Roldán, A., (2011). Evidence of differences between the communities of arbuscular mycorrhizal fungi colonizing galls and roots of *Prunus persica* infected by the root-knot nematode *M. incognita*. *Applied and Environmental Microbiology, 77*, 8656–8661.
107. ZhiPeng, Fayolle, L., Tuinen, D., Chatagnier, O., Li, X., Gianinazzi, S., & Gianinazzi, P. V., (2012). Local and systemic mycorrhiza-induced protection against the ectoparasitic nematode *Xiphinema index* involves priming of defense gene responses in grapevine. *Journal of Experimental Botany, 63*, 3657–3672.
108. Cameron, D. D., Andrew, L. N., Saskia, C. M., Van, W., & Jurriaan, T., (2013). Mycorrhiza induced resistance: More than the sum of its parts? *Trends in Plant Science, 18*, 539–545.
109. Fortin, J. A., Bécard, G., Declerck, S., Dalpé, Y., St. Arnaud, M., Coughlan, A. P., & Piché, Y., (2002). Arbuscular mycorrhiza on root-organ cultures. *Canadian Journal of Botony, 80*, 1–20.

CHAPTER 5

Unraveling the Mechanism of Microbial Biostimulants for Controlling Pests and Diseases

DAISY SENAPOTY[1], LOPAMUDRA GIRI[1], and PUPPALA SRI SARANYA[2]

[1]*Department of Plant Pathology, Assam Agricultural University, Jorhat, Assam, India*

[2]*Department of Entomology, Assam Agricultural University, Jorhat, Assam, India*

ABSTRACT

Pests and diseases are reported to cause devastating losses in agricultural crops worldwide. Various approaches have been pursued to prevent, mitigate, and control these diseases. In addition to good management practices, the main challenge in controlling plant diseases is the use of chemical fertilizers and pesticides, which not only contribute to environmental pollution but also harm beneficial microorganisms in the soil. Therefore, it is crucial to focus on developing eco-friendly alternative strategies for managing plant diseases. The use of microbial bio-control agents has proven to be one of the most effective approaches in disease management. These bio-agents, such as *Trichoderma viride, Pseudomonas fluorescens, Bacillus subtilis, Beauveria bassiana,* etc., have been successfully employed in the management of various plant pathogens and pests. The intricate interactions between these microbes and plants help suppress diseases by stimulating ISR (induced systemic resistance) and SAR (systemic acquired resistance), while also promoting plant growth through activities like nitrogen regulation, IAA (indole acetic acid) production, phosphorous solubilization, and siderophore production.

Microbial Biostimulants: Biorational Pesticides for Management of Plant Pathogens.
Faheem Ahmad and Rakesh Pandey (Eds.)

5.1 INTRODUCTION

Microbial biostimulants are considered an important tool for modern agriculture as part of an integrated crop management (ICM) system. This approach aims to make agriculture more sustainable and resilient. Plants face various environmental stresses, both biotic and abiotic, in their natural habitats. These stresses have led to an increase in diseases and pest attacks, resulting in a decline in crop production. This has raised significant concerns for crop producers and scientists worldwide.

To combat these diseases and pests, many growers heavily rely on chemical fungicides and insecticides. However, the excessive use and misuse of these agrochemicals have caused environmental damage. As a result, there has been a shift in attitudes towards a more sustainable and eco-friendly approach. One alternative is the use of microbial stimulant-based biological control, which can help reduce pollution and the detrimental environmental impact caused by synthetic chemicals.

5.2 TYPES OF MICROBIAL BIOSTIMULANTS

Microbial biostimulants can be of various types which include fungi, bacteria, and actinomycetes.

5.2.1 FUNGI

An eukaryotic, spore-bearing, achlorophyllous, heterotrophic organism that generally reproduces either sexually or asexually, containing chitin or cellulose or both in its cell wall, along with many other organic molecules, and exhibiting an absorptive mode of nutrition. Many fungal agents, such as *Trichoderma* spp., *Glomus* spp., *Pochonia chlamydosporia, Paeciliomyces lilacinus, Ampelomyces quisqualis,* etc., have been found to be effective and beneficial microbial biostimulants.

5.2.2 BACTERIA

These are minute, rigid, unicellular prokaryotic organisms that are devoid of chlorophyll (achlorophyllous) and most commonly reproduce by transverse binary fission. The microbial biostimulants that fall under this category include *Pseudomonas* spp. and *Bacillus* spp., among others.

5.2.3 ACTINOMYCETES

These are a group of unicellular filamentous gram-positive bacteria that form a branching network of filaments and produce spores. They are mostly aerobic but can also be anaerobic. They include *Streptomyces and Corynebacterium,* among others (Table 5.1).

TABLE 5.1 Types of Microbial Biostimulants

	Fungal	
Pathogen	**Biocontrol Agent**	**Target Organisms/Disease**
	Trichoderma viridae	Soil-borne fungal pathogens
	Trichoderma harzianum	Grey mold, *Rhizoctonia*,and *Sclerotina* spp.
	Trichoderma polysporum	Root rot diseases
	Glomus mosseae	Wilt diseases
	Saccharomyces cerevisiae	*Sclerophoma pithyophila* (wood rot fungi) and *Phyllosticta citricarpa* cause citrus black spots.
Pest	*Beauveria bassiana*	Termites, thrips, whiteflies, and aphids
	Metarhizium anisopliae	Caterpillar and locusts.
	Bacterial	
Pathogen	*Bacillus subtilis*	*Fusarium* spp., *Pythium* spp., *Rhizoctonia* spp.
	Bacillus amyloliquefaciens	*Mucor* spp., *Sclerotina sclerotiorum*
	Bacillus licheniformis	*Magnaporthe grisea*
	Bacillus velezensis	*Botrytis cinerea*
	Pseudomonas fluorescens	Soil-borne pathogens and take all diseases of wheat.
	Pseudomonas aeruginosa	*Botrytis cinerea* in tomato.
Pest	*Bacillus thuringiensis* var. *kurstaki.*	Caterpillars of lepidopteran pests, gypsy moth, cabbage looper, etc.
	Bacillus thuringiensis var. *israelensis*	Mosquitos, black flies, fungus gnat larvae.
	Bacillus thuringiensis var. *sandiego and Bacillus thuringiensis* var. *tenebrionis.*	Colorado potato beetle, leaf-eating beetles.
	Bacillus popilliae	Japanese beetle larvae.
	Actinomycetes	
Pathogen	*Streptomyces* spp.	*Xanthomonas* spp., *Ralstonia solanacearum*
Pest	*Corynebacterium glutamicum*	Colorado potato beetle

5.3 CHARACTERISTICS OF MICROBIAL BIOSTIMULANTS

- The microbial biostimulants used should be environmentally friendly and should not cause any harm to other beneficial microorganisms.
- They should exhibit an antagonistic nature, such as antibiosis, competition, parasitism, and predation, when employed against specific pathogens and pests.
- They should be able to reproduce rapidly when subjected to large-scale production in the form of bioformulations.
- They should not cause any harm to the host plant.
- They should not have any residual effect in the soil.
- They should have resistant or virulent genes to the pathogen in order to control the disease caused by it.

5.4 DEFENSE MECHANISMS INVOLVED BY MICROBIAL BIOSTIMULANTS IN PLANTS

Plants and pathogens are found to interact with a wide range of microorganisms throughout their lives. These interactions can have a significant impact on plant health in various ways. To understand the mechanisms involved in biological control, it is useful to know the different ways in which organisms interact. Several of these interactions are discussed in the following subsections.

5.4.1 ANTIBIOSIS

Antibiosis is a biological interaction between two or more organisms in which chemical substances generated by one can slow or kill the growth of the other, which can be a pest or pathogenic microbe. It is an antagonistic mechanism that occurs when one organism interacts with a metabolic product produced by another. The term "antibiosis" was coined by Vuillemin in 1889. Several mechanisms are involved in antibiosis, such as the release of antibiotics, organic compounds, toxins, and various hydrolytic enzymes like beta-xylosidase, chitinase, pectin methylesterase, β-1,3-glucanase, etc. The mode of action includes inhibition of cell wall synthesis, inhibition of protein synthesis (Translation), alteration of cell membranes, inhibition of nucleic acid synthesis, antimetabolite activity, etc. It has been found that the rhizosphere microbial population destroys the pathogen by disintegration of the glycosidic linkages in its cell wall. They also help in the induction of the host plant's negative impact on the biology of the insects and their offspring by mediating both chemical and morphological plant defenses, which results in the death of early instars, reduced size or weight, protracted immature stage growth, reduced adult longevity and fertility, and death in the prepupal or pupal stage.

5.4.1.1 EXAMPLES

One of the most common pests of grain sorghum is the sorghum shoot fly, *Atherigona soccata.* The most important parasitoids used in controlling the pest are egg parasitoids like *Trichogramma chilonis* and *Trichogrammatoidea simmondsi,* as well as larval parasitoids like *Neotrichoporoides nyemitawus* [1]. Antibiosis in sorghum shoot fly results in the retardation of growth and development of the insect, prolonged larval and pupal periods, and poor emergence of adults on resistant genotypes.

In disease management, antibiosis is generally mediated by the production of low molecular weight antibiotics by the antagonist, which can inhibit the growth of the pathogen. Antibiotics play a crucial role in biocontrol by both fungi and bacteria, with a strong correlation between antibiotic production and biocontrol efficiency. Purified antibiotics produced by biocontrol agents mimic their effects, while antibiotic-deficient mutants exhibit less biocontrol activity compared to wild types (WT). Fungal isolates like *Gliocladium* and *Trichoderma* are commonly reported for their production of antibiotics in biological control. Gliotoxin, produced by *Gliocladium virens,* is a key factor in its biocontrol activity against *Pythium ultimum* and *Rhizoctonia solani*[2].

5.4.2 HYPERPARASITISM

Parasitism refers to a direct competitive interaction between two organisms in which one organism obtains nutrients from the other. Hyperparasitism is the parasitic behavior of one species on another species in which the host is also a parasite, such as a parasitic insect, pathogen, or specifically a parasitoid. This type of interaction is common among fungi, although hyperparasitism in bacteria has only been recorded very rarely. *Bdellovibrio bacteriovorus,* a predatory bacterium, has been discovered to have the rare ability to feed on the cytoplasm of other gram-negative bacteria. There are two types of hyperparasitism: obligate and facultative. Obligate hyperparasitism is seen where the organisms have evolved to infect other parasites and in the case of facultative hyperparasites, also known as generalist or opportunistic parasites, they infect a variety of hosts, some of which are parasitic themselves. Hyperparasitism plays a critical role in the evolution and epidemiology of microbial pathogens [3].

According to theory, it has been found that the existence of hyperparasite-infected pathogen strains can influence the stability of trophic [4] and also alter the conditions under which there is a successful invasion of the pathogen [5].

5.4.2.1 EXAMPLE

The hyperparasite *Cryptophonectria hypovirus*-1 (CHV1) parasitizes *Cryphonectria parasitica,* which causes chestnut blight disease. It reduces the pathogen's growth rate and virulence to the host by altering the genetic structure of *C. parasitica* populations. Additionally, it allows tree host populations to recover to a near disease-free state [6].

In the case of insects, hyperparasitism is a highly evolved behavior that is restricted to three orders: Hymenoptera, Diptera, and Coleoptera. For example, in the case of *Alloxysta victrix,* an endophagous hyperparasitoid, the female lays eggs using her ovipositor into the thin exoskeleton of an already parasitized aphid. The eggs hatch only after the primary larva has completely devoured the aphid internally, ultimately killing it [7].

5.4.3 VOLATILE ORGANIC COMPOUNDS (VOC) ANTAGONISM

Volatile organic compounds (VOC) are the primary defense mechanisms of plants, enhancing their resistance and tolerance against herbivorous insects, pathogens, and environmental stresses. These compounds are considered to be chemicals that have potent effects on microbes and can also act as allelopathic agents, reducing pathogen and pest attacks while interacting with plant hormones to regulate plant growth.

Microbial biological control agents (MBCA), particularly bacteria, play a significant role in various mechanisms that protect plants from diseases and pests. The secretion of VOC is one of these mechanisms, and it serves an important function in biological control. VOCs have the ability to diffuse over long distances, limiting fungal proliferation and inducing systemic resistance.

5.4.3.1 EXAMPLE

It has been found that the volatile compound methanol, which is emitted from the leaves of the plant during cell division and cell wall expansion, acts as a primary stimulus when released from damaged tobacco leaves. This compound helps enhance resistance against the pathogenic bacterium *Ralstonia solanacearum*. In in-vitro conditions, it was observed that the volatile compound linalool moderately inhibits the citrus post-bloom fruit drop caused by *Colletotrichum acutatum* [8].

The glasshouse whitefly (*Trialeurodes vaporariorum*) is a common and persistent pest of horticultural and ornamental plants, especially tomatoes (*Solanum lycopersicum*). Analysis of tomato herbivore-induced plant volatiles (HIPVs) led to the selection of methyl salicylate (MeSA) as the plant elicitor. Applying MeSA to uninfested tomato plants successfully inhibits whitefly population development and enhances yield by 11% [9].

5.4.4 MYCOPARASITISM

A mycoparasite is an organism that has the ability to parasitize fungi. Depending on the type of interaction with their host, mycoparasites can be grouped into two types: biotrophic or necrotrophic. Mycoparasitism can be defined as a direct antagonistic mechanism between an endophyte and a pathogen, in which the endophytic microbes feed on the plant pathogen. There are four important steps involved in mycoparasitism activity: targeting the location, recognition of the host, contact and penetration into the host, and nutrient acquisition from the host. In soil, there are many microorganisms that, when they come in contact with plants, act as entomopathogens. These microorganisms can be effectively utilized as biocontrol agents or biopesticides against a variety of agricultural pests. Haustoria are specialized structures employed by hyperparasites to utilize pathogen's cells as a source of nutrients. For mycoparasitism to occur, two interacting species should be in close proximity to each other. There are approximately 100 described species of *Trichoderma,* which are widely distributed throughout the world due to their rapid growth, ability to utilize different substrates, and tolerance to different contaminants and environmental conditions [10].

5.4.4.1 EXAMPLE

It has been observed that *Ampelomyces quisqualis* inhibits and reduces the growth of powdery mildew fungi by parasitizing both its conidia and hyphae.

In another case, the fungi *Coniothyrium minitans* have been found to destroy sclerotia of *Sclerotinia* spp., thus controlling the pathogen [11].

Certain species of Trichoderma have been found to control the insect species cotton jassid, specifically *Amrasca bigutulla bigutulla,* by releasing toxic secondary metabolites [12]. Additionally, *T. viridae* controls the population of *Bombyx mori* by producing antifeedant compounds like chitinases [13].

5.4.5 LYSIS BY EXTRACELLULAR ENZYMES

Lysis can be defined as a process in which the disintegration of a cell takes place by rupturing the cell wall or membrane. It plays an important role in many biological contexts, including the digestion of food by an organism, immune defense against bacteria and viruses, viral replication, and biofilm development. Lysis enzymes help bacteria, viruses, and animal cells invade other cells. There are many lytic enzymes such as chitinases, β-1,3-glucanases, proteases, etc., which activate plant defense mechanisms. These enzymes are usually extracellular, low molecular weight, and highly stable. In humans, the enzyme lysozyme is found in saliva, which lyses bacteria. Additionally, antibiotics like penicillin synthesize an enzyme called β-lactamase that attacks the β-lactam ring of bacteria, causing destruction of their cell wall and resulting in the death of the bacteria. Hence, penicillin is widely used in controlling bacterial disease. Similarly, bacteriophages (viruses that infect bacteria) produce lytic enzymes that rupture the bacterial cell wall and permit the transfer of viral particles (virions) into bacterial host cells. Chitinases are regarded as an effective tool for the complete degradation of mycelia or conidial walls of phytopathogenic fungi.

5.4.5.1 EXAMPLES

Trichoderma is one of the potent biocontrol agents that causes localized lysis of the cell wall of many phytopathogenic fungi at the point of contact between the antagonistic and the host [17]. The cell wall of many fungi consists of chitin or fibrils of β-glucan embedded in the protein matrix. This suggests that the lysis of their hyphae during parasitism is likely affected by β-glucanases, chitinases, and proteases [18]. The species of Trichoderma produces extracellular hydrolytic enzymes that degrade the cell walls of phytopathogens, thus being responsible for the direct attack against the pathogens.

The role of *Bacillus thuringiensis* (Bt) cry genes in controlling Lepidopteran pests has gained worldwide importance. They control insect pests like *Helicoverpa armigera* and *Plutella xylostella* by releasing chitinases such as ChiA and Chi96023 5-459 [19, 20]. Similarly, the fungal biocontrol agent *Beauveria bassiana* produces chitinase enzymes like BbChit1 and chimeric BbChit1, which help in controlling insect pests like *Galleria mellonella* and *Myzus persicae* [21].

5.4.6 MYCOVIRUS-MEDIATED CROSS-PROTECTION (MMCP)

Mycoviruses are viruses that infect fungi, also known as mycophages. The majority of mycoviruses have double-stranded RNA (dsRNA) genomes and isometric particles. However, approximately 30% have positive-sense, single-stranded RNA (+ssRNA) genomes. These viruses have a unique characteristic of lacking an extracellular phase in their life cycles, relying solely on internal mechanisms for transmission. Mycophages are primarily transmitted through spores (vertical transmission to the progeny) and hyphal anastomosis (horizontal transmission between different fungal isolates) [22].

Mycoviruses from plant pathogens have gained significant attention as potential biocontrol tools. They can induce hypovirulence (reduced virulence) in their host fungi when incorporated into them. The successful application of these viruses in disease management greatly depends on their ability to spread in the natural population of pathogens. Since most viruses lack an extracellular replication phase, researchers have developed various transformation and transfection techniques. These techniques include cDNA infectious clones [23], dsRNA [24], *in vitro*-transcribed RNA transcripts [25], and purified virions. These advancements have accelerated the identification of the host and the viral factors involved in the interaction between mycoviruses and their hosts.

There are four basic criteria that must be fulfilled by a mycovirus in order to act as a biocontrol agent. They are:

1. The mycovirus must reduce the virulence of the pathogen and cause hypovirulence by suppressing at least one of the phases in the disease cycle of the fungal pathogen: The mechanism can target both the infection or reproduction process of the pathogen, and this kind of suppression process, either alone or in combination, is found to successfully disturb the disease epidemic by reducing the severity of the disease.

2. It must be transmissible into the fungal pathogen that causes the disease: The majority of mycoviruses do not possess an extracellular phase and can only be transmitted between fungal individuals through hyphal anastomosis (fusion) and cytoplasmic exchange. However, extracellular viral particle transmission is possible, although it appears to be an exception.
3. Must be able to overcome the mechanism of cellular defense induced against the virus infection by the fungi: RNA silencing is an excellent antiviral defense mechanism demonstrated by fungi upon viral infection. To overcome this, the viral RNA within the fungal host is recognized by the mycovirus and broken down into small RNA fragments, ultimately degrading it and suppressing the RNA silencing pathway.
4. Able to replicate and spread through the fungal mycelia: Mycelium is a network of hyphae that make up the filamentous thallus of a fungus. Mycovirus can interfere with the growth of mycelium and the infection process of fungi by spreading the virus to actively growing hyphae and inhibiting their growth. This enables horizontal transmission of the virus, while vertical transmission occurs through reproductive structures and spores.

5.4.6.1 EXAMPLES

To control stem rot in rapeseed (*Brassica napus*), the introduction of a potential mycovirus, namely *Sclerotinia sclerotiorum* hypovirulence-associated DNA virus 1 (SsHADV-1), has been proposed [26].

Fusarium is a significant soil-borne fungal pathogen that causes severe diseases such as wilt. The utilization of advanced RNA deep sequencing techniques has led to the discovery of a wide range of mycoviruses in *Fusarium*. To date, mycoviruses from 13 *Fusarium* species have been fully sequenced, including *Fusarium asiaticum, F. boothii, F. circinatum, F. coeruleum, F. globosum, F. graminearum, F. incarnatum, F. langsethiae, F. oxysporum, F. poae, F. pseudograminearum, F. solani,* and *F. virguliforme* [27].

5.4.7 BACTERIOCINS

Bacteriocins are extracellular proteinaceous or peptidic compounds produced by bacteria that have specific morphological and biochemical features. These compounds range from extremely low molecular weight complexes to high molecular weight complexes, and they are produced to inhibit the growth of similar or closely related bacterial strains. Most bacteria are found to produce a wide range of compounds, such as lytic enzymes, metabolic by-products, proteinaceous exotoxins, ribosomally and non-ribosomally synthesized antibiotics, etc. These compounds help defend against microbial competitors and also mediate local population dynamics. Bacteriocins were first discovered by Andre Gratia in 1925 [28]. Many bacteria produce at least one bacteriocin during their lifetime. These bacteriocins are usually synthesized from plasmids, but many are also made from chromosomes at different stages of the bacterial growth life cycle, under different environmental conditions [29].

5.4.7.1 CLASSIFICATION OF BACTERIOCINS

5.4.7.1.1 Bacteriocins Produced by Gram-Negative Bacteria

They produce two types of bacteriocins, and they are *Colicins* and *Microcins*:

1. **Colicins:** These are the antibacterial proteins produced by bacteria that can kill closely related bacterial strains to reduce competition for nutrients and living space in the environment. Colicins are divided into three domains: an amino-terminal translocation (T) domain that is involved in the transfer of information across the outer membrane *via* the translocator protein; a central receptor-binding (R) domain that binds to a bacterial outer membrane receptor; and a carboxy-terminal cytotoxic (C) domain that exhibits antibacterial activity. Simultaneously, specificity immunity proteins are created to inactivate colicins and prevent self-poisoning by colicins produced by the same bacteria [30].

   It has been observed that the presence of the colicins recognition receptors protein and the translocators protein system on the surface of a bacterial outer membrane allows the colicins to enter the bacteria, resulting in its death. These bacteria are referred to as sensitive strains. Bacteria that lack the receptor protein are classified as resistant strains to a specific colicin. Bacteria that do not possess the translocator protein system are known as tolerant strains, while bacteria that produce immunity proteins are called immune strains. Bacterial strains that are resistant, tolerant, or immune to colicins cannot be eliminated by them.

2. **Microcins:** These are low molecular weight hydrophobic antimicrobial precursor peptides that are synthesized by ribosomes consisting of an N-terminal leader peptide and a core peptide. They are mostly produced by Enterobacteriaceae, which exhibit great tolerance to high heat, pH, and protease [31]. These peptides have a variety of bactericidal mechanisms, including pore-forming, nuclease activity such as DNase and RNase functions, and inhibitors of protein synthesis or DNA replication. Microcins do not possess lysis genes and are released outside the bacteria through the Type I ABC (ATP binding cassette) transporter secretion system, which is composed of several proteins [32].

5.4.7.1.2 Bacteriocins Produced by Gram-Positive Bacteria

Bacteriocins produced by gram-positive bacteria show similar characteristics to those of microcins. These are gene-encoded bacteriocins, generally of low molecular weight antimicrobial peptides consisting of less than 60 amino acids. Among gram-positive bacteria, lactic acid bacteria (LAB) are the typical bacteria that produce a variety of bacteriocins of different sizes, structures, physicochemical properties, and inhibitory spectra (Figure 5.1). Due to their large diversity, they are generally divided into three classes:

- **Class I:** These are modified peptides and are lantibiotics.
- **Class II:** These are unmodified peptides and are non-lanthionine.
- **Class III:** These are large proteins that are heat unstable.

5.4.7.1.3 Example

The bacteria *Pseudomonas putida* WCS358 has been reported to suppress and control *Fusarium* wilt of radish by effectively competing for iron through the production of its pseudobactin siderophore [33]. It also helps in controlling the pathogen *Rhizoctonia solani* in cucumber.

Bacillus thuringiensis (Bt) strain HD-2 produces bacteriocin, known as Thuricin, which is widely used to control lepidopteran pests [34]. Similarly, another strain of *Bacillus thuringiensis* sp. *kurstaki* Bn1 produces bacteriocin Bn1, which controls the hazelnut pest *Balaninus nucum* [35].

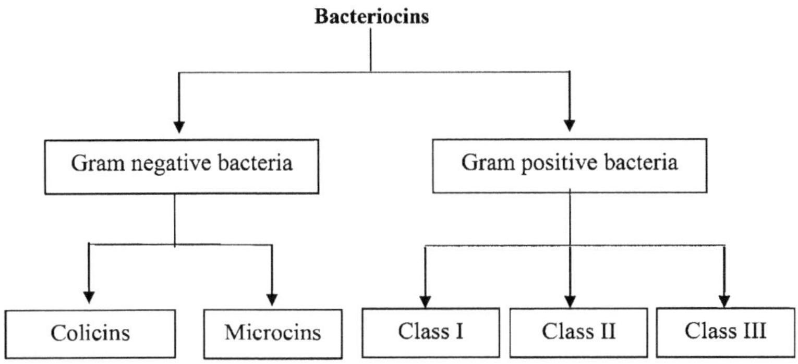

FIGURE 5.1 Flow chart representing bacteriocins produced by both gram-negative and gram-positive bacteria.

5.4.8 INDUCTION OF SYSTEMIC RESISTANCE

It can be of three types. They are discussed in subsections.

5.4.8.1 SYSTEMIC ACQUIRED RESISTANCE (SAR)

"Systemic acquired resistance" refers to the induced resistance that is activated throughout the plant when exposed to certain pathogenic infections, which produce elicitors from virulent, avirulent, or nonpathogenic microbes. It is activated by necrotizing pathogens and an SA-dependent signaling pathway, resulting in increased levels of Salicylic Acid (SA) along with the activation of pathogen-related (PR) proteins. It has been found that the nonpathogenic rhizobacterium *Pseudomonas fluorescens* strain SS101 inhibits the growth of oomycetes and fungal pathogens by inducing plant resistance to certain diseases and pests. *Pseudomonas fluorescens* SS101 induces resistance against bacterial speck disease of tomato caused by *Pseudomonas syringae* pv. tomato, as well as against *Spodoptera exigua,* an herbivorous insect pest [36]."

The plant growth-promoting rhizobacteria (PGPR) *Bacillus cereus AR156* has been observed to induce systemic acquired resistance (SAR) against *Pseudomonas syringae* pv. *tomato,* a leaf pathogen of *Arabidopsis*. This induction was found to be dependent on salicylic acid (SA) and NPR1 (nonexpressor of pathogenesis-related genes), but not on jasmonic acid (JA) and ethylene (ET) pathways [37].

5.4.8.2 INDUCED SYSTEMIC RESISTANCE (ISR)

Plants have a defense mechanism called induced systemic resistance (ISR) that is mainly triggered by infection. Its mode of action relies more on strengthening the physical or chemical barriers of the host plant than on directly eradicating or inhibiting the invading pathogen. The resistance in the plant is induced against several insect herbivores and pathogenic infections, which is mediated by ethylene (ET) and jasmonic acid (JA). There are various other compounds known for inducing resistance, such as proteins with enzymatic activity like xylanases, cellulases, and endo-chitinase. One of the safest modes for ISR and growth promotion in plants was reported as the potential defense responses of plant growth-promoting fungi (PGPF) such as *Trichoderma, Penicillium, Aspergillus,* and *Fusarium* spp. against diseases and pests. The key plant hormones ET and JA regulate ISR during tri-trophic interactions [38]. The main ISR pathway is JA signaling, which defends against herbivorous insects mainly on leaf-chewing pests.

The roots of arabidopsis plants treated with rhizobacteria were observed to show resistance to leaf-feeding insects with the enhanced expression of the JA-dependent gene LOX2 [39].

5.4.8.3 MINDUCED SYSTEMIC TOLERANCE (IST)

The term Induced Systemic Tolerance (IST) was proposed by Yang et al. [40]. It is activated by PGPR-derived elicitors such as phytohormones, volatiles, enzymes, antioxidants (AOX), and exopolysaccharides, which induce physical and chemical changes in crop plants. One of the most common pathways of IST is the synthesis of ACC (1-aminocyclopropane-1-carboxylate) deaminase by plant-associated bacteria. In response to abiotic or biotic stress, plants produce ET, which can limit the expansion of the plant shoot and root, induce chlorosis and leaf abscission, inhibit leaf expansion, or promote epinasty[41].

ACC deaminase, produced by related bacteria, converts ACC, an immediate precursor of ET, into ammonia and α-ketobutyrate. This enzyme decreases the level of stress ET and enhances plant development under stress conditions.

5.5 TYPES OF DEFENSE RESPONSE ACTIVATED IN PLANTS UPON INTERACTION WITH MICROBIAL BIOSTIMULANTS

5.5.1 NITROGEN FIXATION

Plants require nitrogen for their growth and development, as well as for activating defense mechanisms against various abiotic and biotic stresses. An increase in nitrogen can either enhance or reduce the resistance of plants to pathogens, reflecting the infection strategies of the pathogens. Studies have shown that tomato plants exhibit increased resistance against *Fusarium oxysporum* f. sp. *lycopersici* and *Botrytis cinerea* when nitrogen application is increased, indicating the presence of specific and complex pathogen–nitrogen interactions [42]. An increase in nitrogen content contributes to plant defense at every level, whether constitutive or induced. The availability and form of nitrogen not only affect disease development, but also the resistance of plants against diseases. For example, NH^{4+} nutrition has been reported to increase the symptoms of black root rot in sugar beets caused by *Rhizoctonia solani,* as well as *Fusarium* wilt in tomatoes [43]. Nitrogen also activates other defense-associated features, such as an increase in polyamine levels, namely putrescine, spermidine, and spermine production [44]. These polyamine molecules have been found to influence the cell cycle, DNA, and protein synthesis, as well as programmed cell death (PCD) [45], subsequently contributing to PTI (pathogen triggered immunity) and ETI (effector-triggered immunity).

5.5.2 PHOSPHATE SOLUBILIZATION

Phosphorus is considered one of the major nutrients for plants, acting as a vital component of ATP, the energy currency of plant cells, which aids in photosynthesis. Compounds containing phosphorus induce a defense mechanism in plants by inhibiting phosphorylation and competing for phosphate binding sites in phosphorylating enzymes. This leads to alterations in the nucleotide pool of the pathogen, ultimately disrupting its metabolism and inhibiting its growth. Additionally, phosphorus helps activate defensive molecules such as phytoalexins and PR proteins, which directly block pathogens. It also sends systemic alarm signals to nearby non-infected neighboring cells, inducing a defensive response mechanism that includes modifying the cell wall through polysaccharide deposition. Studies have shown that the application of phosphate can confer resistance to infections caused by certain necrotrophic and hemibiotrophic fungal pathogens [46].

5.5.3 SIDEROPHORE PRODUCTION

Siderophores are low molecular-weight organic molecules synthesized by microbes and plants in low or iron-deprived environments. The major function of siderophores is to chelate ferric iron [Fe (III)] from various terrestrial and aquatic environments, making it accessible to microbes and plant cells. The removal of Fe^{3+} by the siderophores from the root region reduces the amount of iron necessary for the growth of pathogens like *Fusarium oxysporum, Pythium ultimum,* and *Fusarium udum,* which cause root rot and wilt diseases of crops [47]. Siderophore-mediated competition for iron deprives pathogens of iron, resulting in a decrease in the incidence of the disease. They are found to be produced by some *Trichoderma* strains, which can restrict the growth and activity of soil-borne pathogens and post-harvest diseases like *Botrytis cinerea,* which causes botrytis bunch rot [48].

Pseudomonas putida WCS358 a plant growth-promoting rhizobacteria (PGPR) produces the siderophore pseudobactin 358 (PSB358), which is effective in the regulation of *Fusarium oxysporum* causing *Fusarium* wilt disease incarnation [49].

5.5.4 PHYTOHORMONE PRODUCTION

Plant hormones play a critical role in growth and development. Several microbial biostimulants (MBs) can produce phytohormones such as auxin, gibberellins, cytokinins, abscisic acid (ABA), and ET, which help the plant develop strength, vitality, and stress tolerance. Inoculation with MB enhances plant development by stimulating the synthesis of lateral roots and root hairs through the production of plant growth-promoting chemicals, enabling plants to absorb more water and nutrients.

Furthermore, MBs directly influence the production of endogenous hormones in plants, affecting plant physiology by altering the phytohormone biosynthesis pathway [50]. In sugarcane, infection with mosaic virus leads to increased expression of the auxin binding protein 1 (ABP1) gene [51]. Similarly, infection with bamboo mosaic virus results in increased ABA content [52].

5.5.5 *HYPERSENSITIVITY*

Hypersensitive response is a type of natural defense mechanism activated by plants in response to various pathogens, aiming to prevent their spread. It is characterized by programmed cell death (PCD) in the region surrounding the infection site on the plant's surface, accompanied by the accumulation of toxic compounds within the dead cells. This restricts the further growth and spread of the pathogens to other parts of the plant. The ability of a pathogen to suppress hypersensitivity plays a critical role in determining whether it can infect a plant or not. Hypersensitivity is a potential target for enhancing the overall resistance of crops to pathogens. Additionally, hypersensitive reactions (HR) such as autophagy, apoptosis, and necrosis can also occur. The HR system is particularly complex to study due to signals from pathogen-associated molecular patterns (PAMPs) and effector-triggered immune pathways, as well as reactive oxygen species (ROS) and phytohormone signals. HR can be classified into structural hypersensitive response and induced hypersensitive response [53].

5.5.6 *REACTIVE OXYGEN SPECIES (ROS)*

Plants use ROS to defend themselves against a range of diseases and pests [54]. The three principal types of ROS are superoxide anion (O_2^-), hydrogen peroxide (H_2O_2), and hydroxyl (OH) radical. These compounds are highly reactive and hazardous, and they can cause oxidative cell death. The rapid accumulation of plant ROS at the site of pathogen attack leads to an oxidative burst, which is detrimental to pathogens directly and can trigger a hypersensitive response (HR) that prevents further spread of biotrophic diseases [55, 56].

In soybean crops infected with the maize earworm *Helicoverpa zea,* there is an occurrence of ROS detoxification, which reduces antioxidant levels but increases harmful oxidation components [57]. Insect feeding triggers oxidative stress responses, which are critical components of plant defense against invading insects. Plants use ROS and local cell death to protect themselves against the phloem sap-feeding green peach aphid. Large concentrations of H_2O_2 and certain other ROS in plants can kill insects directly by damaging their intestinal walls. The death of green peach aphids after eating artificial diets containing H_2O_2 further supports the hypothesis of ROS impacts [58].

- **Advantages:**
  - It is a potential, long-term, cost-effective, and successful method.

- It is an eco-friendly mechanism that enhances soil health, leading to an increase in beneficial soil microorganisms.
- The use of microbial biostimulants helps promote the conservation of the agro-ecosystem and the ecological balance.
- The use of biostimulants enhances the nutrient levels, uptake, and water use efficiency of crops.
- The use of biostimulants increases the crop's tolerance to various biotic and abiotic stresses.
- Improvement of plant health and crop quality, which leads to an increase in harvestable yields.

➢ **Disadvantages:** There are several challenges in the use, selection, and evaluation of microbial biostimulants. The most frequently encountered challenges are as follows:

- The efficiency of a product is strongly dependent on the type of agricultural crop employed, its phenological stage, and its surrounding environment.
- The efficiency of biostimulants will be reduced due to competition between applied microbial strains and the indigenous microbiota for space and nutrients.
- There are several constraints in the development of a new novel microbial bio-stimulant due to complicated business registration procedures.
- A better understanding of microbial activities and interaction processes must be present for the selection of appropriate microbial biostimulants for crops.
- There are only very few accepted protocols.
- The adoption of microbial biostimulants has certain drawbacks, such as lower efficacy and increased environmental sensitivity compared to synthetic products.

5.6 CONCLUSION

Nowadays, microbial-based bioproducts are used under various terms such as biostimulants, bioinoculants, biofertilizers, and biopesticides. They are regarded as key components of ecological sustainability and have significantly increased crop productivity in an environmentally friendly manner. Microbial biostimulants are considered a beneficial agricultural technology and an effective agronomic tool for enhancing crop development and resistance against biotic and abiotic stress. They promote crop growth, health, and quality, which can successfully address the challenges posed by the growing global food demand. However, the progress in the development and practical application of bio-control agents has been somewhat disappointing. Despite more than half a century of intensive research, only a few commercial products have made it to the market, and the initial promise has not been fully realized in practical terms, as bio-control agents have not been able to replace or minimize the use of fungicides and bactericides in crop production systems. To ensure the future success of this novel technique, a more integrated approach is needed, where bio-control agents are seen as part of a comprehensive strategy for disease management. Furthermore, the combination of bio-control agents with different modes of action has proven to be more effective than the application of a single type. Such mixtures can include strains with varying environmental tolerance, thereby expanding the range of conditions in which the formulations will be effective.

KEYWORDS

- **bacteria**
- **biostimulants**
- **environmental tolerance**
- **fungi**
- **microbial bio-agents**
- **mycovirus**
- **nitrogen fixation**
- **phytohormone**

REFERENCES

1. Singh, B. U., & Sharma, H. C., (2002). Natural enemies of sorghum shoot fly, *Atherigona soccata* rondani (*Diptera: Muscidae*). *Biocontrol. Sci. Technol., 12*(3), 307–323.
2. Papavizas, G. C., (1985). *Trichoderma* and *Gliocladium*: Biology, ecology, and potential for biocontrol. *Ann. Rev. Phytopathol., 23*, 23–54.
3. Parratt, S. R., & Laine, A. L., (2016). The role of hyperparasitism in microbial pathogen ecology and evolution. *ISME J., 10*(8), 1815–1822.

4. Holt, R. D., & Hochberg, M. E., (1998). The coexistence of competing parasites. Part II—Hyperparasitism and food chain dynamics. *J. Theor. Biol., 193*(3), 485–495.
5. Taylor, D. R., Jarosz, A. M., Fulbright, D. W., & Lenski, R. E., (1998). The acquisition of hypovirulence in host-pathogen systems with three trophic levels. *Am. Nat., 151*(4), 343–355.
6. Hillman, B. I., & Suzuki, N., (2004). Viruses in the chestnut blight fungus. *Adv. Virus Res., 63*, 423–473.
7. Sullivan, D. J., (1987). Insect hyperparasitism. *Annu. Rev. Entomol., 32*(1), 49–70.
8. Marques, J. P. R., Amorim, L., Silva-Junior, G. J., Spósito, M. B., & Appezzato-Da, G. B., (2015). Structural and biochemical characteristics of citrus flowers associated with defense against a fungal pathogen. *AoB Plants, 7*.
9. Conboy, N. J., McDaniel, T., George, D., Ormerod, A., Edwards, M., Donohoe, P., & Tosh, C. R., (2020). Volatile organic compounds as insect repellents and plant elicitors: An integrated pest management (IPM) strategy for glasshouse whitefly (*Trialeurodes vaporariorum*). *J. Chem. Ecol., 46*(11), 1090–1104.
10. Mishra, N., Khan, S. S., & Sundari, S. K., (2016). Native isolate of *Trichoderma*: A biocontrol agent with unique stress tolerance properties. *World J. Microbiol. Biotechnol., 32*(8), 1–23.
11. Segaran, G., Shankar, S., & Sathiavelu, M., (2022). *Biocontrol Mechanism of Endophytic Microorganisms* (pp. 55–73). AP.
12. Nawaz, A., Gogi, M. D., Naveed, M., Arshad, M., Sufyan, M., Binyameen, M., & Ali, H., (2020). *In vivo* and *in vitro* assessment of *Trichoderma* species and *Bacillus thuringiensis* integration to mitigate insect pests of brinjal (*Solanum melongena* L.). *Egypt J. Biol. Pest. Control., 30*(1), 1–7.
13. Berini, F., Caccia, S., Franzetti, E., Congiu, T., Marinelli, F., Casartelli, M., & Tettamanti, G., (2016). Effects of *Trichoderma viride* chitinases on the peritrophic matrix of *Lepidoptera. Pest Manag., 72*(5), 980–989.
14. Laible, N. J., & Germaine, G. R., (1985). Bactericidal activity of human lysozyme, muramidase-inactive lysozyme, and cationic polypeptides against *Streptococcus sanguis* and *Streptococcus faecalis*: Inhibition by chitin oligosaccharides. *Infect. Immun., 48*(3), 720–728.
15. Ogawara, H., (1981). Antibiotic resistance in pathogenic and producing bacteria, with special reference to beta-lactam antibiotics. *Microbiol. Rev., 45*(4), 591–619.
16. Hasan, S., Gupta, G., Anand, S., & Kaur, H., (2014). Lytic enzymes of *Trichoderma*: Their role in plant defense. *Int. J. Appl. Res. Stud., 3*(2), 1–5.
17. Benhamou, N., & Chet, I., (1993). Hyphal interactions between *Trichoderma harzianum* and *Rhizoctonia solani*: Ultrastructure and gold cytochemistry of the mycoparasitic process. *Phytopathol., 83*, 1062–1071.
18. Kuhn, P. J., Trinci, A. P., Jung, M. J., Goosey, M. W., & Copping, L. G., (2012). *Biochemistry of Cell Walls and Membranes in Fungi*. Springer Science and Business Media.
19. Ni, H., Zeng, S., Qin, X., Sun, X., Zhang, S., Zhao, X., & Li, L., (2015). Molecular docking and site-directed mutagenesis of a *Bacillus thuringiensis* chitinase to improve chitinolytic, synergistic lepidopteran-larvicidal and nematicidal activities. *Int. J. Biol. Sci., 11*(3), 304.
20. Liu, X., Williams, C. E., Nemacheck, J. A., Wang, H., Subramanyam, S., Zheng, C., & Chen, M. S., (2010). Reactive oxygen species are involved in plant defense against a gall midge. *Plant Physiol., 152*(2), 985–999.

21. Fang, W., Leng, B., Xiao, Y., Jin, K., Ma, J., Fan, Y., & Pei, Y., (2005). Cloning of *Beauveria bassiana* chitinase gene Bbchit1 and its application to improve fungal strain virulence. *AEM.*, *71*(1), 363–370.
22. Pearson, M. N., Beever, R. E., Boine, B., & Arthur, K., (2009). Mycoviruses of filamentous fungi and their relevance to plant pathology. *Mol. Plant Pathol.*, *10*(1), 115–128.
23. Choi, G. H., & Nuss, D. L., (1992). Hypovirulence of chestnut blight fungus conferred by an infectious viral cDNA. *Sci.*, *257*(5071), 800–803.
24. Stanway, C. A., & Buck, K. W., (1984). Infection of protoplasts of the wheat take-all fungus, *Gaeumannomyces graminis* var. *tritici*, with double-stranded RNA viruses. *J. Gen. Virol.*, *65*(11), 2061–2065.
25. Chen, B., Choi, G. H., & Nuss, D. L., (1994). Attenuation of fungal virulence by synthetic infectious hypovirus transcripts. *Sci.*, *264*(5166), 1762–1764.
26. Xie, J., & Jiang, D., (2014). New insights into mycoviruses and exploration for the biological control of crop fungal diseases. *Annu. Rev. Phytopathol.*, *52*, 45–68.
27. Li, P., Bhattacharjee, P., Wang, S., Zhang, L., Ahmed, I., & Guo, L., (2019). Mycoviruses in *Fusarium* species: An update. *Front. Cell. Infect. Microbiol.*, *9*, 257.
28. Gratia, A., (1925). Sur un remarquable exemple d'antagonisme entre deux souches de coilbacille. *CR Seances Soc. Biol. Fil.*, *93*, 1040–1041.
29. Daw, M. A., & Falkiner, F. R., (1996). Bacteriocins: Nature, function, and structure. *Micron.*, *27*(6), 467–479.
30. Kleanthous, C., (2010). Swimming against the tide: Progress and challenges in our understanding of colicin translocation. *Nat. Rev. Microbiol.*, *8*(12), 843–848.
31. Rebuffat, S., (2012). Microcins in action: Amazing defense strategies of Enterobacteria. *Biochem. Soc. Trans.*, *40*(6), 1456–1462.
32. Duquesne, S., Petit, V., Peduzzi, J., & Rebuffat, S., (2007). Structural and functional diversity of microcins, gene-encoded antibacterial peptides from enterobacteria. *Microb. Physiol.*, *13*(4), 200–209.
33. De Boer, M., Bom, P., Kindt, F., Keurentjes, J. J., Van, D. S. I., Van, L. L. C., & Bakker, P. A., (2003). Control of *Fusarium* wilt of radish by combining *Pseudomonas putida* strains that have different disease-suppressive mechanisms. *Phytopathology*, *93*(5), 626–632.
34. Favret, M. E., & Yousten, A. A., (1989). Thuricin: The bacteriocin produced by *Bacillus thuringiensis*. *J. Invertebr. Pathol.*, *53*(2), 206–216.
35. Ugras, S., Sezen, K., Kati, H., & Demirbağ, Z., (2013). Purification and characterization of the bacteriocin thuricin Bn1 produced by *Bacillus thuringiensis* subsp kurstaki Bn1 isolated from a hazelnut pest. *J. Microbiol. Biotechnol.*, *23*(2).
36. Van, W. S. C., Van, D. E. S., & Pieterse, C. M., (2008). Plant immune responses triggered by beneficial microbes. *Curr. Opin. Plant Biol.*, *11*(4), 443–448.
37. Niu, D., Wang, X., Wang, Y., Song, X., Wang, J., Guo, J., & Zhao, H., (2016). *Bacillus cereus* AR156 activates PAMP-triggered immunity and induces a systemic acquired resistance through a NPR1-and SA-dependent signaling pathway. *Biochem. Biophys.*, *469*(1), 120–125.
38. Shavit, R., Ofek-Lalzar, M., Burdman, S., & Morin, S., (2013). Inoculation of tomato plants with rhizobacteria enhances the performance of the phloem feeding insect *Bemisia tabaci*. *Front. Plant Sci.*, *4*, 306. doi: 10.3389/fpls.2013. 00306.

39. Pangesti, N., Pineda, A., Dicke, M., & Van, L. J. J. A., (2015a). Variation in plant-mediated interactions between rhizobacteria and caterpillars: Potential role of soil composition. *Plant Biol., 17*, 474–483.
40. Yang, J., Kloepper, J. W., & Choong-Min, R., (2009). Rhizosphere bacteria help plants tolerate abiotic stress. *Trends Plant Sci., 14*(1).
41. Carlson, R., Tugizimana, F., Steenkamp, P. A., Dubery, I. A., Hassen, A. I., & Labuschagne, N., (2020). Rhizobacteria-induced systemic tolerance against drought stress in *Sorghum bicolor* (L.) Moench. *Microbiol. Res., 232*, 126388.
42. Hoffland, E., Jeger, M. J., & Van, B. M. L., (2000). Effect of nitrogen supply rate on disease resistance in tomato depends on the pathogen. *Plant Soil, 218*(1), 239–247.
43. Borrero, C., Trillas, M. I., Delgado, A., & Avilés, M., (2012). Effect of ammonium/nitrate ratio in nutrient solution on control of *Fusarium* wilt of tomato by *Trichoderma asperellum* T34. *Plant Pathol., 61*(1), 132–139.
44. Gupta, K. J., Brotman, Y., Segu, S., Zeier, T., Zeier, J., Persijn, S. T., & Mur, L. A., (2013). The form of nitrogen nutrition affects resistance against *Pseudomonas syringae* pv. *phaseolicola* in tobacco. *J. Exp. Bot., 64*(2), 553–568.
45. Tiburcio, A. F., Altabella, T., Bitrián, M., & Alcázar, R., (2014). The roles of polyamines during the lifespan of plants: From development to stress. *Planta, 240*(1), 1–18.
46. Torregrosa, B. V., Bundo, M., Cardoso, H. M., Pages, M. B., Chiou, T. J., Flors, V., & San, S. B., (2021). *Phosphate-Induced Resistance to Pathogen Infection in Arabidopsis*. bioRxiv.
47. Loper, J. E., (1988). Role of fluorescent siderophore production in biological control of *Pythium ultimum* by a *Pseudomonas fluorescens* strain. *Phytopathology, 78*(2), 166–172.
48. Chet, I., & Inbar, J., (1994). Biological control of fungal pathogens. *Appl. Biochem. Biotechnol., 48*(1), 37–43.
49. Lemanceau, P., Bakker, P. A., De Kogel, W. J., Alabouvette, C., & Schippers, B., (1992). Effect of pseudobactin 358 production by *Pseudomonas putida* WCS358 on suppression of *fusarium* wilt of carnations by non-pathogenic *Fusarium oxysporumFo47. AEM., 58*(9), 2978–2982.
50. Ruzzi, M., & Aroca, R., (2015). Plant growth-promoting rhizobacteria act as biostimulants in horticulture. *Sci. Hortic., 196*, 124–134.
51. Leng, P., Ji, Q., Asp, T., Frei, U. K., Ingvardsen, C. R., Xing, Y., Studer, B., Redinbaugh, M., et al., (2017). Auxin binding protein 1 reinforces resistance to sugarcane mosaic virus in maize. *Mol. Plant, 10*, 1357–1360.
52. Alazem, M., Lin, K. Y., & Lin, N. S., (2014). The abscisic acid pathway has multifaceted effects on the accumulation of Bamboo mosaic virus. *Mol. Plant Microbe Interact., 27*, 177–189.
53. Thakur, A., Verma, S., Reddy, V. P., & Sharma, D., (2019). Hypersensitive responses in plants. *Agric. Rev., 40*(2), 113–120.
54. Mittler, R., Vanderauwera, S., Gollery, M., & Breusegem, F. V., (2004). Reactive oxygen gene network of plants. *Trends Plant Sci., 9*, 490–498.
55. Lamb, C., & Dixon, R. A., (1997). The oxidative burst in plant disease resistance. *Annu. Rev. Plant Physiol. Plant Mol. Biol., 48*, 251–275.
56. Heath, M. C., (2000). Hypersensitive response-related death. *Plant Mol Biol., 44*, 321–334.

57. Bi, J. L., & Felton, G. W., (1995). Foliar oxidative stress and insect herbivory: Primary compounds, secondary metabolites, and reactive oxygen species as components of induced resistance. *J.Chem. Ecol., 21*, 1511–1530.
58. Goggin, F. L., & Fischer, H. D., (2021). Reactive oxygen species in plant interactions with aphids. *Front. Plant Sci., 12*, 811105.

CHAPTER 6

Rhizobacteria as a Potential Bioagent Against Root-Galling Disease in Vegetable Crops

MARWA M. EL-DERINY,[1] DINA S. S. IBRAHIM,[1] and NABIL S. FARAG[2]

[1]*Nematode Diseases Research Department, Plant Pathology Research Institute, Agricultural Research Center (ARC), Giza, Egypt*

[2]*Bacterial Diseases Research Department, Plant Pathology Research Institute, Agricultural Research Center (ARC), Giza, Egypt*

ABSTRACT

Rhizobacteria have gained international recognition and approval for their agricultural benefits over the past decade, particularly for their diverse modes of action that contribute to increased crop yield. This is in response to the growing demand for sustainable agriculture, which emphasizes the importance of growth and environmental preservation, as well as the need to reduce reliance on synthetic chemical solutions. Both the biopesticide market and the number of studies on microbial control of plant diseases through direct and/or indirect antagonistic processes are expanding.

Rhizobacteria can effectively reduce the impact of root-knot nematode (RKN) on plants. Direct antagonism includes predation, competition for resources, the release of nematicidal metabolites and semiochemicals, and niche restriction. However, the outcomes of direct antagonism may vary due to unidentified endogenous and external factors that could hinder rhizobacteria colonization of plant roots.

On the other hand, induced systemic resistance (ISR) can lead to indirect antagonistic effects by priming entire plants to better combat various biotic

Microbial Biostimulants: Biorational Pesticides for Management of Plant Pathogens.
Faheem Ahmad and Rakesh Pandey (Eds.)

and abiotic constraints. This results in faster and/or stronger protective responses (adaptation), enhancing their potential as biocontrol agents. Several studies have utilized rhizobacterial inoculants to evaluate different strains of rhizobacteria for their antagonistic efficacy against root-galling disease, considering soil microbial activity and rhizosphere bacterial populations.

This literature review provides an overview of rhizobacteria-based biological control of economically significant root galling disease in vegetable crops.

6.1 INTRODUCTION

The main culprit behind galling root disease is the root-knot nematode (RKN), *Meloidogyne* spp., one of the most significant economic diseases affecting many plants. Microscopically tiny roundworms known as RKN are among the most prevalent animals on the planet and are members of the phylum Nematoda. The moisture coating that covers soil particles and plant roots is home to the RKN. RKN accesses root tissues using protruding stylets or a mouth spear [1]. This particular nematode species is endoparasitic and can infect plant roots. It can be found worldwide and is known to severely damage crops that are crucial to the economy. There are over 90 species of RKN known, but just four, *M. arenaria, M. javanica, M. hapla,* and *M. incognita* together with a few newly discovered species like *M. chitwoodi* and *M. enterolobii* are responsible for the majority of crop damage [2]. Other nematodes can cause harm to plants; however, the amount lost varies from nation to nation depending on the kind of crop infected. Because they often attack the underground portions of plants and live in soil, RKN are more challenging to control than other pests. Agriculture did not suddenly develop and spread to every corner of the globe; rather, it was the outcome of an evolutionary process that gradually transformed plants from independent, wild progenitors into completely dependent, domesticated cultivars while also enhancing agro profitability. The assumption that soil is the foundation of agriculture is supported by the link between people, the environment, and food supplies. It emphasizes the significance of microbes in maintaining our agricultural and natural ecosystems. Since ancient times, people have used soil bacteria to promote plant development. More recently, scientists researched rhizosphere root colonization and proposed that soil bacteria have the ability to transform atmospheric N_2 into forms that plants can use [3]. The introduction of legumes on cultivated ground increased the soil's fertility. The availability of nutrients in the soil has been increased by using a variety of microorganisms. The growing plant's rhizosphere is impacted by a variety of microorganisms that congregate in, on, and within the recently grown roots. They interact with the plant roots in both positive and detrimental ways. Rhizobacteria are the specific bacteria that live in plant roots.

Plant growth-promoting rhizobacteria, or PGPRs, are another name for rhizobacteria. Numerous researchers have assessed the ability of rhizobacteria to promote plant growth [4, 5]. Rhizobacteria have also been defined differently by several researchers. PGPR refers to soil bacteria that colonize plant roots after being injected into seeds and promote plant growth [6]. PGPR, as described, is a broad set of soil bacteria that are crucial in the interactions between soil and plants in the rhizosphere [7]. These interactions have a variety of effects on plant production and growth. The growth and yields of legumes and nonlegume crops have been improved with PGPR, which involves symbiotic and free-living relationships.

6.2 RHIZOBACTERIA FOR NUTRIENT ACQUISITION

Soils with dynamic microbial ecosystems and high levels of organic matter typically require less fertilizer compared to other soil types that are conventionally maintained [8]. Rhizosphere bacteria play a crucial role in controlling nutrient acquisition and utilization, nutrient transformation, and crop sustainability [9]. For example, when applying organic nitrogen sources, the activity of bulk microbial communities in soils is often taken into consideration. Research on the phytomicrobiome is beginning to uncover specific plant-microbe relationships that directly enhance plant nutrition [10]. Microbes that assist plants in acquiring nutrients (biofertilizers) perform various actions, including increasing root access to nutrients, solubilizing phosphorus, fixing nitrogen, producing HCN, and producing siderophores [11]. Therefore, managing microbial activity presents a significant opportunity to supply crops with the necessary nutrients.

The majority of agricultural soils are rich in phosphorus (P), and the majority of it is in insoluble forms. It has been revealed that some Rhizobacteria species create a phosphatase enzyme capable of releasing P bound to certain other elements [12, 13]. To enhance the availability of native soil P, crops are often treated with phosphate rock derived from a few large deposits. Additionally, phosphorus-solubilizing bacteria can assist plants in accessing non-labile phosphorus by liberating it from its refractory forms.

Inorganic P complexed with Fe, Ca, or Al can be dissolved by organic acids or H+ ions released by phosphorus-solubilizing bacteria. Similarly, reactive P can be released from organic substances by phytase, an enzyme created by microorganisms that solubilize phosphorus. The notion that HCN indirectly improves P availability by chelating metals and retaining these geological materials has recently been refuted [14]. Initially, HCN production by rhizobacteria was assumed to stimulate plant growth by reducing pathogens. By utilizing these substances to convert metal species into immobile forms or chelate them for mobility, phosphorus-solubilizing bacteria can minimize metal toxicity and transport these compounds into plant tissues, providing additional opportunities for phytoextraction [15]. Separate research has also discovered microorganisms that can dissolve phosphate, such as the genera *Bacillus* sp. and *Pseudomonas* sp., which may also stimulate the growth of tomato plants in a greenhouse or field [16].

The symbiotic relationship between legumes and rhizobia, which involves nitrogen fixation, has been extensively studied and utilized. In this interaction, legumes provide rhizobia with reduced carbon (C) and create an anaerobic environment necessary for nitrogenase activity, while rhizobia provide legumes with physiologically accessible nitrogen (N). Both legumes and rhizobia undergo significant changes in this relationship. Rhizobia transform from rod-shaped cells to forked, nitrogen-fixing bacteroids as a result of the legume developing a specialized structure called a nodule to accommodate them [17]. According to various assessments, ranging from 20 to 22 Tg N per year [18] to as high as 40 Tg N per year, rhizobial nitrogen fixation plays a significant role in providing nitrogen to agricultural systems worldwide [19]. Nitrogen-fixing bacteria in tomatoes have been shown to greatly enhance plant growth [20].

The commercialization of free-living nitrogen-fixing bacteria, such as *Bacillus polymyxa*, *Azoarcus* sp., *Diazotrophicus* sp., *Burkholderia* sp., *Azotobacter* sp., *Gluconacetobacter* sp., *Herbaspirillum* sp., and specifically *Azospirillum* sp., has gained increasing attention since the early 21st century [21]. Compared to rhizobia, these free-living diazotrophs are capable of supplying nitrogen to a wider range of crop plants [22, 23]. It has been demonstrated that certain bacteria can enhance nitrogen absorption in plants and improve nitrogen use efficiency, even without nitrogen fixation. This is likely due to the fact that larger roots enable plants to access more soil [10].

Other essential nutrients, such as iron (Fe) and zinc (Zn), may limit crop yields. Fe, despite being abundant in some soils, is often inaccessible to plants. To enhance Fe availability, many bacterial species produce siderophores or organic acids [24, 25]. Additionally, siderophores function by depriving harmful microorganisms of Fe [25, 26]. It has been shown that several strains of Zn-mobilizing bacteria can enhance Zn uptake and consequently increase crop productivity. Although the exact mechanisms of zinc-mobilizing organisms are still unknown, they likely produce compounds and organic acids similar to phosphorus-solubilizing microorganisms and iron-mobilizing organisms [27].

6.3 RECENT INCREASES IN VEGETABLE LOSSES BY ROOT GALLING DISEASE

It has been estimated that plant-parasitic nematodes (PPN) generate $80 billion in damage annually [1]. *Meloidogyne incognita* is the most damaging pathogen to crops worldwide, causing a $100 billion annual economic loss [28]. This polyphagous pest is estimated to cause average yield losses of 10% in vegetable crops and 30% in highly susceptible crops such as tomatoes, eggplants, and melons. *Meloidogyne* species negatively affect numerous agricultural crops globally, with significant yield losses in sub- and tropical agriculture [29]. The potential yield loss of RKN on different tomato cultivars has been extensively studied, ranging from 25% to 100% [30]. Both the quantity and quality of marketable outputs are affected. In addition to direct losses from nematode attacks, injured roots may not effectively use water and fertilizer, resulting in irrigation and fertilizer losses [31]. RKN also exacerbate the damage caused by other plant pathogens, impacting global food production [32, 33]. Therefore, it is urgently necessary for all countries to implement efficient and long-lasting management programs to control RKN. However, due to their small size and subterranean attacks, integrated RKN control methods are limited compared to other insect and animal pests. The aboveground symptoms of RKN sometimes resemble those of other diseases, and environmental root-limiting factors can delay accurate detection and appropriate care [34, 35].

6.4 CURRENT MANAGEMENT METHOD FOR ROOT GALLING DISEASE

Tolerant/resistant cultivars, crop rotation, and nematicides are the main components of the present integrated pest management (IPM) initiative aimed at RKN. However, these techniques remain ineffective, expensive, and/or result in various unanticipated ecological and adverse social effects [36, 37]. The use of nematicides such as milbemectin, Oxamyl, ethoprophos, and fluensulfone is currently the most popular method for controlling RKN infections and field infestations [38–40]. Nematicides have proven to be an effective management tool, but they are costly to purchase and use, pose risks to safety, and have negative environmental impacts [35]. To minimize nematode population density, crop rotation has long been promoted as an alternative to traditional cultural tactics; however, it may not be a practical choice for all crops and farmers [41].

It is now more common to find resistant cultivars, which prevent RKN from reproducing as vigorously as with susceptible cultivars but often result in lower yields [42]. Consequently, some organizations have advocated for the introduction of tolerant cultivars that do not reduce galling root diseases but have minimal impact on production potential [43]. However, the main drawback of resistant cultivars is that they increase the RKN field population density. Therefore, when used as a component of an IPM system, the best results are achieved when using tolerant cultivars. Alternative approaches must therefore be considered for the efficient, long-term management of RKN. One new alternative is biological control, which involves beneficial microbes like Rhizobacteria that can mitigate the effects of RKN, other biotic invaders, and abiotic pressures. By promoting the growth and development of many plant species, they can also assist the plant [44].

6.5 CURRENT UTILITY OF RHIZOBACTERIA AS BIOLOGICAL CONTROL AGENTS (BCA) AGAINST ROOT-KNOT NEMATODE (RKN)

Comparing biological control to conventional approaches, recent research has suggested that it might be a more affordable way to manage RKN. Rhizobacteria, in particular, have become potential BCA options [45, 46]. Rhizobacteria are non-pathogenic bacteria that have been demonstrated to encourage plant development and growth through direct and indirect processes under stress-free and stressful conditions [47]. Rhizobacteria are described by the direct mechanism as bio-fertilizers that generate organic compounds that improve soil nutrient uptake or accelerate plant development. Examples of indirect processes in rhizobacteria-dependent biocontrol include Fe chelators (also known as siderophores), exterior cell wall disintegrating enzymes (such as glucanase and chitinase), and the development of antibiotics that may solubilize the pathogen cell wall [48]. According to this hypothesis, several Rhizobacteria, frequently belonging to the genera *Pseudomonas, Bacillus, Agrobacterium,* and *Paenibacillus,* have been shown to decrease root-knot nematode population through a variety of mechanisms, including direct mechanisms like parasitism, processing of antinematicidal metabolites and hydrolytic enzymes (Table 6.1), or indirect mechanisms like inducing systemic resistance (Table 6.2) [46].

6.5.1 DIRECT ANTAGONISM OF RHIZOBACTERIA FOR ROOT KNOT NEMATODE

Extracellular enzymes such as protease and chitinase can be produced by rhizobacteria [49]. Chitin, a linear polymer composed of widely distributed 1,4-N-acetyl-D-glucosamine (GlcNAc) monomers, can be hydrolytically broken down by chitinase. It has been suggested that this enzyme can lyse the chitin-based outer structure of nematodes [50]. Chitinase activity in plants leads to resistance against plant pathogens. Chitinase enzyme activity has been reported to inhibit *M. incognita* [51, 52]. Rhizobacteria capable of producing chitinase and protease enzymes can lyse the stylet of *Meloidogyne* spp. nematodes *in vitro* [53]. In addition to chitinase, protease is another enzyme that contributes to reducing the population of *Meloidogyne* spp. in the soil. According to reports, the protease enzyme can break down RKN cuticles and protein-filled eggs [54, 55].

The growth and reproduction of RKN can also be inhibited by direct antagonistic interactions with rhizobacteria, primarily through predatory behavior and the release of toxins or hydrolytic enzymes such as glucanases, lipases, 2,4-diacetyl phloroglucinol (DAPG), and hydrogen cyanide (HCN) (Table 6.1). The predatory bacterium *Pasteuria penetrans* attacks and colonizes the body cavity of RKN. Once adhered to the cuticle layer of RKN, spores of this species begin to produce germination tubes that penetrate and rupture the phytonematode cell membrane [56].

Furthermore, various rhizobacteria from the *Bacillus* and *Pseudomonas* families produce nematicidal substances. For example, certain strains of *P. fluorescens* reduce the density of phytonematodes in soil or *in vitro* by producing DAPG, a phenolic antiphytopathogenic metabolite [57].

The wild-type (WT) bacterium can inhibit the motility of juvenile PCNs by up to 85%, compared to mutant strains that lack DAPG production. However, the correct functioning and morphology of mitochondria in RKN (*M. javanica* and *M. incognita*) are negatively affected by the volatile chemical HCN produced by *P. aeruginosa* and *P. fluorescens* strain CHA0 [58, 59]. Interestingly, two secondary metabolites, Pyoluteorin and DAPG, released by *P. fluorescens* strain CHA0, significantly reduce the population density of *M. javanica* in a soil environment and *in vitro*, respectively [60, 61].

The exogenous administration of *P. penetrans* showed considerable reductions in *M. javanica* population density by an average of 97.5% [62]. Some *Pasteuria* species, though, have been turned into products for sale. As a result, these products seem to only be effective against one or two specific species, which limits their usefulness in real-world situations. Producers frequently seek out products that offer a greater variety of PPN defensive responses and disease resistance.

Several hydrolytic enzymes produced by rhizobacteria can inhibit the development of cellular organelles, as well as the synthesis and stability of cell walls and membranes [63]. It is believed that *Corynebacterium pauro-metabolu,* chitinases, and certain proteolytic enzymes produced by *B. megaterium, B. firmus, B. cereus, B. licheniformis,* and *B. subtilis* are responsible for suppressing the growth of various RKN (Table 6.1).

However, recent attempts to commercialize rhizobacteria for RKN control have only achieved limited success [64]. The inconsistent performance of rhizobacteria in controlling phytonematodes in the field has been a major concern, likely due to various external and internal factors that restrict rhizobacteria root colonization. Factors such as other plants, diverse rhizospheric metabolites, and various rhizospheric organisms contribute to this issue [65]. Therefore, it is necessary to explore ways to enhance and maximize the ability of rhizobacteria to function as biological control agents (BCA). Numerous studies are now focusing on the potential effectiveness of rhizobacteria's indirect antagonism against plant parasitic nematodes as a practical alternative to strengthen plant defenses against phytonematodes, other pathogens, and abiotic constraints, while also promoting plant growth and development.

TABLE 6.1 A List of Rhizobacteria That Have Direct Antagonistic Effects on Root Knot Nematode

PGPR	Plant	Molecules or Modes	References
Bacillus amyloliquefaciens FZB 42	–	Secondary metabolites	[66]
Bacillus cereus	Tomato	Protease, sphingosine, secondary metabolites, antibiotic production, chitinase, IAA.	[67–69]
Bacillus licheniformis	Tomato, eggplant	Chitinase, protease	[70–72]
Bacillus firmus	Cucumber, tomato,	Secondary metabolites, Sep 1 protease.	[73–78]
Bacillus subtilis	Lettuce, tomato,	Secondary metabolites, hydrolytic enzymes, lipopeptide antibiotics, IAA.	[69, 79–83]
Bacillus pumilus L1	Tomato	Protease, chitinase, IAA	[69, 84, 85]
Bacillus thuringiensis	Tomato	Thuringiensin (β-exotoxin), *Bt* crystal protein (toxin protein).	[86–88]
Bacillus megaterium	Tomato	Protease, secondary metabolites, extracellular hydrolytic enzymes, IAA.	[69, 89–91]
Bacillus coagulans	Tomato, cotton	Hydrolytic enzymes.	[92–94]
Corynebacterium paurometabolus	–	Hydrogen sulfide, chitinase.	[95]
Pseudomonas aeruginosa	Tomato	Hydrogen cyanide (HCN)	[58, 96]
Pseudomonas fluorescens	Groundnut	Indole acetic acid, antibiotic.	[97, 98]
Pseudomonas putida GR12-2	–	Synthesis of phytohormones indole-acetic acid.	[99]
Pseudomonas fluorescens Wood1R	–	Antibiotic 2,4-diacetyl phloroglucinol DAPG.	[61]
Pseudomonas fluorescens CHA0	Tomato, mungbean	DAPG, HCN, extracellular protease, pyoluteorin.	[59, 60, 100]
Pseudomonas stutzeri	Mungbean	HCN	[101]
Pasteuria penetrans	Tomato and cucumber.	Predation	[56, 62, 102, 103]
Paenibacillus illinoisensis KJA-424	–	Chitinase	[104]
Paenibacillus castaneae	Tomato	Antibiotics and secondary metabolites, gelatinase, and chitinase.	[105–107]
Serratia marcescens	–	Volatile metabolites, prodigiosin.	[108]

6.5.2 INDIRECT ANTAGONISM, RHIZOBACTERIA'S INDUCED SYSTEMIC RESISTANCE (ISR) TO THE ROOT-KNOT NEMATODE (RKN)

The term "PGPR-mediated priming" refers to the process of equipping "whole" plants more effectively by inducing quicker and/or more powerful defense reactions (adaptation) on a surface exposed to various biotic and abiotic challenges, within environmental limits. Indirect inhibition of root-knot nematode by Rhizobacteria occurs through ISR, also known as induced systemic resistance. ISR provides plants with non-specific, long-lasting protection and broad-spectrum disease resistance (defense responses) against numerous pathogenic microorganisms, pests, and insect herbivores, such as phytonematodes [109]. In fact, ISR induced by various *Pseudomonas* spp. and *Bacillus* spp., including *B. amyloliquefaciens* and *B. subtilis,* has been shown to suppress plant parasitic nematodes in radish (*Raphanus sativus* L.), cucumber (*Cucumis sativus* L.), and tomato (*Solanum lycopersicum* L.) [109] (Table 6.2).

However, the molecular processes that underlie its manifestations remain obscure. Due to diminished PGPR-induced priming in mutant plants with defects in JA/ET biosynthesis or signaling, the majority of studies to date have focused on the possible functional significance of JA and ET hormones in the development of ISR. Exogenous JA administration may also show systemic resistance to *M. incognita, M. graminicola,* and *M. javanica* in a variety of plant species [110, 111]. However, neither the root colonization by rhizobacteria nor the induction of JA/ET-responsive gene expression in leaves appeared to enhance the creation within systemic (leaf) tissues of (JA) or (ET) [112, 113]. ISR may instead enhance the activity of JA/ET-responsive genes. ISR-permitted VSP, PDF1.2, and HEL are induced more quickly and/or more intensely than JA/ET-responsive genes when other biotic stressors afterward attacked plants that had been exposed to or treated with rhizobacteria (such as microbial diseases, herbivores, or insects) [113, 114].

Nevertheless, some findings from past research supported the opposite. Rhizobacteria's root inoculations quickly sparked JA synthesis and systemic (leaf) tissues' signaling transductions, which in consequence intensified the whole plant's level of defense reactions [115]. Several JA-responsive and biosynthetic genes, including ChiB, PDF1.2, PAL, HEL, LOX, and VSP, were rapidly upregulated in systemic (leaf) tissue across a variety of agricultural plants after *S. marcescens* strain 90-166 and *P. fluorescens* WCS417r were injected into the plants [116]. After that, JA synthesis works in conjunction by using abscisic acid (ABA) signaling to activate certain transcription factors that respond to stress, including OCP3 (overexpressor of cationic peroxidase 3) as well as MYB60, which ultimately regulates stomatal closure and decreases plant growth and yield [117]. The reduction of JA's ability to promote plant growth and development may be explained by these signaling and metabolic cascades [118]. Indeed, exogenous JA application typically inhibits root growth [109, 119, 120], which suggests that (i) ISR uses alternative and/or extra signaling and/or metabolic mechanisms to enhance plant disease resistance without lowering yield potential (in addition to JA-responsive processes); or (ii) resource allocations could be the subject of competition between ISR and PGPR-induced JA signals.

TABLE 6.2 A List of Rhizobacteria That Have Root Knot Nematode Indirect Antagonistic Effects

PGPR	Plant	Target PPN	Modes	References
Bacillus amyloliquefaciens (syn. *Bacillus velezensis*)	Tomato, paper,	*Meloidogyne incognita*	ISR and systemic acquired resistance (SAR).	[133, 135, 139–142]
Bacillus aryabhattai P-3	Tomato	*Meloidogyne incognita*	ISR	[143]
Bacillus cereus	Tomato	*Meloidogyne javanica, Meloidogyne incognita*	ISR	[132, 133]
Bacillus mojavensis	Chinese cabbage	*Meloidogyne incognita*	ISR	[133, 138]
Bacillus pasteurii	–	*Meloidogyne incognita*	ISR	[116, 133]
Bacillus pumilus	–	*Meloidogyne incognita*	ISR and SAR	[123, 131, 133, 134, 136, 137]
Bacillus sphaericus	–	*Meloidogyne incognita*	ISR	[133]
Bacillus subtilis	–	*Meloidogyne arenaria, Meloidogyne incognita, Meloidogyne javanica*	ISR and SAR	[123, 133–136]
Bacillus thuringiensis	Tomato	*Meloidogyne incognita*	ISR	[131, 144]
Pseudomonas dimunita	Tomato	*Meloidogyne* spp.	ISR	[128]
Pseudomonas fluorescens	Tomato, potato	*Meloidogyne incognita, Meloidogyne javanica*	ISR and SAR	[59, 98, 121–127]
Pseudomonas putida	Tomato	*Meloidogyne incognita*	ISR	[122, 127, 129, 130]
Serratia marcescens	Tomato	*Meloidogyne incognita*	ISR	[129–131]

6.5.3 SALICYLIC ACID (SA) POTENTIAL FUNCTIONS IN INDIRECT CONFLICT, SYSTEMIC ACQUIRED RESISTANCE (SAR) OF RHIZOBACTERIA AGAINST PHYTONEMATODES

It is generally understood that pathogenic microbes produce systemic acquired resistance (SAR), but rhizobacteria do not. However, recent data suggests that rhizobacteria could promote SAR [145, 146]. The catch is that cellular mechanisms are needed for induced systemic resistance (ISR) and SAR that are mutually antagonistic, with jasmonic acid (JA) and salicylic acid (SA) signaling for ISR and SAR, respectively [109, 145]. This initially contrasts with the involvement of SAR and SA signaling in the indirect antagonism of rhizobacteria. Furthermore, a few investigations have shown that certain *Bacillus* species, like *B. pumilus* and *B. subtilis,* can cause plant SA accumulations and signaling [152], leading researchers to hypothesize that an ISR mode mimics or shares SAR processes. For example, *B. amyloliquefaciens* MBI600 was found to activate defense that depends on SA but not JA and reduce *tomato spotted wilt virus* (TSWS) disease incidence [135]. Notably, research conducted independently by other teams utilizing related rhizobacteria strains, including *B. amyloliquefaciens, B. mycoides, B. pumilus,* and *B. subtilis,* found an activation of JA signaling [147]. Although the reason for this discrepancy is unknown, it could be due to various experimental circumstances or even experimental mistakes. Alternatively, timing variations could cause rhizobacteria to initiate JA and SA signaling, respectively. This finding implies that systemic resistance caused by rhizobacteria may be mediated by complex and/or different pathways [46]. Additional research is required to fully understand the combined or antagonistic activities of several plant hormone signals, including JA, SA, ethylene (ET), and abscisic acid (ABA), in transmitting rhizobacteria's activation of the defense system.

6.5.4 SIGNALING INTERACTIONS BETWEEN SALICYLIC ACID (SA) AND JASMONIC ACID (JA)

As mentioned, there is crosstalk between JA and SA signals, which can have either antagonistic or synergistic effects [148, 149]. Antagonism between JA and SA has been observed in tomatoes, as reported in one of the early studies on their interaction. It was observed that the SA formulation of aspirin significantly reduced gene transcription associated with JA [150]. Similarly, *P. syringae*-induced JA activation in tomatoes inhibited the expression of

SA marker genes [151]. Furthermore, exogenous JA injection resulted in decreased SA activity and elevated JA signaling [152]. A closer examination of this antagonistic relationship revealed that SA signaling prevents further JA action by targeting the promoter region downstream of JA-responsive genes [153]. Plants benefit from energy conservation through the antagonistic connection between these two hormones (JA and SA), optimizing their defense mechanisms when faced with a single pathogen [148, 149].

Later research proposed that the interaction between hormones is dependent on their concentration. When JA or SA levels are excessively high, there is an antagonistic interaction; however, when those levels are lower, the two hormones work in concert [154]. The precise concentration at which the two hormones become antagonistic and stop working together is still unknown.

Detection of any modulation in the JA and SA pathways by pathogenic microbes is a crucial component of this type of study because the goal of biological control research is to optimize the BCA to be as effective and efficient against the target pathogen. This, in turn, can help determine the best strategy to provide the target pathogen with the best possible protection and defense. When selecting the optimal BCA for each management plan, it is crucial to keep this in mind.

6.6 CURRENT APPLICATIONS OF RHIZOBACTERIA AGAINST ROOT KNOT NEMATODE ON VEGETABLE CROPS

Further investigation is currently underway to explore additional monitoring and control methods for *Meloidogyne* spp. infested vegetable crops. The focus is on utilizing biocontrol agents, specifically rhizobacteria, as potential alternatives to chemical nematicides. This shift is due to stricter regulations on the use of chemical nematicides and growing environmental concerns.

The rhizosphere plays a crucial role in protecting roots from pathogen invasion, serving as the first line of defense. Therefore, suitable rhizobacteria for biocontrol agents are those capable of thriving in the rhizosphere. Four *Pseudomonas* strains (*P. putida* (NRRL B-8), *P. fluorescens* (DSM 50090), *P. fluorescens* (NRRL B-23932), and *P. fluorescens* – 348) have shown promising results in improving tomato plant development and reducing root galling. This led to the conclusion that Pseudomonas species can induce systemic resistance in plants against pathogens [127].

In another study, rhizobacteria promoting plant growth significantly reduced the presence of RKN (root-knot nematodes) egg masses and root galling on tomato crops. This resulted in increased crop yield [155]. *Pseudomonas fluorescens* was found to effectively reduce the severity of root damage and root knots caused by *Meloidogyne* spp. infection in tomato plants [156].

A separate study demonstrated that a biological substance from rhizobacteria could suppress the root system of tomato and carrot plants [157]. The level of infection in the roots was closely correlated with increased tomato production in soils treated with a bionematicide formula derived from *Bacillus* and *Pseudomonas.* Plants treated with the bionematicide formula at a 2% dose exhibited less overall root knotting and less severe root damage compared to the control plants [128]. Additionally, RKN in carrot plants might be successfully controlled by a liquid mixture of molasses mixture and the antagonistic bacterium *P. fluorescens* [158]. Second-stage juveniles of *M. incognita* greatly decreased the infection of tomato roots with *P. fluorescens* and *B. licheniformis* [159]. *Bacillus aryabhattai* had a 41% control impact on *M. incognita* in the pot experiment [143]. *Bacillus halotolerans, B. toyonensis, B. oceanisediminis, B. kochii, B. pseudomycoides, B. pumilus, B. cereus,* and *P. aeruginosa* were found to be efficient at controlling tomato infected with *M. incognita* [160]. Additionally, *P. castaneae* had a substantial impact on the index of root-galling, egg mass, and shoot dry weight of tomato [106]. The impact of *S. marcescens* treatment on nematode-infected tomato plants was studied, this species induced polyphenol oxidase (PPO) and 1,3-glucanase activity as well as systemic resistance in the plants., which in turn boosted the expression of the plants' defense mechanisms and promoted the creation of osmoprotectants like sugars and amines [161]. By reducing the development of galls, restricting nematode reproduction, and hatching and eliminating young worms by releasing toxins, *Bacillus* has been demonstrated to elicit suppressive activity against *M. hapla*-infected lettuce [162]. In tomato plants, *Pseudomonas* sp. responded to *M. incognita* by producing protease enzymes, which led to the induction of defenses towards second-stage juveniles [163]. According to the outcomes of field tests, *B. vietnamiensis* B418 had a 71.15% control efficacy against root-knot nematode-infected watermelon. Microbiome assemblage study revealed that B418 might compete with other bacteria and encourage fungal variety in the rhizosphere because it led to a significant rise in the fungal community and a significant drop in the bacterial community [164]. The chosen isolates of *B. subtilis, B. pumilus, B. cereus,* and *B. megaterium* have a substantial influence on plant root morphology and nutrient uptake, according to research by Devindrappa et al. [69]. These isolates are extremely nematicidal and have a variety of plant growth promotion (PGP) properties. When compared to untreated seeds, these microorganisms were said to boost seed germination and seedling vigor index by altering the auxin pathways in the host plant. To prepare the seed for radicle emergence, the completion of pre-germination metabolic processes was accelerated [165]. Tomato plants infected with root-knot nematodes showed increased levels of their secondary metabolite profiles after exposure to *P. aeruginosa* and *B. gladioli* [166]. The rhizosphere rhizobacteria may also protect the host root from damage caused by pathogens. As biotechnology, molecular biology, and bioinformatics advance, new methods and studies will provide better guidance for the development of more efficient strategies for the biological management of root-knot nematodes.

6.7 CONCLUSION

The complex biological balance in the soil, which eventually contributes to the modification of agroecosystems and achieving more favorable circumstances for plant development and health, is essential for agricultural production. For example, rhizobacteria play a crucial role in maintaining the soil ecosystem's health, helping to control nematodes and other pests. The ancient and extensive coevolution has led to interactions between plants and the phytomicrobiome. We should expect to discover a wide range of new, occasionally unexpected interactions that benefit crops and, consequently, global food production, as evolution is pragmatic, random, and relentless. The phytomicrobiome holds enormous potential for novel and sustainable crop management techniques. However, there is still much work to be done, and our understanding of this potential remains minimal.

The largest agricultural corporations in the world are attempting to expand their crop protection industry, particularly in the field of biological control products. Therefore, extension education, product marketing, creative business management, and in-depth research will be essential for the future success of enterprises producing microbial inoculants, specifically rhizobacteria. Improved fermentation and formulation procedures are needed to effectively utilize rhizobacteria strains for the biological management of phytonematodes, which cause significant economic losses to crop plants. As agriculture moves towards sustainability, the potential of microbes as biocontrol agents in farming becomes increasingly evident. However, it is important to be practical when it comes to warnings. Currently, there are not many commercial biocontrol products made from rhizobacteria with nematicidal properties that are widely used in agriculture. Most of the research conducted so far has been in controlled environments or pots, therefore extensive field investigations are necessary to determine the potential of rhizobacteria. A concerted effort will be required to highlight the benefits of rhizobacteria biocontrol agents to farmers and raise awareness about environmentally friendly alternatives.

KEYWORDS

- **antagonistic bacterium**
- **biocontrol agents**
- **crop protection industry**
- **mode of action**
- **phytonematodes**
- **rhizobacteria**
- **root galling nematode**

REFERENCES

1. Nicol, J. M., Turner, S. J., Coyne, D. L., Den, N. L., Hockland, S., & Maafi, Z. T., (2011). Current nematode threats to world agriculture. In: Jones, J., Gheyse, G., & Ffenoll, C., (eds.), *Genomics and Molecular Genetics of Plant-Nematode Interaction* (pp. 21–43). Springer;Berlin/Heidelberg, Germany.
2. Moens, M., Perry, R. N., & Starr, J. L., (2009). *Meloidogyne* species – a diverse group of novel and important plant parasites. In: Perry, R. N., Moens, M., & Starr, J. L., (eds.). *Root-Knot Nematodes* (pp. 1–17). Wallingford, UK, CAB International.
3. Hellriegel, H., & Wilfarth, H., (1888). *Untersuchungen uber die Stickstoffnahrung der Gramineen und Leguminosen* (p. 234). Beilageheft zu der Zeitschrift des Vereins fur Rubenzucker-Industrie Deutschen Reichs.
4. Backer, R., Rokem, J. S., Iangumaran, G., Lamont, J., Praslickova, D., Ricci, E., Subramanian, S., & Smith, D. L., (2018). Plant growth-promoting rhizobacteria: Context, mechanisms of action, and roadmap to commercialization of biostimulants for sustainable agriculture. *Front. Plant Sci., 9*, 1473. doi: 10.3389/fpls.2018.01473.
5. Sidhu, H. S., (2018). Potential of plant growth-promoting rhizobacteria in the management of nematodes: A review. *Journal of Entomology and Zoology Studies*, *6*(3), 1536–1545.

6. Kloepper, J. W., & Schroth, M. N., (1978). Plant growth-promoting rhizobacteria on radishes. *Proceedings of the 4th International Conference on Plant Pathogenic Bacteria INRA* (Vol. 2, pp. 879–882). Angers, France.
7. Bishnoi, U., (2015). PGPR interaction: An ecofriendly approach promoting the sustainable agriculture system. *Advances in Botanical Research, 75*, 81–113.
8. Bender, S. F., Wagg, C., & Van, D. H. M. G., (2016). An underground revolution: Biodiversity and soil ecological engineering for agricultural sustainability. *Trends Ecol. Evol., 31*, 440–452. doi: 10.1016/j.tree.2016.02.016.
9. Prasad, M., Chaudhary, M., Choudhary, M., Kumar, T., & Jat, L., (2017). Rhizosphere microorganisms towards soil sustainability and nutrient acquisition. *Agriculturally Important Microbes for Sustainable Agriculture* (pp. 31–49). doi: 10.1007/978-9811055898_2.
10. Beattie, G. A., (2015). Microbiomes: Curating communities from plants. *Nature, 528*, 340, 341. doi: 10.1038/nature16319.
11. Pii, Y., Mimmo, T., Tomasi, N., Terzano, R., Cesco, S., & Crecchio, C., (2015). Microbial interactions in the rhizosphere: Beneficial influences of plant growth-promoting rhizobacteria on nutrient acquisition process: A review. *Biol. Fertil. Soils., 51*, 403–415. doi: 10.1007/s00374-015-0996-1.
12. Cabugao, K. G., Timm, C. M., Carrell, A. A., Childs, J., Lu, T. Y. S., Pelletier, D. A., Weston, D. J., & Norby, R. J., (2017). Root and rhizosphere bacterial phosphatase activity varies with tree species and soil phosphorus availability in Puerto Rico tropical forest. *Front. Plant Sci., 8*, 1834. doi: 10.3389/fpls.2017.01834.
13. Divjot, K., Rana, K. L., Tanvir, K., Yadav, N., Yadav, A. N., Kumar, M., Kumar, V., et al., (2021). Biodiversity, current developments, and potential biotechnological applications of phosphorus solubilizing and-mobilizing microbes: A review. *Pedosphere, 31*(1), 43–75. doi: 10.1016/S1002-0160(20)60057-1.
14. Rijavec, T., & Lapanje, A., (2016). Hydrogen cyanide in the rhizosphere: Not suppressing plant pathogens, but rather regulating availability of phosphate. Front. *Microbiol., 7*, 1785. doi: 10.3389/fmicb.2016.01785.
15. Ahemad, M., (2015). Phosphate-solubilizing bacteria-assisted phytoremediation of metalliferous soils: A review. *3 Biotech., 5*, 111–121. doi: 10.1007/s13205-014-0206-0.
16. Zhang, J., Wang, P., Fang, L., Zhang, Q. A., Yan, C., & Chen, J., (2017). Isolation and characterization of phosphate-solubilizing bacteria from mushroom residues and their effect on tomato plant growth promotion. *Polish J. Microbiol., 66*(1), 57–65. doi: 10.5604/17331331.1234993.
17. Oke, V., & Long, S. R., (1999). Bacteroid formation in the rhizobium–legume symbiosis. *Curr. Opin. Microbiol., 2*, 641–646. doi: 10.1016/S1369-5274(99)00035-1.
18. Herridge, D. F., Peoples, M. B., & Boddey, R. M., (2008). Global inputs of biological nitrogen fixation in agricultural systems. *Plant Soil, 311*, 1–18. doi: 10.1007/s11104-008-9668-3.
19. Galloway, J. N., Townsend, A. R., Erisman, J. W., Bekunda, M., Cai, Z., Freney, J. R., et al., (2008). Transformation of the nitrogen cycle: Recent trends, questions, and potential solutions. *Science, 320*, 889–892. doi: 10.1126/science.1136674.
20. Yavarian, S., Jafari, P., Akbari, N., & Feizabadi, M. M., (2021). Selective screening and characterization of plant growth-promoting bacteria for growth enhancement of tomato, Lycopersicon esculentum. *Iranian J. Microbiol., 13*(1), 121–129. doi: 10.18502/ijm.v13i1.5502.

21. Vessey, J. K., (2003). Plant growth promoting rhizobacteria as biofertilizers. *Plant Soil, 255*, 571–586. doi: 10.1023/A:1026037216893.
22. Adesemoye, A. O., & Kloepper, J. W., (2009). Plant–microbes interactions in enhanced fertilizer-use efficiency. *Appl. Microbiol. Biotechnol., 85*, 1–12. doi: 10.1007/s00253-009-2196-0.
23. Kuan, K. B., Othman, R., Abdul, R. K., & Shamsuddin, Z. H., (2016). Plant growth-promoting rhizobacteria inoculation to enhance vegetative growth, nitrogen fixation and nitrogen remobilization of maize under greenhouse conditions. *PloS One, 11*(3), e0152478. https://doi.org/10.1371/journal.pone.0152478.
24. Archana, G., Buch, A., & Kumar, G. N., (2012). Pivotal role of organic acid secretion by rhizobacteria in plant growth promotion. In: Satyanarayana, T., & Johri, B., (eds.), *Microorganisms in Sustainable Agriculture and Biotechnology*. Springer, Dordrecht., https://doi.org/10.1007/978-94-007-2214-9_3.
25. Ahmed, E., & Holmstrom, S. J., (2014). Siderophores in environmental research: Roles and applications. *Microb. Biotechnol., 7*, 196–208. doi: 10.1111/1751-7915.12117.
26. Saha, M., Sarkar, S., Sarkar, B., Sharma, B. K., Bhattacharjee, S., & Tribedi, P., (2016). Microbial siderophores and their potential applications: A review. *Environ. Sci. Pollut. Res., 23*, 3984–3999. doi: 10.1007/s11356-015-4294-0.
27. Hafeez, F. Y., Abaid-Ullah, M., & Hassan, M. N., (2013). Plant growth-promoting rhizobacteria as zinc mobilizers: A promising approach for cereals biofortification. In: Maheshwari, D. K., Saraf, M., & Aeron, A., (eds.), *Bacteria in Agrobiology: Crop Productivity* (pp. 217–235). Berlin: Springer.
28. Trudgill, D. L., & Blok, V. C., (2001). Apomictic, polyphagous root-knot nematodes: Exceptionally successful and damaging biotrophic root pathogens. *Annu Rev Phytopathol., 39*, 53–77.
29. Sikora, R. A., & Fernandez, E., (2005). Nematode parasites of vegetables. In: Luc, M., Sikora, R. A., & Bridge, J., (eds.). *Plant-Parasitic Nematodes in Subtropical and Tropical Agri-culture* (2nd edn., pp. 319–391). Wallingford, UK, CAB International.
30. Seid, A., Fininsa, C., Mekete, T., Decraemer, W., & Wesemael, W. M. L., (2015). Tomato (*Solanum lycopersicum*) and root-knot nematodes (*Meloidogyne* spp.)-a century-old battle. *Nematology, 17*, 995–1009. https://doi.org/10.1163/15685411-00002935.
31. Mitkowski, N. A., & Abawi, G. S., (2003). Root-knot nematodes. *The Plant Health Instructor.* doi: 10.1094/PHI-I-2003-0917-01.
32. Kassie, Y., (2019). Status of root-knot nematode (*Meloidogyne* species) and *fusarium* wilt (*Fusarium oxysporum*) disease complex on tomato (*Solanum lycopersicum* L.) in the Central Rift Valley, Ethiopia. *Agricultural Sciences, 10*, 1090–1103. doi: 10.4236/as.2019.108082.
33. El-Deriny, M. M., Hammad, E. A., & Essa, T. A., (2022). Potentiality of *Trichoderma* species against *Fusarium oxysporum* f. sp. *cucumerinum* and *Meloidogyne javanica* disease complex in cucumber plants. *Egypt. J. Agronematol., 21*(1), 34–58. doi: 10.21608/EJAJ.2022.222316.
34. Bernard, G. C., Egnin, M., & Bonsi, C., (2017). The impact of plant-parasitic nematodes on agriculture and methods of control. In: Shah, M. M., & Mohammod, M., (eds.), *Nematology—Concepts, Diagnosis and Control* (pp. 121–151). IntechOpen*; London,* UK.
35. Elhady, A., Giné, A., Topalovic, O., Jacquiod, S., Sørensen, S. J., Sorribas, F. J., Heuer, H., & Castagnone-Sereno, P., (2017). Microbiomes associated with infective stages of

root-knot and lesion nematodes in soil. *PLoS One*, *12*, e0177145. doi: 10.1371/journal.pone.0177145.
36. Pimentel, D., (2005). Environmental and economic costs of the application of pesticides, primarily in the United States. *Environ. Dev. Sustain.*, *7*, 229–252. doi: 10.1007/s10668-005-7314-2.
37. Mitiku, M., (2016). Plant-parasitic nematodes and their management: A review. *Agric. Res. Technol. Open Access J.*, *16*, 555980. doi: 10.19080/ARTOAJ.2018.16.55580.
38. Calvo-Araya, J. A., (2017). A novel nematicide for the control of root-knot nematodes in tomato crop. *Int. J. Phytopathol.*, *6*(2), 35–40.
39. Jones, J. G., Kleczewski, N. M., Desaeger, J., Meyer, S. L. F., & Johnson, G. C., (2017). Evaluation of nematicides for southern root-knot nematode management in lima bean. *Crop Protection, 96*, 151–157. https://doi.org/10.1016/j.cropro.2017.02.015.
40. Talavera-Rubia, M., Vela-Delgado, M. D., & Verdejo-Lucas, S., (2020). Nematicidal efficacy of milbemectin against root-knot nematodes. *Plants (Basel, Switzerland)*, *9*(7), 839. https://doi.org/10.3390/plants9070839.
41. Onkendi, E. M., Kariuki, G. M., Marais, M., & Moleleki, L. N., (2014). The threat of root-knot nematodes (*Meloidogyne* spp.) in Africa: A review. *Plant Pathology, 63*, 727–737.
42. Vyska, M., Cunniffe, N., & Gilligan, C., (2016). The trade-off between disease resistance and crop yield: A landscape-scale mathematical modeling perspective. *J. R. Soc. Interface, 13*, 20160451. doi: 10.1098/rsif.2016.0451.
43. Mukhtar, T., Hussain, M. A., Kayani, M. Z., & Aslam, M. N., (2014). Evaluation of resistance to root-knot nematode (*Meloidogyne incognita*) in okra cultivars. *Crop Protection, 56*, 25–30. https://doi.org/10.1016/j.cropro.2013.10.019.
44. De Vrieze, J., (2015). The littlest farmhands. *Science, 349*, 680–683. doi: 10.1126/science.349.6249.680.
45. Amer-Zareen, A. M. J., Abid, M., Gowen, S. R., & Kerry, B. R., (2004). Management of root-knot nematode (*Meloidogyne javanica*) by biocontrol agents in two crop rotations. *International Journal of Biology and Biotechnology, 1*, 67–73.
46. Subedi, P., Gattoni, K., Liu, W., Lawrence, K. S., & Park, S. W., (2020). Current utility of plant growth-promoting rhizobacteria as biological control agents towards plant-parasitic nematodes. *Plants (Basel)*, *9*(9), 1167. doi: 10.3390/plants9091167.
47. Lugtenberg, B., & Kamilova, F., (2009). Plant-growth-promoting rhizobacteria. *Annu. Rev. Microbiol.*, *63*, 541–556. doi: 10.1146/annurev.micro.62.081307.162918.
48. Sultana, F., & Hossain, M. M., (2022). Assessing the potentials of bacterial antagonists for plant growth promotion, nutrient acquisition, and biological control of Southern blight disease in tomato. *PLOS One*, *17*(6), e0267253. https://doi.org/10.1371/journal.pone.0267253.
49. Wiratno, W., Syakir, M., Sucipto, I., & Pradana, A. P., (2019). Isolation and characterization of endophytic bacteria from roots of *Piper nigrum* and their activities against *Fusarium oxysporum* and *Meloidogyne incognita.Biodiversitas, 20*(3), 682–687. doi: 10.13057/biodiv/d200310.
50. Jha, S., & Modi, H., (2018). Statistical optimization of chitinase production by *Streptomyces rubiginosus* SP24 and efficacy of purified chitinase to control root-knot nematode infection in *Vigna radiata* under controlled conditions. *Chem Biol Technol Agric.*, *5*(1), 1–13. doi: 10.1186/s40538-018-0133-0.

51. Gupta, R., Singh, A., Ajayakumar, P., & Pandey, R., (2017). Chitinolytic microbes confer *Meloidogyneincognita* resistance and augment secondary metabolites in *Bacopa monnieri* (L.) Pennell. *Arch Phytopathol Pflanzenschutz, 50*(3, 4), 178–196. doi: 10.1080/03235408.2017.1285522.
52. Kassab, S., Eissa, M., Badr, U., Ismail, A., Abdel, R. A., & Soliman, G., (2017). Nematicidal effect of a wild type of *Serratia marcescens* and its mutants against *Meloidogyne incognita* juveniles. *Egypt J. Agronematol., 16*(2), 95–114. doi: 10.21608/ejaj.2017.56204.
53. Safni, I., Lubis, K., Tantawi, A. R., & Murthi, S., (2018). Isolation and characterization of rhizobacteria for biological control of root-knot nematodes in Indonesia. *J ISSAAS, 24*, 67–81.
54. De Souza, G. A., De Freitas, S. F. E., Morgan, T., Sufiate, B. L., Tavares, G. P., Braga, F. R., Monteiro, T. S. A., et al., (2017). Enhanced production of *Monacrosporium thaumasium* protease and destruction action on root-knot nematode *Meloidogyne javanica* eggs. *Rhizosphere, 3*, 13–15. doi: 10.1016/j.rhisph.2016.12.001.
55. Gomes, E., Soares, F., Souza, D., Lima, L., Sufiate, B., Ferreira, T., & Queiroz, J., (2019). Role of *Synadenium grantii* latex proteases in nematicidal activity on *Meloidogyne incognita* and *Panagrellus redivivus*. *Braz J Biol., 79*(4), 665–668. doi: 10.1590/1519-6984.188129.
56. Bishop, A., (2011). *Pasteuria penetrans* and its parasitic interaction with plant parasitic nematodes. In: Logan, N. A., & De Vos, P., (eds.), *Endospore-Forming Soil Bacteria* (pp. 181–201). Springer; Berlin/Heidelberg, Germany.
57. Cronin, D., Moenne-Loccoz, Y., Fenton, A., Dunne, C., Dowling, D. N.,& O'gara, F., (1997). Role of 2,4-diacetyl-phloroglucinol in the interactions of the biocontrol *Pseudomonad* strain F113 with the potato cyst nematode *Globodera rostochiensis*. *Appl. Environ. Microbiol., 63*, 1357–1361. doi: 10.1128/AEM.63.4.1357-1361.1997.
58. Siddiqui, I. A., & Ehteshamul-Haque, S., (2001). Suppression of the root rot–root knot disease complex by *Pseudomonas aeruginosa* in tomato: The influence of inoculum density, nematode populations, moisture, and other plant-associated bacteria. *Plant Soil., 237*, 81–89. doi: 10.1023/A:1013313103032.
59. Siddiqui, I., & Shaukat, S., (2003). Suppression of root-knot disease by *Pseudomonas fluorescens* CHA0 in tomato: Importance of bacterial secondary metabolite, 2,4-diacetylpholoroglucinol. *Soil Biol. Biochem., 35*, 1615–1623. doi: 10.1016/j.soilbio.2003.08.006.
60. Hamid, M., Siddiqui, I. A., & Shahid, S. S., (2003). Improvement of *Pseudomonas fluorescens* CHA0 biocontrol activity against root-knot nematode by the addition of ammonium molybdate. *Lett. Appl. Microbiol., 36*, 239–244. doi: 10.1046/j.1472-765X.2003.01299.x.
61. Timper, P., Koné, D., Yin, J., Ji, P., & McSpadden, G. B. B., (2009). Evaluation of an antibiotic-producing strain of *Pseudomonas fluorescens* for suppression of plant-parasitic nematodes. *J. Nematol., 41*, 234–240.
62. Bhuiyan, S. A., Garlick, K., Anderson, J. M., Wickramasinghe, P., & Stirling, G. R., (2018). Biological control of root-knot nematode on sugarcane in soil naturally or artificially infested with *Pasteuria penetrans*. *Australas. Plant Pathol., 47*, 45–52. doi: 10.1007/s13313-017-0530-z.

63. Maksimov, I. V., Abizgil'dina, R. R., & Pusenkova, L. I., (2011). Plant growth promoting rhizobacteria as alternative to chemical crop protectors from pathogens (review). *Appl. Biochem. Microbiol., 47*, 333–345. doi: 10.1134/S0003683811040090.
64. Ramamoorthy, V., Viswanathan, R., Raguchander, T., Prakasam, V., & Samiyappan, R., (2001). Induction of systemic resistance by plant growth promoting rhizobacteria in crop plants against pests and diseases. *Crop Prot., 20*, 1–11. doi: 10.1016/S0261-2194(00)00056-9.
65. Babalola, O. O., (2010). Beneficial bacteria of agricultural importance. *Biotechnol. Lett., 32*, 1559–1570. doi: 10.1007/s10529-010-0347-0.
66. Chen, X. H., Vater, J., Piel, J., Franke, P., Scholz, R., et al., (2006). Structural and functional characterization of three polyketide synthase gene clusters in *Bacillus amyloliquefaciens* FZB 42. *J. Bacteriol., 188*, 4024–4036.
67. Oka, Y., Chet, I., & Spiegel, Y., (1993). Control of the root-knot nematode *Meloidogyne javanica* by *Bacillus cereus*. *Biocontrol. Sci. Technol., 3*, 115–126. doi: 10.1080/09583159309355267.
68. Gao, H., Qi, G., Yin, R., Zhang, H., Li, C., & Zhao, X., (2016). *Bacillus cereus* strain S2 shows high nematicidal activity against *Meloidogyne incognita* by producing sphingosine. *Sci. Rep., 6*:28756. doi: 10.1038/srep28756.
69. Devindrappa, M., Kamra, A., Grover, M., et al., (2022). Nematicidal rhizobacteria with plant growth-promoting traits associated with tomato in root-knot infested polyhouses. *Egypt J Biol Pest Control., 32*, 51. https://doi.org/10.1186/s41938-022-00539-1.
70. Siddiqui, Z. A., & Husain, S. I., (1991). Studies on the biological control of root-knot nematode. *Curr. Nematol., 2*, 5–6.
71. Siddiqui, Z. A., & Mahmood, I., (1992). Biological control of root-rot disease complex of chickpea caused by *Meloidogyne incognita* Race 3 and *Macrophomina phaseolina*. *Nematol. Medit., 20*, 199–202.
72. El-Nagdi, W. M. A., Abd-El-Khair, H., Soliman, G. M., Ameen, H. H., & El-Sayed, G. M., (2019). Application of protoplast fusants of *Bacillus licheniformis* and *Pseudomonas aeruginosa*on *Meloidogyne incognita* in tomato and eggplant. *Middle East J. Appl. Sci., 9*, 622–629.
73. Giannakou, I. O., Karpouzas, D. G., & Prophetou-Athanasiadou, D., (2004). A novel non-chemical nematicide for the control of root-knot nematodes. *Appl. Soil Ecol., 26*, 69–79. doi: 10.1016/j.apsoil.2003.09.002.
74. Mendoza, A. R., Kiewnick, S., & Sikora, R. A., (2008). *In vitro* activity of *Bacillus firmus* against the burrowing nematode *Radopholus similis*, the root-knot nematode *Meloidogyne incognita* and the stem nematode *Ditylenchus dipsaci*. *Biocontrol Sci. Technol., 18*, 377–389. doi: 10.1080/09583150801952143.
75. Terefe, M., Tefera, T., & Sakhuja, P. K., (2009). Effect of a formulation of *Bacillus firmus* on root-knot nematode *Meloidogyne incognita* infestation and the growth of tomato plants in the greenhouse and nursery. *J. Invertebr. Pathol., 100*, 94–99. doi: 10.1016/j.jip.2008.11.004.
76. Terefe, M., Tefera, T., & Sakhuja, P. K., (2012). Biocontrol (formulation of *Bacillus firmus* (BioNem)) of root-knot nematode, *Meloidogyne incognita* on tomato plants in the field. *Ethiop. J. Agric. Sci., 22*, 102–116.
77. Xiong, J., Zhou, Q., Luo, H., Xia, L., Li, L., Sun, M., & Yu, Z., (2015). Systemic nematicidal activity and biocontrol efficacy of *Bacillus firmus* against the root-knot

nematode *Meloidogyne incognita*. *World J. Microbiol. Biotechnol., 31*, 661–667. doi: 10.1007/s11274-015-1820-7.
78. Geng, C., Nie, X., Tang, Z., Zhang, Y., Lin, J., Sun, M., & Peng, D., (2016). A novel serine protease, Sep1, from *Bacillus firmus* DS-1 has nematicidal activity and degrades multiple intestinal-associated nematode proteins. *Sci Rep., 6*, 25012. doi: 10.1038/srep25012.
79. Gautam, A., Siddiqui, Z., & Mahmood, I., (1995). Integrated management of *Meloidogyne incognita* on tomato. *Nematol. Mediterr., 23*, 245–248.
80. Prakob, W., Nguen-Hom, J., Jaimasit, P., Thanunchai, J., & Chaisuk, P., (2009). Biological control of lettuce root-knot disease by the used of *Pseudomonas aeruginosa*, *Bacillus subtilis* and *Paecilomyces lilacinus*. *J. Agric. Technol., 13*, 179–191.
81. Kavitha, P. G., Jonathan, E. L., & Nakkeeran, S., (2012). Effects of crude antibiotic of *Bacillus subtilis* on hatching of eggs and mortality of juveniles of *Meloidogyne incognita*. *Nematol. Mediterr., 40*, 211–215.
82. Basyony, A. G., & Abo-Zaid, G. A., (2018). Biocontrol of the root-knot nematode, *Meloidogyne incognita*, using an eco-friendly formulation from *Bacillus subtilis*, lab and greenhouse studies. *Egypt. J. Biol. Pest Control., 28*, 87. doi: 10.1186/s41938-018-0094-4.
83. De Mazzuchelli, R. C. L., Mazzuchelli, E. H. L., & De Araujo, F. F., (2020). Efficiency of *Bacillus subtilis* for root-knot and lesion nematodes management in sugarcane. *Biol. Control., 143*, 104185. doi: 10.1016/j.biocontrol.2020.104185.
84. Lee, Y. S., & Kim, K. Y., (2016). Antagonistic potential of *Bacillus pumilus* L1 against root-knot nematode, *Meloidogyne arenaria*. *J. Phytopathol., 164*, 29–39. doi: 10.1111/jph.12421.
85. Forghani, F., & Hajihassani, A., (2020). Recent advances in the development of environmentally benign treatments to control root-knot nematodes. *Front. Plant Sci., 11*, 1125. doi: 10.3389/fpls.2020.01125.
86. Wei, J. Z., Hale, K., Carta, L., Platzer, E., Wong, C., Fang, S. C., & Aroian, R. V., (2003). *Bacillus thuringiensis* crystal proteins that target nematodes. *Proc. Natl. Acad. Sci. USA, 100*, 2760–2765. doi: 10.1073/pnas.0538072100.
87. Mohammed, S., Saedy, M., Enan, M., Ibrahim, N. E., Ghareeb, A., & Moustafa, S., (2008). Biocontrol efficiency of *Bacillus thuringiensis* toxins against root-knot nematode, *Meloidogyne incognita*. *J. Cell Mol. Biol., 7*, 57–66.
88. Ramalakshmi, A., Sharmila, R., Iniyakumar, M., et al., (2020). Nematicidal activity of native *Bacillus thuringiensis* against the root-knot nematode, *Meloidogyne incognita* (Kofoid and White). *Egypt J. Biol. Pest Control., 30*, 90. https://doi.org/10.1186/s41938-020-00293-2.
89. Kloepper, J. W., & Beauchamp, C. J., (1992). A review of issues related to measuring of plant roots by bacteria. *Can. J. Microbiol., 38*, 1219–1232. doi: 10.1139/m92-202.
90. Padgham, J. L., & Sikora, R. A., (2007). Biological control potential and modes of action of *Bacillus megaterium* against *Meloidogyne graminicola* on rice. *Crop Prot., 26*, 971–977. doi: 10.1016/j.cropro.2006.09.004.
91. Huang, Y., Xu, C., Ma, L., Zhang, K., Duan, C., & Mo, M., (2010). Characterization of volatiles produced from *Bacillus megaterium* YFM3. 25 and their nematicidal activity against *Meloidogyne incognita*. *Eur. J. Plant Pathol., 126*(3), 417–422.
92. Ambo, P. B. N., Ethiopia, E. A., Serfoji, P., Rajeshkumar, S., & Selvaraj, T., (2010). Management of root-knot nematode, *Meloidogyne incognita* on tomato cv Pusa ruby

by using vermicompost, AM fungus, *Glomus aggregatum* and mycorrhiza helper bacterium, *Bacillus coagulans*. *J. Agric. Sci. Technol., 6*, 37–45.
93. Serfoji, P., Smithra, P., Saravavanan, K., & Durai, R. K., (2013). Plant growth promotion and management of root-knot nematode *Meloidogyne incognita* through glomus aggregatum and *Bacillus coagulans* and vermicomposting in tomato. *Int. J. Pharm. Biol. Arch., 4*, 532–536.
94. Xiang, N., Lawrence, K., & Donald, P., (2018). Biological control potential of plant growth-promoting rhizobacteria suppression of *Meloidogyne incognita* on cotton and *Heterodera glycines* on soybean: A review. *J. Phytopathol., 166*, 449–458. doi: 10.1111/jph.12712.
95. Mena, J., & Pimentel, E., (2002). Mechanism of action of *Corynebacterium pauronetabolum* strain C-924 on nematodes. *Nematology., 4*, 287.
96. Singh, P., & Siddiqui, Z. A., (2010). Biocontrol of root-knot nematode *Meloidogyne incognita* by the isolates of *Pseudomonas* on tomato. *Arch. Phytopathol. Plant Protect., 43*, 1423–1434. doi: 10.1080/03235400802536857.
97. Meena, B., Radhajeyalakshmi, R., Marimuthu, T., Vidhyasekaran, P., Sabitha, D., & Velazhahan, R., (2000). Induction of pathogenesis related proteins, phenolics and phenylalanine ammonia lyase in groundnut by *Pseudomonas fluorescens*. *Zeitschrift fir Pflanzenkrantheiten and Pflanzeschutz, 107*, 514–527.
98. Patricia, T., Daouda, K., Jingfang, Y., Pingsheng, J., Brian, B., & Mcspadden, G., (2009). Evaluation of an antibiotic-producing strain of *Pseudomonas fluorescens* for suppression of plant-parasitic nematodes. *Journal of Nematology, 41*(3), 234–240.
99. Xie, H., Pasternak, J. J., & Glick, B. R., (1996). Isolation and characterization of mutants of the plant growth promoting rhizobacterium *Pseudomonas putida* GR12-2 that over produce indoleacetic acid. *Curr. Microbiol., 32*, 67–71.
100. Siddiqui, I. A., Haas, D., & Heeb, S., (2005). Extracellular protease of *Pseudomonas fluorescens* CHA0, a biocontrol factor with activity against the root-knot nematode *Meloidogyne incognita. Appl. Environ. Microbiol., 71*, 5646–5649. doi: 10.1128/AEM.71.9.5646-5649.2005.
101. Khan, M. R., Mohidin, F. A., Khan, U., & Ahamad, F., (2016). Native *Pseudomonas* spp. suppressed the root-knot nematode in *in vitro* and *in vivo*, and promoted the nodulation and grain yield in the field grown mungbean. *Biol. Control., 101*, 159–168. doi: 10.1016/j.biocontrol.2016.06.012.
102. Dube, B., & Smart, G. C., (1987). Biological control of *Meloidogyne incognita* by *Paecilomyces lilacinus* and *Pasteuria penetrans*. *J. Nematol., 19*, 222–227.
103. Kokalis-Burelle, N., (2015). *Pasteuria penetrans* for control of *Meloidogyne incognita* on tomato and cucumber, and *M. arenaria* on snapdragon. *Journal of Nematology, 47*(3), 207–213.
104. Jung, W. J., Jung, S. J., An, K. N., Jin, Y. L., Park, R. D., Kim, K. Y., Shon, B. K., & Kim, T. H., (2002). Effect of chitinase-producing *Paenibacillus illinoisensis* KJA-424 on egg hatching of root-knot nematode (*Meloidogyne incognita*). *J. Microbiol. Biotechnol., 12*(6), 865–871.
105. Bakengesa, J. A., (2016). *Potential of Paenibacillus spp. as a Biocontrol Agent for Root-Knot Nematodes (Meloidogyne spp.)* (p. 31). Thesis in Master of Science in Nematology, Univ. of Gent.

106. Cetintas, R., Kusek, M., & Fateh, S. A., (2018). Effect of some plant growth-promoting rhizobacteria strains on root-knot nematode, *Meloidogyne incognita*, on tomatoes. *Egypt J. Biol. Pest Control, 28*, 7. https://doi.org/10.1186/s41938-017-0008-x.
107. AbdelRazek, G. M., & Yaseen, R., (2020). Effect of some rhizosphere bacteria on root-knot nematodes. *Egypt. J. Biol. Pest Control, 30*, 140. https://doi.org/10.1186/s41938-020-00340-y.
108. Rahul, S., Chandrashekhar, P., Hemant, B., Chandrakant, N., Laxmikant, S., & Satish, P., (2014). Nematicidal activity of microbial pigment from *Serratia marcescens*. *Nat. Prod. Res., 28*, 1399–1404. doi: 10.1080/14786419.2014.904310.
109. Pieterse, C. M., Van, W. S. C., Van, P. J. A., Knoester, M., Laan, R., Gerrits, H., Weisbeek, P. J., & Van, L. L. C., (1998). A novel signaling pathway controlling induced systemic resistance in *Arabidopsis*. *Plant Cell., 10*, 1571–1580. doi: 10.1105/tpc.10.9.1571.
110. Fujimoto, T., Tomitaka, Y., Abe, H., Tsuda, S., Futai, K., & Mizukubo, T., (2011). Expression profile of jasmonic acid-induced genes and the induced resistance against the root-knot nematode (*Meloidogyne incognita*) in tomato plants (*Solanum lycopersicum*) after foliar treatment with methyl jasmonate. *J. Plant Physiol., 168*, 1084–1097. doi: 10.1016/j.jplph.2010.12.002.
111. Nahar, K., Kyndt, T., De Vleesschauwer, D., Höfte, M., & Gheysen, G., (2011). The jasmonate pathway is a key player in systemically induced defense against root knot nematodes in rice. *Plant Physiol., 157*, 305–331. doi: 10.1104/pp.111.177576.
112. Hubbard, K. E., Nishimura, N., Hitomi, K., Getzoff, E. D., & Schroeder, J. I., (2010). Early abscisic acid signal transduction mechanisms: Newly discovered components and newly emerging questions. *Genes Dev., 24*, 1695–1708. doi: 10.1101/gad.1953910.
113. Lee, S. C., & Luan, S., (2012). ABA signal transduction at the crossroad of biotic and abiotic stress responses. *Plant Cell Environ., 35*, 53–60. doi: 10.1111/j.1365-3040.2011.02426.x.
114. Daszkowska-Golec, A., & Szarejko, I., (2013). The molecular basis of ABA-mediated plant response to drought. In: Vahdati, K., & Leslie, C., (eds.), *Abiotic Stress—Plant Responses and Applications in Agriculture*. IntechOpen; Rijeka, Croatia. Available online: https://www.intechopen.com/books/abiotic-stress-plant-responses-and-applications-in-agriculture/the-molecular-basis-of-aba-mediated-plant-response-to-drought (accessed on 9 January 2023).
115. Der Ent, S. V., Hulten, M. V., Pozo, M. J., Czechowski, T., Udvardi, M. K., Pieterse, C. M. J., & Ton, J., (2009). Priming of plant innate immunity by rhizobacteria and β-aminobutyric acid: Differences and similarities in regulation. *New Phytol., 183*, 419–431. doi: 10.1111/j.1469-8137.2009.02851.x.
116. Ryu, C. M., Farag, M. A., Hu, C. H., Reddy, M. S., Wei, H. X., Paré, P. W., & Kloepper, J. W., (2003). Bacterial volatiles promote growth in *Arabidopsis*. *Proc. Natl. Acad. Sci. USA.,100*, 4927–4932. doi: 10.1073/pnas.0730845100.
117. Lim, C. W., Baek, W., Jung, J., Kim, J. H., & Lee, S. C., (2015). Function of ABA in stomatal defense against biotic and drought stresses. *Int. J. Mol. Sci., 16*, 15251–15270. doi: 10.3390/ijms160715251.
118. Huang, H., Gao, H., Liu, B., Qi, T., Tong, J., Xiao, L., Xie, D., & Song, S., (2017). *Arabidopsis* MYB24 regulates jasmonate-mediated stamen development. *Front. Plant Sci., 8*, 1525. doi: 10.3389/fpls.2017.01525.
119. Feys, B., Benedetti, C. E., Penfold, C. N., & Turner, J. G., (1994). *Arabidopsis* mutants selected for resistance to the phytotoxin coronatine are male sterile, insensitive to

methyl jasmonate, and resistant to a bacterial pathogen. *Plant Cell, 6*, 751–759. doi: 10.2307/3869877.

120. Mosblech, A., Thurow, C., Gatz, C., Feussner, I., & Heilmann, I., (2011). Jasmonic acid perception by COI1 involves inositol polyphosphates in *Arabidopsis thaliana. Plant J., 65*, 949–957. doi: 10.1111/j.1365-313X.2011.04480.x.
121. Leeman, M., Van, P. J. A., Den, O. F. M., Heinsbroek, M., Bakker, P. A. H. M., & Schippers, B., (1995). Induction of systemic resistance by *Pseudomonas fluorescens* in radish cultivars differing in susceptibility to *Fusarium* wilt, using a novel bioassay. *Eur. J. Plant Pathol., 101*, 655–664. doi: 10.1007/BF01874869.
122. Krechel, A., Faupel, A., Hallmann, J., Ulrich, A., & Berg, G., (2002). Potato-associated bacteria and their antagonistic potential towards plant-pathogenic fungi and the plant-parasitic nematode *Meloidogyne incognita* (Kofoid & White) Chitwood. *Can. J. Microbiol., 48*, 772–786.
123. Choudhary, D. K., Prakash, A., & Johri, B. N., (2007). Induced systemic resistance (ISR) in plants: Mechanism of action. *Indian J. Microbiol., 47*, 289–297. doi: 10.1007/s12088-007-0054-2.
124. Tian, B. O., Yang, J., & Zhang, K. Q., (2007). Bacteria used in the biological control of plant-parasitic nematodes: Populations, mechanisms of action and future prospects. *FEMS Microbiology & Ecology, 61*, 197–213.
125. De Vleesschauwer, D., Djavaheri, M., Bakker, P. A. H. M., & Höfte, M., (2008). *Pseudomonas fluorescens* WCS374r-induced systemic resistance in rice against *Magnaporthe oryzae* is based on pseudobactin-mediated priming for a salicylic acid-repressible multifaceted defense response. *Plant Physiol., 148*, 1996–2012. doi: 10.1104/pp.108.127878.
126. Saikia, S. K., Tiwari, S., & Pandey, R., (2013). Rhizospheric biological weapons for growth enhancement and *Meloidogyne incognita* management in *Withania somnifera* cv. *Poshita. Biol. Control., 65*, 225–234. doi: 10.1016/j.biocontrol.2013.01.014.
127. ElSayed, I. A., & Edrees, N. O., (2014). Potency evaluation of *Pseudomonas* strains against root-knot nematode infecting tomato. *International Journal of Advanced Research, 2*(8), 602–608.
128. Asyiah, I. N., Prihatin, J., Hastuti, A. D., Winarso, S., Widjayanthi, L., Nugroho, D., Firmansyah, K., & Pradana, A. P., (2021). Cost-effective bacteria-based bionematicide formula to control root-knot nematode *Meloidogyne* spp. in tomato plants.*Biodiversitas, 22*(6), 3256–3264. doi: 10.13057/biodiv/d220630.
129. Mohamed, Z. K., El-Sayed, S. A., Radwan, T. E. E., & Abd El-Wahab, G. S., (2009). Potency evaluation of *Serratia marcescens* and *Pseudomonas fluorescens* as biocontrol agents for root-knot nematodes in Egypt. *J. Appl. Sci. Res., 4*(1), 93–102.
130. Almaghrabi, O. A., Massoud, S. I., & Abdelmoneim, T. S., (2013). Influence of inoculation with plant growth promoting rhizobacteria (PGPR) on tomato plant growth and nematode reproduction under greenhouse conditions. *Saudi J. Biol. Sci., 20*, 57–61. doi: 10.1016/j.sjbs.2012.10.004.
131. Zhang, S., Moyne, A. L., Reddy, M. S., & Kloepper, J. W., (2002). The role of salicylic acid in induced systemic resistance elicited by plant growth-promoting rhizobacteria against blue mold of tobacco. *Biol. Control., 25*, 288–296. https://doi.org/10.1016/S1049-9644(02)00108-1.
132. De Halfeld-Vieira, B. A., Vieira, Jr. J. R., Da Romeiro, R. S., Silva, H. S. A., & Baracat-Pereira, M. C., (2006). Induction of systemic resistance in tomato by the autochthonous

phylloplane resident *Bacillus cereus*. *Pesqui. Agropecu. Bras., 41*, 1247–1252. doi: 10.1590/S0100-204X2006000800006.
133. Xiang, N., Lawrence, K. S., Kloepper, J. W., Donald, P. A., McInroy, J. A., & Lawrence, G. W., (2016). Biological control of *Meloidogyne incognita* by spore-forming plant growth-promoting rhizobacteria on cotton. *Plant Dis., 101*, 774–784. doi: 10.1094/PDIS-09-16-1369-RE.
134. Kavitha, J., Jonathan, E. I., & Umamaheswari, R., (2007). Field application of *Pseudomonas fluorescens*, *Bacillus subtilis* and *Trichoderma viride* for the control of *Meloidogyne incognita* (Kofoid and White) Chitwood on sugarbeet. *J. Biol. Control., 21*, 211–215.
135. Ryu, C. M., Farag, M. A., Hu, C. H., Reddy, M. S., Kloepper, J. W., & Paré, P. W., (2004). Bacterial volatiles induce systemic resistance in *Arabidopsis*. *Plant Physiol., 134*, 1017–1026. doi: 10.1104/pp.103.026583.
136. Lastochkina, O., Pusenkova, L., Yuldashev, R., Babaev, M., Garipova, S., Blagova, D., Khairullin, R., & Aliniaeifard, S., (2017). Effects of *Bacillus subtilis* on some physiological and biochemical parameters of *Triticum aestivum* L. (wheat) under salinity. *Plant Physiol. Biochem., 121*, 80–88. doi: 10.1016/j.plaphy.2017.10.020.
137. Bargabus, R., Zidack, N., Sherwood, J., & Jacobsen, B., (2004). Screening for the identification of potential biological control agents that induce systemic acquired resistance in sugar beet. *Biol. Control Biol. Control., 30*, 342–350. doi: 10.1016/j.biocontrol.2003.11.005.
138. Liu, K., Garrett, C., Fadamiro, H., & Kloepper, J. W., (2016). Induction of systemic resistance in Chinese cabbage against black rot by plant growth-promoting rhizobacteria. *Biol. Control., 99*, 8–13. doi: 10.1016/j.biocontrol.2016.04.007.
139. Burkett-Cadena, M., Kokalis-Burelle, N., Lawrence, K. S., Van, S. E., & Kloepper, J. W., (2008). Suppressiveness of root-knot nematodes mediated by rhizobacteria. *Biol. Control., 47*, 55–59. doi: 10.1016/j.biocontrol.2008.07.008.
140. Choudhary, D. K., & Johri, B. N., (2009). Interactions of *Bacillus* spp. and plants with special reference to induced systemic resistance (ISR). *Microbiol. Res., 164*, 493–513. doi: 10.1016/j.micres.2008.08.007.
141. Li, J., Zou, C., Xu, J., Ji, X., Niu, X., Yang, J., Huang, X., & Zhang, K. Q., (2015). Molecular mechanisms of nematode-nematophagous microbe interactions: Basis for biological control of plant-parasitic nematodes. *Annu. Rev. Phytopathol., 53*, 67–95. doi: 10.1146/annurev-phyto-080614-120336.
142. Xie, S., Jiang, H., Ding, T., Xu, Q., Chai, W., & Cheng, B., (2018). *Bacillus amyloliquefaciens* FZB42 represses plant MiR846 to induce systemic resistance via a jasmonic acid-dependent signaling pathway. *Mol. Plant Pathol., 19*, 1612–1623. doi: 10.1111/mpp.12634.
143. Du, J., Gao, Q., Liu, Z., Li, C., Song, X., Xu, R., et al., (2020). Plant rhizosphere growth-promoting bacterium with root-knot nematode inhibition and its effect on the tomato rhizosphere microbial community structure. *J. Bacteriol. Mycol., 7*(8), 1159.
144. Zuckerman, B. M., Dicklow, M. B., & Acosta, N., (1993). A strain of *Bacillus thuringiensis* for the control of plant-parasitic nematodes. *Biocontrol Sci. Technol., 3*, 41–46. https://doi.org/10.1080/09583159309355257.
145. Vlot, A. C., Klessig, D. F., & Park, S. W., (2008). Systemic acquired resistance: The elusive signal(s). *Curr. Opin. Plant Biol., 11*, 436–442. doi: 10.1016/j.pbi.2008.05.003.

146. Yu, Y., Gui, Y., Li, Z., Jiang, C., Guo, J., & Niu, D., (2022). Induced systemic resistance for improving plant immunity by beneficial microbes. *Plants (Basel, Switzerland), 11*(3), 386. https://doi.org/10.3390/plants11030386.
147. Liu, Y., Lai, Q., & Shao, Z., (2018). Genome analysis-based reclassification of *Bacillus weihenstephanensis* as a later heterotypic synonym of *Bacillus* mycoides. *Int. J. Syst. Evol. Microbiol., 68*, 106–112. doi: 10.1099/ijsem.0.002466.
148. Glazebrook, J., (2005). Contrasting mechanisms of defense against biotrophic and necrotrophic pathogens. *Ann. Rev. Phytopathol., 43*, 205–227. doi: 10.1146/annurev.phyto.43.040204.135923.
149. Gimenez-Ibanez, S., & Solano, R., (2013). Nuclear jasmonate and salicylate signaling and crosstalk in defense against pathogens. *Front. Plant Sci., 4*, 72. doi: 10.3389/fpls.2013.00072.
150. Leon-Reyes, A., Van, D. D. D., De Lange, E. S., Delker, C., Wasternack, C., Van, W. S. C., Ritsema, T., & Pieterse, C. M., (2010). Salicylate-mediated suppression of jasmonate-responsive gene expression in *Arabidopsis* is targeted downstream of the jasmonate biosynthesis pathway. *Planta, 232*(6), 1423–1432. https://doi.org/10.1007/s00425-010-1265-z.
151. Uppalapati, S. R., Ishiga, Y., Wangdi, T., Kunkel, B. N., Anand, A., Mysore, K. S., & Bender, C. L., (2007). The phytotoxin coronatine contributes to pathogen fitness and is required for suppression of salicylic acid accumulation in tomato inoculated with *Pseudomonas syringae* pv. tomato DC3000. *Mol. Plant Microbe Interact., 20*, 955–965. doi: 10.1094/MPMI-20-8-0955.
152. Tamaoki, D., Seo, S., Yamada, S., Kano, A., Miyamoto, A., Shishido, H., Miyoshi, S., et al., (2013). Jasmonic acid and salicylic acid activate a common defense system in rice. *Plant Signal. Behav., 8*, 24260. doi: 10.4161/psb.24260.
153. Van, D. D. D., Leon-Reyes, A., Koornneef, A., Van, V. M. C., Rodenburg, N., Pauwels, L., Goossens, A., Körbes, A. P., Memelink, J., Ritsema, T., et al., (2013). Salicylic acid suppresses jasmonic acid signaling downstream of SCFCOI1-JAZ by targeting GCC promoter motifs via transcription factor ORA59. *Plant Cell, 25*, 744–761. doi: 10.1105/tpc.112.108548.
154. Mur, L. A. J., Kenton, P., Atzorn, R., Miersch, O., & Wasternack, C., (2006). The outcomes of concentration-specific interactions between salicylate and jasmonate signaling include synergy, antagonism, and oxidative stress leading to cell death. *Plant Physiol., 140*, 249–262. doi: 10.1104/pp.105.072348.
155. Kokalis-Burelie, N., & Dickson, D. W., (2003). Effects of soil fumigants and bioyieldtm on root knot nematode incidence and yield of tomato. *Proc. Int. Res. Conf. Methyl Bromide Alternatives and Emissions Reductions, 50*, 1–50.3.
156. Nishantha, K., Jayasiri, H., Herath, H., Nugaliyadde, M., Dissanayake, M., & De Costa, D. M., (2018). Antagonistic effect of eight Sri Lankan isolates of *Pseudomonas fluorescens* on *Meloidogyne incognita* in tomato, *Lycopersicon esculentum. Int J Environ Agric Biotech., 3*(6), 1980–1986. doi: 10.22161/ijeab/3.6.4.
157. Viljoen, J. J., Labuschagne, N., Fourie, H., & Sikora, R. A., (2019). Biological control of the root-knot nematode *Meloidogyne incognita* on tomatoes and carrots by plant growth-promoting rhizobacteria. *Trop Plant Pathol., 44*(3), 284–291. doi: 10.1007/s40858-019-00283-2.

158. Nagachandrabose, S., (2018). Liquid bioformulations for the management of root-knot nematode, *Meloidogyne hapla* that infects carrot. *Crop Prot., 114*, 155–161. doi: 10.1016/j.cropro.2018.08.022.
159. Colagiero, M., Rosso, L., & Ciancio, A., (2017). Diversity and biocontrol potential of bacterial consortia associated to root-knot nematodes. *Biological Control.*
160. Liu, G., Lin, X., Xu, S., Liu, G., Liu, F., & Mu, W., (2020). Screening, identification, and application of soil bacteria with nematicidal activity against root-knot nematode (*Meloidogyne incognita*) on tomato. *Pest Management Science.*
161. Abd-Elgawad, M. M., & Kabeil, S. S., (2012). Biological control of *Meloidogyne incognita* by *Trichoderma harzianum* and *Serratia marcescens* and their related enzymatic changes in tomato roots. *African Journal of Biotechnology., 11*, 16247–16252.
162. Chen, J., Abawi, G. S., & Zuckerman, B. M., (2000). Efficacy of *Bacillus thuringiensis, Paecilomyces marquandii*, and *Streptomyces costaricanus* with and without organic amendments against *Meloidogyne hapla* infecting lettuce. *Journal of Nematology., 32*, 70–77.
163. Sharma, I. P., & Sharma, A. K., (2017). Effective control of root-knot nematode disease with *Pseudomonad* rhizobacteria filtrate. *Rhizosphere, 3*, 123–125.
164. Liu, M., Philp, J., Wang, Y., et al., (2022). Plant growth-promoting rhizobacteria *Burkholderia vietnamiensis* B418 inhibits root-knot nematode on watermelon by modifying the rhizosphere microbial community. *Sci. Rep., 12*, 8381. https://doi.org/10.1038/s41598-022-12472-2.
165. Sukumar, P., Legue, V., Vayssières, A., Martin, F., Tuskan, G. A., & Kalluri, U. C., (2013). Involvement of auxin pathways in modulating root architecture during beneficial plant–microorganism interactions. *Plant Cell Environ., 36*, 909–919.
166. Khanna, K., Jamwal, V. L., Sharma, A., Gandhi, S. G., Ohri, P., Bhardwaj, R., Al-Huqail, A. A., et al., (2019). Evaluation of the role of rhizobacteria in controlling root-knot nematode (RKN) infection in *Lycopersicon esculentum* plants by modulation in the secondary metabolite profiles. *AoB Plants.*

CHAPTER 7

The Role of Endophytic Bacteria in Plant Immune Responses and Plant Disease Control

NABIL RADOUANE,[1,2] HAITAM LAHMAMSI,[1,3] JIHANE KENFAOUI,[1,2] IKRAM LEGRIFI,[1,2] AISSAM DAABOUB,[1] KHADIJA GOURA,[1,4] SAID EZRARI,[1,2] HAJAR EL-HAMSS,[1] and RACHID LAHLALI[1]

[1]*Department of Plant Protection, Phytopathology Unit, Ecole Nationale d'Agriculture de Meknès, BPS 40, Meknès, Morocco*

[2]*Laboratory of Functional Ecology and Environmental Engineering, Sidi Mohamed Ben Abdellah University, Route d'Imouzzer, Fez, Morocco*

[3]*Laboratory of Microbial Biotechnology and Bioactive Molecules, Sidi Mohamed Ben Abdellah University, Route d'Imouzzer, Fez, Morocco*

[4]*Laboratory of Biotechnology and Valorization of Phyto-Resources, Faculty of Sciences, Moulay Ismail University, Avenue Zitoun, Meknes, Morocco*

ABSTRACT

Agricultural development must keep pace with the rapid and continuous evolution of the global population, which has led to increased use of chemical-based inputs for enhancing productivity and controlling emerging diseases in various crops. However, these chemical inputs pose significant risks to the environment and human health. Endophytic bacteria (EB) have gained significant importance as a biological control method due to their numerous advantages. EB are a category of microorganisms that live symbiotically within plants and promote crop growth through various mechanisms.

Microbial Biostimulants: Biorational Pesticides for Management of Plant Pathogens.
Faheem Ahmad and Rakesh Pandey (Eds.)

Additionally, these bacteria play other roles, such as interacting with plants and causing significant changes in plant physiology. EB establish different types of relationships with plants, including mutualism and commensalism, and apart from promoting plant growth (PGP), they also contribute to nitrogen fixation in the soil, phosphate solubilization, and phytohormone signaling, which help plants cope with various environmental stresses. This chapter provides a comprehensive discussion on the capabilities of endophytic bacteria in promoting plant growth, interacting with their hosts, studying defense mechanisms against plant pathogens, and mitigating environmental stresses.

7.1 INTRODUCTION

The UN forecasts that the global human population will reach 9.7 billion people by the year 2050 and increase by an additional 1.5 billion people or more in the subsequent five decades after 2050. During these two future periods, the population will increase by about 32% and 53%, respectively, from approximately 7.3 billion people in 2015 [1]. Under the pressure of these important demographic forces, it is hard to be optimistic about the future, given the limited quantity of natural resources, climate change, depleted reserves of water primarily needed for irrigation, and secondary ecological factors, which only magnify this challenge even further. The quest to increase crop yields has already led to an over-utilization of the soil's ecological system, making it difficult to adopt a similar traditional agricultural policy to address this menace. In addition, activities that are primarily anthropogenic, with the example of over application of agricultural-based chemicals and monocultures, are a major concern since they disrupt the functioning of the ecological system [2].

To meet these challenges, the current agricultural system must be reshaped to minimize such negative impacts on the ecosystem. One of the most interesting solutions to achieve this objective is the utilization of endophytic bacteria (EB), which is capable of conferring a plethora of advantages to the host-acting plant, including the promotion of growth and deterrence against pathogenic organisms [3]. According to Hallmann [4], endophytic bacteria (EB) include bacteria that have been isolated mainly from tissues of plants that were previously surface-sterilized but do not pose any damage to their host plants. The aforementioned bacteria were identified in all of the studied plant species [4].

Therefore, a plant devoid of endophytes is considered a rare occurrence in natural environments [5]. They belong to a specific category of rhizobacteria that have adapted themselves to invade their plant hosts [6]. Numerous research studies have documented the various impacts of EB on the health and growth of crops. They can act directly by promoting the acquisition of essential macro-nutrients such as nitrogen, iron, and phosphorus, and/or by modulating the levels of phytohormones like cytokinin, auxin, and gibberellin [3]. Indirectly, they prevent plant damage following infection with a phytopathogen through a variety of defense mechanisms, such as the secretion of lytic enzymes, toxin inhibition, antibiosis, competition for resources and space, and conferring immunity in plants or initiating systemic resistance [3, 7, 8].

When applied externally, potato endophytes can activate both the host plant's basal MAMP (microbe-associated molecular pattern)-triggered immunity and inducible defense mechanism [9]. They can assist the host in tolerating stressful situations and even elicit allelopathic effects against secondary competitive species of crops [10]. According to Li et al. [11], 54 bacterial endophytes showed significant inhibition against *Agrobacterium tumefaciens* when they tested 305 endophytes for their antagonistic activity. de Almeida et al. [12] revealed that *Bacillus* and *Burkholderia* isolates were highly effective in combating the phytopathogens *Phomopsis sojae, Sclerotinia sclerotiorum,* and *Rhizoctonia solani in vitro* [12]. The endophytic bacterium *Pseudomonas synxantha* DLS65, first isolated from kiwifruit tissue, was tested against *M. fructicola* and *M. fructigena.* The results showed that *P. synxantha* DLS65 was able to completely suppress the mycelial growth of *Monilinia fructicola* and *Monilinia fructigena* in Petri dishes [13].

7.2 ENDOPHYTIC BACTERIA (EB) DIVERSITY

To gain a more comprehensive understanding of bacterial endophyte diversity, numerous studies have investigated bacterial endophytes in various plant species. These include prairie plants, agronomic crops, plants that thrive in harsh environments, perennial and wild plants [14, 15]. Additionally, several endophytes have been identified in rice [16], medicinal plants [17], tomato [18], citrus [19], potato [20], chickpea [21], and others.

Bacterial endophytes have been previously isolated from different parts of plants, including the rhizosphere and the phyllosphere [22]. These parts encompass seeds, roots, tubers, stems, leaves, fruits, ovules, and internal plant tissues such as vascular tissue, intercellular space, and periderm [8, 23]. Among these, roots have been found to harbor the highest abundance of bacterial endophytes compared to other tissues [24].

Santoyo et al. [3] reported that proteobacteria, including the Classes α, β, and γ-proteobacteria are the leading predominant phylum in the diversity analysis of endophytes [3]. An evaluation of endophytic communities of bacteria belonging to tomato leaves revealed that proteobacteria are the leading majority in terms of representation, followed closely by *Actinobacteria, Planctomycetes, Verrucomicrobia,* and *Acidobacteria* [25]. By estimation, an approximated 129 types of bacterial endophyte species have a representation exceeding 54 genera previously identified from a plethora of host plants. The important reported bacterial endophytes belong to the genera *Arthrobacter, Bacillus, Burkholderia, Enterobacter, Methylobacterium, Microbacterium, Micrococcus, Paenibacillus, Pantoea, Phyllobacterium, Pseudomonas, Rhanella, Rhodanobacter, Sphingomonas,* and *Stenotrophomonas* [3].

The diverse nature of endophytic bacterial communities present in crops is driven by many abiotic and biotic factors. Several studies report that the genotype, host plant stages, root exudation, and secondary metabolites can greatly determine the kind of endophytic bacteria harbored [26]. Not only that, the soil type, climatic conditions, presence of pollutants, moisture, soil pH, and temperature can also influence these communities [26]. The approach embraced in studying the aforementioned bacteria is another important factor influencing their diversity. According to Hallmann et al. [4], the concentration, nature, and even duration of treatment for a sterilizing agent utilized in recovering bacteria can greatly impact this [4].

7.3 PLANT IMMUNE RESPONSE WITHIN ENDOPHYTIC BACTERIA (EB)

Endophytic bacteria (EB) improve immunity and plant resistance to biotic stresses by secreting elicitors. Additionally, these bacteria enhance abiotic stress tolerance to factors such as drought, cold, and salt resistance, supporting host plants in surviving severe conditions (Figure 7.1) [27].

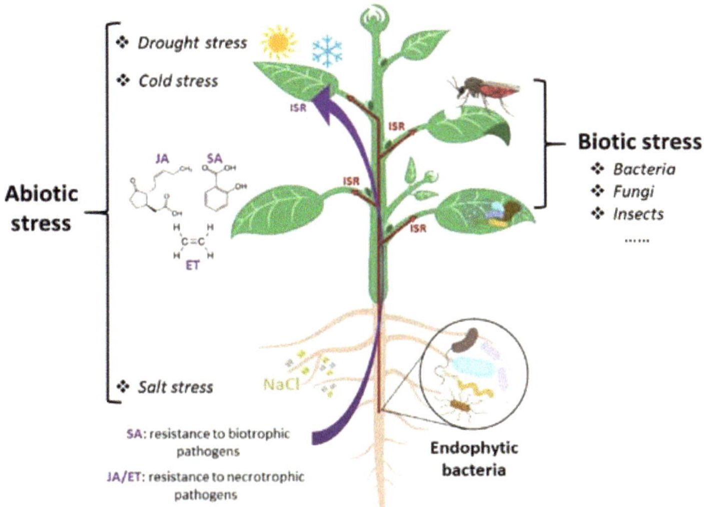

FIGURE 7.1 Endophytic bacteria increase resistance to biotic and abiotic stresses by inducing systemic resistance via one or combined SA, JA, and ET signaling pathways, which boost the plant's immune system, making it more capable of fighting disease.

7.3.1 MECHANISMS OF INTERACTIONS AGAINST BIOTIC STRESS

Endophytic bacteria (EB) promote plant resistance to phytopathogens by inducing defensive mechanisms, a mechanism known as Induced Systemic Resistance (ISR) [28, 29]. This provides evidence that the interactions between EB and the host trigger immune responses in plants, similar to those observed with phytopathogens. Subsequently, these microorganisms colonize plant tissues by evading defensive responses, as observed with *Bacillus* and *Pseudomonas* spp. [30]. Various compounds, including siderophores, N-acyl-homoserine lactones, salicylic acid (SA), jasmonic acid (JA), antibiotics, volatiles (e.g., acetoin), and lipopolysaccharides, all play crucial roles in inducing ISR. Indeed, plant defense mechanisms and protection against biotic stresses are associated with ISR [31]. While various EB-stimulated ISR through SA, plant hormones like ethylene (ET) and JA play crucial regulatory functions in the signaling pathways involved in the induction of ISR [32]. The resistance to biotrophic infections is regulated by the SA mechanism, whereas necrotrophic pathogens are regulated by the JA/ET pathway [33, 34]. Endophytic bacteria regulate plant diseases by increasing plant resistance, which has garnered significant attention and interest. Plants colonized by endophytes often exhibit higher immunity against pathogens compared to uninfected plants. The strain 89B-61 of *Pseudomonas fluorescens,* an endophytic bacterium, was the first to be described, demonstrating the induction of ISR (induced systemic resistance) to protect cucumbers against anthracnose disease [35]. Targeting endophytes to specific parts of the host plant significantly reduces the disease index, even when pathogens affect other regions of the plant [36]. Previous studies have shown that endophytic bacteria defend plants from soil-borne fungi by promoting systemic resistance in their above-ground parts. For example, *Arabidopsis thaliana* inoculated with *B. pumilus* SE34 exhibited lower disease severity and reduced symptoms associated with cucumber mosaic virus [37]. The induction of systemic resistance was facilitated by plant signaling compounds such as jasmonic acid (JA) and ethylene (ET) [37]. The endophyte *Streptomyces albidoflavus* OsiLf-2 demonstrated strong antagonistic activity against rice blast disease caused by *Magnaporthe oryzae,* and triggered host plant resistance through the upregulation of JA and salicylic acid (SA) expression [38]. Similarly, the presence of *Methylobacterium* sp. IMBG290 improved potato plant resistance to the disease caused by *Pectobacterium atrosepticum.* The mode of resistance was associated with changes in the nature of the local endophytic population. Modifications that affect the endophytic community have been connected to disease resistance, suggesting that the endophytes play a crucial part in disease suppression [9].

7.3.2 MECHANISMS OF INTERACTIONS AGAINST ABIOTIC STRESS

Many extreme conditions, such as ecological constraints and stresses brought on by living organisms, have an impact on plant development and health. Plants can withstand abiotic stress in two ways: (i) by activating response mechanisms immediately after exposure [39]; and (ii) by producing biochemical compounds that act as anti-stress agents in the extracellular environment [30].

Subramanian et al. [40] described that tomato plants inoculated with *P. frederiksbergensis* OS261 and *P. vancouverensis* OB155 performed better under cold stress (10–12°C). Compared to the control plants, there was less membrane loss and higher antioxidant activity. Cold adaptation genes (LeCBF1 and LeCBF3) were also activated in the treated plants [40]. Similarly, *B. phytofirmans* PsJN improves grapevine cold resistance by modifying carbohydrate metabolism and photosynthetic activity associated with tolerance to cold stress [41]. The presence of this bacteria in the host aided the organism's ability to adapt to lower temperatures, reducing cell damage, boosting photosynthesis, and causing an accumulation of molecules related to cold stress, including starch, proline, and phenolic compounds.

Indeed, extracellular bacteria have been shown to improve plant resistance to drought stress [42, 43]. Sugarcane inoculated with *Gluconacetobacter diazotrophicus* isolate PAL5 becomes more resistant to drought stress than control plants through drought-responsive gene expression and ethylene (ET) and abscisic acid (ABA) hormone pathways [44]. Moreover, these bacteria have been effectively employed to counteract the harmful impacts of salts and improve plant development under stressful circumstances. Additionally, endophytes provide positive effects in plants such as detoxification, nutrient acquisition, osmotic adjustment, and phytohormone modulation, which help mitigate the consequences of salt stress [45]. This reveals that endophytes can identify physiological effects on plants and modify the expression of genes to adapt to changes in their environment. Bacterial endophytes can be developed as protective agents under harsh climatic conditions, as they can impact the physiological responses of plants to stresses [46].

7.4 ENDOPHYTIC BACTERIA (EB) AS PLANT DISEASE CONTROL AGENT

Biological agents are known as effective substitutes for chemical and cultural control against various plant diseases. Recently, the use of bacterial endophytes as natural agents to manage plant infections has emerged as a crucial aspect in phytopathological studies [47]. The application of bacterial endophytes as natural control agents is an eco-friendly and non-toxic approach, with sustainable implementations for human health, agriculture, and the environment. These microorganisms reside inside plant tissues, primarily in the intercellular space, providing protection from environmental stress and forming a symbiotic association with their host crops without causing pathogenic effects. This association helps promote crop growth and yield performance [48]. Bacterial endophytes have been reported in different hosts across virtually all documented plant species.

Several bacterial endophytes produce secondary metabolites, which have been isolated, extracted, and characterized in various studies [49]. Some of these bioactive compounds can inhibit the development of plant pathogens, suppress plant diseases, and enhance systemic resistance through the production of proteins, particularly pathogenesis-related proteins (PRPs), as well as defense enzymes that slow down or inhibit the growth of disease-causing microorganisms [50].

Diverse studies have been conducted on the antibiosis mechanism, which is the capacity of bacteria to release secondary metabolites as volatile compounds, siderophores, hydrogen cyanide (HCN), and antibiotics possessing antibacterial, insecticidal, and antifungal properties [51]. These antimicrobial substances are mostly derived from the genus Streptomyces [52].

Species of the *Bacillus* genus are considered major producers of compounds that are effective against the development of various phytopathogens [53]. In this context, *Bacillus* species have been reported to produce abundant peptide antibiotics [54]. The lipopeptides have been evaluated as a powerful substance to deal with different types of pathogen agents. Similar to another study, lipopeptides have been extracted from *Bacillus* spp. isolated from yellow dent corn and Indian popcorn, showing antifungal activity against the fungus *Fusarium moniliforme* [55]. In addition, a study has suggested that *Burkholderia* sp. possesses a high ability to secrete secondary metabolites including pyrrolnitrin; this substance has an effective antagonist activity to control plants against phytopathogen infections [56, 57].

Siderophores are substances produced by bacteria when there is a deficiency in iron concentration in the soil, as iron is an essential element for the viability of microorganisms. Eventually, the siderophores secreted by EB solubilize this iron in the soil due to their high affinity for the substrate and their ability to absorb it. The siderophore activity of these EB can limit the acquisition of iron by all microorganisms present in the soil, thus preventing the proliferation of undesirable agents in the surrounding area due to low iron concentration [58, 59]. Siderophores represent an important mechanism of EB that could be involved in inhibiting phytopathogens and managing plant diseases. Several studies have shown the ability of EB to inhibit the development of phytopathogens through the secretion of siderophores [60]. This activity was demonstrated in a study that evaluated the effect of the endophytic bacterium *P. fluorescens* as a biological control agent (BCA) against the pathogen *E. carotovora*, the causative agent of potato soft rot disease [61]. Various derivatives of siderophores, such as catecholets, carboxylate, and hydroxymate, are secreted by different bacteria. For example, *Bacillus* spp. isolated from potatoes was found to produce two major types of siderophores, salicylate and catecholets. *Bacillus* spp. tested in this study showed a strong inhibitory effect against the causal agent of banana wilt disease [62] (Table 7.1).

TABLE 7.1 List of Endophytic Bacteria with Biocontrol Potentials

Plant	Endophytic Bacteria	Disease	References
Cucumber	*Bacillus* sp.	Downy mildew	[63]
Tomato	*Staphylococcus epidermidis* *Pseudomonas mossellii* *Pseudomonas fluorescence* *Pseudomonas brassicacearum* *P. mossellii* *P. fluorescence*	*R. solanacearum* (bacterial wilt disease).	[64]
Grapevine	*Bacillus subtilis*	Crown gall *Agrobacterium tumefaciens*	[65]
Rice	*Bacillus subtilis* *B. amyloliquefaciens* (FZB 24), EPB 9, EPB10, EPCO 29, and EPCO 78	Bacterial leaf blight, sheath blight.	[66]
Rice	*Pseudomonas* spp.	Rice blast fungus	[67]
Sugar beet	*Bacillus subtilis*	Cercospora leaf spot	[68]
Cacao	*Bacillus subtilis*	Damping off	[69]
Banana	*Pseudomonas putida* *Acromobacter* spp. *Rhizobium* spp. *Bacillus flexustwo* *Bacillus cereus* *Pseudomonas putida*	*Fusarium* wilt	[70]
Strawberry	*Bacillus amyloliquefaciens* strain S13-3	Anthracnose	[71]
Wheat	*Bacillus megaterium* (BM1), *Bacillus subtilis* (BS43, BSM0, and BSM2)	Fusarium head blight	[72]
Coffea arabica L. *Coffea robusta* L.	*Bacillus lentimorbus* *Bacillus cereus*	Leaf rust	[73]
Ribbon plant	*P. fluorescens* Pf1 *P. fluorescens*	*Alternaria* leaf blight	[74]
Chili pepper	*Bacillus pseudomycoidesNBRC 101232* *B. mycoidesstrain 273* *B. thuringiensisATCC 10792*	Bacterial wilt (*Ralstonia syzigii*) *Fusarium* wilt	[75]
Rice	*B.amyloliquefaciens* strains A1, A3, A13 *B. methylotrophicus* strain A2 *B. subtilis* strain A15 *Bacilli* D29 and H8	Bacterial leaf blight	[76]

7.5 ENDOPHYTIC BACTERIA (EB) AS PGPR

As described earlier, endophytic bacteria (EB) constitute a bacterial group that can colonize the internal tissues of several higher vascular plants. Unlike phytopathogens, they do not pose any threat to the plants. On the contrary, bacterial endophytes have been shown to possess a synergistic association with the plant host. The plant provides a rich environment for the bacteria, and in return, it benefits from a variety of beneficial plant traits offered by endophytes, as shown in Figure 7.2 [3, 77].

EB could be opportunistic and enter the plant occasionally to profit from a highly nutritious environment and lack of competition (generally from wounds, germinating radicles, stomates, or after a nematode invasion of the roots) [78, 79], or they can be competent, meaning they can actively enter the plant host and are equipped with a group of genes that ensure invasion, colonization, and the maintenance of plant-endophyte association [24]. The main traits employed by this type of bacteria are chemotaxis, cell-wall degrading enzymes, lipopolysaccharide production, and motility [46].

In real conditions, cropped soils generally do not contain all the essential micro- and macro-nutrients for plant growth, in addition to the omnipresence of soil-borne pathogens that threaten plant health and productivity, such as *Phytophthora, Pythium,* and *Fusarium* ssp. [46]. As a solution to manage these problems, several strategies have been employed, mainly the chemical approach by the use of commercial pesticides and fertilizers; to compensate for the shortage of macro- and micro-elements in the soil and/or to suppress phytopathogens. However, the hazards posed by these chemicals and their direct impact on human health, the environment, and soil fertility make it necessary to limit their use and find alternative solutions [80].

EB resides among the most promising and eco-friendly alternatives. These bacteria do not necessarily inhibit pathogens; however, they enhance the growth of plants and physiology through several indirect and direct mechanisms. The former mechanisms, as described above, are generally ensured by suppressing phytopathogens and stimulating plant immune responses [81]. In this section, we will give an overview of the direct mechanisms, which comprise nutrient acquisition by providing the plant with nutritious compounds (e.g., soluble phosphorus, nitrogen, etc.), the secretion of phytohormones in addition to the ET biosynthesis inhibition to mitigate various abiotic and biotic stresses, etc. [82].

7.5.1 PHOSPHATE SOLUBILIZATION

Among the essential macronutrients for crop growth and physiology, soluble phosphate ($H_2PO_4^-$ and HPO_4^{2-} ions) constitutes a limiting factor in many enzymatic reactions and also in the development of young plantlets [83]. However, its availability in soil is generally low since most phosphate forms present in soil are not soluble and therefore cannot be absorbed by the plant. In addition, the plant assimilates only 25% of the phosphorus fertilizers, while the rest is transformed into insoluble forms (calcium-bound P) after complexation with soil [84].

EB and many other phosphate solubilizing bacteria (PSB) are a real ally for the plant host, thanks to their potential to increase the availability of soluble phosphorus forms by using several mechanisms, mainly by increasing acidity through organic acid production, which dissolves the insoluble precipitate directly or by chelating calcium ions that release the soluble P forms [85]. Some bacteria also use enzymatic mechanisms, participating in the mineralization of organic P by secreting acid phosphatase [86].

This feature plays a pivotal role in enhancing crop growth, thereby increasing yields and productivity. Numerous studies have documented the isolation of phosphate-solubilizing EB and their potential use as plant inoculants. Recently, 17 EB isolated from *Brassica napus* plants exhibited significant phosphate solubilizing activity, reaching 456 mg•L^{-1}. The isolate 1L6 was identified as *Bacillus pumilus* [87]. Researchers were able to isolate 10 EB from cultivated tomato plants with phosphate solubilizing activity, along with other PGP and antagonistic activities [88]. The ability of two endophytes, *Pseudomonas brenneri* LJ215 and *Paenibacillus glycanilyticus* LJ121, as plant growth promoters under seminatural and field conditions, was also assessed in white lupin [89].

7.5.2 NITROGEN FIXATION

Nitrogen fixation is another way that bacterial endophytes could improve plant growth in poor or nitrogen-deficient soils. This property enables endophytic bacteria to convert atmospheric nitrogen (N_2) into ammonia (NH_4^+), which can be easily assimilated by the host plant through the expression of nitrogenase enzyme synthesis genes [86]. This provides the plant with the necessary nitrogenous compounds essential for its growth, thereby enhancing plant physiology and increasing crop yields [90].

It is important to note that this activity is more efficient in endophytic bacteria than in other epiphytic or free rhizobial bacteria. Due to their presence inside plant tissue, endophytes can rapidly and efficiently transfer assimilable nitrogen to the plant [91]. Additionally, nitrogen fixation by endophytes is enhanced in conditions where they are protected from external factors such as UV radiation, temperature fluctuations, and high oxygen pressure, which can inhibit nitrogen fixation [92].

Many nitrogen-fixing endophytic strains are already used as plant inoculum. For example, an increase in plant biomass was reported after inoculation with strains like *Azospirillum brasilense, Gluconacetobacter-diazotrophicus,* and *Burkholderia* spp. The benefits of nitrogen fixation on plant growth promotion (PGP) were assessed by Pham et al. [92] in greenhouse conditions through a comparison between rice plantlets inoculated with *Pseudomonas stutzeri* strain A15, a nitrogenase-deficient mutant, and non-inoculated plantlets [92].

7.5.3 ACC DEAMINASE/SIDEROPHORE PRODUCTION AND PHYTOREMEDIATION

EB is capable of improving plant growth through stress mitigation. ET functions as a plant phytohormone that plays several roles in plant growth and metabolism modulation [78]. However, an increase in ET amounts inside the plant occurs during stress conditions (e.g., metals-contaminated soils) which could eventually lead to plant growth inhibition [43]. Some EB can stop massive ET production by cleaving its immediate precursor, 1-aminocyclopropane-1-carboxylate (ACC), into α-ketobutyrate and ammonia; the cleavage is done by ACC deaminase produced by EB (Figure 7.2) [93].

Indeed, strains *Pantoea agglomerans, Ralstonia* sp., and *Pseudomonas thivervalensis* isolated from Cu-tolerant plants were shown to exhibit ACC deaminase activity, enhanced rape plants growth, and Cu uptake under cupper-contaminated soils [94]. In a parallel study, Rezvani Borujeni et al. [95] asserted that the potential use of several ACC deaminase *Pseudomonas* strains as efficient poly-aromatic hydrocarbons (PAHs) decontaminants, with a PAHs removal percentage reaching 74.9% [95]. It was also proven that inoculation with ACC deaminase-producing bacteria improved the plant's adaptive properties towards heavy metal stress [96].

On the other hand, siderophores are minute, low molecular weight (<10 kDa) proteaceous compounds secreted by several rhizobia and endophytes, presenting high affinity to ferric ions Fe3+ and considered as chelating iron molecules [97]. Based on their chemical structure, siderophores are subdivided into three families – catecholate (phenolate), carboxylate, and hydroxamate. According to the previous definition, the main role of siderophores is iron scavenging; however, several studies have reported their ability to also transport many other metallic ions like V, Cr, Pb, Mo, and Co [98, 99].

These properties of siderophores make them an attractive tool for scientists to use in the decontamination of heavy metal-polluted soils. EB was proven to mainly secrete catechol-type siderophores [100], while *Streptomyces* sp. has the capacity to produce three different types of siderophores. Additionally, it was proven that the addition of Cd^+ in the medium elevated siderophore production, indicating their implication in Cd^+ decontamination.

Interestingly, a synergistic effect is observed when these two compounds are applied in combination. Indeed, the inoculation of *Helianthus annuus* seeds by a consortium of high ACC deaminase-producing *Pseudomonas* strain and siderophore-producing *Bacillus aerophilus* showed better tolerance to metal contamination and plant growth traits compared to the sole use of each bacterium [101].

7.5.4 INDOLE-3-ACETIC ACID (IAA)/ABSCISIC ACID (ABA) PRODUCTION AND DROUGHT TOLERANCE

Auxins are a group of phytohormones that regulate plant growth and many physiological activities. IAA is the primary plant auxin that is generally produced by plant-associated microbes and is relatively abundant in EB represented by *Rhizobium, Microbacterium,* and *Mycobacterium* genera [102]. It plays a key role in gene regulation and the modulation of cellular responses to different stresses [43].

Several researchers have reported the implication of auxin expression in drought tolerance, as it induces lateral rooting and hairing, thus increasing the exchange surface area. This, in turn, enhances water and nutrient uptake and ultimately helps the plant to overcome drought [103]. Among the over 30 Aux/IAA genes implicated in auxin signaling identified in rice, the OsIAA6 gene has been proven to be highly expressed in drought conditions. This significantly ameliorates rice drought tolerance by regulating auxin biosynthesis [104].

Abscisic acid, or ABA, is a weak acid and one of the low molecular weight phytohormones produced by plants. Similar to auxins, it plays various roles in plant growth, physiology, and response to abiotic stress [105]. Under normal conditions, ABA synthesis is generally low. However, osmotic stress triggers the expression of ABA synthesis genes, resulting in increased ABA production and its transportation through vascular tissues to the leaves. Once in the leaves, ABA induces stomatal closure to reduce evapotranspiration [106].

Furthermore, ABA strengthens the root system, allowing plants to dig deeper into the soil and increase their surface area for optimal moisture uptake [107]. It also enhances antioxidant enzyme activities and helps maintain relative water content by storing compatible solutes [106].

Despite the well-established literature on ABA and its mechanisms of action, little is known about the ability of EB to synthesize it. For example, Forchetti et al. [108] reported the isolation of ABA-producing endophytic strains *Bacillus pumilus* and *Achromobacter xiloxidans* from sunflower soil samples. In another study, the authors isolated the ABA-producing endophytic *Bacillus amyloliquefaciens* RWL-1 and assessed its capacity to enhance drought resistance in rice plants under normal and saline conditions [108, 109].

In summary, EB is a better option than PGPR for PGP, stress alleviation, and phytoremediation, as they can effectively communicate with the host by directly secreting several signaling and beneficial molecules inside plant tissues and vascular systems. Additionally, their presence inside the plant provides better protection against various biotic and abiotic stresses, while also reducing nutrient competition with other bacteria (Table 7.2).

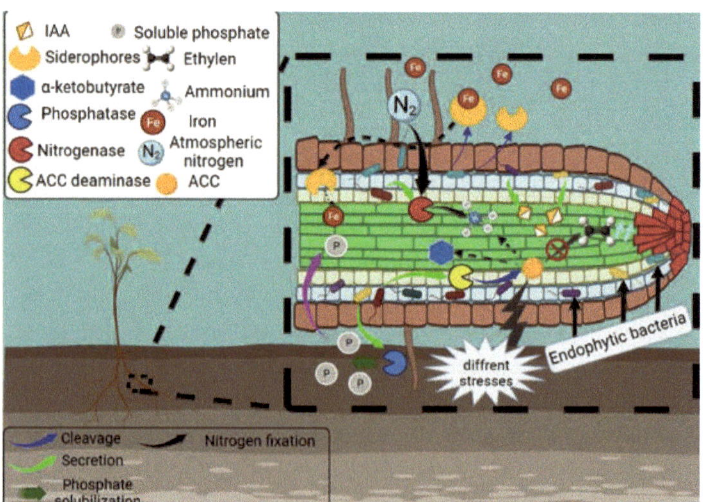

FIGURE 7.2 Illustration of the different PGP traits exhibited by endophytic bacteria.

TABLE 7.2 Main PGP Traits Exhibited by Endophytic Bacteria on Their Host Plants

Main PGP Traits	Endophytic Bacteria Species	Host Plant	References
Siderophores production	–	*Festuca rubra* and *Brassica napus*	[100]
IAA and siderophores production.	*Pseudomonas* sp.; *Acitenobacter brumalii*, *Enterobacter asburiae*, *Sinorhizobium meliloti*	Maize	[110]
Nitrogen fixation	*Paenibacillus polymyxa*	Lodgepole pine	[111]
Phosphate solubilization	*Bacillus megaterium* and *Ochrobactrum pseudogrignonense*	Millet	[112]
Zinc and phosphate solubilization, IAA production.	*Bacillus*, *Pseudomonas*, *Methylobacterium*, *Rhizobium*, *Paenibacillus*, and *Rummeliibacillus*	*Handroanthus impetiginous*	[113]
phosphate solubilization, ammonia, IAA, amylase, gelatinase, sidero-phores, and catalase production.	*Pseudomonas rhodesiae*, *Pantoea ananatis*, *Bacillus cereus*, *Enterobacter asburiae*, *Staphylococcus sciuri*, *Burkholderia cenocepacia*, and *Ralstonia pickettii*	Rice	[114]
Auxin production and solubilize phosphate.	*Curtobacterium* sp., *Bacillus amyloliquefaciens*, *Methylobacterium* sp., and *Microbacterium* sp.	*Urochloa ramosa*	[115]
Calcium phosphate solubilization, synthesis of IAA and siderophores.	*B. velezensis*, *B. aryabhattai*, *Klebsiella pneumonia*,and *B. cereus*	Cassava	[116]
ACC deaminase and IAA production.	*Bacillus mojavensis* PRN2	*Pisum sativum*	[117]
IAA, geberilins, ABA, and ACC deaminase production.	*Curtobacterium* sp. SAK1	*Artemisia princeps*	[118]
IAA production, pellicle formation, ACC deaminase, phosphate solubilization, protease, lipase, cellulase, and amylase production.	*Bacillus*, *Staphylococcus*,and *Kocuria*	*Ammodendron bifolium*	[119]
Produced indole-acetic amino acid (IAA), 6 produced siderophore, 27 showed 1-aminocyclopropane-1-carboxylic acid (ACC) deaminase activity, 37 dissolved precipitated tricalcium phosphate, 36 fixed nitrogen, and 12 hydrolyzed starch.	*Pantoea*, *Enterobacter*, *Bacillus*,*Klebsiella*, *Tetragenococcus*, *Massilia*, *Raoultella*, *Serratia*,and *Pectobacterium*	*Nypa fruticans*	[120]

TABLE 7.2 *(Continued)*

Main PGP Traits	Endophytic Bacteria Species	Host Plant	References
ACC deaminase and phosphate and potassium solubilization, nitrogen fixation.	*Bacillus*, *Stenotrophomonas*, *Pseudomonas*, *Microbacterium*,and *Curtobacterium*	*Helianthus tuberosus*	[121]
Mineral phosphate solubilization, N-2-fixation, ACC deaminase, IAA, and siderophores production.	*Bacillus microbacterium*	*Oryza rufipogon Griff*	[122]
Siderophore, (IAA), and (ACC) deaminase production	*Pseudomonas aeruginosa* L10	*Phragmites australis*	[123]
ACC deaminase activity, siderophore, and IAA production, heavy metal detoxification, and P-solubilization.	*Serratia marcescens*	*Pteris vittate*	[124]
ACC deaminase, IAA, exopolysaccharide production.	*Bacillus aquimaris*, *Micrococcus luteus*, *Bacillus methylotrophicus*	*Helianthus tuberosus*	[125]
ACC deaminase	*Pseudomonas putida* PIR3Cand *Roultella terrigena* PCM8	Soybean	[126]
Biofilm formation, hydrolytic enzymes secretion like cellulase, protease, and amylase, phosphate solubilization, nitrogen fixation, IAA, siderophore, ammonia, and 1-aminocyclopropane-1-carboxylic acid deaminase production.	*Stenotrophomonas* sp. and *Pseudomonas* sp.	*Camelliasinensis*	[127]
Cell wall-degrading enzymes (protease and cellulase), siderophores, and non-volatile metabolites.	*Bacillus velezensis*	*Fraxinus hupehensis*	[128]
Phosphate solubilization, IAA, siderophores, lytic enzymes production.	*Bacillus megaterium* DZK1BH	*Lotus corniculatus* and *Dactylis glomerata*	[129]
Solubilization of phosphate from hydroxyapatite, production of siderophores, and nitrogen fixation.	*Pseudomonas protegens*	The Mexican giant maize of Jala landrace	[130]

7.6 APPLICATION OF ENDOPHYTIC BACTERIA (EB)

In sustainable agriculture, endophytes are frequently employed as biofertilizers, biocontrol agents, and inoculants to alleviate abiotic constraints such as drought, salinity, cold, and soil pH variations. Although omics and other technologies have advanced, the prospective applications of endophytes are still largely unknown [131]. Recently, the application of endophytes in medicine, agriculture, and other fields has increased. These applications include enhancing crop growth and wellness through the secretion of phytohormones or biofertilizers, fostering tolerance to biotic and abiotic stresses, enhancing plant immunity, and suppressing disease. These applications are also associated with the release of new compounds, secondary metabolites, siderophores, herbicides, and pesticides such as nitrogenous or other ionic fertilizers [17].

In agriculture, horticulture, agroforestry, and phytoremediation, plant growth is promoted by using endophytic bacteria (EB) which operate similarly to rhizosphere bacteria. Plant growth-promoting endophytes (PGPEs) are a type of bio-fertilizer that enhances the host's ability to absorb nutrients, increases production, reduces fertilizer costs, and creates an eco-safe environment by minimizing nitrogen leaching. Moreover, endophytic microbes assist plants in thriving under challenging conditions such as salinity, nutrient stress, trace metal stress, temperature stress, or drought. They also aid in the growth of plants in contaminated environments by breaking down harmful compounds. Additionally, they produce various growth-stimulating hormones like indole acetic acid (IAA), gibberellic acid (GA), ABA, and cytokinin, which play a crucial role in plant growth and mitigating biotic stress [132, 133]. To enhance the effectiveness of biological control methods, appropriate and relevant application techniques for EB must be selected. Researchers have explored different strategies, including seed treatment, soil trenching, stem injection, and foliage spraying, and have found that each method has varying levels of efficacy [134].

In this study, scientists investigated a seed inoculation approach to enhance the survival and effectiveness of bacterial inoculations. The goal was to extend the shelf lives of seeds and improve their compatibility with fungicides commonly used. Crop Genetics International Ltd. developed a method for seed inoculation by introducing the endophytic bacterium *Achromobacter xylosoxidans* to the seeds. This approach reduced the symptoms of rice blast disease by promoting the synthesis of enzymes associated with plant defense, resulting in significant increases in rice output [134]. Another investigation discovered an endophytic bacterium called *Bacillus safensis* strain ZY16, which exhibited the ability to produce biosurfactants, degrade hydrocarbons, tolerate salt, and enhance plant development [135].

Furthermore, *Bacillus amyloliquefaciens* S13-3, an endophyte renowned for producing antibiotics, has also been found to inhibit strawberry anthracnose (*Colletotrichum gloeosporioides*) when applied topically [71].

According to Sundaramoorthy et al. [136], the use of talc-based bioformulation of *P. fluorescens* (Pf1) and *B. subtilis* resulted in a 17-30% reduction in Fusarium wilt in chili peppers compared to the control (EPCO16 and EPC5 strains) [136]. The incidence of Fusarium wilt in cv. Grand Naine banana was significantly reduced by applying these bacterial mixtures of *Bacillus* and *Pseudomonas* to the soil [70].

On the other hand, the secretion of lytic enzymes like cellulases and chitinases by EB enables them to be used for phytoremediation of environmental pollutants or the management of fungal diseases. Additionally, their broad host range allows for a variety of applications as they are sources of novel biomolecules for the biochemical and pharmaceutical industries. In recent times, EB has also been used to create nanoparticles for both industrial and medical applications [17]..

In the pharmaceutical industry, researchers claimed that endophytes were responsible for approximately half of all novel medications introduced between 1981 and 2010. Insecticides, antioxidants (AOX), antibacterial agents, anticancer, and antidiabetic substances are just a few of the applications. Many anticancer and antibacterial metabolites have multiple targets in plants, animals, and human diseases, making them widely applicable in medical and veterinary medicine. The majority of these applications are eco-safe [17].

Biopesticides, medicines, detergents, and secondary metabolites are biologically active substances produced by microorganisms and used in the production of various industrial items, including food and food supplements, biofuels, and pharmaceuticals. This method of production has been successfully employed in various studies. For example, Ntabo et al. identified EB from mangrove plants grown in Kenya and demonstrated that they secrete enzymes with potential economic utility, such as proteases, pectinases, cellulases, amylases, and chitinases. The Phragmites australis root-associated endophytic *Pseudomonas aeruginosa* L10 effectively decomposed hydrocarbons and produced a biosurfactant [17]. Baker et al. investigated *Coffea arabica* L. EB and discovered their ability to degrade caffeine, which could be utilized for decaffeinating beverages. A comprehensive understanding of EB can greatly assist researchers in utilizing them more efficiently and incorporating them to enhance the quality and yield of medicinal-based plants [137].

7.7 CONCLUSION AND PERSPECTIVES

Increasingly, the plant defense system is being considered as a comprehensive immune system, in which the plant's response to abiotic/biotic stress is the outcome of a complex interaction involving microbes. In this review paper, we have outlined the mechanisms of action exerted by diverse bacterial endophytes to enhance plant defense. Based on this role, we have discussed some studies that demonstrate the ability of these bacteria to control plant diseases. The studies discussed in this paper may provide insights on developing future strategies to control plant pathogens and gain a deeper understanding of the various interactions between endophytes, plants, and the entire microbial community.

Given the immense diversity of bacterial endophytes, we recommend exploring wild plants that may harbor specialized endophytes capable of combating abiotic and biotic stresses. These bacteria have the potential to significantly improve plant performance, which is essential for sustainable agricultural production. Furthermore, endophytes may serve as agents for component production, contributing to increased production yield and promoting more sustainable agriculture.

ACKNOWLEDGMENTS

This study was financially supported by the Phytopathology Unit of the Department of Plant Protection and Environment at Ecole Nationale d'Agriculture de Meknès (Morocco).

CONFLICTS OF INTEREST

No potential conflict of interest was reported by the authors.

KEYWORDS

- **abiotic stress**
- **biological control**
- **biotic stress**
- **endophytic bacteria**
- **immune response**
- **phosphate solubilization**
- **plant growth-promoting rhizobacteria**
- **plant disease control agent**

REFERENCES

1. FAO, (2018). *The Future of Food and Agriculture – Alternative Pathways to 2050* (p. 228). Food and Agriculture Organization of the United Nations. Rome, Italy.
2. Bishnoi, U., (2015). *PGPR Interaction: An Ecofriendly Approach Promoting the Sustainable Agriculture System*. Elsevier Ltd.
3. Santoyo, G., Moreno-Hagelsieb, G., Del Carmen Orozco-Mosqueda, M., & Glick, B. R., (2016). Plant growth-promoting bacterial endophytes. *Microbiological Research, 183*, 92–99. doi: 10.1016/j.micres.2015.11.008.
4. Hallmann, J., (1997). *Bacterial Endophytes in Agricultural Crops,43*(10), 895–914. cdnsciencepub.com. doi: 10.1139/m97-131.
5. Partida-Martínez, L. P., & Heil, M., (2011). The microbe-free plant: Fact or artifact? *Frontiers in Plant Science, 2*, 1–16. doi: 10.3389/fpls.2011.00100.
6. Reinhold-Hurek, B., & Hurek, T., (1998). Life in grasses: Diazotrophic endophytes. *Trends in Microbiology, 6*(4), 139–144. doi: 10.1016/S0966-842X(98)01229-3.
7. Solanki, M. K., Yandigeri, M. S., Kumar, S., Singh, R. K., & Srivastava, A. K., (2019). Co-inoculation of different antagonists can enhance the biocontrol activity against *Rhizoctonia solani* in tomato. *Antonie van Leeuwenhoek, International Journal of General and Molecular Microbiology, 112*(11), 1633–1644. doi: 10.1007/s10482-019-01290-8.
8. Pandey, P. K., Samanta, R., & Yadav, R. N. S., (2019). Inside the plant: Addressing bacterial endophytes in biotic stress alleviation. *Archives of Microbiology, 201*(4), 415–429. doi: 10.1007/s00203-019-01642-y.
9. Ardanov, P., Ovcharenko, L., Zaets, I., Kozyrovska, N., & Pirttilä, A. M., (2011). Endophytic bacteria enhancing growth and disease resistance of potato (*Solanum tuberosum* L.). *Biological Control, 56*(1), 43–49. doi: 10.1016/j.biocontrol.2010.09.014.
10. Cipollini, D., Rigsby, C. M., & Barto, E. K., (2012). Microbes as targets and mediators of allelopathy in plants. *Journal of Chemical Ecology, 38*(6), 714–727. doi: 10.1007/s10886-012-0133-7.
11. Li, Q., Guo, R., Li, Y., et al., (2019). Insight into the bacterial endophytic communities of peach cultivars related to crown gall disease resistance. *Applied and Environmental Microbiology, 85*(9). doi: 10.1128/AEM.02931-18.
12. De Almeida, L. K. B., Carpentieri-Pipolo, V., Fira, D., et al., (2018). Screening of bacterial endophytes as potential biocontrol agents against soybean diseases. *Journal of Applied Microbiology, 125*(5), 1466–1481. doi: 10.1111/JAM.14041.
13. Aiello, D., Restuccia, C., Stefani, E., Vitale, A., & Cirvilleri, G., (2019). Postharvest biocontrol ability of *Pseudomonas synxantha* against *Monilinia fructicola* and *Monilinia fructigena* on stone fruit. *Postharvest Biology and Technology, 149*, 83–89. https://core.ac.uk/download/pdf/195755122.pdf (accessed on 9 January 2023).
14. Nair, D. N., & Padmavathy, S., (2014). Impact of endophytic microorganisms on plants, environment, and humans. *The Scientific World Journal, 2014*. doi: 10.1155/2014/250693.
15. Zinniel, D. K., Lambrecht, P., Harris, N. B., et al., (2002). Isolation and characterization of endophytic colonizing bacteria from agronomic crops and prairie plants. *Applied and Environmental Microbiology, 68*(5), 2198–2208. doi: 10.1128/AEM.68.5.2198-2208.2002.

16. Ali, M., Ali, Q., Sohail, M. A., et al., (2021). Molecular sciences diversity and taxonomic distribution of endophytic bacterial community in the rice plant and its prospective state key laboratory for managing biotic and chemical threats to the quality and safety of agro-products. *Int. J. Mol. Sci, 22*, 10165. doi: 10.3390/ijms221810165.
17. Wu, W., Chen, W., Liu, S., et al., (2021). Beneficial relationships between endophytic bacteria and medicinal plants. *Frontiers in Plant Science, 12*, 1–13. doi: 10.3389/fpls.2021.646146.
18. Sharma, A., Kaushik, N., Sharma, A., et al., (2021). Screening of tomato seed bacterial endophytes for antifungal activity reveals lipopeptide producing *Bacillus siamensis* strain NKIT9 as a potential bio-control agent. *Frontiers in Microbiology, 12*, 1–17. doi: 10.3389/fmicb.2021.609482.
19. Mushtaq, S., Shafiq, M., Ashraf, T., Haider, M. S., Ashfaq, M., & Ali, M., (2019). Characterization of plant growth promoting activities of bacterial endophytes and their antibacterial potential isolated from citrus. *Journal of Animal and Plant Sciences, 29*(4), 978–991.
20. Liu, J. M., Wang, S. S., Zheng, X., et al., (2020). Antimicrobial activity against phytopathogens and inhibitory activity on solanine in potatoes of the endophytic bacteria isolated from potato tubers. *Frontiers in Microbiology, 11*, 1–16. doi: 10.3389/fmicb.2020.570926.
21. Brígido, C., Singh, S., Menéndez, E., et al., (2019). Diversity and functionality of culturable endophytic bacterial communities in chickpea plants. *Plants, 8*(2), 1–21. doi: 10.3390/plants8020042.
22. Senthilkumar, M., Anandham, R., Madhaiyan, M., Venkateswaran, V., & Sa, T., (2011). Endophytic bacteria: Perspectives and applications in agricultural crop production. *Bacteria in Agrobiology: Crop Ecosystems*, 61–96. doi: 10.1007/978-3-642-18357-7_3.
23. Maggini, V., Mengoni, A., Gallo, E. R., et al., (2019). Tissue specificity and differential effects on *in vitro* plant growth of single bacterial endophytes isolated from the roots, leaves and rhizospheric soil of *Echinacea purpurea*. *BMC Plant Biology, 19*(1), 1–9. doi: 10.1186/s12870-019-1890-z.
24. Rosenblueth, M., & Martínez-Romero, E., (2006). Bacterial endophytes and their interactions with hosts. *Molecular Plant-Microbe Interactions, 19*(8), 827–837. doi: 10.1094/MPMI-19-0827.
25. Romero, F. M., Marina, M., & Pieckenstain, F. L., (2014). The communities of tomato (*Solanum lycopersicum* L.) leaf endophytic bacteria, analyzed by 16S-ribosomal RNA gene pyrosequencing. *FEMS Microbiology Letters, 351*(2), 187–194. doi: 10.1111/1574-6968.12377.
26. Papik, J., Folkmanova, M., Polivkova-Majorova, M., Suman, J., & Uhlik, O., (2020). The invisible life inside plants: Deciphering the riddles of endophytic bacterial diversity. *Biotechnology Advances, 44*, 107614. doi: 10.1016/j.biotechadv.2020.107614.
27. Oukala, N., Aissat, K., & Pastor, V., (2021). Bacterial endophytes: The hidden actor in plant immune responses against biotic stress. *Plants*, 1–24.
28. Jha, Y., (2019). Endophytic bacteria as a modern tool for sustainable crop management under stress. *Soil Biology*, 203–223.
29. Pozo, M. J., & Azcón-Aguilar, C., (2007). Unraveling mycorrhiza-induced resistance. *Current Opinion in Plant Biology, 10*(4), 393–398. doi: 10.1016/j.pbi.2007.05.004.

30. Eid, A. M., Salim, S. S., Hassan, S. E. D., Ismail, M. A., & Fouda, A., (2019). Role of endophytes in plant health and abiotic stress management. In: *Microbiome in Plant Health and Disease* (pp. 119–144). Springer.
31. Gwinn, K. D., (2018). Bioactive natural products in plant disease control. In: *Studies in Natural Products Chemistry* (Vol. 56, pp. 229–246). Elsevier B.V.
32. Pieterse, C. M. J., Zamioudis, C., Berendsen, R. L., Weller, D. M., Van, W. S. C. M., & Bakker, P. A. H. M., (2014). Induced systemic resistance by beneficial microbes. *Annual Review of Phytopathology, 52.*
33. Portieles, R., Xu, H., Yue, Q., et al., (2021). Heat - killed endophytic bacterium induces robust plant defense responses against important pathogens. *Scientific Reports, 11*, 1–14. doi: 10.1038/s41598-021-91837-5.
34. Firdous, J., Lathif, N. A., Mona, R., & Muhamad, N., (2019). Endophytic bacteria and their potential application in agriculture: A review. *Indian Journal of Agriculture, 53*(1), 1–7. doi: 10.18805/IJARe.A-366.
35. Kloepper, J. W., & Ryu, C., (2006). Bacterial endophytes as elicitors of induced systemic resistance. *Microbial Root Emdophytes, 9.*
36. Xia, Y., Liu, J., Chen, C., et al., (2022). The multifunctions and future prospects of endophytes and their metabolites in plant disease management. *Microorganisms, 10*, 1–19.
37. Ryu, C., Farag, M. A., Hu, C., Reddy, M. S., Kloepper, J. W., & Pare, P. W., (2004). Bacterial volatiles induce systemic resistance in *Arabidopsis*. *Plant Physiology, 134*(3), 1017–1026. doi: 10.1104/pp.103.026583.with.
38. Gao, Y., Zeng, X. D., Ren, B., et al., (2019). Antagonistic activity against rice blast disease and elicitation of host-defense response capability of an endophytic *Streptomyces albidoflavus* OsiLf-2.*Plant Pathology*., 259–271. doi: 10.1111/ppa.13118.
39. Meena, K. K., Sorty, A. M., Bitla, U. M., et al., (2017). Abiotic stress responses and microbe-mediated mitigation in plants: The omics strategies. *Frontiers in Plant Science, 8*, 172.
40. Subramanian, P., Mageswari, A., Kim, K., Lee, Y., & Sa, T., (2015). Psychrotolerant endophytic *Pseudomonas* sp. strains OB155 and OS261 induced chilling resistance in tomato plants (*Solanum lycopersicum* Mill.) by activation of their antioxidant capacity. *Molecular Plant-Microbe Interactions, 28*(10), 1073–1081.
41. Fernandez, O., Theocharis, A., Bordiec, S., et al., (2012). Burkholderia phytofirmans PsJN acclimates grapevine to cold by modulating carbohydrate metabolism. *Molecular Plant-Microbe Interactions, 25*(4), 496–504.
42. Paul, D., & Lade, H., (2014). Plant-growth-promoting rhizobacteria to improve crop growth in saline soils: A review. *Agron. Sustain. Dev., 34*, 737–752. doi: 10.1007/s13593-014-0233-6.
43. Ullah, A., Nisar, M., Ali, H., et al., (2019). Drought tolerance improvement in plants: An endophytic bacterial approach. *Applied Microbiology and Biotechnology, 103*(18), 7385–7397. doi: 10.1007/s00253-019-10045-4.
44. Vargas, L., Santa, B. A. B., Mota, F. J. P., et al., (2014). Drought tolerance conferred to sugarcane by association with *Gluconacetobacter diazotrophicus*: A transcriptomic view of hormone pathways. *PLoS One, 9*(12), e114744.
45. Vaishnav, A., Shukla, A. K., Sharma, A., Kumar, R., & Choudhary, D. K., (2019). Endophytic bacteria in plant salt stress tolerance: Current and future prospects. *Journal of Plant Growth Regulation, 38*(2), 650–668. doi: 10.1007/s00344-018-9880-1.

46. Liu, H., Carvalhais, L. C., Crawford, M., et al., (2017). Inner plant values: Diversity, colonization, and benefits from endophytic bacteria. *Frontiers in Microbiology, 8*, 1–17. doi: 10.3389/fmicb.2017.02552.
47. Soylu, S., Soylu, E. M., Kurt, S., & Ekici, O. K., (2005). Antagonistic potentials of rhizosphere-associated bacterial isolates against soil-borne diseases of tomato and pepper caused by *Sclerotinia sclerotiorum* and *Rhizoctonia solani*. *Journal of Biological Sciences, 9*(1), 43–48.
48. Dutta, D., Puzari, K. C., Gogoi, R., & Dutta, P., (2014). Brazilian archives of biology and technology endophytes: Exploitation as a tool in plant protection. *Arch. Biol. Technol., 57557*(5), 621–629.
49. Strobel, G., Daisy, B., Castillo, U., & Harper, J., (2004). Natural products from endophytic microorganisms. *Journal of Natural Products, 67*(2), 257–268. doi: 10.1021/np030397v.
50. Zachow, C., Jahanshah, G., De Bruijn, I., et al., (2015). The novel lipopeptide poaeamide of the endophyte *Pseudomonas poae* RE∗1-1-14 is involved in pathogen suppression and root colonization. *Molecular Plant-Microbe Interactions, 28*(7), 800–810. doi: 10.1094/MPMI-12-14-0406-R.
51. Gupta, G. N., Srivastava, S., Khare, S. K., & Prakash, V., (2014). Role of phosphate solubilizing bacteria in crop growth and disease management. *Journal of Pure and Applied Microbiology, 8*(1), 461–474.
52. Demain, A. L., & Sanchez, S., (2009). *Microbial Drug Discovery: 80 Years of Progress*, 5–16. doi: 10.1038/ja.2008.16.
53. Meena, K. R., & Kanwar, S. S., (2015). *Lipopeptides as the Antifungal and Antibacterial Agents: Applications in Food Safety and Therapeutics, 2015.*
54. Tabbene, O., Slimene, I. B., & Bouabdallah, F., (2009). *Production of Anti-methicillin-Resistant Staphylococcus Activity from Bacillus Subtilis sp. Strain B38 Newly Isolated from Soil,* 407–419. doi: 10.1007/s12010-008-8277-1.
55. Gond, S. K., Bergen, M. S., Torres, M. S., & Jr, J. F. W., (2014). Endophytic *Bacillus* spp. produce antifungal lipopeptides and induce host defense gene expression in maize. *Microbiological Research.* doi: 10.1016/j.micres.2014.11.004.
56. Rika, F. N. B., Aris, T. W., & Nurita, T. M., (2014). Control activity of potential antifungal-producing *Burkholderia* sp. in suppressing *Ganoderma boninense* growth in oil palm. *Asian Journal of Agricultural Research, 8*, 259–268.
57. Hammer, P. E., Hill, D. S., Lam, S. T., Pe, K. V., Ligon, J. M., (1997). Four genes from *Pseudomonas fluorescens* that encode the biosynthesis of pyrrolnitrin. *Applied and Environmental Microbiology, 63*(6), 2147–2154.
58. Rodrigues, G. M., & Smith, I., (2003). Mechanisms of iron regulation in mycobacteria: Role in physiology and virulence. *Molecular Microbiology, 47*(6), 1485–1494.
59. Beneduzi, A., Ambrosini, A., & Passaglia, L. M. P., (2012). *Plant Growth-Promoting Rhizobacteria (PGPR): Their Potential as Antagonists and Biocontrol Agents, 4*, 1044–1051.
60. Husen, E., (2003). Screening of soil bacteria for plant growth promotion activities *in vitro*. *Indonesian Journal of Agricultural Science, 4*(1), 27–31.
61. Algeblawi, A., & Adam, F., (2013). Biological control of *Erwinia carotovora* subsp. Carotovora by *Pseudomonas fluorescens*, *Bacillus subtilis*, and *Bacillus thuringiensis*. *Int. J. Chem. Environ. Biol. Sci., 1*(5), 770–771.

62. Kesaulya, H., Hasinu, J. V., & Tuhumury, G. N. C., (2018). Potential of *Bacillus* spp produces siderophores in suppressing the wilt disease of banana plants. *IOP Conference Series: Earth and Environmental Science, 102*(1). doi: 10.1088/1755-1315/102/1/012016.
63. Sun, Z. B., Yuan, X. F., Zhang, H., Wu, L. F., Liang, C., & Feng, Y. J., (2013). Isolation, screening, and identification of antagonistic downy mildew endophytic bacteria from cucumber. *European Journal of Plant Pathology, 137*(4), 847–857. doi: 10.1007/s10658-013-0293-5.
64. Nawangsih, A. A., Damayanti, I. K. A., Wiyono, S., & Kartika, J. G., (2011). Selection and characterization of endophytic bacteria as biocontrol agents of tomato bacterial wilt disease. *HAYATI Journal of Biosciences, 18*(2), 66–70. doi: 10.4308/hjb.18.2.66.
65. Ferrigo, D., Causin, R., & Raiola, A., (2017). Effect of potential biocontrol agents selected among grapevine endophytes and commercial products on crown gall disease. *BioControl, 62*(6), 821–833. doi: 10.1007/s10526-017-9847-3.
66. Nagendran, K., Karthikeyan, G., Faisal, P. M., Raveendran, M., Prabakar, K., & Raguchander, T., (2013). Management of bacterial leaf blight disease in rice with endophytic bacteria. *World Applied Sciences Journal, 28*(12), 2229–2241. doi: 10.5829/idosi.wasj.2013.28.12.2009.
67. Omoboye, O. O., Oni, F. E., Batool, H., Yimer, H. Z., De Mot, R., & Höfte, M., (2019). *Pseudomonas* cyclic lipopeptides suppress the rice blast fungus *Magnaporthe oryzae* by induced resistance and direct antagonism. *Frontiers in Plant Science, 10*, 1–17. doi: 10.3389/fpls.2019.00901.
68. Collins, D. P., & Jacobsen, B. J., (2003). Optimizing a *Bacillus subtilis* isolate for biological control of sugar beet cercospora leaf spot. *Biological Control, 26*(2), 153–161. https://www.sciencedirect.com/science/article/abs/pii/S1049964402001329 (accessed on 9 January 2023).
69. Melnick, R. L., (2010). *Endophytic Bacillus spp. of Theobroma cacao: Ecology and Potential for Biological Control of Cacao Diseases*. Doctoral Dissertation.
70. Raman, T., & Muthukathan, G., (2015). Field suppression of *Fusarium* wilt disease in banana by the combined application of native endophytic and rhizospheric bacterial isolates possessing multiple functions. *Phytopathologia Mediterranea, 54*(2), 241–252. doi: 10.14601/Phytopathol.
71. Yamamoto, S., Shiraishi, S., & Suzuki, S., (2015). Are cyclic lipopeptides produced by *Bacillus amyloliquefaciens* S13-3 responsible for the plant defense response in strawberry against *Colletotrichum gloeosporioides*? *Letters in Applied Microbiology, 60*(4), 379–386.
72. Pan, D., Mionetto, A., Tiscornia, S., & Bettucci, L., (2015). Endophytic bacteria from wheat grain as biocontrol agents of *Fusarium* graminearum and deoxynivalenol production in wheat. *Mycotoxin Research, 31*(3), 137–143. doi: 10.1007/s12550-015-0224-8.
73. Franco, S. H., Sandro, A. S. H., Soares De, M. I., Vieira, N. F., & Bettiol, W., (2006). Bioprospecting endophytic bacteria for biological control of coffee leaf rust bioprospecção de bactérias endofíticas como agentes de biocontrole da ferrugem do cafeeiro. *Sci. Agric. Piracicaba, 63*(1), 32–39.
74. Muthukumar, A., & Venkatesh, A., (2013). *A New Record of Leaf Blight of Ribben Plant Caused by Alternaria Alternata in India, 2*(3), 228–230.
75. Yanti, Y., Warnita, Reflin, & Busniah, M., (2018). Indigenous endophyte bacteria ability to control *Ralstonia* and *Fusarium* wilt disease on chili pepper. *Biodiversitas, 19*(4), 1532–1538. doi: 10.13057/biodiv/d190446.

76. El-shakh, A. S. A., Kakar, K. U., Wang, X., et al., (2015). Controlling bacterial leaf blight of rice and enhancing the plant growth with endophytic and rhizobacterial *Bacillus* strains. *Toxicological and Environmental Chemistry, 97*(6), 766–785. doi: 10.1080/02772248.2015.1066176.
77. Shcherbakov, A. V., Bragina, A. V., Kuzmina, E. Y., et al., (2013). Endophytic bacteria of sphagnum mosses as promising objects of agricultural microbiology. *Microbiology (Russian Federation),82*(3), 306–315. doi: 10.1134/S0026261713030107.
78. Hardoim, P. R., Van, O. L. S., & Elsas, J. D. V., (2008). Properties of bacterial endophytes and their proposed role in plant growth. *Trends in Microbiology, 16*(10), 463–471. doi: 10.1016/j.tim.2008.07.008.
79. Rajkumar, M., Ae, N., & Freitas, H., (2009). Endophytic bacteria and their potential to enhance heavy metal phytoextraction. *Chemosphere, 77*(2), 153–160. doi: 10.1016/j.chemosphere.2009.06.047.
80. Aktar, W., Sengupta, D., &Chowdhury, A., (2009). Impact of pesticides use in agriculture: Their benefits and hazards. *Interdisciplinary Toxicology, 2*(1), 1–12. doi: 10.2478/v10102-009-0001-7.
81. Yadav, A., (2017). Exploring the potential of endophytes in agriculture: A minireview. *Advances in Plants & Agriculture Research, 6*(4). doi: 10.15406/apar.2017.06.00221.
82. Singh, M., Kumar, A., Singh, R., & Pandey, K. D., (2017). Endophytic bacteria: A new source of bioactive compounds. *3 Biotech, 7*(5). doi: 10.1007/s13205-017-0942-z.
83. Puente, M. E., Li, C. Y., & Bashan, Y., (2009). Endophytic bacteria in cacti seeds can improve the development of cactus seedlings. *Environmental and Experimental Botany, 66*(3), 402–408. doi: 10.1016/j.envexpbot.2009.04.007.
84. Bhattacharyya, P. N., & Jha, D. K., (2012). Plant growth-promoting rhizobacteria (PGPR): Emergence in agriculture. *World Journal of Microbiology and Biotechnology, 28*(4), 1327–1350. doi: 10.1007/s11274-011-0979-9.
85. Nautiyal, C. S., Bhadauria, S., Kumar, P., Lal, H., Mondal, R., & Verma, D., (2000). Stress induced phosphate solubilization in bacteria isolated from alkaline soils. *FEMS Microbiology Letters, 182*(2), 291–296. doi: 10.1016/S0378-1097(99)00605-9.
86. Afzal, I., Shinwari, Z. K. Z. K., Sikandar, S., & Shahzad, S., (2019). Plant beneficial endophytic bacteria: Mechanisms, diversity, host range and genetic determinants. *Microbiological Research, 221*, 36–49. doi: 10.1016/j.micres.2019.02.001.
87. Lipková, N., Medo, J., Artimová, R., et al., (2021). Growth promotion of rapeseed (*Brassica napus* L.) and blackleg disease (*Leptosphaeria maculans*) suppression mediated by endophytic bacteria. *Agronomy, 11*(10). doi: 10.3390/agronomy11101966.
88. Romero, F. M., Marina, M., & Pieckenstain, F. L., (2016). Novel components of leaf bacterial communities of field-grown tomato plants and their potential for plant growth promotion and biocontrol of tomato diseases. *Research in Microbiology, 167*(3), 222–233. doi: 10.1016/j.resmic.2015.11.001.
89. Ferchichi, N., Toukabri, W., Boularess, M., Smaoui, A., Mhamdi, R., & Trabelsi, D., (2019). Isolation, identification, and plant growth promotion ability of endophytic bacteria associated with lupine root nodule grown in Tunisian soil. *Archives of Microbiology, 201*(10), 1333–1349. doi: 10.1007/s00203-019-01702-3.
90. Govindarajan, M., Balandreau, J., Kwon, S. W., Weon, H. Y., & Lakshminarasimhan, C., (2008). Effects of the inoculation of *Burkholderia vietnamensis* and related endophytic diazotrophic bacteria on grain yield of rice. *Microbial Ecology, 55*(1), 21–37. doi: 10.1007/s00248-007-9247-9.

91. Lata, R. K., Divjot, K., & Nath, Y. A., (2019). Endophytic microbiomes: Biodiversity, ecological significance, and biotechnological applications. *Research Journal of Biotechnology, 14*(10), 142–162.
92. Pham, V. T. K., Rediers, H., Ghequire, M. G. K., et al., (2017). The plant growth-promoting effect of the nitrogen-fixing endophyte *Pseudomonas stutzeri* A15. *Archives of Microbiology, 199*(3), 513–517. doi: 10.1007/s00203-016-1332-3.
93. Gamalero, E., & Glick, B. R., (2015). Bacterial modulation of plant ethylene levels. *Plant Physiology, 169*(1), 13–22. doi: 10.1104/pp.15.00284.
94. Zhang, Y. F., He, L. Y., Chen, Z. J., Wang, Q. Y., Qian, M., & Sheng, X. F., (2011). Characterization of ACC deaminase-producing endophytic bacteria isolated from copper-tolerant plants and their potential in promoting the growth and copper accumulation of *Brassica napus*. *Chemosphere, 83*(1), 57–62. doi: 10.1016/j.chemosphere.2011.01.041.
95. Rezvani, B. S., Khavazi, K., Asgharzadeh, A., & Rezvani, B. I., (2018). Use of bacterial acc deaminase to increase oil (especially poly aromatic hydrocarbons) phytoremediation efficiency for maize (*Zea mays*) seedlings. *International Journal of Phytoremediation, 20*(5), 476–482. doi: 10.1080/15226514.2017.1374330.
96. Grobelak, A., Kokot, P., Światek, J., Jaskulak, M., & Rorat, A., (2018). Bacterial ACC deaminase activity in promoting plant growth on areas contaminated with heavy metals. *Journal of Ecological Engineering, 19*(5), 150–157. doi: 10.12911/22998993/89818.
97. Ghosh, S. K., Bera, T., & Chakrabarty, A. M., (2020). Microbial siderophore – A boon to agricultural sciences. *Biological Control, 144*, 104214. doi: 10.1016/j.biocontrol.2020.104214.
98. Bellenger, J. P., Wichard, T., Kustka, A. B., & Kraepiel, A. M. L., (2008). Uptake of molybdenum and vanadium by nitrogen-fixing soil bacterium using siderophores. *Nature Geoscience, 1*(4), 243–246. doi: 10.1038/ngeo161.
99. Braud, A., Jézéquel, K., Bazot, S., & Lebeau, T., (2009). Enhanced phytoextraction of an agricultural Cr- and Pb-contaminated soil by bioaugmentation with siderophore-producing bacteria. *Chemosphere, 74*(2), 280–286. doi: 10.1016/j.chemosphere.2008.09.013.
100. Grobelak, A., & Hiller, J., (2017). Bacterial siderophores promote plant growth: Screening of catechol and hydroxamate siderophores. *International Journal of Phytoremediation, 19*(9), 825–833. doi: 10.1080/15226514.2017.1290581.
101. Kumar, A., Tripti, Maleva, M., Bruno, L. B., & Rajkumar, M., (2021). Synergistic effect of ACC deaminase producing *Pseudomonas* sp. TR15a and siderophore producing *Bacillus aerophilus* TR15c for enhanced growth and copper accumulation in *Helianthus annuus* L. *Chemosphere, 276*, 130038. doi: 10.1016/j.chemosphere.2021.130038.
102. Tsavkelova, E. A., Cherdyntseva, T. A., Klimova, S. Y., Shestakov, A. I., Botina, S. G., & Netrusov, A. I., (2007). Orchid-associated bacteria produce indole-3-acetic acid, promote seed germination, and increase their microbial yield in response to exogenous auxin. *Archives of Microbiology, 188*(6), 655–664. doi: 10.1007/s00203-007-0286-x.
103. Spaepen, S., & Vanderleyden, J., (2011). Auxin and plant-microbe interactions. *Cold Spring Harbor Perspectives in Biology, 3*(4), a001438.
104. Jung, H., Lee, D. K., Choi, Y. Do., & Kim, J. K., (2015). OsIAA6, a member of the rice Aux/IAA gene family, is involved in drought tolerance and tiller outgrowth. *Plant Science, 236*, 304–312. doi: 10.1016/j.plantsci.2015.04.018.
105. Dong, T., Park, Y., & Hwang, I., (2015). Abscisic acid: Biosynthesis, inactivation, homoeostasis and signaling. In: Guilfoyle, T., & Hagen, G., (eds.), *Essays in Biochemistry*, (Vol. 58, pp. 29–48). doi: 10.1042/bse0580029.

106. Ullah, A., Manghwar, H., Shaban, M., et al., (2018). Phytohormones enhanced drought tolerance in plants: A coping strategy. *Environmental Science and Pollution Research, 25*(33), 33103–33118. doi: 10.1007/s11356-018-3364-5.
107. Vysotskaya, L. B., Korobova, A. V., Veselov, S. Y., Dodd, I. C., & Kudoyarova, G. R., (2009). ABA mediation of shoot cytokinin oxidase activity: Assessing its impacts on cytokinin status and biomass allocation of nutrient-deprived durum wheat. *Functional Plant Biology, 36*(1), 66–72. doi: 10.1071/FP08187.
108. Forchetti, G., Masciarelli, O., Alemano, S., Alvarez, D., & Abdala, G., (2007). Endophytic bacteria in sunflower (*Helianthus annuus* L.): Isolation, characterization, and production of jasmonates and abscisic acid in culture medium. *Applied Microbiology and Biotechnology, 76*(5), 1145–1152. doi: 10.1007/s00253-007-1077-7.
109. Shahzad, R., Khan, A. L., Bilal, S., Waqas, M., Kang, S. M., & Lee, I. J., (2017). Inoculation of abscisic acid-producing endophytic bacteria enhances salinity stress tolerance in *Oryza sativa. Environmental and Experimental Botany, 136*, 68–77. doi: 10.1016/j.envexpbot.2017.01.010.
110. Sandhya, V., Shrivastava, M., Ali, S. Z., & Sai Shiva, K. P. V., (2017). Endophytes from maize with plant growth promotion and biocontrol activity under drought stress. *Russian Agricultural Sciences, 43*(1), 22–34. doi: 10.3103/s1068367417010165.
111. Puri, A., Padda, K. P., & Chanway, C. P., (2015). Can a diazotrophic endophyte originally isolated from lodgepole pine colonize an agricultural crop (corn) and promote its growth? *Soil Biology and Biochemistry, 89*, 210–216. doi: 10.1016/j.soilbio.2015.07.012.
112. Silva, U. C., Cuadros-Orellana, S., Silva, D. R. C., et al., (2021). Genomic and phenotypic insights into the potential of rock phosphate solubilizing bacteria to promote millet growth *in vivo. Frontiers in Microbiology, 11*, 1–17. doi: 10.3389/fmicb.2020.574550.
113. Yarte, M. E., Gismondi, M. I., Llorente, B. E., & Larraburu, E. E., (2022). Isolation of endophytic bacteria from the medicinal, forestal and ornamental tree *Handroanthus impetiginosus. Environmental Technology (United Kingdom), 43*(8), 1129–1139. doi: 10.1080/09593330.2020.1818833.
114. Penido, A., Mendes, P., Campos, I., & Mendes, L., (2013). *Malaysian Journal of Microbiology, 9*(2), 166–175. doi: 10.1017/CBO9781107415324.004.
115. Verma, S. K., & White, J. F., (2018). Indigenous endophytic seed bacteria promote seedling development and defend against fungal disease in browntop millet (*Urochloa ramosa* L.). *Journal of Applied Microbiology, 124*(3), 764–778. doi: 10.1111/jam.13673.
116. Ferreira, S. Da. C., Nakasone, A. K., Do Nascimento, S. M. C., et al., (2021). Isolation and characterization of cassava root endophytic bacteria with the ability to promote plant growth and control the *in vitro* and *in vivo* growth of *Phytopythium* sp. *Physiological and Molecular Plant Pathology, 116*. doi: 10.1016/j.pmpp.2021.101709.
117. Maheshwari, R., Bhutani, N., & Suneja, P., (2020). Isolation and characterization of ACC deaminase producing endophytic *Bacillus mojavensis* PRN2 from *Pisumsativum. Iranian Journal of Biotechnology, 18*(2), 11–20. doi: 10.30498/IJB.2020.137279.2308.
118. Khan, M. A., Asaf, S., Khan, A. L., et al., (2019). Alleviation of salt stress response in soybean plants with the endophytic bacterial isolate *Curtobacterium* sp. SAK1. *Annals of Microbiology, 69*(8), 797–808. doi: 10.1007/s13213-019-01470-x.
119. Zhu, Y., & She, X., (2018). Evaluation of the plant-growth-promoting abilities of endophytic bacteria from the psammophyte *Ammodendron bifolium. Canadian Journal of Microbiology, 64*(4), 253–264. doi: 10.1139/cjm-2017-0529.

120. Fernando, T. C., & Cruz, J. A., (2019). Profiling and biochemical identification of potential plant growth-promoting endophytic bacteria from *Nypa fruticans*. *Philippine Journal of Crop Science, 44*(2), 77–85. doi: 10.1093/nq/s2-X.255.387-g.
121. Khamwan, S., Boonlue, S., riddech, N., Jogloy, S., & Mongkolthanaruk, W., (2018). Characterization of endophytic bacteria and their response to plant growth promotion in *Helianthus tuberosus* L. *Biocatalysis and Agricultural Biotechnology, 13*, 153–159. doi: 10.1016/j.bcab.2017.12.007.
122. Zhang, Z., Liu, T., Zhang, X., et al., (2021). Cultivable endophytic bacteria in seeds of dongxiang wild rice and their role in plant-growth promotion. *Diversity, 13*(12):665.
123. Wu, T., Xu, J., Xie, W., et al., (2018). *Pseudomonas aeruginosa* L10: A hydrocarbon-degrading, biosurfactant-producing, and plant-growth-promoting endophytic bacterium isolated from a reed (*Phragmites australis*). *Frontiers in Microbiology, 9*, 1–12. doi: 10.3389/fmicb.2018.01087.
124. Wang, L., Lin, H., Dong, Y., He, Y., & Liu, C., (2018). Isolation of vanadium-resistance endophytic bacterium PRE01 from *Pteris vittata* in stone coal smelting district and characterization for potential use in phytoremediation. *Journal of Hazardous Materials, 341*, 1–9. doi: 10.1016/j.jhazmat.2017.07.036.
125. Namwongsa, J., Jogloy, S., Vorasoot, N., Boonlue, S., Riddech, N., & Mongkolthanaruk, W., (2019). Endophytic bacteria improve root traits, biomass and yield of *Helianthus tuberosus* L. under normal and deficit water conditions. *Journal of Microbiology and Biotechnology, 29*(11), 1777–1789. doi: 10.4014/jmb.1903.03062.
126. Simarmata, R., Rohman, S., & Simanjuntak, P., (2018). Amelioration of salt tolerance in soybean (*Glycine max*. L) by plant-growth promoting endophytic bacteria produce 1-aminocyclopropane-1-carboxylase deaminase. *Annales Bogorienses, 22*(2), 81–93.
127. Hazarika, S. N., Saikia, K., Borah, A., & Thakur, D., (2021). Prospecting endophytic bacteria endowed with plant growth promoting potential isolated from camellia sinensis. *Frontiers in Microbiology, 12*, 1–18. doi: 10.3389/fmicb.2021.738058.
128. Zheng, T. W., Liu, L., Nie, Q. W., Hsiang, T., Sun, Z. X., & Zhou, Y., (2021). Isolation, identification, and biocontrol mechanisms of endophytic bacterium D61-A from *Fraxinus hupehensis* against *Rhizoctonia solani*. *Biological Control, 158*, 104621. doi: 10.1016/j.biocontrol.2021.104621.
129. Knežević, M., Berić, T., Buntić, A., et al., (2021). Potential of root nodule nonrhizobial endophytic bacteria for growth promotion of *Lotus corniculatus* L. and *Dactylis glomerata* L. *Journal of Applied Microbiology, 131*(6), 2929–2940. doi: 10.1111/jam.15152.
130. Rios-Galicia, B., Villagómez-Garfias, C., La Vega-Camarillo, D., et al., (2021). The Mexican giant maize of Jala landrace harbors plant-growth-promoting rhizospheric and endophytic bacteria. *3 Biotech, 11*(10), 1–16.
131. Verma, H., Kumar, D., Kumar, V., et al., (2021). The potential application of endophytes in management of stress from drought and salinity in crop plants. *Microorganisms, 9*(8), 1729.
132. Muhae-Ud-Din, G., Ali, M. A., Naveed, M., et al., (2018). Consortium application of endophytic bacteria and fungi improves grain yield and physiological attributes in advanced lines of bread wheat. *Turkish Journal of Agriculture – Food Science and Technology, 6*(2), 136. doi: 10.24925/turjaf.v6i2.136-144.1416.
133. ALKahtani, M. D. F., Fouda, A., Attia, K. A., et al., (2020). Isolation and characterization of plant growth promoting endophytic bacteria from desert plants and their application

as bioinoculants for sustainable agriculture. *Agronomy, 10*(9), 1–18. doi: 10.3390/agronomy10091325.
134. Latha, P., Karthikeyan, M., & Rajeswari, E., (2019). Endophytic bacteria: Prospects and applications for the plant disease management. *Plant Health Under Biotic Stress: Microbial Interactions, 2*, 1–50.
135. Wu, T., Xu, J., Liu, J., et al., (2019). Characterization and initial application of endophytic *Bacillus safensis* strain ZY16 for improving phytoremediation of oil-contaminated saline soils. *Frontiers in Microbiology, 10*. doi: 10.3389/fmicb.2019.00991.
136. Sundaramoorthy, S., Raguchander, T., Ragupathi, N., & Samiyappan, R., (2012). Combinatorial effect of endophytic and plant growth promoting rhizobacteria against wilt disease of *Capsicum annum* L. caused by *Fusarium solani*. *Biological Control, 60*(1), 59–67.
137. Baker, S., Sahana, S., Rakshith, D., Kavitha, H. U., Kavitha, K. S., & Satish, S., (2012). Biodecaffeination by endophytic *Pseudomonas* sp. isolated from *Coffee arabica* L. *Journal of Pharmacy Research, 5*(7), 3654–3657.

CHAPTER 8

Exploring the Role of *Piriformospora indica* and *Pichia anomala* in the Management of Plant Pathogen

MOHINI YADAV, VIKASH KUMAR, and SAURABH C. SAXENA

Department of Biochemistry, Central University of Haryana, Mahendragarh, Haryana, India

ABSTRACT

Biotic stress is a condition in which the basic metabolism of plants is affected by pathogens such as viruses, bacteria, and fungi. Plant pathogens vary greatly in size, from viruses to insects that cause infection in our crops. Pathogens cause a variety of changes in plants, including a reduction in biomass, premature flowering, synthesis of defensive compounds, and many more. Several signaling pathways are involved in plant defense against biotic stress caused by pathogens. Phytohormones like jasmonic acid (JA), salicylic acid (SA), ethylene (ET), and ABA play a major role in overcoming biotic stress. Pathogenesis-related (PR) proteins are important for plant resistance to infections, so when plants are attacked by pathogens, the level of PR protein is highly expressed. Pesticides have traditionally been used to control pathogens; however, they are expensive and have adverse effects on human health and the environment. Pesticides can cause bio-magnification by entering into the food chain, reducing soil fertility, and causing loss of biodiversity. In agriculture, the use of biostimulants for controlling pathogens and plant diseases can be a suitable alternative to pesticides. Beneficial microbes, their genes, and their products, such as metabolites, are used in biological control to minimize the adverse effects of plant diseases and stimulate positive plant responses. Therefore, this chapter focuses on the fungi

Microbial Biostimulants: Biorational Pesticides for Management of Plant Pathogens.
Faheem Ahmad and Rakesh Pandey (Eds.)

Piriformospora indica (*P. indica*) and *Pichia anomala* as biostimulants. *P. indica* is an artificially cultivable fungus that shows adaptability by directly influencing plant hormone-signaling pathways during mutualism, allowing it to colonize a broad range of plant species. Through signal transduction, it can provide both physiological and pathological defenses against viral and fungal plant disorders. Stress tolerance can be conferred by *P. indica*-mediated activation of the antioxidant defense system and expression of stress-tolerant genes. *Pichia anomala* is an ascomycetous yeast that was the first to be discovered to have potent antifungal properties. It also acts as a bio-control agent, as its killer toxin could be used as an antimicrobial, inhibiting different molds in a wide range of environments. This chapter examines the impact of the biostimulants *Piriformaspora indica* and *Pichia anomala* on plant defense mechanisms and their potential use for controlling plant diseases as biocontrol agents.

8.1 INTRODUCTION

Biotic stress is a condition in which the basic metabolism of plants is affected by pathogens such as fungi, bacteria, and viruses. Plant diseases have consistently been documented as causing destruction to food and fiber crops throughout recorded history. There have been notable instances of epidemics over the past 200 years that have had a significant impact on the social development of humanity. In 1845, when the late blight fungus *Phytophthora infestans* completely decimated Ireland's potato crop, it was the most devastating pandemic the country had ever experienced [1]. Plant pathogens affect plants by inhibiting their growth and development. Previous use of pesticides has been employed to prevent harmful pathogens, but they are costly and have detrimental effects on living organisms and the environment. Biostimulants offer an eco-friendly and viable alternative to pesticides. A biostimulant is an organic substance or microbe provided to plants to enhance crop quality, stress resistance, and productivity [2]. They reduce the need for fertilizers while promoting plant growth, water tolerance, and stress resistance [3]. They are environmentally safe and have positive effects on plant development. They improve plant metabolism and physiology, particularly under stressful conditions [4]. Many beneficial microbes, along with their genes and products such as metabolites, are utilized as biological controls to minimize the negative impacts of plant diseases and stimulate positive plant responses. Therefore, in this chapter, we focus on the fungus

Piriformospora indica (*P. indica*) and *Pichia anomala* as biostimulants. *P. indica* is a cultivable phytopromotional, biotrophic mutualistic root endosymbiont that mimics the functions of normal arbuscular mycorrhizal fungi [5]. This fungus can colonize the roots of a wide range of higher plants, providing them with various benefits including nutrient absorption, stress resistance, and growth stimulation [6]. Numerous roles of this fungus have been described, such as biological hardening during the transplantation of micro-propagated plantlets, activation of the antioxidant defense system, and expression of genes related to stress tolerance [7]. The ascomycetous yeast *Pichia anomala* was the first to exhibit strong antifungal properties. Its lethal toxin can be used as an antibacterial agent and inhibits various molds in different environmental conditions, making it a potential biocontrol agent. It protects fruits and grains from harmful microbes after harvest [8].

8.2 PLANT PATHOGEN DERIVED CROP LOSS

Plant diseases are spreading worldwide, posing a significant threat to agricultural commodities. The current scenario highlights the serious status of global food security, which can be improved by protecting crops against diseases caused by agricultural pests. Crop loss refers to the decrease in both quantity and quality of yield, and it can be measured by the gap between achievable and actual production levels [9]. Pathogen-caused losses have both direct and indirect impacts, with various features that may have immediate or long-term effects. Direct yield losses, caused by pathogens, weeds, and animals, account for 20-40% of the overall loss in agricultural productivity worldwide. Global pre-harvest losses associated with plant diseases amount to 9-15% of the annual yield [10]. Invasive plant infections, such as fungi, are estimated to cause $21 billion in annual agricultural losses in the US [11]. In India, insect pests are responsible for reducing crop yields in important food and cash crops by 15% to 20% [12]. Successful plant-pathogen contact leads to a full infection, known as a compatible interaction. To evaluate and control crop loss, a multidisciplinary approach is required, considering the interactions between pathogens (fungi, viruses, bacteria, and nematodes) and other biotic and abiotic components [9]. Pathogens interfere with processes and disrupt the products of photosynthesis and other physiological processes, causing crop losses [13]. The negative effects of infection on host plants are primarily due to the disruption of host metabolic processes, rather than competition for host metabolites, except in cases of diseases like rusts and smuts when the pathogen biomass is high [13]. Invasive insect species pose a growing threat to Indian agricultural products, resulting in significant negative economic impacts on agriculture. Since 1889, there have been reports of 24 different insect species invading India [14]. The evolution of pesticide resistance in various insect species is a major concern for applied evolutionary biologists. It is estimated that a total of 954 pest species have developed resistance to different types of pesticides [15]. Fungal diseases also pose a serious threat to the most important crops, causing significant yield losses across various crops. The 1943 Bengal famine in India, which was brought on by the destruction of rice by *Helminthosporium oryzae*, and the destruction of wheat by stem rust (*Puccinia graminis* f. sp. *tritici*), are two well-known cases of significant crop loss [16]. Numerous economically significant crops are now susceptible to viruses, which can cause severe crop losses [17]. The effective control of threatening plant virus illnesses in some crops is challenging due to the diversity of virus strains, their vectors, and environmental conditions [18]. Among the more than 200 plant pathogenic bacterial species known to affect plant health, *Agrobacterium, Xanthomonas, Pseudomonas, Ralstonia,* and other genera are considered the most significant [19]. Plant parasitic nematodes, especially *Meloidogyne* species that cause root knots, significantly reduce the productivity of vegetable crops under intensive farming [20]. The concept of using microorganisms as pathogen biocontrol is part of the modern plant protection solution approach [21]. Estimating disease-related losses is a crucial step in any plant protection program, as it is difficult to address a problem that has not been adequately described [22]. Without estimating crop losses, it would be impossible to establish plant protection action thresholds or minimize the need for periodic pesticide applications.

8.3 BIOSTIMULANTS

Biostimulants are organic materials or microbes that are administered to crops to improve crop quality, nutrition efficiency, and stress resistance. They have the ability to decrease the demand for fertilizers while promoting plant growth, water tolerance, and stress resistance [3] (Figure 8.1). Biostimulants are beneficial even at low concentrations as they encourage the proper functioning of vital plant processes, resulting in maximum yields and output. In order to survive and thrive, plants must face various pressures and develop effective adaptive strategies to escape or tolerate the negative effects [4].

Biostimulants arise from natural resources, and they can:

- Soil supplements act to enhance the structure of soil, which in turn promotes plant responsiveness [23].
- When used in moderation, it encourages the growth and development of plants [23].
- They help in enhancing the effectiveness of plant nutrients [24].

Chemical fertilizers and insecticides are used to fertilize crops and control pests, allowing agricultural systems to increase their output capacity [25]. The widespread utilization of insecticides and fertilizers has had a severe impact on the atmosphere as well as public health. Crop resistance to stress has also decreased due to excessive manufacturing techniques and the improper use of pesticides [26]. The employment of plant-useful microbes is an appealing way to avoid complex chemical-based techniques that further contribute to ecologically safe and sustainable agriculture.

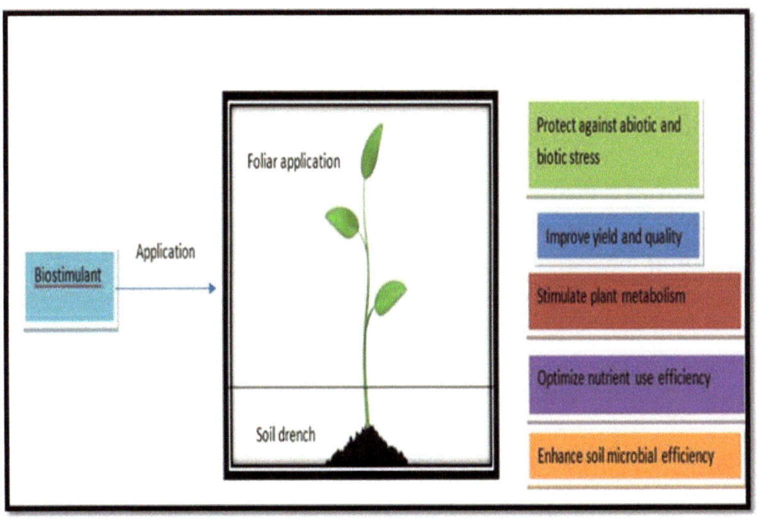

FIGURE 8.1 Effect of biostimulants on plant growth and productivity.

Plant growth promotion (PGP), biological disease control, increased crop output, and improved quality are all examples of beneficial microbial effects [25]. Research has been conducted to develop bio-fertilizers as effective designed products to replace chemical fertilizers [24]. The most beneficial microbial inoculants are fungi and bacteria that promote plant growth and are classified as bio-fertilizers or biocontrol agents depending on their activity [27]. They are, in fact, manufactured products containing microbes that enhance plant nutrient levels by substituting soil nutrients, increasing nutrient availability, improving access to nutrients for plants, or producing specialized compounds [28].

8.3.1 TYPES OF BIOSTIMULANTS

Biostimulants are safe for the environment and have favorable impacts on plant development. They improve plant metabolism and physiology [29]. The categories of the most popular biostimulants are discussed in subsections (Figure 8.2).

8.3.1.1 HUMIC SUBSTANCES (HS)

They occur naturally in soil and are produced by the microbial activity of the soil as well as the decomposition of living matter. Humic substances (HS) are groups of heterogeneous molecules that were initially divided into fulvic acids, humic acids, and humins according to their solubility and molecular weights [30]. Humic compounds influence the structural, molecular, and physical characteristics of soil and have long been identified as significant contributions to soil fertility. The majority of biostimulant effects relate to improving root nutrition through various processes [29]. One of them is the enhanced absorption of macro- and micronutrients as a result of the soil's higher cation exchange capacity and enhanced phosphorus availability due to HS obstructing Ca-P precipitation in the polyanionic HS-containing soil. The activation of cell membrane H+-ATPases, which transforms energy made available by ATP hydrolysis into a transmembrane electrochemical potential used for the transport of nitrate and other minerals, is another important factor of HS in root feeding [31]. Proton pumping via plasma membrane ATPases promotes food uptake in addition to the enlargement of cells and the development of organs [32]. It increases invertase and respiratory activity by supplying carbon substrates. In seedlings of maize grown hydroponically, it has been demonstrated that high-molecular-mass HS increases the functioning of crucial enzymes in this process, indicating that HS may modulate stress response [33, 34].

Treatments with humic compounds have been shown to aid in plant root growth and development through upregulation of the antioxidant system and metal chelation [35]. This is demonstrated by improved water and nutrient uptake, as well as increased resilience to environmental challenges [36].

8.3.1.2 SEAWEED EXTRACT

Seaweeds are a diverse group of brown, red, and green multicellular, macroscopic marine algae. It is believed that there are over 9,000 different types of seaweed, which play a vital role in the marine coastal environment [4]. Seaweeds are an important source of nutrients for fertilizers and organic materials. They contribute to enhancing plant growth through xylem hydraulic conductance, stomatal regulation, photosynthetic activity, abiotic stress tolerance, and resistance to fungus and viruses. As a result, they can increase crop yield and productivity [4]. Seaweeds used for producing biostimulants contain cytokinins, auxins, and other hormone-like compounds [37]. Additionally, they contain complex polysaccharides such as laminarin, fucoidan, alginates, as well as plant hormones, and various active minerals and chemical substances [38].

8.3.1.3 AMINO ACIDS

Amino acids and peptide combinations are created by chemically and enzymatically hydrolyzing agricultural by-products, such as animal wastes like collagen and crop residues [24]. Because they serve as the building blocks for proteins and peptides, amino acids are crucial chemicals in plant processes. Some bioactive peptides can change critical physiological processes in plants because they bind to the cell receptors that regulate various metabolic activities [39]. There are two sources of this kind of amino acids and peptides: an animal source that is obtained through the high-temperature hydrolysis of collagen and vegetal hydrolysates that are obtained through the low-temperature enzymatic hydrolysis of collagen. They help in the growth of plants by lowering the accumulation of ROS, helping in osmoprotection and metal chelation [40].

8.3.1.4 BENEFICIAL MICROORGANISMS

Microorganisms are divided into two main families: beneficial bacteria and beneficial fungi. Mycorrhizal fungi (MF) coexist with plants in a mutualistic symbiosis. Over 90% of all plant species have symbiotic relationships with the heterogeneous group of taxa known as MF [25]. Through an external net of hyphae, the mycorrhizal fungus penetrates the roots of plants and provides them with water and minerals collected from the soil, while the plant supplies the microorganism with organic compounds and sugar. The plant's root surface is multiplied by 100-1,000, increasing its ability to absorb nutrients and water [41].

Beneficial bacteria cluster together and settle in the rhizosphere. They primarily influence how plants function nutritionally, how they absorb nutrients, and how effectively they can survive abiotic stress. Rhizobium and rhizobacteria that encourage plant growth are the primary items in the market [42].

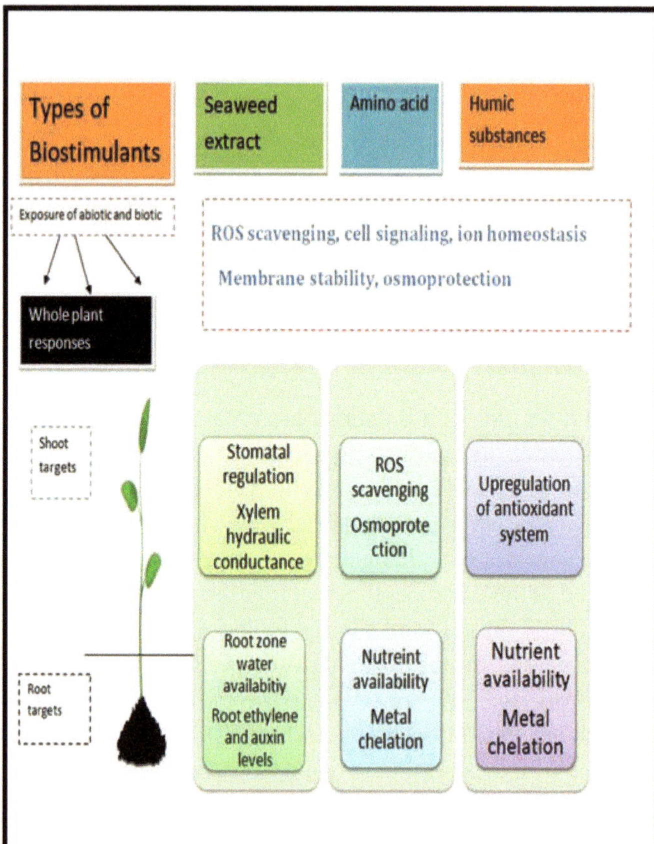

FIGURE 8.2 Major bio-stimulatory processes focused on non-microbial biostimulants upon interaction with plants.

8.4 *PIRIFORMOSPORA INDICA* (*P. INDICA*) AS A BIOSTIMULANT

P. indica is a root-colonizing endophytic fungus that colonizes a lot of hosts without showing any host specificity that permits plants to thrive under harsh physical and nutritional challenges [5]. It was initially discovered in association with xerophytic plants in the Indian Thar Desert Sharma [2]. It colonizes a wide range of dicots and monocots, as well as bryophytes, pteridophytes, and gymnosperms. Electron microscopy and molecular methods based on 18S rRNA sequences have been used to determine the taxonomic position of *P. indica*, revealing that it belongs to the Hymenomycetes of the Basidiomycota [5]. The fungus forms pear-shaped, auto-fluorescent chlamydospores in the cortical zone of colonized roots and the rhizosphere, but it does not penetrate the aerial endodermis of the plants [43]. The fungus aids in nutrient uptake, prepares plants to tolerate water stress, temperature stress, and salt stress, and provides systemic resistance to heavy metal ions, toxins, pests, and pathogens [44].

Phytohormones have an impact on how plants grow and develop, as well as how they interact with other organisms like endophytic fungi. Plant growth and development, resistance to abiotic stressors, and interactions with mutualistic fungi are all impacted by phytohormones like auxin, cytokinin, ethylene (ET), salicylic acid (SA), abscisic acid (ABA), gibberellic acid and jasmonates [6]. Phytohormones, which are related to defense, regulate root colonization, which is necessary for the symbiotic relationship to remain stable. A systemic resistance response involving SA, ET, ABA, and jasmonates is widely used to explain higher stress tolerance, and their interactions with *P. indica* have just recently been discovered [45]. Sirrenberg et al. [70] were the first to show that in liquid culture *P. indica* produces IAA and that the synthesis of fungal auxin influences the growth of the root. Auxin levels vary significantly among plant species colonized by *P. indica,* and these differences have been directly measured or estimated from the auxin-responsive gene expression [46]. A large-scale microarray study of *Arabidopsis* roots colonized by *P. indica* revealed that auxin content and gene expression related to auxin in the whole root remained unaltered in response to *P. indica* [46]. In plant-microbe interactions, the phytohormones JA and SA mediate defense responses and systemic signal propagation, with SA primarily involved in the response to biotrophic microorganisms and JA to necrotrophic microbes [45]. Vahabi et al. [71] demonstrated that SA and jasmonate levels increased when *Arabidopsis* roots were placed close to *P. indica* before any physical contact occurred. The symbiotic association between *P. indica* and *Arabidopsis* showed that modest stressors such as limited food availability, exposure to salt or heavy metals, light and osmotic stress, and pathogen infections improve plant growth in symbiosis [46]. Akum et al. [72] studied that *P. indica* effector protein PIIN 08944, which encourages root colonization and mutualistic symbiosis when expressed ectopically in *Arabidopsis* and barley, aided root colonization by interfering with the SA-mediated basal immune response. In addition to its effect on biomass, *P. indica* in barley defends its host from fungal infections and salt stress. It protects against *Fusarium graminearum,* which causes root rot in barley roots [47]. *P. indica* stimulates JA-dependent defense responses to regulate secondary colonization in distal root regions, demonstrating systemic signaling, while induced systemic resistance (ISR) is typically thought to be dependent on SA [45]. When plants are colonized by *P. indica,* the stimulation of jasmonates and JA-responsive genes by infection of *A. brassicae* is greatly reduced, implying that the colonized plants suppress the immunity of the host because they are stronger and healthier and require less JA-dependent defense for protection [6]. *Gaeumannomyces graminis,* a seed and virulent root pathogen, was used to challenge *P. indica.* Pathogen development was completely stopped in an experiment when the fungus was permitted to grow early, and the pathogen was introduced later in the center of the solidified nutritional medium [7]. The growth of the pathogen was also entirely inhibited by *P. indica* culture filtrate. The combination of *P. indica* and *Aspergillus sydowii* had similar inhibitory effects [43]. *P. indica* minimized the deleterious impact of *F. culmorum, Pseudocercosporella herpotrichoides, and Blumeriagraminis f. sp. tritici* on winter wheat growing in pot cultures [48]. *P. indica* helps the commercially useful chickpea plant (*Cicer arietinum*) grow and defends it from *Botrytis cinerea.* It was shown that *P. indica* colonized chickpea plants significantly boosted biomass and root growth when compared to plants infected with *B. cinerea* and chickpea plants grown without *P. indica* [49]. Systemically elevated antioxidants (AOX) and reduction in powdery mildew disease were observed in *P. indica*-infested plants [50]. Therefore, considering the above-mentioned research outputs (Figure 8.3), it can be safely concluded that *P. indica* acts as a potential bio-fertilizer and bio-control agent in plants challenged with different pathogens.

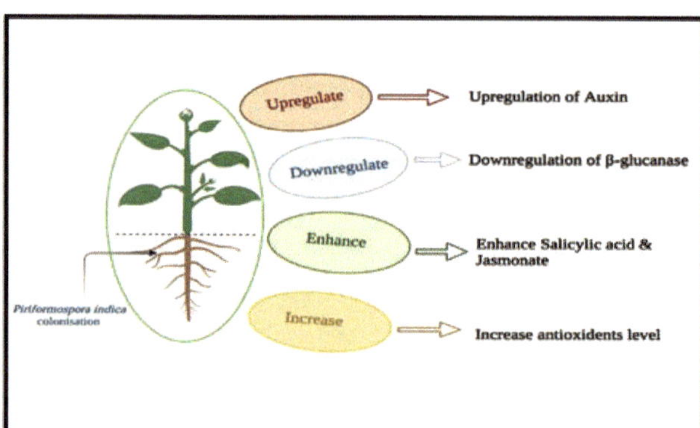

FIGURE 8.3 *P. indica* induced response during colonization (programmed cell death) and biocontrol of pathogens (increased level of salicylic acid, jasmonate, and different antioxidants).

8.4.1 MECHANISM OF ACTION OF PIRIFORMOSPORA INDICA (P. INDICA)

Plant stress tolerance is induced by the symbiotic association of fungi and is mediated by two mechanisms:

- When exposed to stress, it may trigger the action of host stress-responsive systems, directing the plants to either avoid or reduce that stress [51]; and
- Synthesis of antistress biochemicals [45].

One explanation for the induced resistance was the increased antioxidant production in shoots and roots of *P. indica*-colonized plants [50, 52]. Additionally, it was demonstrated that after *P. indica* inoculation, several genes known to be involved in plant defense mechanisms were systemically induced [50]. According to some findings, *P. indica's* bio-protective activity may be mediated by the host plants' systemic production of antioxidant defense [53]. *P. indica* mimics AMF (arbuscular mycorrhizal fungi) and has a strong association with many valuable plants from an economic and medicinal perspective [54]. *P. indica* spores grow and penetrate chickpea roots within 5 to 7 days. The amount of the inoculum and length of co-cultivation are both important factors in the colonization of the roots [55]. A maximum of 65% colonization by *P. indica* in chickpea roots was observed within 30 days of interaction. During *P. indica's* colonization of the chickpea plants, the GST (glutathione-S-transferase) activity in the shoot and root showed many fold increments [56]. Antibiosis assay reveals that *P. indica* is not involved in any chemical or antibiotic secretion to inhibit *B. cinerea*[55]. Additionally, it was discovered that in the maize plant *P. indica*, colonization caused a reduced growth of the pathogen *Fusarium* without any antibiotic secretion [53]. Barley roots with *P. indica* infestations were resistant to *Fusarium graminearum,* and the plant's PR gene expression was investigated. The outcomes show that the endophyte inhibits PR protein accumulation [47]. A study looked at *P. indica's* ability to manage a disease called onion leaf blight, which is caused by the necrotrophic fungal infection *Stemphylium vesicarium* [57]. The role of *P. indica* colonization in signaling oxidative stress and its protective impact against peroxidative damage was analyzed by analyzing changes in malondialdehyde (MDA) and H_2O_2 concentration [49]. It was also demonstrated that *P. indica* administration alters the enzyme activity of catalase (CAT), peroxidase (POD), phenylalanine ammonia-lyase, and superoxide dismutase (SOD) during infection [49]. Expression study based on qPCR of defense-related genes *AcLOX1, AcLOX2, AcGST, AcPAL1, AcWRKY1,* and *AcWRKY70* and *AcCHI* gave additional evidence of *P. indica's* capacity to induce onion systemic response [49]. Studies in the Arabidopsis plant showed that to cause systemic resistance, the fungus only needs NPR1 localized in the cytosol, not the nuclear region [58]. The decrease in powdery mildew was shown to be caused by a mechanism similar to systemic-induced resistance that depends on jasmonate signaling in Arabidopsis plants with *P. indica* colonization [59]. *P. indica* appears to cause plants to undergo programmed cell death (PCD), much like several other pathogens, however, unlike these pathogens, the endophyte's ability to colonize the root depends on the plant's cell death program [47]. When the number of dead cells in the root reaches a specific threshold due to excessive inoculum and early-on excessive colonization, it may negatively affect plant growth and development [43].

8.5 *PICHIA ANOMALA* AS BIOSTIMULANT

Boysen et al. [73] were the first to note that the ascomycetous yeast *Pichia anomala* had a high antifungal activity. It is highly competitive and can inhibit various microorganisms. It is often regarded as a biopreservation agent due to its antimicrobial properties [8]. According to Zhao et al. [74], *P. anomala* (strain FY-102) which is isolated from apple surfaces has a strong antagonistic effect on *Botrytis cinerea.* Regarding apples, it exhibits strong and consistent antagonistic activity against *P. expansum* as well.

Pichia anomala is often mentioned as a protector of postharvest fruits and grains against wound pathogens and *Penicillium roquefertii* [8]. However, like any other biological control agent (BCA), this yeast must meet various standards related to biological, technical, and toxicological qualities in order to become an economically viable alternative to chemical control. The biocontrol effectiveness of *P. anomala* strain K was evaluated in orchard trials over two consecutive growing seasons to determine its suitability for orchard use [60]. *P. anomala* strain K could potentially be a significant component of integrated systems that also incorporate chemical or physical treatments to effectively combat post-harvest diseases [60]. *P. anomala* strain J121 effectively prevented the most significant spoilage mold, *P. roqueforti,* in grains stored airtight in temperate regions, regardless of the level of air leakage investigated [61].

P. anomala, either with chitosan or alone, significantly enhanced the enzymatic activity of CAT, polyphenol oxidase (PPO), POD, and phenylalanine

(PAL) in table grapes compared to the control [62]. Treatment of table grapes with *P. anomala* upregulated the genes responsible for these important disease defense enzymes, as confirmed by RT-qPCR (reverse transcription-quantitative polymerase chain reaction) results [62]. A recent study demonstrated that incubating *P. anomala* with chitosan greatly improved yeast enzymatic activities such as CAT and β-1,3-glucanase, while reducing MDA accumulation [62]. CAT has been known to effectively detoxify reactive oxygen species (ROS) generated at wound sites in plants. Lowering MDA content also helps minimize oxidative damage to yeast cells [62]. Enzymes like 1,3-glucanase enhance the antagonistic ability of yeasts and cause cell lysis in pathogens by destroying their cell walls [61]. POD enzyme has been observed to regulate H_2O_2 levels in the fruit cell wall, which is crucial for the cross-linking of phenolic compounds in response to various external stressors such as wounds, pathogen attacks, and environmental stress [62]. In a study, chitosan (1% w/v) induced *P. anomala* treatment significantly increased POD activity in grapes. The highest concentration of total phenolic compounds was found in *P. anomala*-induced chitosan-treated table grapes [63]. Phenolic compounds increase the post-harvest shelf life of fruits and reduce stresses during post-harvest storage. When plants are exposed to abiotic and biotic stress, such as heat stress, fungal, bacterial, and viral infections, an increase in the PAL (phenylalanine ammonia-lyase) enzyme may lead to a higher accumulation of phenolic chemicals in the fruit [61]. A recent study also demonstrates that post-harvest treatment of table grapes with chitosan induced by *P. anomala* can significantly increase the accumulation of total flavonoids in the fruit, which plays a crucial role in enabling the fruit to combat post-harvest diseases and other biotic stress [62]. The incubation period before pathogen inoculation also influences the antagonistic action. In Citrus, it was observed that the longer the time between applying the antagonist and administering the fungal spores, the higher the level of protection [60]. The pathogen inoculation and the concentration ratio of antagonist-to-pathogen also affect the antagonistic action of *P. anomala*. The design of the yeast formulation sometimes enables a stronger and more stable potency of the antagonist [8].

8.5.1 MECHANISM OF PICHIA ANOMALA

Pichia anomala exhibits antibacterial activity against various phylogenetic classes of microorganisms. The ability of Strain K to inhibit the growth of *Botrytis cinerea* on fruits is attributed to the yeast's ß-1.3 glucanase activity

[64]. This fungus has been found to restrict the growth of *Aspergillus* and *Penicillium* species, as well as unexpectedly inhibiting *Enterobacteriaceae* [65]. Instead of competing for nutrients, it has been shown that the suppression of filamentous fungi is primarily linked to oxygen scavenging and the production of ethyl acetate [64]. The involvement of killer proteins in the inhibition of bacteria is still not fully understood [8]. Various strains of *P. anomala* have demonstrated killer functions against Gram-negative bacteria, filamentous fungi, and other yeasts. Killer proteins exhibit a remarkable variation in molecular weights, ranging from 8 to 300 kilodaltons, which is likely due to varying degrees of glycosylation [65]. Anti-idiotypic antibodies that mimic the effects of killer toxins have been developed since the killer proteins themselves are too large to be used in therapeutic treatments.

When cultivated on different cereal grains, *Pichia* has shown significant biocontrol efficacy against a variety of molds [8]. In one study, *P. anomala* outperformed 60 different yeast species in its ability to inhibit the growth of *Penicillium roqueforti* in a test tube simulation of an airtight grain silo. Numerous yeasts reached similar levels of growth as *P. anomala* without exerting any influence on mold growth [66].

8.6 FUTURE PROSPECTIVE OF *PIRIFORMOSPORA INDICA* (*P. INDICA*)AND *PICHIA ANOMALA*

P. indica is a new class of axenically cultivable, root-colonizing, plant-promoting fungi, which is a beneficial agent for controlling plant root diseases [43]. The establishment and survival of micropropagated rootlets are also facilitated by this fungus [43]. The symbiotic nature of *P. indica*, as well as potential pre and post-root colonization scenarios, the molecular basis of root penetration, signaling between the plant and fungus, and ecological dispersal, are the main areas that require further exploration. It would also be interesting to investigate two additional aspects of *P. indica's* GSH (glutathione) metabolism: first, the detoxification of various pesticides and other xenobiotics that rely on GSH, catalyzed by the enzyme GST, and second, the GSH-dependent detoxification of heavy metals carried out by substances known as phytochelatins in several plant families [43]. If *P. indica* proves highly effective in detoxifying xenobiotics and heavy metals, this information could be utilized to cultivate *P. indica*-dependent plants in regions where soil contamination poses environmental issues. This approach would help eliminate excess xenobiotics from the soil, thereby improving atmospheric sustainability.

Pichia anomala can suppress dangerous microbes in various environments, making it an effective biocontrol agent. In addition to its antimicrobial impact on feed grain, yeasts are also known for providing protein and vitamins, thereby enhancing the quality of feed [67]. *P. anomala* has the ability to convert nitrate and ammonium (NH_3) substances into proteins, which can be utilized as feed due to its capacity to consume a wide range of nitrogen compounds [43]. Furthermore, it exhibits significant phytase activity [65]. Further research on the physiology and genetics of *P. anomala* is necessary to fully understand its role in different environments. Yeasts are used as model organisms to study the regulation of essential metabolic activities, and further research on this respiratory yeast could expand its applications [66]. Anti-idiotypic antibodies have been developed using the killer proteins from *P. anomala* as a matrix, and they have shown effectiveness against microbial pathogens [65]. This is an intriguing strategy, and further studies may lead to alternative antibacterial treatments for both humans and animals [65].

8.7 CONCLUSION

In today's world, people are becoming more health-conscious and thus have heightened concerns about the use of pesticides and insecticides in crop plants. The most effective alternative for disease suppression in plants appears to be the use of BCA (Biological Control Agents). Bio-agents prevent diseases in plants while posing no environmental risks. Over the past 10 years, the usage of biostimulants in agriculture has significantly increased, mainly due to their diverse qualities [68]. Agricultural crops treated with biostimulants have been found to be more resistant to environmental changes. To ensure the optimal application of these biostimulants, it is important to develop an understanding of plant characteristics and techniques that have the greatest impact on preventing biotic stress. Furthermore, more emphasis should be placed on gaining a thorough knowledge of the functional mechanisms of biostimulants. It is crucial to recognize that different crop systems respond to plant biostimulants (PBs) in various ways, and favorable effects are often observed in specific crop species and under carefully controlled laboratory or greenhouse conditions [69].

P. indica is an endophytic fungus that demonstrates adaptability in colonizing plant species by directly manipulating the transmission of plant hormones.

Through signal transduction, it also induces both local and systemic resistance to several plant diseases. P. indica performs various functions, including nutrient uptake, disease resistance, stress tolerance, and growth promotion. The use of this fungus enables the successful hardening of micro-propagated plantlets after transplantation. Numerous medicinal plants infested with P. indica have been found to enhance the production of beneficial secondary metabolites [51]. P. anomala exhibits antibacterial activity against various phylogenetic classes of microbes, thereby positively influencing plant growth and development.

KEYWORDS

- **amino acids**
- **beneficial microorganisms**
- **biocontrol agents**
- **biostimulants**
- **humic substances**
- ***Pichia anomala***
- ***Piriformospora indica***
- **plant pathogen**

REFERENCES

1. Goss, E. M., Tabima, J. F., Cooke, D. E., Restrepo, S., Fry, W. E., Forbes, G. A., Fieland, V. J., et al., (2014). The Irish potato famine pathogen Phytophthora infestans originated in central Mexico rather than the Andes. *Proceedings of the National Academy of Sciences, 111*(24), 8791–8796.
2. Sharma, P., Kharkwal, A. C., Abdin, M. Z., & Varma, A., (2014). *Piriformospora indica* improves micropropagation, growth and phytochemical content of *Aloe vera* L. plants. *Symbiosis, 64*(1), 11–23.
3. Traon, D., Amat, L., Zotz, F., & Du Jardin, P., (2014). *A Legal Framework for Plant Biostimulants and Agronomic Fertilizer Additives in the EU.* Report to the European Commission, DG Enterprise & Industry.
4. Sharma, H. S., Fleming, C., Selby, C., Rao, J. R., & Martin, T., (2014). Plant biostimulants: A review on the processing of macroalgae and use of extracts for crop management to reduce abiotic and biotic stresses. *Journal of Applied Phycology, 26*(1), 465–490.

5. Varma, A., Sherameti, I., Tripathi, S., Prasad, R., Das, A., Sharma, M., Bakshi, M., et al., (2012). 13 symbiotic fungus *Piriformospora indica*. In: *Fungal Associations* (pp. 231–254). Springer.
6. Pozo, M. J., López-Ráez, J. A., Azcón-Aguilar, C., & García-Garrido, J. M., (2015). Phytohormones as integrators of environmental signals in the regulation of mycorrhizal symbioses. *New Phytologist*, *205*(4), 1431–1436.
7. Narayanasamy, P., (2013). Mechanisms of action of fungal biological control agents. In: *Biological Management of Diseases of Crops* (pp. 99–200). Springer.
8. Jijakli, H., (2010). *Pichia Anomala in Biocontrol for Fruits: 20 Years of Fundamental and Practical Researches*. In Meeting International sur les applications de P. Anomala.
9. Chiarappa, L., Chiang, H. C., & Smith, R. F., (1972). Plant pests and diseases: Assessment of crop losses: A multidisciplinary scientific approach is required to solve an ancient agricultural problem. *Science*, *176*(4036), 769–773.
10. Bag, M. K., Yadav, M. K., & Mukherjee, A. K., (2017). Changing disease scenario with special emphasis on false smut of rice. *SATSA Mukhapatra*, *21*, 219–224.
11. Fisher, M. C., Gurr, S. J., Cuomo, C. A., Blehert, D. S., Jin, H., Stukenbrock, E. H., Stajich, J. E., et al., (2020). Threats posed by the fungal kingdom to humans, wildlife, and agriculture. *MBio*, *11*(3), e00449–20.
12. Oerke, E. C., Dehne, H. W., Schönbeck, F., & Weber, A., (2012). *Crop Production and Crop Protection: Estimated Losses in Major Food and Cash Crops*. Elsevier.
13. Lucas, J. A., (2009). *Plant Pathology and Plant Pathogens*. John Wiley & Sons.
14. Kumar, J., Murali-Baskaran, R. K., Jain, S. K., Sivalingam, P. N., Dixit, A., Mallikarjuna, J., & Ghosh, P. K., (2021). Biotic stresses of agricultural crops in India: Re-visiting national status and mitigation strategies. *Current Science*, *120*(2), 264.
15. Kole, R. K., Roy, K., Panja, B. N., Sankarganesh, E., Mandal, T., & Worede, R. E., (2019). Use of pesticides in agriculture and emergence of resistant pests. *Indian J. Anim. Hlth*, *58*(2), 53–70.
16. Savary, S., Bregaglio, S., Willocquet, L., Gustafson, D., D'Croz, D. M., Sparks, A., Castilla, N., et al., (2017). Crop health and its global impacts on the components of food security. *Food Security*, *9*(2), 311–327.
17. Vurro, M., Bonciani, B., & Vannacci, G., (2010). Emerging infectious diseases of crop plants in developing countries: Impact on agriculture and socio-economic consequences. *Food Security*, *2*(2), 113–132.
18. Jones, R. A. C., (2014). Plant virus ecology and epidemiology: Historical perspectives, recent progress and future prospects. *Annals of Applied Biology*, *164*(3), 320–347.
19. Mansfield, J., Genin, S., Magori, S., Citovsky, V., Sriariyanum, M., Ronald, P., Dow, M. A. X., et al., (2012). Top 10 plant pathogenic bacteria in molecular plant pathology. *Molecular Plant Pathology*, *13*(6), 614–629.
20. Blair, B. L., & Stirling, G. R., (2007). The role of plant-parasitic nematodes in reducing yield of sugarcane in fine-textured soils in Queensland, Australia. *Australian Journal of Experimental Agriculture*, *47*(5), 620–634.
21. Heydari, A., & Pessarakli, M., (2010). A review on biological control of fungal plant pathogens using microbial antagonists. *Journal of Biological Sciences*, *10*(4), 273–290.
22. James, W. C., (1990). Estimated losses of crops from plant pathogens. *CRC Handbook of Pest Management*, *1*, 5–50.

23. Juwarno, J., Nugroho, H., Hardiyati, T., & Yuniaty, A., (2020). Molecular profiles of five salinity-resistant soybean ${$glycine max (L.) Merr.$}$ cultivars. *Molekul, 15*(3), 184–190.
24. Calvo, P., Nelson, L., & Kloepper, J. W., (2014). Agricultural uses of plant biostimulants. *Plant and Soil, 383*(1), 3–41.
25. Hayat, R., Ali, S., Amara, U., Khalid, R., & Ahmed, I., (2010). Soil beneficial bacteria and their role in plant growth promotion: A review. *Annals of Microbiology, 60*(4), 579–598.
26. Malik, K. A., Mirza, M. S., Hassan, U., Mehnaz, S., Rasul, G., Haurat, J., Bally, R., & Normand, P., (2002). The role of plant-associated beneficial bacteria in rice-wheat cropping system. *Biofertilizers in Action* (pp. 73–83). Rural Industries Research and Development Corporation, Canberra.
27. Vassilev, N., Eichler-Löbermann, B., Flor-Peregrin, E., Martos, V., Reyes, A., & Vassileva, M., (2017). Production of a potential liquid plant bio-stimulant by immobilized *Piriformospora indica* in repeated-batch fermentation process. *AMB Express, 7*(1), 1–7.
28. Malusa, E., & Vassilev, N., (2014). A contribution to set a legal framework for biofertilizers. *Applied Microbiology and Biotechnology, 98*(15), 6599–6607.
29. Du Jardin, P., (2015). Plant biostimulants: Definition, concept, main categories, and regulation. *Scientia Horticulturae, 196*, 3–14.
30. Asli, S., & Neumann, P. M., (2010). Rhizosphere humic acid interacts with root cell walls to reduce hydraulic conductivity and plant development. *Plant and Soil, 336*(1), 313–322.
31. Lamar, R., Olk, D. C., Mayhew, L., & Bloom, P. R., (2013). Evaluation of a proposed standardized analytical method for the determination of humic and fulvic acids in commercial products. In: *Functions of Natural Organic Matter in Changing Environment* (pp. 1071–1073). Springer.
32. Jindo, K., Martim, S. A., Navarro, E. C., Pérez-Alfocea, F., Hernandez, T., Garcia, C., Aguiar, N. O., & Canellas, L. P., (2012). Root growth promotion by humic acids from composted and non-composted urban organic wastes. *Plant and Soil, 353*(1), 209–220.
33. Olivares, F. L., Aguiar, N. O., Rosa, R. C. C., & Canellas, L. P., (2015). Substrate biofortification in combination with foliar sprays of plant growth promoting bacteria and humic substances boosts production of organic tomatoes. *Scientia Horticulturae, 183*, 100–108.
34. Schiavon, M., Pizzeghello, D., Muscolo, A., Vaccaro, S., Francioso, O., & Nardi, S., (2010). High molecular size humic substances enhance phenylpropanoid metabolism in maize (*Zea mays* L.). *Journal of Chemical Ecology, 36*(6), 662–669.
35. Trevisan, S., Francioso, O., Quaggiotti, S., & Nardi, S., (2010). Humic substances biological activity at the plant-soil interface: From environmental aspects to molecular factors. *Plant Signaling & Behavior, 5*(6), 635–643.
36. Nardi, S., Pizzeghello, D., Schiavon, M., & Ertani, A., (2016). Plant biostimulants: Physiological responses induced by protein hydrolyzed-based products and humic substances in plant metabolism. *Scientia Agricola, 73*, 18–23.
37. Hamza, B., & Suggars, A., (2001). Biostimulants: Myths and realities. *Turf Grass Trends, 8*, 6–10.
38. Battacharyya, D., Babgohari, M. Z., Rathor, P., & Prithiviraj, B., (2015). Seaweed extracts as biostimulants in horticulture. *Scientia Horticulturae, 196*, 39–48.

39. Nacry, P., Bouguyon, E., & Gojon, A., (2013). Nitrogen acquisition by roots: Physiological and developmental mechanisms ensuring plant adaptation to a fluctuating resource. *Plant and Soil*, *370*(1), 1–29.
40. Halpern, M., Bar-Tal, A., Ofek, M., Minz, D., Muller, T., & Yermiyahu, U., (2015). The use of biostimulants for enhancing nutrient uptake. *Advances in Agronomy*, *130*, 141–174.
41. Vessey, J. K., (2003). Plant growth promoting rhizobacteria as biofertilizers. *Plant and Soil*, *255*(2), 571–586.
42. Bonfante, P., & Genre, A., (2010). Mechanisms underlying beneficial plant–fungus interactions in mycorrhizal symbiosis. *Nature Communications*, *1*(1), 1–11.
43. Varma, A., Singh, A., Sahay, N. S., Sharma, J., Roy, A., Kumari, M., Rana, D., et al., (2001). *Piriformospora Indica*: An axenically culturable mycorrhiza-like endosymbiotic fungus. In: *Fungal Associations* (pp. 125–150). Springer.
44. Varma, A., Chordia, P., Bakshi, M., & Oelmüller, R., (2013). Introduction to sebacinales. In: *Piriformospora Indica* (pp. 3–24). Springer.
45. Xu, L., Wu, C., Oelmüller, R., & Zhang, W., (2018). Role of phytohormones in *Piriformospora indica*-induced growth promotion and stress tolerance in plants: More questions than answers. *Frontiers in Microbiology*, *9*, 1646.
46. Vadassery, J., Ritter, C., Venus, Y., Camehl, I., Varma, A., Shahollari, B., Novák, O., et al., (2008). The role of auxins and cytokinins in the mutualistic interaction between *Arabidopsis* and *Piriformospora indica*. *Molecular Plant-Microbe Interactions*, *21*(10), 1371–1383.
47. Deshmukh, S. D., & Kogel, K. H., (2007). *Piriformospora Indica* protects barley from root rot caused by *Fusariumgraminearum*. *Journal of Plant Diseases and Protection*, *114*(6), 263–268.
48. Khalid, M., Rahman, S., & Huang, D., (2019). Molecular mechanism underlying *Piriformospora indica*-mediated plant improvement/protection for sustainable agriculture. *Acta Biochimica et Biophysica Sinica*, *51*(3), 229–242.
49. Roylawar, P., Khandagale, K., Randive, P., Shinde, B., Murumkar, C., Ade, A., Singh, M., et al., (2021). *Piriformospora indica* primes onion response against *Stemphylium* leaf blight disease. *Pathogens*, *10*(9), 1085.
50. Waller, F., Achatz, B., Baltruschat, H., Fodor, J., Becker, K., Fischer, M., Heier, T., et al., (2005). The endophytic fungus *Piriformospora indica* reprograms barley to salt-stress tolerance, disease resistance, and higher yield. *Proceedings of the National Academy of Sciences*, *102*(38), 13386–13391.
51. Gill, S. S., Gill, R., Trivedi, D. K., Anjum, N. A., Sharma, K. K., Ansari, M. W., Ansari, A. A., et al., (2016). *Piriformospora indica*: Potential and significance in plant stress tolerance. *Frontiers in Microbiology*, *7*, 332.
52. Serfling, A., Wirsel, S. G., Lind, V., & Deising, H. B., (2007). Performance of the biocontrol fungus *Piriformospora indica* on wheat under greenhouse and field conditions. *Phytopathology*, *97*(4), 523–531.
53. Kumar, M., Yadav, V., Tuteja, N., & Johri, A. K., (2009). Antioxidant enzyme activities in maize plants colonized with *Piriformospora indica*. *Microbiology*, *155*(3), 780–790.
54. Singh, A., Sharma, J., Rexer, K. H., & Varma, A., (2000). Plant productivity determinants beyond minerals, water, and light: *Piriformospora indica*–a revolutionary plant growth promoting fungus. *Current Science*, 1548–1554.

55. Narayan, O. P., Verma, N., Singh, A. K., Oelmüller, R., Kumar, M., Prasad, D., Kapoor, R., et al., (2017). Antioxidant enzymes in chickpea colonized by *Piriformospora indica* participate in defense against the pathogen botrytis cinerea. *Scientific Reports*, *7*(1), 1–11.
56. Gomaa, M., & Dawood, M. F., (2021). Ecotoxicological impacts of arsenic on plants and its remediation approaches. In: *Heavy Metal Toxicity in Plants* (pp. 207–230). CRC Press.
57. Hay, F., Stricker, S., Gossen, B. D., McDonald, M. R., Heck, D., Hoepting, C., Sharma, S., & Pethybridge, S., (2021). Stemphylium leaf blight: A re-emerging threat to onion production in Eastern North America. *Plant Disease*, *105*(12), 3780–3794.
58. Kinkema, M., Fan, W., & Dong, X., (2000). Nuclear localization of NPR1 is required for activation of PR gene expression. *The Plant Cell*, *12*(12), 2339–2350.
59. Stein, E., Molitor, A., Kogel, K. H., & Waller, F., (2008). Systemic resistance in *Arabidopsis* conferred by the mycorrhizal fungus *Piriformospora indica* requires jasmonic acid signaling and the cytoplasmic function of NPR1. *Plant and Cell Physiology*, *49*(11), 1747–1751.
60. Lahlali, R., Serrhini, M. N., & Jijakli, H., (2004). Efficacy assessment of *Candida oleophila* (strain O) and *Pichia anomala* (strain K) against major postharvest diseases of citrus fruits in Morocco. *Comm. Appl. Biol. Sci. Ghent University*, *69*(4).
61. Schnürer, J., & Jonsson, A., (2011). *Pichia anomala* J121: A 30-year overnight near success biopreservation story. *Antonie Van Leeuwenhoek*, *99*(1), 5–12.
62. Godana, E. A., Yang, Q., Zhao, L., Zhang, X., Liu, J., & Zhang, H., (2021). *Pichia anomala* induced with chitosan triggers defense response of table grapes against post-harvest blue mold disease. *Frontiers in Microbiology*, 1684.
63. Hu, W., Godana, E. A., Xu, M., Yang, Q., Dhanasekaran, S., & Zhang, H., (2021). Transcriptome characterization and expression profiles of disease defense-related genes of table grapes in response to *Pichia anomala* induced with chitosan. *Foods*, *10*(7), 1451.
64. Walker, G. M., (2011). *Pichia anomala*: Cell physiology and biotechnology relative to other yeasts. *Antonie Van Leeuwenhoek*, *99*(1), 25–34.
65. Passoth, V., Olstorpe, M., & Schnürer, J., (2011). Past, present, and future research directions with *Pichia Anomala*. *Antonie Van Leeuwenhoek*, *99*(1), 121–125.
66. Passoth, V., Fredlund, E., Druvefors, U. Ä., & Schnürer, J., (2006). Biotechnology, physiology and genetics of the yeast *Pichia anomala*. *FEMS Yeast Research*, *6*(1), 3–13.
67. Olstorpe, M., Borling, J., Schnürer, J., & Passoth, V., (2010). *Pichia anomala* yeast improves feed hygiene during storage of moist crimped barley grain under Swedish farm conditions. *Animal Feed Science and Technology*, *156*(1, 2), 47–56.
68. Albrecht, U., (2019). Plant biostimulants: Definition and overview of categories and effects: HS1330, 5/2019. *EDIS*, *2019*(3).
69. Van, O. M. J., Pepe, O., De Pascale, S., Silletti, S., & Maggio, A., (2017). The role of biostimulants and bioeffectors as alleviators of abiotic stress in crop plants. *Chemical and Biological Technologies in Agriculture*, *4*(1), 1–12.
70. Sirrenberg, A., Göbel, C., Grond, S., Czempinski, N., Ratzinger, A., Karlovsky, P., & Pawlowski, K., (2007). *Piriformospora indica* affects plant growth by auxin production. *Physiologia Plantarum*, *131*(4), 581–589.
71. Vahabi, K., Sherameti, I., Bakshi, M., Mrozinska, A., Ludwig, A., Reichelt, M., & Oelmüller, R., (2015). The interaction of *Arabidopsis* with *Piriformospora indica*

shifts from initial transient stress induced by fungus-released chemical mediators to a mutualistic interaction after physical contact of the two symbionts. *BMC Plant Biology, 15*(1), 1–15.

72. Akum, F. N., Steinbrenner, J., Biedenkopf, D., Imani, J., & Kogel, K. H., (2015). The *Piriformospora indica* effector PIIN_08944 promotes the mutualistic Sebacinalean symbiosis. *Frontiers in Plant Science, 6*, 906.
73. Boysen, M. E., Stina, B., & Johan, S., (2000). Effect of the biocontrol yeast *Pichia anomala* on interactions between *Penicillium roqueforti*, *Penicillium carneum*, and *Penicillium paneum* in moist grain under restricted air supply. *Postharvest Biol. Technol., 19*, 173–179. doi: 10.1016/S0925-5214(00) 00086-7.
74. Zhao, L., Zhang, H., Li, J., Cui, J., Zhang, X., & Ren, X., (2012). Enhancement of biocontrol efficacy of *Pichia carribbica* to postharvest diseases of strawberries by addition of trehalose to the growth medium. *International Journal of Molecular Sciences, 13*(3), 3916–3932.

CHAPTER 9

Fungal Endophytes in Plants: Friends or Betrayers?

DAISY SENAPOTY,[1] V. BHUVANESWARI,[1] R. K. MOHANA PRADEEP,[2] and MATETI GAYITHRI[1]

[1]*Department of Plant Pathology, Assam Agricultural University, Jorhat, Assam, India*

[2]*Department of Plant Pathology, Tamil Nadu Agricultural University, Coimbatore, Tamil Nadu, India*

ABSTRACT

Fungal endophytes constitute the majority of microorganisms in nature, residing inside plant tissues in a mutually beneficial relationship with their hosts. These endophytes share food and nutrients with the plants, enhancing the plants' immune response to combat biotic and abiotic stress. They regulate phytohormone levels, sequester ferric ions, solubilize macronutrients, produce antibiotics and lytic enzymes, all of which contribute to systemic resistance. Fungal endophytes have shown potential as biocontrol agents, effectively competing with phytopathogens. However, their association with plants can be neutral or even pathogenic, depending on the host, endophyte genotype, and climatic conditions. Various factors, such as host age, nutrition, plant type, habitat, and light intensity, can disrupt the endophyte-symbiotic continuum, causing them to switch to a saprophytic or pathogenic phase. This chapter provides an overview of the diverse mechanisms by which fungal endophytes alleviate plant stress and adapt to different lifestyles.

Microbial Biostimulants: Biorational Pesticides for Management of Plant Pathogens.
Faheem Ahmad and Rakesh Pandey (Eds.)

9.1 INTRODUCTION

An increasing global population and urbanization of agricultural lands both necessitate high production per unit area. Though scientists have developed high-yielding varieties and modern technologies to cope with farmers in terms of yield, the plants are prone to pests and diseases due to the "climate crisis." Pests and diseases are estimated to cause 20 to 40% of global crop production loss each year. Every year, invasive insects cost the global economy $70 billion and plant diseases around $220 billion [1]. Hence, the management of diseases is inevitable to ensure food security and safety in the future. Many tactics and techniques are accessible for disease management, including cultural, physical, chemical, and host plant resistance [2]. Among them, synthetic fungicide application is widely adopted among the farming community because of their excellence in the mode of action, efficacy, and effectiveness in disease control to avoid yield losses. Fungicides have become an intrinsic part of Asian agriculture, with overuse of fungicides and bactericides accounting for 56.5% of the world's consumption in crop protection [3–5]. In India, the massive use of fungicides recorded in recent decades ranged from 13.05 kg/ton to 20.09 kg/ton [1]. The widespread use of fungicides in agriculture also raises health and environmental concerns. These agrochemicals are much more expensive, non-eco-friendly, and harm ecosystems and beneficial soil microbes [6]. It increases plant susceptibility towards the establishment of new races, pathogen resistance development, and, most notably, agrochemical deposition on post-harvest produce [7-9]. From the aforementioned statements, the employment of the rhizosphere and phyto-microbiome has become a greater strategy to combat pathogens for long-term sustainable plant disease management [10]. The microbial community assists the plant in acquiring nutrients, stress tolerance, increased fitness against biotic stress, and growth promotion. Similarly, the plant provides space for food and nutrients to the microbe's population. Endophytic fungi are one among the Phyto microbiome with a symbiotic association, which largely colonize to favor the plants for resistance against various pathogens through their multiple modes of action such as secretion of lytic enzymes, antibiosis, competition for ecological niche, siderophore production, and induced systemic resistance (ISR). The subtle interaction between the host and endophytes revolves around mutualism, commensalism, and symbiotic relationships depending on their colonization and mode of nutrition. The environmental factors and unbalanced nutrients between the host and microbes decide their mode of living, either as endophytes or pathogens. It's a complex phenomenon where different endophytes interact differently with their host. Omics technologies and next-generation sequencing enlighten the physical, biochemical, and genetic basis of endophyte reciprocity with plants and host responses to microbes, which is highly remarkable. Henceforth, better knowledge of the functional role and modulation of host immune response against biotic and abiotic stress is of utmost importance. This chapter discusses how fungal endophytes furnish increased fitness of plants to combat various biotic, environmental, and edaphic stresses and further debates the lifestyle switching from symbiotic to parasitic state.

9.2 FUNGAL ENDOPHYTES: "THE SECRET GUARDIAN OF PLANTS"

A multitude of microorganisms colonize plants either intercellularly or intracellularly in an endosymbiotic manner to perform various functions related to plant health and survival, without causing visible symptoms [11]. Endophyte colonization is widespread in all parts of the plant, including stems, roots, leaves, petioles, buds, flowers, seeds, and fruits. However, most of them exhibit tissue specificity, host recurrence, spatial heterogeneity, and high biodiversity within the plant community [12–14]. The localization of the endophyte community on plant parts strongly fluctuates with seasonal and climatic conditions, with roots being more highly colonized than leaves during summer rather than winter. *Colletotrichum* and *Diaporthe* colonization are commonly found in matured leaves, while *Trichoderma* and *Fusarium* occupy roots, and *Allophoma* populates the stem. Therefore, endophytes rarely move systematically within the tissues [15, 16]. Throughout evolution, plants have recruited a plethora of microorganisms into their rhizosphere due to the close connection between roots and the soil environment. Various root exudates secreted by plants, such as strigolactones (SLs), flavonoids, benzoxazinoids (BXs), and coumarins, perform a crucial role in engaging beneficial microbes under nutrient starvation conditions in plants [17–19]. When plants succumb to extremely unfavorable conditions, such as high temperature, drought, salinity, metal toxicity, cold stress, and water-logged conditions, endophytes are known to perform various functions to mitigate and overcome the stress by producing numerous chemical metabolites (e.g., secondary metabolites and growth hormones) and activating stress regulatory pathways. Under biotic stress conditions, fungal endophytes compete with pathogens by secreting lytic enzymes and antimicrobial compounds, alerting the plant's defensive pathways, and activating the ISR.

9.3 MODULATING IMMUNE RESPONSES OF THE HOST AGAINST BIOTIC STRESS

9.3.1 PRODUCTION OF ANTIBIOTIC COMPOUNDS

The potential of fungal endophytes to produce antifungal, antibacterial, antiviral, insecticidal, and nematicidal activities against fungi, bacteria, viruses, insect pests, and nematodes is constantly studied for *in-vivo* commercial application. The direct antagonism of fungal endophytes against detrimental phytopathogens was examined through dual culture assay [20] and volatile compound assay [21]. Furthermore, the identification of antimicrobial compounds is done through GC-MS-TD (gas chromatography mass spectrometry and thermal desorption), FT-IR (Fourier transform infrared), and HPLC (high-performance liquid chromatography) methods. Endophytic fungi have a broad range of sources for the production of secondary metabolic compounds such as alkaloids, phenols, isocoumarins, lactones, terpenoids, phenylpropanoids, lignans, and quinone steroids [22]. *Fusarium oxysporum,* an endophyte isolated from *Dendrobium lindleyi* orchid, was effective against *Fusarium* spp. pathogen due to the production of coumarin compounds [23]. Gluconic acid, phenyl mercury tri ethanol ammonium (NH_3) lactate (similar to puratized fungicide), hydrazine, and benzaldehyde metabolites were produced by *Trichoderma longibrachiatum* endophyte from paddy and showed inhibitory activity against *Rhizoctonia solani.* The same endophyte secretes different chemical compounds, namely nicotinic acid, oxalic acid, 1 HBT, and picolinic acid, when co-cultured with *Magnaporthe grisea* [24]. Benzothiazole, 4-methylquinazoline, and benzyl alcohol are volatile organic compounds (VOC) secreted by *Xylariaceae* and *Alternaria* sp. isolate 2 showed diminishing growth activity against *Colletotrichum acutatum,* respectively, which were recovered from disinfected leaves of *Olea europaea* [25]. N-hexane and chloroform culture filtrates of *Cephalosporium* sp. and *Fusarium* solani endophytes exhibited enhanced nematicide activity against the root-knot nematode (RKN) *Meloidogyne javanica* within 48 hours, showing 100% death of juveniles. *Cephalosporium* produces octadecanoic acid, heptadecanoic acid, methyl 10-trans, 12-cis-octadecadienoate, and Fusarium solani secretes 1,1,2-trichloro-3,3,3-trifluoro, 1-Propene antimicrobial compounds that help in disease and insect pest management in agriculture [26]. Sometimes, the secondary metabolic compounds are not independently manufactured by endophytes alone. The endophytes, during colonization and adverse effects of biotic stress, chemically interact with the plant, leading to the modification of metabolic pathways of plants, thus improving the production of antimicrobial compounds.

9.3.2 SECRETION OF LYTIC ENZYMES

The hidden function behind the production of extracellular hydrolases by fungal endophytes on plants is to obtain nutrition from their host by hydrolyzing food substances and inducing a defense response against invasive pathogens [27–29]. There are two categories of extracellular enzymes: hydrolases, such as chitinases, amylases, lipases, proteases, pectinases, β-1,4-glucanase, cellulases, xylanases, asparaginases, and oxidoreductases; and PODs and laccase, which fall under the oxidoreductases type [30]. The maize endophyte *Acremonium zeae* NRRL 6415 hydrolyzes arabinogalactan and utilizes cell wall components for penetration [31]. *Antennariella placitae* degrades the cell wall of *Ustilaginoidea virens* false smut fungi of rice by secreting the chitinase enzyme [32]. Phosphatases, cellulases, and glucosidases produced by *Penicillium* and *Preussia* endophytes demonstrate competition with pathogens through limited colonization of specific plant cells [33]. Endophytes isolated from various parts of *Baliospermum montanum* revealed that 83% of them secrete amylase and 79% produce cellulase, which breaks down complex carbohydrate substances as substrates and activates the cell wall-mediated defense pathway, respectively. However, they did not exhibit laccase activity [34]. Medicinal plants colonized with fungal endophytes have shown significant evidence of laccase production [35]. Grasses inhabited by *Epichloae* secrete β-1,6-glucanases within the apoplast, providing nutrients for their own survival and degrading the cell wall of antagonistic pathogens encountered within the plant system [36]. The *E. festucae* E437 endophytic strain demonstrated strong endoglucanase and protease activity in the broth medium, but it did not exhibit significant lytic activity on *Drechslera erythrospila* when co-cultured with the pathogen [37]. The diverse characteristics of fungal endophytes primarily depend on the substrate of the medium, incubation temperature, relative humidity, and pH of the medium under *in vitro* conditions. In *in vivo* conditions, the host habitat, tissue specificity, soil fertility, and climatic factors determine the potential secretion of lytic enzymes. Biotechnological genome editing tools, such as CRISPR (clustered regularly interspaced short palindromic repeats)-Cas9, can be employed to enhance the production of strain-specific extracellular enzymes, which can be utilized in agriculture, industry, and medicine.

Instead of strain-specific approaches, the concept of one strain with multiple enzymes (OSME) is gaining more attention nowadays to obtain multiple benefits from fungal endophytes.

9.3.3 PRODUCTION OF SIDEROPHORES

Fe is an essential micronutrient required for living organisms, including animals, plants, and microbes. The low molecular weight compound siderophores have a high affinity towards iron. In low iron conditions, endophytic microbes chelate the iron to maintain the homeostasis of cellular respiration in plants. During pathogen infection, a rivalry between the host and endophytes occurs for iron acquisition. Several studies have revealed that mammalian pathogenic microbes require siderophores for their virulence [38]. Similarly, phytopathogenic bacteria and fungi utilize siderophores for their multiplication in hosts and to facilitate infection. Oide et al. [39] demonstrated that nonribosomal peptide synthetase *NPS6* is a critical component for extracellular siderophore biosynthesis, which is a virulence determinant of the *Cochliobolus heterostrophus* maize pathogen. The mutant *nps6* showed reduced virulence and hypersensitivity to hydrogen peroxide (H_2O_2). External supplementation of iron leads to the restoration of tolerance to H_2O_2 by the mutants. Fungal endophytes solubilize ferric iron (Fe3+) from the environment and transport it for plant cell growth under iron-limited conditions. Competition for iron is a feasible mechanism to protect the host from phytopathogens. *Lasiodiplodia pseudotheobromae* endophytic fungi compete for iron with *Fusarium oxysporum* and *Sclerotium rolfsii* [40]. Siderophores produced by endophytes also exhibit antibacterial activity. Haselwandter & Winkelmann [41] reported that the dark septate endophytic fungi (DSEF) diffuse ferricrocin, an important hydroxamate-type siderophore, while colonizing the root system. Ferrirubine is a novel siderophore produced by *Phialocephala fortinii* DSEF [42]. Intracellular and extracellular production of fusarinine and ferricrocin-type siderophores by the foliar endophyte of perennial ryegrass *Epichloe festucae* [43]. Pine and labrador tea-specific foliar endophytes produce ferricrocin, which has antibacterial and antioxidant activity. Ferricrocin prevents the Fenton's reaction and plays a significant role in cytoprotection and antioxidant activity. Ferrous ions react with H_2O_2 and result in ferric ions, reactive oxygen free radicals, and hydroxide ion deposition in plants. On one hand, ROS accumulation assists the compatible interaction between endophytes and their host, and on the other hand, it creates a barrier for pathogens incompatible with the host in iron sequestration [44].

The relative proportion of siderophore production ranges from maximum to minuscule in winter and summer, respectively, in *Vitis vinifera*. This is due to the freezing soil temperature, which implies the unavailability of iron to both endophytes and plants. Therefore, endophytes actively recruit iron from the soil through increased siderophore production [45].

9.3.4 PHOSPHORUS AND ZINC SOLUBILIZATION

Phosphorus is an indispensable macronutrient for plants, required in larger quantities. P is a non-available nutrient for plant health, present in organic, inorganic, and insoluble forms due to its chemical nature. Interestingly, a few endophytic strains solubilize it from unavailable sources to make it available for plant absorption [46]. Endophytes mobilize phosphorus either by secreting organic acids or by ejecting protons or through enzymatic degradation. In calcareous soil, phosphorus precipitated by calcium leads to lower solubility of P. In acidic soil, iron and aluminum utilize free phosphorus and form aluminum and iron phosphates. pH plays a significant role in phosphorus availability to plants. *Penicillium sclerotiorum,* a foliar endophyte of tea, resulted in an increase in soluble phosphate through the secretion of different organic acids like malic, oxalic, and gluconic acids, thereby reducing the pH of the medium. Insoluble potassium aluminum silicate salts, hydroxyapatite, and zinc oxides are solubilized into soluble minerals by endophytes and utilized by plants [47]. Fungal endophytes of *Ephedra pachyclada,* such as *Penicillium crustosum* and *Penicillium caseifulvum,* significantly improved the root length and fresh weight of maize seeds when compared with un-inoculated plants. It is due to the enhanced phosphorus solubilization and ammonia production of these endophytes [48]. *Trichoderma longibrachiatum* S12 endophyte recovered from soybean manifested improved phosphorus and zinc solubilization index and assisted in elevating nutrient uptake [49]. Conidia of fungal strains relatively move greater distances in soil and air. Upon repeated subculturing, the mineral solubilizing potential of the fungal hyphae is not lost, unlike endophytic bacteria (EB). DSEF colonizing roots deter phytopathogens by exclusively reducing the carbon availability in the rhizospheric environment. Secretion of melanin coated on the root cells diffuses multiple secondary metabolites thereby inhibiting herbivore attacks and altering host performances. In extreme environmental conditions, the DSE replaces the mycorrhizal associations with roots and mineralizes the immobile N, P, S, and Zn into forms that are available to plants [50].

9.3.5 PHYTOHORMONE PRODUCTION

Endophytes are capable of producing phytohormones such as Auxin (IAA), gibberellin, and cytokinin, which play a significant role in root and shoot development. IAA performs various growth-promoting activities, including cell division and tropism. Endophytic fungi produce L-tryptophan, which acts as a precursor for IAA biosynthesis. However, the metabolic pathway used by them for IAA production has not been elucidated yet. *Phoma glomerata* and *Penicillium* sp. are endophytic fungi that produce IAA and GA [51]. By producing phytohormones, endophytes help plants mitigate abiotic stress. *A. fumigatus* TS1 and *F. proliferatum* BRL1 (endophytic fungi) obtained from the roots of *Oxalis corniculata* have been shown to increase chlorophyll content, biomass production, and root and shoot length [52, 53]. Inoculated plants with the fungal endophyte Ld-03 exhibited maximum root and shoot length compared to control plants. Endophytes produce gibberellins (GA1 and GA3) from Acetyl-CoA through the mevalonic acid (MVA) pathway. Increased levels of GA1, GA3, GA4, and GA7 were identified in the hyphal extract of *P. citinum* IR-3-3 compared to the original GA producer *Gibberella fujikuroi*. The application of culture filtrate from an endophytic isolate of *Cladosporium* sp. resulted in remarkable growth of cucumber plants due to gibberellin production [54].

9.3.6 INDUCED SYSTEMIC RESISTANCE (ISR)

Endophytes trigger ISR through phytohormones, such as jasmonic acid (JA) and ET, which pave the way for PR protein assimilation. PR proteins, like chitinase and glucanase, are involved in the cell wall degradation of pathogens. The NPR1 gene is required for the JA/ET-dependent pathway, which is cytosolic-mediated in ISR and nucleus-mediated in the case of systemic acquired resistance (SAR). *Fusarium solani*, an endophyte obtained from tomato roots, induces ISR against the foliar pathogen *Septoria lycopersici* through the expression of PR5 and PR7 in roots [55]. In contrast to the usual pathway, induced resistance mediated by fungal endophyte against *Fusarium wilt* is independent of the signaling agents' JA/ET pathways [56]. The PAMP molecules (Pathogen-associated molecular patterns) chitin, flagellin, and VOCs are the elicitors that trigger the ISR pathway. Pattern recognition receptors (PRRs) have evolved to identify these elicitors, triggering WRKY53 gene expression and MAPK signaling, which result in

PAMP-triggered immunity (PTI). *P. indica* modifies the immune response of plants through ISR against *Fusarium* in a wide range of crops including barley, wheat, maize, and tomato. The pre-inoculation of *Piriformosa indica* before infection of *Blumeria graminis* in *Hordeum vulgare* potentially reduces the development of symptoms, activating ISR [57]. Endophytes passively invade the plants as the host is incapable of recognizing the MAMPs (cell wall of endophyte) which are composed of beta-glucan binding lectin due to modification in the FGB1 gene [58]. Evolutionarily, the endophytic strains negatively regulate the effectors responsible for virulence-like pathogens to establish colonization. The host's defensive nature is rapidly enhanced by the ISR pathway. *Beauveria bassiana* treated tomato and cotton seeds induced resistance against *Rhizoctonia solani* and *Pythium myriotylum* [59].

9.4 ALLEVIATION OF HOST FITNESS AGAINST ABIOTIC STRESS

The diverse metabolic and genetic strategies were governed by host-associated endophytic fungi to increase their fitness against various abiotic stresses [60, 61]. The current and improved understanding of the mechanisms of physical and biochemical changes in endophyte and non-endophyte colonized plants (NENC) were discussed. The overall performance of endophyte colonized and non-colonized plants against various abiotic stresses is given in Figure 9.1.

9.4.1 AN ELEVATION IN TOTAL BIOMASS

An increase in biomass content has generally been estimated as one of the indicators for hosts to cope with salinity and drought stress. Several studies have been conducted to increase the total biomass in various plant species, including *Brassica juncea*[62], *Solanum lycopersicum*[63], *Glycine max*[64], *Zea mays*[65], *Oryza sativa*[66], *Chenopodium quinoa*[67], *Nicotiana benthamiana*[68], and *Triticum aestivum*[69], by using ENC compared to NENC host plants.

The total biomass was also assessed based on plant growth analysis, including Relative Growth Rate, Net Assimilation Rate, and Leaf Area Ratio in plants. Hou et al. [70] found that DSE alleviates salt stress in *Artemesia ordosica* by increasing shoot and root biomass compared to NDSEC host plants.

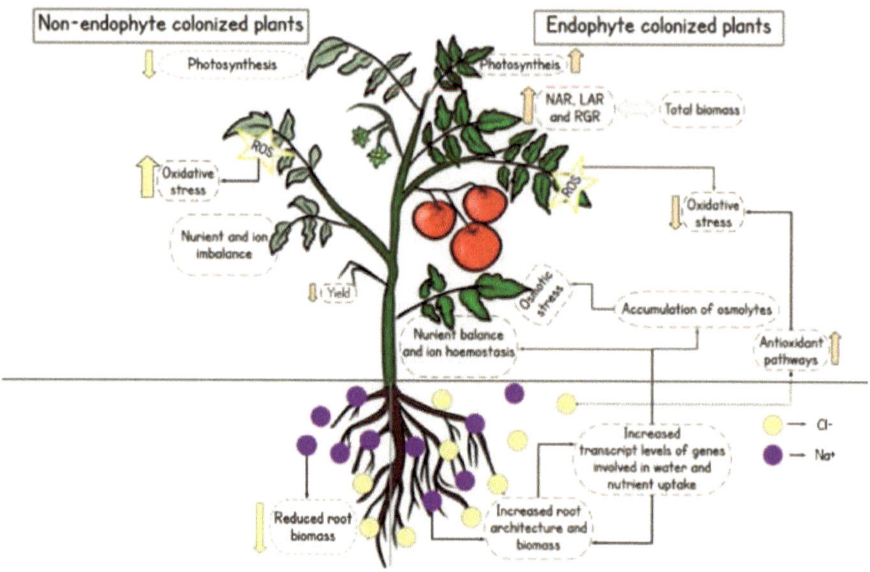

FIGURE 9.1 Beneficial effects of symbiotic-endophytic fungi in plants under abiotic stress conditions, summarized from this part of the chapter.

Note: In general, stress reduces biomass in non-endophyte colonized plants on the left. Endophyte-colonized plants, however, improve root biomass by modifying their root architecture through increased transcription levels of genes that govern nutrient and water uptake. Additionally, the increased transcription levels of genes enhance the accumulation of osmolytes under osmotic stress, thereby maintaining nutrient and ionic homeostasis Na^+ (Sodium) and Cl^- (Chloride) channels will increase their antioxidant pathways (producing scavenging enzymes) and reduce oxidative stress (ROS). In non-endophyte colonized plants, they act in a reverse manner with an increase in oxidative stress. Endophytic fungi increase the photosynthetic activity through various mechanisms and increase NAR, LAR, and RGR of plants. Ultimately, this leads to an increase in yield and total biomass of the endophyte colonized plants compared to non-endophytic plants.

9.4.2 MANIPULATION OF ROOT ARCHITECTURE

RSA (Root System Architecture) is the primary plant part affected by various abiotic stresses, such as salt, drought, flooding, heat, metal, and other soil factors [71, 72]. It significantly alters and influences root architecture, including primary and lateral root growth, root hairs, and surface area [73]. RSA plays a crucial role in the absorption of water and mineral nutrients from non-saline areas, contributing to improved plant health and biomass

[74, 75]. RSA regulates salt acquisition and translocation, enhancing tolerance to salt stress [76], while faster growth with finer root diameter aids in nutrient allocation to mitigate drought [77, 78].

In ENC (Endophytic Fungi-Colonized) rice, maize, and barley crops, the length and biomass of roots were increased compared to NENC (Non-Endophytic Fungi-Colonized) crops, helping to mitigate salinity [79–81]. The beneficial effects of endophytic fungi-mediated RSA for enhanced water and mineral uptake under saline conditions need to be further explored.

9.4.3 NUTRIENT ACQUISITION AND IONIC HOMEOSTASIS

Nutrient acquisition is one of the major challenges faced by terrestrial plants during various abiotic stresses [82]. Endophytes assist the plants by efficient mobilization and solubilization of insoluble macronutrients such as nitrogen (N), phosphorus (P), and potassium (K) into soluble form for easy accessibility to plants to maintain desired ratios Na^{+}: K^{+}, Na^{+}: Ca^{2+} and Ca^{2+}: Mg^{2+}[83, 84]. In the past few decades, endophytic fungi have been assessed for mobilization and solubilization of macro- and micronutrients [85–87]. Moreover, a few studies revealed that DSE (*Periconia macrospinosa*) increased its efficiency of intake of organic nutrients for tomato growth and its development [86]. Similar results, in the uptake of macro- and micronutrients, were regulated by phytohormone-producing *Galactomyces geotrichum* in soybean [88]. The endophytic fungus (*P. indica*) elevated the elemental concentration of P and Zn within the roots and shoots of rice seeds [66]. Numerous research disclosed that root and seed endophytes are more intricate in nutrient acquisition than foliar endophytes, Chen et al. [89] found that foliar endophytes stimulate the symbiotic association of roots with other microbes for improving nutrient acquisition.

Ionic homeostasis plays a crucial role in the maintenance of turgor pressure, enzyme activity, photosynthesis, and stomatal activity in plants [90]. A high Na^{+}: K^{+} ratio acts as an indicator of plants suffering from stress, which consequently disrupts the cellular metabolic pathways of plants [91]. Plants have evolved with a variety of antiporters, membrane proteins, ionic sodium and potassium transporters, cationic channels, aquaporins, and vacuolar ATPases [92] to achieve ionic homeostasis. Abdelaziz et al. [91] elucidated the upregulation of genes encoding high-affinity K^{+} transporter 1 (HKT1) and inward-rectifying K+ channels KAT1 and KAT2 for maintaining Na^{+}: K^{+} homeostasis in ENC than NENC *A. thaliana* plants.

9.4.4 OXIDATIVE STRESS

Abiotic stresses commonly interfere with normal cellular functions and components, such as membrane disruption, lipid peroxidation, cell death, and modification of proteins and DNA, due to the overproduction of reactive oxygen species (ROS) including hydroxyl (OH-), superoxide (O_2^-) radicals, and H_2O_2 in crop plants [93–95]. Plants have evolved two major systems, enzymatic and non-enzymatic antioxidative systems, to cope with adverse consequences and maintain redox homeostasis. These systems include enzymes like ascorbate peroxidase (APX), catalase (CAT), glutathione reductase (GR), superoxide dismutase (SOD), monodehydroascorbate reductase (MDHAR), dehydroascorbate reductase (DHAR), and ascorbic acid (AsA), as well as compounds like carotenoids, glutathione (GSH), and osmoregulatory substances [93, 96]. The levels of reactive oxygen species (ROS) and antioxidants (AOX) may vary depending on the cultivar and stress duration [97]. Baltruschat et al. [98] conducted experiments on *Serendipita indica* and barley root colonization, and found elevated levels of antioxidant enzymes (CAT, GR, APX, and DHAR) as well as the AsA system to neutralize free radicals [99]. Similar elevation of antioxidant enzymes was reported in rice plants colonized by *S. indica*. An endophyte, *Trichoderma longibrachiatum* T6 (ENC), colonizes wheat plants and enhances antioxidative defense enzymes such as peroxidase (POD), SOD, and CAT, as well as the genes encoding defense enzymes against O_2^- and H_2O_2 radicals, compared to NENC plants [100]. *S. indica* colonized plant roots contain elevated concentrations of ascorbate and altered CAT, GR, APX, MDHAR, and DHAR in comparison with NENC plants. These alterations yield considerable results with a reduction in lipid peroxidation of leaves [79]. Zhang, Gan, & Xu [100] reported a 15% reduction of malondialdehyde (MDA) in ENC salt stress plants compared to NENC plants. MDA is a product of lipid peroxidation and generally acts as an indicator for cell membrane injury caused by oxidative stress.

9.4.5 PHOTOSYNTHESIS

Abiotic stresses hinder photosynthesis, resulting in extensive losses in terms of crop productivity. Salt stress [101] and drought [102] have been identified as major factors affecting photosynthesis. Salt stress has a multifarious effect on photosynthesis, degrading D_1 and D_2 proteins of Photosystem II [103], reducing the activity of enzymes synthesizing photosynthetic pigments, limiting CO_2 availability and assimilation, and affecting RuBisCO [104]. Drought resistance in most photosynthetic apparatus is achieved through efficient mechanisms such as paraheliotropism (leaf movement), minimizing non-photochemical quenching (NPQ) [105], and thermal dissipation of excitons trapped in PSII. Plants exposed to water stress experience a decrease in the net CO_2 concentration in the chloroplast due to a decrease in the photochemical product of PSII, leading to increased thermal dissipation of excitons in PSII, reduced nitrate reductase and sucrose phosphate synthase activity, and elevated levels of ribulose 1,5-bisphosphate oxygenase activity [106]. ENC plants have been reported to exhibit lower NPQ and efficient conversion of light into chemical energy compared to NENC plants [107]. Endophytic symbiosis utilizes these mechanisms to combat the negative effects of salt stress on photosynthetic activity [75, 108]. Magnesium (Mg^{2+}) is an essential macronutrient involved in photosynthesis, protein and nucleic acid synthesis, and enzyme activation [109]. ENC plants improve Mg^{2+} uptake to stabilize plastid integrity and enhance photosynthetic efficiency [75, 110, 111]. Kumar et al. [112] reported that ENC plants (*P. indica*) maintain a high antioxidative ascorbate-glutathione (GSH) defense priming, leading to enhanced photosynthetic efficiency and maintenance of plastid integrity during salt stress. Furthermore, ENC plants repair the disrupted D1/D2/Cytb 559 complex in PSII by accumulating glycine betaine during salt stress, maintaining the integrity of PSII compared to NENC crops [113].

9.4.6 PHYTOHORMONAL REGULATION

The induction of phytohormones, especially GA's, played a significant role in mitigating salt and drought stress by fungal endophytes [51, 79, 114]. Khan et al. [115] discovered that ENC (*Aspergillus fumigatus* LH02) soybean plants alleviated salt stress by regulating GAs, specifically GA4, GA9, and GA12, in comparison to NENC soybean plants. Similar results were obtained by Waqas et al. [51], where ENC, namely *Phoma glomerata* LWL2 and *Penicillium* sp. LWL3, cucumber plants alleviated both physiologically active GA's, such as GA1, GA3, GA4, and inactive GA7 in LWL2, followed by two active GA1 and GA3 in LWL3, respectively. Furthermore, it also regulated IAA, ABA production, and JA, SA signaling to mitigate both salt and drought stress conditions compared to NENC cucumber plants.

9.5 LIFESTYLE SWITCHING FROM ENDOPHYTE TO PARASITE AND VICE-VERSA

The lifestyle of endophytic fungi on host plants is ambiguous and can express multiple modes of living, including mutualism, commensalism, parasitism, and saprophytic states. In a mutualistic relationship, both partners benefit from each other by exchanging nutrients and energy resources. Commensalism occurs when one partner benefits from the other organism without depriving the associated partner. Parasitism occurs when the microbes damage the host to some extent in order to uptake nutrients and establish their populations. In a saprophytic state, the microbes reside on dead or decomposed matter, and the death of the host is not caused by the organism.

Many economically important diseases are caused by endophytes that asymptomatically colonize host plants [116] and exhibit latent infection [117–120]. Endophytes such as *Cladosporium musae*, *Pestalotiopsis* sp., *C. gloeosporioides, Deightoniella torulosa, Guignardia cocoicola, Periconiella musae,* and *Cordana musae* have been recovered from *Musa acuminata* [121]. Since most of them are reported as potential banana pathogens, these endophytes were tested for pathogenicity on banana leaves by spraying conidial suspensions (SU). Among them, *Deightoniella torulosa* caused leaf spots with necrotic brown pale lesions, while the others did not, indicating that pathogens have a hidden life in a host as endophytes [122].

Endophytes such as *Botryosphaeria dothidea, Neofusicoocum, Cytospora, Diaporthe*, and *Pestalotiopsis* spp. have been isolated from fruit tree shoots. When they were re-inoculated on their respective hosts, a few endophytes produced sunken cankers and lesions, clearly indicating their latent pathogenic nature [123]. Under certain circumstances, the microbes may act as endophytes or become pathogenic. The different lifestyles of endophytes on host plants depend on the host, the virulence of the endophytes, and abiotic factors. For a list of endophytes that switch their lifestyle to a pathogenic state, refer to (Table 9.1).

9.5.1 HOST FACTORS

Switching of lifestyles occurs not only in the endophyte genome but also in the respective partner. The host plays an equal role in adopting the symbiotic-plant continuum. When the host completes its life cycle and senesces, the colonized endophytes do not receive sufficient nutrition from the host. Host nutrition is a major driving factor in shifting endophytes to necrotrophs or saprobes [142, 143]. The endophyte is an evolutionarily unstable trait that

TABLE 9.1 List of Switching Over Lifestyle of Endophyte to Parasite State

Fungus	Isolated From	Parasitic	Mutualistic	Factors Affecting Lifestyle Switching	References
Colletotrichum magna	Cucurbits	Cucurbit plants	Non-cucurbit plants.	Path-1 mutation in single locus. REMI mutation in single locus.	[124, 125]
Colletotrichum acutatum	Strawberry	Strawberry	Vicia, Conyza, eggplant, tomato, bean	Host specificity	[126]
Botrytis cinerea	Hybrid polyanthus	Hybrid polyanthus	Hybrid polyanthus	Host age, flowering stage.	[127]
Neotyphodium starri	*Festuca arizonica*	Arizona fescue	Perennial ryegrass Tall fescue	Host genotype	[128]
Neotyphodium coenophialum	*Lolium perenne* (perennial rye grass)	Perennial rye grass	Perennial rye grass	Flowering stage	[129]
Discula quercina	Oak trees	Oak trees	Oak trees	Host senescence	[130]
Diplodia mutila	*Iriartea deltoidea*	Young seedlings	Mature plants	Light induces the hydrogen peroxide production.	[131]
Lasiodiplodia theobromae	*Terminalia* spp.	*Terminalia arjuna*	*Terminalia catappa*	Latent pathogen under unfavorable biotic and abiotic stress.	[120]
	Terminalia spp.	*Terminalia tomentosa*	*Terminalia mantaly*		
Lasiodiplodia parva	*Terminalia* spp.		*Terminalia catappa*		
Endomelanconiopsis endophytica		*Terminalia bellarica*	*Terminalia mantaly*		
		Theobroma cocoa	*Terminalia ivorensis*		
		Theobroma cocoa	*Terminalia superba*		
Epichloe festucae	*Lolium perenne* (perennial rye grass)	Perennial rye grass	Perennial rye grass	Disruption of Nox complex (NoxA, NoxR, and RacA).	[132, 133]
Lasiodiplodia theobromae	*Adansonia gregorii* (Baobab)	Baobab trees	Baobab trees	Abiotic factors – temperature and dry period, latent pathogen.	[134]

TABLE 9.1 *(Continued)*

Fungus	Isolated From	Parasitic	Mutualistic	Factors Affecting Lifestyle Switching	References
Harpophora oryzae	*Oryza granulata* (Chinese wild paddy)	Rice	Rice	Transposable elements	[135]
Colletotrichum tofieldiae	*Arabidopsis thaliana*	*Arabidopsis thaliana*	*Arabidopsis thaliana*	Phosphate availability in soil.	[136]
Fusarium solani	*Phragmites australis* (common reed)	Paddy	Paddy	Salinity	[137]
Alternaria alternata	*Vitis vinefera* (grapevine)	Grapevine	Grapevine	Reduction in carbohydrate level in leaves.	[138]
Leptosphaeria biglobosa	*Brassica napus*	*Brassica napus*	*Brassica napus*	Mycovirus – ds RNA quadrivirus	[139]
Sclerotinia sclerotiorum	*Brassica napus*(rapeseed)	Rapeseed	Rapeseed	*Sclerotinia sclerotiorum*hypo-virulence associated DNA virus 1 (SsHADV-1)	[140]
Pestalotiopsis theae	*Camellia sinensis* (tea)	Tea	Tea	Mycovirus – Pestalotiopsis theae chrysovirus (PtCV1)	[141]

shares a common clade with pathogenic *Guignardia, Fusarium, Colletotrichum,* and *Phomopsis* to saprophytes [144, 145]. During the senescence of *Magnolia liliifera* leaves, associated endophytes including *Leptosphaeria, Phomopsis,* and *Colletotrichum* isolates secrete various cell wall degrading enzymes. In the early stage of senescence, the aforementioned endophytes produce xylanase, β-mannanase (hemicellulose breakdown), and finally laccase and polygalacturonase, which are secreted to digest lignin and pectin, respectively [146].*Phomopsis liquidambari* B3 endophyte colonizes symbiotically in rice and *A. thaliana* generally lives as saprophytes performing elevation of carbon and nitrogen cycle by decaying the litter fall [147]. The density of endophyte colonization in conifer trees increases with the age of needles. It is proposed that if the density of colonization reaches a maximum threshold, the needles eventually die. Generally, the threshold never reaches a high level until the onset of natural senescence. In due course of time, the colonized endophytic fungi initiate spore production to shift to a saprophytic lifestyle as the plant ages [148]. The ecological fitness of the endophytes is modified according to the host genotype.

9.5.2 ENDOPHYTE GENETIC FACTORS

Disruption or mutation in a single locus of the endophyte genome alters the endophyte's nature to a pathogenic lifestyle without affecting host specificity [149, 150]. Mutation in the NoxA gene of *Epichloe festucae*-infected plants observed stunted growth and premature senescence of leaves. The gene expression is usually upregulated for coding NADPH (nicotinamide adenine dinucleotide phosphate) complex, essential for cellular differentiation and polarized growth of endophyte hyphae. This complex encodes the ROS production of endophytes, which helps restrict proliferative and polarized growth in the shoot apical meristem and activates the extension of leaf primordium. Mutant *noxA* is defective in ROS production due to the single copy of inserted plasmid, thereby unrestrictedly proliferating the cells and not switching to an expansion of intercalary cells, resulting in stunted growth [151]. Fungal ROS production strictly maintains the symbiotic endophyte-plant interaction. Mutation in components of the ROS-producing Nox complexes, such as NoxR and RacA, led to the loss of apical dominance, ultimately breaking the mutualistic relationship between fungi and plants [151]. The stress-activated-mitogen-activated protein kinase (MAPK) is responsible for inducing signal transduction pathways in response to various environmental cues. Eaton et al. [150] investigated the disruption in the fungal MAPK gene *sakA* in relation to the symbiosis of *Epichloe festucae* with perennial ryegrass. High-throughput mRNA sequencing reveals the negative regulation of large clusters of genes encoding secondary metabolite production and the positive regulation of several sets of transporter and hydrolytic genes in the *sakA* mutant. Drastic reduction in apical dominance and lateral root growth due to the downregulation of auxin biosynthesis and upregulation of auxin influx transporter genes switches the intercellular growth of hyphae to vasculature colonization. It increases plant biomass and collapses the symbiotic association, resulting in precocious senescence, and the plant eventually dies. UV-induced mutation in a single locus converts the most destructive *Colletotrichum magna* pathogen infecting cucurbits into non-pathogenic and performs a mutualistic endophyte lifestyle without host specificity [152, 124]. In addition, the *pth-1 mutant* colonizes tissue-specific manner and further activates the systemic resistance against the wild-type *C. magna, Colletotrichumorbiculare,* and *Fusarium oxysporum* [153]. REMI (restriction enzyme-mediated integration) transformants with hygromycin-resistant plasmids of *Colletotrichum magna* mutants were generated by disrupting genes in a single locus [125]. Both REMI and UV mutants were 'locked' into one lifestyle (mutualistic or commensal) and couldn't switch over to the pathogenic phase [154]. The aforementioned shreds of evidence prove that disruption in genes converts the pathogen to non-pathogenic. This mechanism has attracted scientists to reduce the virulence of pathogens and utilize them in plant disease control. The hypervirulence of *Leptosphaeria biglobosa* induces systemic acquired resistance against *Lephtosphaeria maculans* with the infection of mycovirus. There is a dramatic alteration in amino acid metabolism, upregulation of WRKY genes, and carbohydrate metabolism due to mycovirus-infected *L. biblobosa* inoculation on plants [139]. A potential necrotrophic pathogen, *Sclerotinia sclerotiorum,* was converted into mutualistic colonization by mycovirus (SsHADV-1) infection in rapeseed. The mycovirus downregulates the key virulence factors of the pathogen and suggests that they shape the origin of endophytes [140]. *Pestalotiopsis thea* chrysovirus-1 (PtCV-1) modulates the pathogenic life of *P. thea* into a symbiotic association with tea [141].

9.5.3 ABIOTIC FACTORS

Environment is the major influencing dependent variable for disease development in the presence of a virulent strain of pathogen and a susceptible host plant. The endophytes switch lifestyles from symbiotic to parasitic continuum, relying on abiotic causes. They provide increased stress tolerance under varying climatic conditions. A total of 120 fungal endophytes were acquired from seed pods, inflorescence, and stems of symptomless *Arabidopsis thaliana* plants in a natural environment, where 48% of the isolates were *Leptosphaeria maculans.* The plants grown in tissue-cultured media and inoculated with *L. maculans* showed a harmful effect on the root growth of seedlings. On the contrary, the endophyte colonized plants (ENC) flowered earlier, and the inflorescence length was maximum compared to un-inoculated plants. Under stressed environments, the inflorescence is induced sooner in the plants. This significantly deciphers the artificial conditions (laboratory), enabling stress on the plant and a favorable situation for the endophyte to become pathogenic. This confirms the delicate balance between endophyte virulence and host defense, modifying the fine-line of the endophyte to the necrotrophic pathogen. In the natural environment, high precipitation enables L. maculans to effectively colonize the host plant. Inoculation with root endophytes, namely *Geniculosporium* sp. 9725, *Stagonospora* sp. 9722, and *Phoma* sp., resulted in a severe reduction in root length and the appearance of necrotic lesions under laboratory conditions [156].

Alvarez-Loayza et al. [131] reported that high light intensity triggers the *Diplodia mutila* endophytic fungus to switch into a necrotrophic pathogen by causing lesions on the young plants of *Iriartea deltoidea. D. mutila* resembles an endophyte under shady understory, and when the plants are exposed to light, the pathogen produces melanin and H_2O_2. High light activates H_2O_2 biosynthesis by the endophyte through the glyoxylate pathway, whereas melanin imposes the accumulation of ROS in plants, ultimately causing the death of cells by hypersensitivity. Matured plants are not affected by *D. mutila,* even after exposure to high light intensity. Moreover, it showed increased resistance to herbivore infestation (stem borer – Coccotrypes sp.). These findings reveal that endophyte virulence is unambiguously plant age and light-dependent.

Lasiodiplodia theobromae, an endophytic fungus, switches to pathogenic due to extreme climatic conditions like prolonged dry spells and high temperatures in the North West of Australia. Since *L. theobromae* has a broad host range, they colonize the *Adansonia gregorii* (Baobab trees) asymptomatically and resemble a latent pathogen [134].

The symbiotic relationship between oak trees and the endophytes *Apiognomonia quercina/Discula quercina* becomes unbalanced under distressful climate conditions. This imbalance generates conflicts in the symbiotic interaction between both partners, subsequently causing the endophyte to switch from a quiescence state to a neutral state, and eventually becoming antagonistic [130]. In another specific case, the endemic endophyte *Colletotrichum tofieldiae* solubilizes phosphate in *A. thaliana* plants under phosphorus-limited conditions, while the ubiquitous endophyte *Piriformosa indica* mobilizes phosphate in both high and low phosphate soil [155]. High phosphorus levels in the soil stimulate the pathogenic lifestyle of endophytes. In this context, the host PSR (Phosphate Starvation System) regulates the root colonization of *C. tofieldiae,* which is necessary for macronutrient uptake from the soil. The plant growth promotion (PGP) mechanism relies on PEN2 indole glucosinolate biosynthesis, which plays a crucial role in maintaining the mutualistic association between the endophyte and the host [136]. *Fusarium solani,* under low salt stress, colonizes rice plants, enhancing germination percentage and seedling length, while high salinity inhibits root growth in young rice seedlings [137].

9.6 CHALLENGES IN THE COMMERCIALIZATION OF FUNGAL ENDOPHYTES AS BIOCONTROL AGENTS

Though fungal endophytes provide significant benefits to plants, such as plant growth promotion (PGP) and increased resistance to phytopathogens and environmental stress, there are several limitations to their commercialization and large-scale application.:

- Majority of the endophytes are dormant pathogens.
- Most of the fungal endophytes are latent pathogens.
- High cost and tissue-specificity limit their application to a broad range of plants.
- The mutualistic plant interaction is evolutionarily highly unstable due to the prevalent environment, host phenotype, genotype, and soil nutrient availability.
- Assessing the mutualistic colonization of the fungal endophyte in relation to a different host and climatic conditions is highly impractical.
- Natural selection and mutation can modify the endophyte lifestyle to a parasitic state.
- Compatible relationship of the host with endophytes needs to be well studied.
- The beneficial characteristics of endophytes, explored during host colonization, may not be displayed under *in vitro* conditions due to potential coevolution of endophytes with the host through genetic recombination.

- Difficulty in optimizing favorable growth conditions for fungal endophytes.
- Co-existence of fungal endophytes with their native phytomicrobiomes to be elucidated.

9.7 CONCLUSION

From the various studies conducted so far, the mutualistic association between fungal endophytes and plants has been well defined in terms of their beneficial aspects and adaptation to each other under different biotic and environmental cues. The multipartite mode of action exhibited by endophytes has piqued our interest in understanding the intricate symbiotic intimacy with their hosts. The internal dynamics, such as mutations in single loci and transposons, as well as external pressures like competition with other microbes for niche and host response, play a crucial role in determining the endophyte lifestyle. Despite the high demand for microbial-based biofertilizers and biocontrol products in the current market, there are still several bottlenecks that need to be addressed regarding the underlying mechanisms of endophyte-plant interactions. The application of innovative biotechnological tools will enhance our understanding of endophyte colonization in plants, particularly in terms of growth promotion and immune system modification to combat biotic and abiotic stress. By discovering the molecular components utilized by endophytes during their interactions and signal transduction, we can gain insights into their co-existence and evolutionary strategies. Ultimately, the bioprospecting of fungal endophytes offers a promising alternative for sustainable crop production and maintaining plant health.

KEYWORDS

- **defense mechanism**
- **endophyte genetic factors**
- **fungal endophytes**
- **ionic homeostasis**
- **lifestyle switching over**
- **nutrient acquisition**
- **plant health**

REFERENCES

1. FAOSTAT, (2020). https://www.fao.org/faostat/en/#home (accessed on 9 January 2023).
2. Maloy, O. C., (2005). Plant disease management. *The Plant Health Instructor*.
3. Poole, N. F., & Arnaudin, M. E., (2014). The role of fungicides for effective disease management in cereal crops. *Can. J. Plant Pathol., 36*(1), 1–11.
4. Thind, T. S., (2017). Role of fungicides in crop health management: Prospects and challenges. *Develop. Fungal Biol App. Mycol.*, 433–447.
5. Ons, L., Bylemans, D., Thevissen, K., & Cammue, B., (2020). Combining biocontrol agents with chemical fungicides for integrated plant fungal disease control. *Microorganisms., 8*(12), 1930.
6. Gikas, G. D., Parlakidis, P., Mavropoulos, T., & Vryzas, Z., (2022). Particularities of fungicides and factors affecting their fate and removal efficacy: A review. *Sustainability., 14*, 4056.
7. Wightwick, A., Walters, R., Allinson, G., Reichman, S., & Menzies, S. R. A. N., (2010). *Environmental Risks of Fungicides Used in Horticultural Production Systems, Fungicides.* InTech., Shanghai.
8. Zubrod, J. P., Bundschuh, M., Arts, G., Brühl, C. A., Imfeld, G., Knäbel, A., Payraudeau, S., et al., (2019). An overlooked pesticide classes. *Environ. Sci. Technol., 53*(7), 3347–3365.
9. Singh, V. K., Singh, R., Kumar, A., & Bhadouria, R., (2021). Current status of plant diseases and food security. *Food Security and Plant Disease Management Book* (pp. 19–35).
10. Tariq, M., Khan, A., Asif, M., Khan, F., Ansari, T., Shariq, M., & Siddiqui, M. A., (2020). Biological control: A sustainable and practical approach for plant disease management. *Acta Agric. Scand. – B Soil Plant Sci., 70*(6), 507–524.
11. Compant, S., Samad, A., Faist, H., & Sessitsch, A., (2019). A review on the plant microbiome: Ecology, functions, and emerging trends in microbial application. *J. Adv. Res., 19*, 29–37.
12. Huang, W. Y., Cai, Y. Z., Hyde, K. D., Corke, H., & Sun, M., (2008). Biodiversity of endophytic fungi associated with 29 traditional Chinese medicinal plants. *Fungal Diversity*.
13. Xia, Y., Sahib, M. R., Amna, A., Opiyo, S. O., Zhao, Z., & Gao, Y. G., (2019). Culturable endophytic fungal communities associated with plants in organic and conventional farming systems and their effects on plant growth. *Sci. Rep., 9*(1), 1–10.
14. Kiti, H. M., Cosmas, N. M., Josiah, O. O., Guyo, P. M., & Cromwell, M. K., (2021). Diversity of mangrove fungal endophytes from selected mangrove species of coastal Kenya. *Int. J. Mar. Sci., 20*(1), 125–136.
15. Wearn, J. A., Sutton, B. C., Morley, N. J., & Gange, A. C., (2012). Species and organ specificity of fungal endophytes in herbaceous grassland plants. *J. Ecol., 100*(5), 1085–1092.
16. Fang, K., Miao, Y. F., Chen, L., Zhou, J., Yang, Z. P., Dong, X. F., & Zhang, H. B., (2019). Tissue-specific and geographical variation in endophytic fungi of *Ageratina adenophora* and fungal associations with the environment. *Front. Microbiol., 10*, 2919.
17. Schlemper, T. R., Leite, M. F., Lucheta, A. R., Shimels, M., Bouwmeester, H. J., Van, V. J. A., & Kuramae, E. E., (2017). Rhizobacterial community structure differences among sorghum cultivars in different growth stages and soils. *FEMS Microbiol. Ecol., 93*, 8.

18. Yu, P., He, X., Baer, M., Beirinckx, S., Tian, T., Moya, Y. A., & Hochholdinger, F., (2021). Plant flavones enrich rhizosphere *Oxalobacteraceae* to improve maize performance under nitrogen deprivation. *Nat. Plants, 7*(4), 481–499.
19. Cotton, T. E., Petriacq, P., Cameron, D. D., Meselmani, M. A., Schwarzenbacher, R., Rolfe, S. A., & Ton, J., (2019). Metabolic regulation of the maize rhizobiome by benzoxazinoids. *ISME J., 13*(7), 1647–1658.
20. Dennis, C., & Webster, J., (1971). Antagonistic properties of species groups of *Trichoderma* III. hyphal interaction *Trans. Br. Mycol. Soc., 57*, 363–369.
21. Rahmansyah, M., (2013). Endophytic fungi isolated from mangrove plant and have antagonism role against *Fusarium* wilt. *J. Agric. Biol. Sci., 8*(3), 251–257.
22. Radic, N., & Strukelj, B., (2012). Endophytic fungi: The treasure chest of antibacterial substances. *Phytomedicine, 19*, 1270–1284.
23. Bungtongdee, N., Sopalun, K., Laosripaiboon, W., & Iamtham, S., (2019). The chemical composition, antifungal, antioxidant and antimutagenicity properties of bioactive compounds from fungal endophytes associated with Thai orchids. *J. Phytopathol., 167*, 56–64.
24. Sornakili, A., Thankappan, S., Sridharan, A. P., Nithya, P., & Uthandi, S., (2020). Antagonistic fungal endophytes and their metabolite-mediated interactions against phytopathogens in rice. *Physiol. Mol. Plant Pathol., 112*, 501–525.
25. Landum, M. C., Felix, Mdo. R., Alho, J., Garcia, R., Cabrita, M. J., Rei, F., Carla, M. R., & Varanda, C. M., (2016). Antagonistic activity of fungi of *Olea europaea* L. against *Colletotrichum acutatum*. *Microbiol. Res., 183*, 100–108.
26. Farhat, H., Urooj, F., Tariq, A., Sultana, V., Ansari, M., Ahmad, V. U., & Haque, S. E., (2019). Evaluation of antimicrobial potential of endophytic fungi associated with healthy plants and characterization of compounds produced by endophytic *Cephalosporium* and *Fusarium solani*. *Biocatal. Agric. Biotechnol., 18*, 10104.
27. Pavithra, N. L., Sathish, K., & Ananda, K., (2012). Antimicrobial and enzyme activity of endophytic fungi isolated from Tulsi. *J. Pharm. Biomed. Sci., 16*, 12–18.
28. Desire, M. H., Bernard, F., Forsah, M. R., Assang, C. T., & Denis, O. J., (2014). Enzymes and qualitative phytochemical screening of endophytic fungi isolated from *Lantana camara* Linn. leaves. *J. Appl. Biol. Biotech.*, 2.
29. Uzma, F., Konappa, N. M., & Chowdappa, S., (2016). Diversity and extracellular enzyme activities of fungal endophytes isolated from medicinal plants of Western Ghats, Karnataka. *Egypt. J. Basic Appl. Sci., 3*(4), 335–342.
30. Raghav, D., Jyoti, A., Siddiqui, A. J., & Saxena, J., (2021). Plant-associated endophytic fungi as potential bio-factories for extracellular enzymes: Progress, challenges, and strain improvement with precision approaches. *J. App. Microbiol.*
31. Bischoff, K. M., Wicklow, D. T., & Jordan, D. B., (2009). Extracellular hemicellulolytic enzymes from the maize endophyte *Acremonium zeae*. *Curr. Microbiol., 58*, 499–503.
32. Andargie, M., Congyi, Z., & Yun, Y., (2017). Identification and evaluation of potential bio-control fungal endophytes against *Ustilagonoidea virens* on rice plants. *World J. Microbiol. Biotechnol., 33*, 120.
33. Khan, A. L., Al-Harrasi, A., Al-Rawahi, A., Farsi, Z., Al-Mamari, A., Waqas, M., Asaf, S., et al., (2016). Endophytic fungi from frankincense tree improves host growth and produces extracellular enzymes and indole acetic acid. *PLoS One, 11*(6), e0158207.
34. Jagannath, S., Konappa, N., Lokesh, A., Bhuvaneshwari, Dasegowda, T., Udayashankar, A. C., Chowdappa, S., et al., (2021). Bioactive compounds guided diversity of endophytic

fungi from *Baliospermum montanum* and their potential extracellular enzymes. *Anal. Biochem., 1*, 614–114024.
35. Toghueo, R. M. K., Zabalgogeazcoa, I., De Aldanaand, B. V., & Boyom, F. F., (2017). Enzymatic activity of endophytic fungi from the medicinal plants *Terminalia catappa, Terminalia mantaly* and *Cananga odorata South Afr. J. Bot., 109*, 146–153.
36. Moy, M., Li, H. M., Sullivan, R., White, J. F. Jr., & Belanger, F. C., (2002). Endophytic fungal beta-1, *6*-glucanase expression in the infected host grass. *Plant Physiol., 130*(3), 1298–1308.
37. Niones, J. T., & Takemoto, D., (2014). An isolate of *Epichloe festucae*, an endophytic fungus of temperate grasses, has growth inhibitory activity against selected grass pathogens. *J. Gen. Plant Pathol., 80*, 337–347.
38. Schrettl, M., Bignell, E., Kragl, C., Sabiha, Y., Loss, O., Eisendle, M., Wallner, A., et al., (2007). Distinct roles for intra- and extracellular siderophores during *Aspergillus fumigatus* infection. *PloS Pathog., 3*, 1195–1207.
39. Oide, S., Moeder, W., Krasnoff, S., Gibson, D., Haas, H., Yoshioka, K., & Turgeon, B. G., (2006). NPS6, encoding a nonribosomal peptide synthetase involved in siderophore-mediated iron metabolism, is a conserved virulence determinant of plant pathogenic ascomycetes. *The Plant Cell, 18*(10), 2836–2853.
40. Aramsirirujiwet, Y., Gumlangmak, C., &Kitpreechavanich, V., (2016). Studies on antagonistic effect against plant pathogenic fungi from endophytic fungi isolated from *Houttuynia Cordata* thunb and screening for siderophore and indole-3-acetic acid production. *Asia-Pac. J. Sci. Technol., 21*, 55–66.
41. Haselwandter, K., & Winkelmann, G., (2007). Siderophores of symbiotic fungi. In: Varma, A., & Chincholkar, S. B., (eds.), *Microbial Siderophores, Soil Biology*(Vol. 12, pp. 91–103).
42. Bartholdy, B. A., Berreck, M., & Haselwandter, K., (2001). Hydroxamate siderophore synthesis by *Phialocephala fortinii*, a typical dark septate fungal root endophyte. *BioMetals.*, 33–42.
43. Johnson, L. J., (2008). Iron and siderophores in fungal–host interactions. *Mycol. Res., 112*, 170–183.
44. Kajula, M., Tejesvi, M. V., Kolehmainen, S., Makinen, A., Hokkanen, J., Mattila, S., & Pirttila, A. M., (2010). The siderophore ferricrocin produced by specific foliar endophytic fungi *in vitro*. *Fungal Biol., 114*(2, 3), 248–254.
45. Kulisova, M., Vrublevskaya, M., Lovecka, P., Vrchotova, B., Stranska, M., Kolarik, M., & Kolouchova, I., (2021). Fungal endophytes of *Vitis vinifera*—Plant growth promotion factors. *Agriculture., 11*, 1250.
46. Patle, P., Navnage, N., & Ramteke, P., (2018). Endophytes in plant system: Roles in growth promotion, mechanism, and their potentiality in achieving agriculture sustainability. *Int. J. Chem. Stud., 6*, 270–274.
47. Nath, R., Sharma, G. D., & Barooah, M., (2015). Plant growth promoting endophytic fungi isolated from tea (*Camellia sinensis*) shrubs of Assam, India. *Appl. Ecol. Environ. Res., 13*(3), 877–891.
48. Khalil, A. M. A., Hassan, S. E. D., Alsharif, S. M., Eid, A. M., Ewais, E. E. D., Azab, E., Gobouri, A. A., et al., (2021). Isolation and characterization of fungal endophytes isolated from medicinal plant *Ephedra pachyclada* as plant growth-promoting. *Biomolecules., 11*(2), 140.

49. Sallam, N., Ali, E. F., Seleim, M. A., & Bagy, H. M. K., (2021). Endophytic fungi associated with soybean plants and their antagonistic activity against *Rhizoctonia solani*. *Egypt. J. Biol. Pest Control., 31*(1), 1–9.
50. Mandyam, K., & Jumpponen, A., (2005). Seeking the elusive function of the root-colonizing dark septate endophytic fungi. *Stud. Mycol. Stud., 53*, 173–189.
51. Waqas, M., Khan, A. L., Kamran, M., Hamayun, M., Kang, S. M., Kim, Y. H., & Lee, I. J., (2012). Endophytic fungi produce gibberellins and indole acetic acid and promote host-plant growth during stress. *Molecules, 17*(9), 10754–10773.
52. Bilal, L., Asaf, S., Hamayun, M., Gul, H., Iqbal, A., Ullah, I., & Hussain, A., (2018). Plant growth promoting endophytic fungi *Aspergillus fumigatus* TS1 and *Fusarium proliferatum* BRL1 produce gibberellins and regulates plant endogenous hormones. *Symbiosis, 76*(2), 117–127.
53. Khan, M. S., Gao, J., Munir, I., Zhang, M., Liu, Y., Xue, J., & Zhang, X., (2021). Characterization of endophytic fungi, *Acremonium* sp., from *Lilium davidii* and analysis of its antifungal and plant growth-promoting effects. *Bio. Med Res. Int.*
54. Humayun, M., Khan, S. A., Khan, A. L., Rehman, G., Kim, Y. H., Iqbal, I., Hussain, J., et al., (2010). Gibberellin production and plant growth promotion from pure cultures of *Cladosporium* sp. MH-6 isolated from cucumber (*Cucumis sativus* L.). *Mycologia., 102*(5), 989–995.
55. Kavroulakis, N. S., Zervakis, G. I., Ehaliotis, C., Haralampidis, K., & Papadopoulou, K. K., (2007). Role of ethylene in the protection of tomato plants against soil-borne fungal pathogens conferred by an endophytic *Fusarium solani* strain. *J. Exp. Bot., 58*, 3853–3864.
56. Constantin, M. E., De Lamo, F. J., Vlieger, B. V., Rep, M., & Takken, F. L., (2019). Endophyte-mediated resistance in tomato to *Fusarium oxysporum* is independent of ET, JA, and SA. *Front. Plant Sci., 10*, 979.
57. Waller, F., Achatz, B., Deshmukh, S., Baltruschat, H., & Kogel, K. H., (2006). Induction of systemic resistance by the root endophytic fungus *Piriformospora indica* in barley. *Plant Cell Physiol., 47*, 96.
58. Wawra, S., Fesel, P., Widmer, H., Timm, M., Seibel, J., Leson, L., & Zuccaro, A., (2016). The fungal-specific β-glucan-binding lectin FGB1 alters cell-wall composition and suppresses glucagon triggered immunity in plants. *Nat. Commun., 7*(1), 1–11.
59. Bamisile, B. S., Dash, C. K., & Akutse, K. S., (2018). Prospects of endophytic fungal entomopathogens as biocontrol and plant growth promoting agents: An insight on how artificial inoculation methods affect endophytic colonization of host plants. *Microbiol. Res., 217*, 34–50.
60. Singh, R., (2014). Microorganism as a tool of bioremediation technology for cleaning environment: A review. *Proc Int. Acad. Ecol. Environ. Sci., 4*(1).
61. Gopalakrishnan, S., Sathya, A., Vijayabharathi, R., Varshney, R. K., Gowda, C. L., & Krishnamurthy, L., (2015). Plant growth promoting rhizobia: Challenges and opportunities. *Biotech., 5*, 355–377.
62. Ahmad, P., Hashem, A., Abd-Allah, E. F., Alqarawi, A. A., John, R., Egamberdieva, D., & Gucel, S., (2015). Role of *Trichoderma harzianum* in mitigating NaCl stress in Indian mustard (*Brassica juncea* L.) through antioxidative defense system. *Front. Plant Sci., 6*, 868.
63. Azad, K., & Kaminskyj, S., (2016). A fungal endophyte strategy for mitigating the effect of salt and drought stress on plant growth. *Symbiosis., 68*, 73–78.

64. Hamayun, M., Hussain, A., Khan, S. A., Kim, H. Y., Khan, A. L., Waqas, M., Irshad, M., et al., (2017). Gibberellins producing endophytic fungus *Porostereum spadiceum* AGH786 rescues growth of salt affected soybean. *Front. Microbiol., 8*, 686.
65. Rho, H., Hsieh, M., Kandel, S. L., Cantillo, J., Doty, S. L., & Kim, S. H., (2018). Do endophytes promote growth of host plants under stress? A meta-analysis on plant stress mitigation by endophytes. *Microb. Ecol., 75*, 407–418.
66. Saddique, M. A. B., Ali, Z., Khan, A. S., Rana, I. A., & Shamsi, I. H., (2018). Inoculation with the endophyte *Piriformospora indica* significantly affects mechanisms involved in osmotic stress in rice. *Rice, 11*, 34.
67. Gonzalez-Teuber, M., Urzua, A., Plaza, P., & Bascunan-Godoy, L., (2018). Effects of root endophytic fungi on response of *Chenopodium quinoa* to drought stress. *Plant Ecol., 219*(3), 231–240.
68. Dastogeer, K. M., Li, H., Sivasithamparam, K., Jones, M. G., & Wylie, S. J., (2018). Fungal endophytes and a virus confer drought tolerance to *Nicotiana benthamiana* plants through modulating osmolytes, antioxidant enzymes and expression of host drought responsive genes. *Environ. Exp. Bot., 149*, 95–108.
69. Zhang, S., Gan, Y., & Xu, B., (2019b). Mechanisms of the IAA and ACC-deaminase producing strain of *Trichoderma longibrachiatum* T6 in enhancing wheat seedling tolerance to NaCl stress. *BMC Plant Biol., 19*, 22.
70. Hou, L., Li, X., He, X., Zuo, Y., Zhang, D., & Zhao, L., (2021). Effect of dark septate endophytes on plant performance of *Artemisia ordosica* and associated soil microbial functional group abundance under salt stress. *Appl. Soil Ecol., 165*, 103998.
71. Remans, T., Thijs, S., Truyens, S., Weyens, N., Schellingen, K., Keunen, E., Gielen, H., & Cuypers, A., (2012). Understanding the development of roots exposed to contaminants and the potential of plant-associated bacteria for optimization of growth. *Ann. Bot., 110*, 239–252.
72. Verma, S. K., Sahu, P. K., Kumar, K., Pal, G., Gond, S. K., Kharwar, R. N., & White, J. F., (2021). Endophyte roles in nutrient acquisition, root system architecture development and oxidative stress tolerance. *J. Appl. Microbiol., 131*(5), 2161–2177.
73. Liu, C. Y., Zou, Y. N., Zhang, D. J., Shu, B., & Wu, Q. S., (2019). Mycorrhizae and tolerance of abiotic stress in *citrus* plants. In: Giri, B., Prasad, R., Wu, Q. S., & Varma, A., (eds.), *Biofertilizers for Sustainable Agriculture and Environment* (Vol. 55, pp. 465–487).
74. Hund, A., Ruta, N., & Liedgens, M., (2009). Rooting depth and water use efficiency of tropical maize inbred lines, differing in drought tolerance. *Plant and Soil., 318*(1), 311–325.
75. Jogawat, A., Saha, S., Bakshi, M., Dayaman, V., Kumar, M., Dua, M., Varma, A., et al., (2013). *Piriformospora indica* rescues growth diminution of rice seedlings during high salt stress. *Plant Signal. Behav., 8*, e26891.
76. Jung, J. K. H. M., & McCouch, S. R. M., (2013). Getting to the roots of it: Genetic and hormonal control of root architecture. *Front Plant Sci., 4*, 186.
77. Comas, L. H., Becker, S. R., Cruz, V. M. V., Byrne, P. F., & Dierig, D. A., (2013). Root traits contributing to plant productivity under drought. *Front. Plant Sci., 4*, 442.
78. Palta, J. A., & Turner, N. C., (2019). Crop root system traits cannot be seen as a silver bullet delivering drought resistance. *Plant and Soil., 439*(1), 31–43.
79. Waller, F., Achatz, B., Baltruschat, H., Fodor, J., Becker, K., Fischer, M., Heier, T., et al., (2005). The endophytic fungus *Piriformospora indica* reprograms barley to salt-stress

tolerance, disease resistance, and higher yield. *Proceedings of the National Academy of Sciences, 102*, 13386–13391.

80. Yun, P., Xu, L., Wang, S. S., Shabala, L., Shabala, S., & Zhang, W. Y., (2018). *Piriformospora indica* improves salinity stress tolerance in *Zea mays* L. plants by regulating Na^+ and K^+ loading in root and allocating K^+ in shoots. *Plant Growth Regul., 86*, 323–331.
81. Kord, H., Fakheri, B., Ghabooli, M., Solouki, M., Emamjomeh, A., Khatabi, B., Sepehri, M., et al., (2019). Salinity-associated microRNAs and their potential roles in mediating salt tolerance in rice colonized by the endophytic root fungus *Piriformospora indica. Funct. Integr. Genomics, 19*, 659–672.
82. White, J. F., Kingsley, K. L., Verma, S. K., & Kowalski, K. P., (2018a). Rhizophagy cycle: An oxidative process in plants for nutrient extraction from symbiotic microbes. *Microorganisms, 6*, 95.
83. Soares, M. A., Li, H. Y., Kowalski, K. P., Bergen, M., Torres, M. S., & White, J. F., (2016). Functional role of bacteria from invasive *Phragmites australis* in promotion of host growth. *Microb. Ecol., 72*, 407–417.
84. Irizarry, I., & White, J. F., (2017). *Bacillus amyloliquefaciens* alters gene expression, ROS production and lignin synthesis in cotton seedling roots. *J. Appl. Microbiol., 124*, 1589–1603.
85. Carvalho, T. L. G., Balsemao-Pires, E., Saraiva, R. M., Ferreira, P. C. G., & Hemerly, A. S., (2014). Nitrogen signaling in plant interactions with associative and endophytic diazotrophic bacteria. *J. Exp. Bot., 65*, 5631–5642.
86. Yakti, W., Kovacs, G. M., Vagi, P., & Franken, P., (2018). Impact of dark septate endophytes on tomato growth and nutrient uptake. *Plant Ecol. Divers., 11*, 637–648.
87. Verma, S. K., Kharwar, R. N., & White, J. F., (2019). The role of seed-vectored endophytes in seedling development and establishment. *Symbiosis, 78*, 107–113.
88. Waqas, M., Kim, Y. H., Khan, A. L., Shahzad, R., Asaf, S., Hamayun, M., Kang, S. M., & Khan, M. A., (2017). Additive effects due to biochar and endophyte application enable soybean to enhance nutrient uptake and modulate nutritional parameters. *J. Zhejiang Univ. Sci. B., 18*, 109–124.
89. Chen, Z., Jin, Y., Yao, X., Chen, T., Wei, X., Li, C., White, J., & Nan, Z., (2020). Fungal endophyte improves survival of *Lolium perenne* in low fertility soils by increasing root growth, metabolic activity, and absorption of nutrients. *Plant Soil, 452*, 185–206.
90. Evelin, H., Devi, T. S., Gupta, S., & Kapoor, R., (2019). Mitigation of salinity stress in plants by arbuscular mycorrhizal symbiosis: Current understanding and new challenges. *Front. Plant Sci., 10*, 470.
91. Abdelaziz, M. E., Kim, D., Ali, S., Fedoroff, N. V., & Al-Babili, S., (2017). The endophytic fungus *Piriformospora indica* enhances *Arabidopsis thaliana* growth and modulates Na^+/K^+ homeostasis under salt stress conditions. *Plant Sci., 263*, 107–115.
92. Grabov, A., (2007). Plant KT/KUP/HAK potassium transporters: Single family–multiple functions. *Ann. Bot., 99*, 1035–1041.
93. Mittler, R., (2002). Oxidative stress, antioxidants, and stress tolerance. *Trends Plant Sci., 7*, 405–410.
94. Miller, G. A. D., Suzuki, N., Ciftci-Yilmaz, S., & Mittler, R. O. N., (2010). Reactive oxygen species homeostasis and signaling during drought and salinity stresses. *Plant Cell Environ., 33*(4), 453–467.

95. Marcińska, I., Czyczyło-Mysza, I., Skrzypek, E., Filek, M., Grzesiak, S., Grzesiak, M. T., & Quarrie, S. A., (2013). Impact of osmotic stress on physiological and biochemical characteristics in drought-susceptible and drought-resistant wheat genotypes. *Acta Physiol. Plant., 35*(2), 451–461.
96. Gupta, S., Schillaci, M., Walker, R., Smith, P. M. C., Watt, M., & Roessner, U., (2021). Alleviation of salinity stress in plants by endophytic plant-fungal symbiosis: Current knowledge, perspectives, and future directions. *Plant Soil., 461*, 219–244.
97. Hasanuzzaman, M., Nahar, K., & Fujita, M., (2013). Plant response to salt stress and role of exogenous protectants to mitigate salt-induced damages. In: Ahmad, P., Azooz, M. M., & Prasad, M. N. V., (eds.), *Ecophysiology and Responses of Plants Under Salt Stress* (pp. 25–87). Springer, New York.
98. Baltruschat, H., Fodor, J., Harrach, B. D., Niemczyk, E., Barna, B., Gullner, G., & Zuccaro, A., (2008). Salt tolerance of barley induced by the root endophyte *Piriformospora indica* is associated with a strong increase in antioxidants. *New Phytol., 180*, 501–510.
99. Bagheri, A. A., Saadatmand, S., Niknam, V., Nejadsatari, T., & Babaeizad, V., (2013). Effect of endophytic fungus, *Piriformospora indica*, on growth and activity of antioxidant enzymes of rice (*Oryza sativa* L.) under salinity stress. *Int. J. Adv. Biol. Biomed. Res., 1*, 1337–1350.
100. Zhang, S., Gan, Y., & Xu, B., (2016). Application of plant-growth-promoting fungi *Trichoderma longibrachiatum* T6 enhances tolerance of wheat to salt stress through improvement of antioxidative defense system and gene expression. *Front. Plant Sci., 7*, 1405.
101. Sudhir, P., & Murthy, S., (2004). Effects of salt stress on basic processes of photosynthesis. *Photosynthetica, 42*, 481–486.
102. Chaves, M., Flexas, J., & Pinheiro, C., (2009). Photosynthesis under drought and salt stress: Regulation mechanisms from whole plant to cell. *Ann. Bot., 103*, 551–560.
103. Jansen, M. A., Gaba, V., Greenberg, B. M., Mattoo, A. K., & Edelman, M., (1996). Low threshold levels of ultraviolet-B in a background of photosynthetically active radiation trigger rapid degradation of the D2 protein of photosystem-II. *Plant J., 9*, 693–699.
104. Seemann, J. R., & Critchley, C., (1985). Effects of salt stress on the growth, ion content, stomatal behavior, and photosynthetic capacity of a salt-sensitive species. *Phaseolus vulgaris* L. *Planta., 164*, 151–162.
105. Lima, N. M. C., Martins, M. D. O., Ferreira-Silva, S. L., & Silveira, J. A. G., (2015). *Jatropha curcas* and *Ricinus communis* display contrasting photosynthetic mechanisms in response to environmental conditions. *Sci. Agr., 72*, 260–269.
106. Cornic, G., & Massacci, A., (1996). Leaf photosynthesis under drought stress. In: *Photosynthesis and the Environment* (pp. 347–366). Springer, Dordrecht.
107. Pehlivan, N., Yesilyurt, A. M., Durmus, N., & Karaoglu, S. A., (2017). *Trichoderma lixii* ID11D seed bio-priming mitigates dose dependent salt toxicity in maize. *Acta Physiol. Plant, 39*, 79.
108. Molina-Montenegro, M. A., Acuna-Rodriguez, I. S., Torres-Diaz, C., & Gundel, P. E., (2018). Root endophytes improve physiological performance and yield in crops under salt stress by up-regulating the foliar sodium concentration. *Bio. Rxiv.*, 435032.
109. Chen, T., Johnson, R., Chen, S., Lv, H., Zhou, J., & Li, C., (2018). Infection by the fungal endophyte *Epichloe bromicola* enhances the tolerance of wild barley (*Hordeum brevisubulatum*) to salt and alkali stresses. *Plant and Soil, 428*(1), 353–370.

110. Johnson, J. M., Alex, T., & Oelmuller, R., (2014). *Piriformospora indica*: The versatile and multifunctional root endophytic fungus for enhanced yield and tolerance to biotic and abiotic stress in crop plants. *J. Trop. Agric., 52*:103–122.
111. Yin, L., Ren, A., Wei, M., Wu, L., Zhou, Y., Li, X., & Gao, Y., (2014). *Neotyphodium coenophialum*-infected tall fescue and its potential application in the phytoremediation of saline soils. *Int. J. Phytoremediation, 16*, 235–246.
112. Kumar, M., Sharma, R., Jogawat, A., Singh, P., Dua, M., Gill, S. S., Trivedi, D. K., et al., (2012). *Piriformospora indica,* a root endophytic fungus, enhances abiotic stress tolerance of the host plant. In: Narendra, T., Sarvajeet, G. S., Tiburcio, A. F., & Tuteja, R., (eds.), *Improving Crop Resistance to Abiotic Stress* (pp. 543–558). Wiley-Blackwell, Weinheim.
113. Rivero, R. M., Mestre, T. C., Mittler, R., Rubio, F., Garcia-Sanchez, F., & Martinez, V., (2014). The combined effect of salinity and heat reveals a specific physiological, biochemical, and molecular response in tomato plants. *Plant Cell Environ., 37*, 1059–1073.
114. Contreras-Cornejo, H. A., Macias-Rodríguez, L., Cortes-Penagos, C., & Lopez-Bucio, J., (2009). *Trichoderma virens,* a plant beneficial fungus, enhances biomass production and promotes lateral root growth through an auxin-dependent mechanism in *Arabidopsis*. *Plant Physiol., 149*, 1579–1592.
115. Khan, A. L., Hamayun, M., Kim, Y. H., Kang, S. M., Lee, J. H., & Lee, I. J., (2011). Gibberellins producing endophytic *Aspergillus fumigatus* sp. LH02 influenced endogenous phytohormonal levels, isoflavonoids production and plant growth in salinity stress. *Process Biochem., 46*(2), 440–447.
116. Petrini, O., (1991). Fungal endophytes of tree leaves. In: Andrews, J. H., &Hirano, S. S., (eds.), *Microbial Ecology of Leaves* (pp. 179–197). Springer-Verlag: New York, USA.
117. Brown, K. B., Hyde, K. D., & Guest, D. I., (1998). Preliminary studies on endophytic fungal communities of *Musa acuminata* species complex in Hong Kong and Australia. *Fungal Divers., 1*, 27–51.
118. Romero, A., Carrion, G., & Rico-Gray, V., (2001). Fungal latent pathogens and endophytes from leaves of *Parthenium hysterophorus* (*Asteraceae*). *Fungal Divers., 7*, 81–87.
119. Slippers, B., Michael, J., & Wingfield, (2007). *Botryosphaeriaceae* as endophytes and latent pathogens of woody plants: Diversity, ecology, and impact. *Fungal Biol. Rev., 21*, 90–106.
120. Begoude, B. A. D., Slippers, B., Wingfield, M. J., & Roux, J., (2011). The pathogenic potential of endophytic *Botryosphaeriaceous* fungi on *Terminalia* species in Cameroon. *For Pathol., 41*, 281–292.
121. Photita, W., Lumyong, S., Lumyong, P., & Hyde, K. D., (2001). Endophytic fungi of wild banana (*Musa acuminata*) at Doi Suthep Pui National Park, in Thailand. *Mycol. Res., 105*, 1508–1513.
122. Photita, W., Lumyong, S., Lumyong, P., McKenzie, E. H. C., & Hyde, K. D., (2004). Are some endophytes of *Musa acuminata* latent pathogens? *Fungal Divers., 16*, 131–140.
123. Sessa, L., Abreo, E., & Lupo, S., (2018). Diversity of fungal latent pathogens and true endophytes associated with fruit trees in Uruguay. *J. Phytopathol., 166*, 633–647.
124. Freeman, S., & Rodriguez, R. J., (1993). Genetic conversion of a fungal plant pathogen to a non-pathogenic, endophytic mutualist. *Science, 260*, 75–86.

125. Redman, R. S., Ranson, J., & Rodriguez, R. J., (1999b). Conversion of the pathogenic fungus*Colletotrichum magna* to a non-pathogenic endophytic mutualist by gene disruption. *Mol. Plant Microbe Interact., 12*, 969–975.
126. Freeman, S., Horowitz, S., & Sharon, A., (2001). Pathogenic and non-pathogenic lifestyles in *Colletotrichum acutatum* from strawberry and other plants. *Phytopathology, 91*, 986–992.
127. Barnes, S. E., & Shaw, M. W., (2002). Factors affecting symptom production by latent *Botrytis cinerea* in *Primula × polyantha. Plant Pathology, 51*, 746–754.
128. Faeth, S. H., & Sullivan, T. J., (2003). Mutualistic asexual endophytes in a native grass are usually parasitic. *Am. Nat., 161*, 310–325.
129. Schardl, C. L., Leuchtmann, A., & Spiering, M. J., (2004). Symbioses of grasses with seed borne fungal endophytes. *Annu. Rev. Plant Biol., 55*, 315–340.
130. Moricca, S., & Ragazzi, A., (2008). Fungal endophytes in Mediterranean oak forests: A lesson from *Discula quercina. Phytopathol., 98*(4), 380–386.
131. Alvarez-Loayza, P., White, J. F. Jr., Torres, M. S., Blslev, H., Kristiansen, T., Svenning, J. C., & Gil, N., (2011). Light converts endosymbiotic fungus to pathogen, influencing seedling survival and niche-space filling of a common tropical tree, *Iriartea deltoidea. PLoS One, 6*, e16386.
132. Eaton, C. J., Cox, M. P., & Scott, B., (2011). What triggers grass endophytes to switch from mutualism to pathogenism? *Plant Sci. J., 180*(2), 190–195.
133. Tanaka, A., Christensen, M. J., Takemoto, D., Park, P., & Scott, B., (2006). Reactive oxygen species play a role in regulating a fungus-perennial ryegrass mutualistic association. *Plant Cell, 18*, 1052–1066.
134. Sakalidis, M. L., Giles, E. S. H., & Treena, I. B., (2011). Endophytes as potential pathogens of the baobab species *Adansonia gregorii*: A focus on the *Botryosphaeriaceae. Fungal Ecol., 4*, 1–14.
135. Xu, X. H., Su, Z. Z., Wang, C., Kubicek, C. P., Feng, X. X., Mao, L. J., Wang, J. Y., et al., (2014). The rice endophyte *Harpophora oryzae* genome reveals evolution from a pathogen to a mutualistic endophyte. *Sci. Rep., 4*(1), 1–9.
136. Hiruma, K., Gerlach, N., Sacristan, S., Nakano, R. T., Hacquard, S., Kracher, B., Neumann, U., et al., (2016). Root endophyte *Colletotrichum tofieldiae* confers plant fitness benefits that are phosphate status dependent. *Cell, 165*, 464–474.
137. Eydoux, L., & Farrer, E. C., (2020). Does salinity affect lifestyle switching in the plant pathogen *Fusarium solani*? *Acc. Microbial., 26*, acmi000114.
138. Ribeiro, T. H. C., Fernandes-Brum, C. N., De Souza, C. R., Dias, F. A. N., Almeida-Junior, O. D., Regina, M. D. A., & Chalfun-Junior, A., (2020).Transcriptome analyses suggest that changes in fungal endophyte lifestyle could be involved in grapevine bud necrosis. *Scientific Reports, 10*(1), 1–13.
139. Shah, U. A., Kotta-Loizou, I., Fitt, B. D. L., & Coutts, R. H. A., (2020). Mycovirus-induced hypervirulence of *Leptosphaeria biglobosa* enhances systemic acquired resistance to *Leptosphaeria maculans* in *Brassica napus. Mol. Plant Microbe Interact., 33*, 98–107.
140. Zhang, H., Xie, J., Fu, Y., Cheng, J., Qu, Z., Zhao, Z., Cheng, S., et al., (2020). A 2-kb myco-virus converts a pathogenic fungus into a beneficial endophyte for *Brassica* protection and yield enhancement. *Mol. Plant, 3*, 1–14.

141. Zhou, L., Li, X., Kotta-Loizou, I., Dong, K., Li, S., Ni, D., Hong, N., Wang, G., & Xu, W., (2021). A mycovirus modulates the endophytic and pathogenic traits of a plant associated fungus. *ISME J., 15*, 1893–1906.
142. Arnold, A. E., Miadlikowska, J., Higgins, K. L., Sarvate, S. D., Gugger, P., Way, A., Hofstetter, V., et al., (2009). A phylogenetic estimation of trophic transition networks for ascomycetous fungi: Are lichens cradles of symbiotrophic fungal diversification? *Syst. Biol., 58*, 283–297.
143. Wang, Z., Johnston, P. R., Yang, Z. L., & Townsend, J. P., (2009). Evolution of reproductive morphology in leaf endophytes. *PLoS One, 4*, e4246.
144. Promputtha, I., Lumyong, S., Dhanasekaran, V., McKenzie, E. H. C., Hyde, K. D., & Jeewon, R., (2007). A phylogenetic evaluation of whether endophytes become saprotrophs at host senescence. *Microb. Eco., 53*(4), 579–590.
145. Delaye, L., Garcia-Guzman, G., & Heil, M., (2013). Endophytes versus biotrophic and necrotrophic pathogens—Are fungal lifestyles evolutionarily stable traits? *Fungal Divers., 60*(1), 125–135.
146. Promputtha, I., Hyde, K. D., McKenzie, E. H. C., Peberdy, J. F., & Lumyong, S., (2010). Can leaf degrading enzymes provide evidence that endophytic fungi becoming saprobes? *Fungal Divers., 41*, 89–99.
147. Zhou, J., Li, X., Huang, P. W., & Dai, C. C., (2018). Endophytism or saprophytism: Decoding the lifestyle transition of the generalist fungus *Phomopsis liquidambari. Microbiol. Res., 206*, 99–112.
148. Sieber, T. N., (2007). Endophytic fungi in forest trees: Are they mutualists? *Fungal Biol. Rev., 21*, 75–89.
149. Takemoto, D., Tanaka, A., & Scott, B., (2006). A p67 (Phox)-like regulator is recruited to control hyphal branching in a fungal-grass mutualistic symbiosis. *Plant Cell, 18*, 2807–2821.
150. Eaton, C. J., Cox, M. P., Ambrose, B., Becker, M., Hesse, U., Schardl, C. L., & Scott, B., (2010). Disruption of signaling in a fungal-grass symbiosis leads to pathogenesis. *Plant Physiol., 153*, 1780–1794.
151. Tanaka, A., Takemoto, D., Hyon, G. S., Park, P., & Scott, B., (2008). NoxA activation by the small GTPase RacA is required to maintain a mutualistic symbiotic association between *Epichloe festucae* and perennial ryegrass. *Mol. Microbiol., 68*, 1165–1178.
152. Freeman, F., Kim, D. S., & Rodriguez, E., (1992). Preparation of 1, 4-diketones and their reactions with bis (trialkyltin) or bis (triphenyltin) sulfide-boron trichloride. *J. Org. Chem., 57*(6), 1722–1727.
153. Redman, R. S., Freeman, S., Clifton, D. R., Morrel, J., Brown, G., & Rodriguez, R. J., (1999). Biochemical analysis of plant protection afforded by a non-pathogenic endophytic mutant of *Colletotrichum magna. Plant Physiol., 119*, 795–804.
154. Rodriguez, R. J., & Redman, R. S., (2008). More than 400 million years of evolution and some plants still can't make it on their own: Plant stress tolerance via fungal symbiosis. *J. Exp. Bot., 59*, 1109–1114.
155. Yadav, V., Kumar, M., Deep, D. K., Kumar, H., Sharma, R., Tripathi, T., Tuteja, N., et al., (2010). A phosphate transporter from the root endophytic fungus *Piriformospora indica* plays a role in phosphate transport to the host plant. *J. Biol. Chem., 285*, 26532–26544.
156. Junker, C., Draeger, S., & Schulz, B. A., (2012). Fine line – endophytes or pathogens in*Arabidopsis thaliana. Fungal Ecol., 5*, 657–662.

CHAPTER 10

Promoting Plant Growth by ACC Deaminase Enzymes Producing Bacterial Endophytes Under Biotic Stress Conditions

MOHAMMAD YASEEN MIR[1] and FADIME KARABULUT[2]

[1]*Center of Research for Development, University of Kashmir, Jammu and Kashmir, India*

[2]*Department of Biology, Firat University, Elazıg, Turkey*

ABSTRACT

Environments in which plants are exposed to phytopathogens grow under biotic stress, and it modulates ethylene (ET) and is used as a signaling molecule in stress tolerance mechanisms. In the presence of these stress conditions, plants can induce higher levels of ET than their threshold values, resulting in reduced plant growth and crop productivity. In this context, bacterial endophytes play an important role in managing soil and plant health, as well as supporting plant growth under biotic stress conditions. Most endophytic bacteria appear to possess the enzyme aminocyclopropane-1-carboxylic acid (ACC) deaminase, which enhances the growth of the host plant. ACC-deaminase is involved in regulating ET levels in plant development. These bacterial endophytes effectively degrade ACC-deaminase and simultaneously stimulate plant growth (such as increasing root length and plant biomass) at low ET levels and under stressful conditions. Additionally, these bacteria play a crucial role in regulating plant physiology through the production of plant growth regulators. Endophytes also act as microorganisms with beneficial mechanisms of action that increase crop yields, prevent,

Microbial Biostimulants: Biorational Pesticides for Management of Plant Pathogens.
Faheem Ahmad and Rakesh Pandey (Eds.)

or even eliminate infections and pathogens. The goal of the researchers should be to uncover beneficial bacteria-plant interactions and manage this partnership. To achieve this, the molecular interaction between plants and endophyte bacteria should be better demonstrated for the production of endophyte bacteria. In the coming years, with increased awareness among manufacturers, more efficient products will be obtained under biotic stress conditions.

10.1 INTRODUCTION

The various stresses that can occur in the soil have a negative impact on the growth and development of plants. Biotic stresses, such as the formation of symbiotic relationships between pathogens and microorganisms, also affect plant growth. Additionally, the normal functioning of a plant's metabolism and life cycle is disrupted under these stressful conditions. The production of ethylene (ET) increases when plants experience a range of developmental, morphological, physiological, and biochemical disorders, which in turn inhibits root growth and overall plant growth [1, 2]. Endophytic bacteria (EB) are present in plant tissues, either intercellularly or intracellularly, and their diversity and distribution vary depending on the tissue type, soil type, and genotype of the host plant. EB does not cause any pathogenic symptoms [3].

Sustainability is greatly aided by plant-microbe interactions, whether they are symbiotic, endophytic, rhizospheric, or phyllospheric. The use of rhizobacteria in agriculture to enhance plant growth and restore health in unfavorable environmental conditions without degrading soil fertility and natural resources is the best example of this dual association that has been documented [2]. By reducing the amount of ET that is produced as a direct result of ACC deaminase stress, plant growth-promoting bacteria (PGPR) can make plants more resistant to the effects of a variety of biotic stress conditions. Rhizosphere: the region around the plant's roots; Phyllosphere: the region around the plant's leaves; or Plant Tissue: the region inside which PGPR develops (endophytes). Solubilizing growth regulators (such as auxin, gibberellins, cytokinin, and ET) or immobilized nutrients and making them available to plants encourages the synthesis of various plant growth factors [4]. Plants growing in stressful environments benefit from the effects of microbes that promote plant growth and have ACC deaminase potential [5, 6]. Despite its pleasant surroundings, ET controls a number of processes involved in plant growth and development. However, understanding the effect of ACC deaminase on plant metabolic activities, enzyme action mechanism, biochemistry, and genetics increases the knowledge of plant-soil-borne pathogen and provides plants with tolerance against various biotic stresses [2].

10.2 BIOTIC STRESS IN PLANTS

PGPR, which has ACC Deaminase activity, offers phytopathogen antagonism as well as resistance to a variety of biotic stresses in the soil. Because of PGPR, ET regulation in plants can, therefore, also be effectively established.

10.2.1 PATHOGENICITY STRESS

Globally, ecosystems and food production are seriously threatened by pathogenic microorganisms. In order to achieve PGPR-mediated biocontrol, host plants must compete for a substrate or an ecological niche, generating systemic resistance (ISR) against various pathogens and allelochemicals [7]. In addition to directly combating pathogens, ACC deaminase bacteria help plants resist pathogen attacks. ISR-inducing rhizobacteria must exhibit a different expression than pathogen elicitors because beneficial rhizobacteria obviously do not harm their hosts or cause local necrosis.

Following vaccination against the pathogen causing less severe disease, the expression of ISR is comparable to systemic acquired resistance (SAR). This results in a decrease in the number of diseased plants. The reduction is attributed to slower pathogen growth and less colonization of induced tissues, which impacts the plant's ability to defend against the pathogen [8]. Analysis of ISR signaling mediated by rhizobacteria has revealed that salicylic acid (SA) is an essential signaling molecule in systemic and locally induced resistance responses, while ethylene (ET) and jasmonic acid (JA) also play important roles. Rhizobacteria are responsible for mediating this type of signaling. Hence, there are similarities in the expression of ISR and SAR, and how each is influenced by different types of biological induction and various defense-related processes [9]. It is worth noting that using PGPR to induce ISR can help prevent diseases caused by pathogens that are weak against ET- and JA-dependent defenses. ACC deaminase bacteria could potentially combat at least one of the two pathogens that infect potatoes, namely *Ralstonia solanacearum* and *Rhizoctonia solani* [8, 10].

10.2.2 ETHYLENE (ET)–IAA INTERACTION

IAA is well known for its ability to activate ACC synthase transcription, but the potential of ET to block IAA transport and signal transduction is less well known. After each stressful event in the plant's life, this feedback loop of ET inhibition of IAA synthesis and/or function reduces the amount of ACC synthase, ACC, and ultimately ET in the plant's system. When ET levels in plant roots are reduced by a PGPB containing ACC deaminase, auxin response factor synthesis is no longer repressed by ET, indirectly promoting plant growth. As a result, PGPR containing ACC deaminase promotes plant growth by lowering ET inhibition and allowing IAA stimulation without negatively affecting plant ET levels or ACC synthase levels [8].

10.2.3 RHIZOBIAL INFECTION IN THE ACTIVITY OF ACC DEAMINASE

There is a lot of evidence to support the idea that ET produced after rhizobia infects legumes inhibits the nodulation process. The most recent research has shown that PGPR with ACC deaminase activity enhances nitrogen fixation and symbiosis in plants by promoting nodulation in legumes through inhibition of ET biosynthesis. The increased nodulation in *Lotus japonicus* was caused by the upregulation of the ACC deaminase activity gene mlr5932 in bacteroids. The discovery of an endophytic ACC deaminase bacterium that has the ability to control nodulation in *Mimosa pudica* has been made. The coinoculation of Bradyrhizobium plus ACC deaminase rhizobacteria was found to increase moong (green lentil) bean nodulation in comparison to *Bradyrhizobium* spp. inoculation alone [8].

10.3 THE ROLE OF ACCD-PGPR IN THE PROCESS OF NODULATION UNDER CONDITIONS OF BIOLOGICAL STRESS

The process of nitrogen fixation, carried out by symbiotic organisms, is essential for sustainable agriculture. It is common for nitrogen-fixing bacteria, most notably rhizobium, to coexist in the roots of legume plants. These bacteria promote the growth of nodules on the roots of legume plants. Nodules respond to various environmental stresses and plant hormones, such as abscisic acid (ABA), auxin, cytokinin, and ethylene, in terms of their development and growth. Stress conditions that accumulate in plant tissues typically trigger ET production [11]. In several leguminous plants, such as

Medicago sativa, Pisum sativum, and *Trifolium repens,* ET is also known to negatively regulate nodule formation. Nodule counts decrease, and fixed nitrogen levels are significantly reduced [12]. Various rhizobium species, including *Sinorhizobium meliloti, Rhizobium leguminosarum,* and *Mesorhizobium loti,* have been found to produce ACC deaminase. It is common knowledge that under both stressful and non-stressful circumstances, rhizobial bacteria that produce ACC deaminase can densely nodulate the roots of associated legumes. Previous understanding suggested that free-living bacteria had a higher level of ACC deaminase activity compared to native rhizobium species. There is, therefore, a substantial body of research demonstrating that pulse products infused with ACCD-PGPR have lower stress ET levels [11]. Bradyrhizobium japonicum and Pseudomonas putida, which produce ACC deaminase, were co-injected into mung beans to increase nodulation. *Mesorhizobium ciceri* was injected with *Serratia proteamaculans* and *Citrobacter koseri* to increase the amount of nodulation [13]. These were instances of how ACC deaminase-producing bacteria were inoculated into legume crops to increase nodulation under non-stressful circumstances, but co-grafting with ACC deaminase-producing bacteria demonstrated better nodulation in legumes under a variety of environmental stresses.

Another method for improving nodulation of rhizobial bacteria in legumes involves overexpressing the ACC deaminase (acdS) gene in rhizobial bacteria that lack ACC deaminase activity. Additionally, the genetically altered Mesorhizobium loti MAFF303099 promotes improved nodule development in Lotus plants by reliably expressing ACC deaminase [14]. ACCD-PGPRs are capable of enhancing bacterial symbiosis in both normal and stressful environments. This can be achieved through co-vaccination assay or by using a transformed Rhizobium strain expressing the acdS gene.

10.3.1 ETHYLENE'S (ET) FUNCTION IN CONTROLLING PLANT PROCESSES AND STRESS REACTIONS

The simplest plant hormone produced by plants, ethylene (ET), controls a number of crucial plant processes. These processes include root elongation, root hair development, fruit ripening, senescence, seed germination, and induction of a variety of biotic stresses such as exposure to nematodes, phytopathogens, organic, and inorganic chemicals [15]. ET produced under these stressful circumstances is known as "stress ethylene" [5]. This type of ET initiates the transcription and subsequent expression of genes, which causes plants to age. The biosynthesis of ET in plants involves the conversion of methionine to S-adenosyl methionine (SAM) by the enzyme SAM synthetase, and SAM is then used as a substrate by ACC synthetase to form ACC. ACC, produced in the previous step, acts as a precursor for the production of ACC oxidase, which is essential for ET production.

10.3.2 SYNERGY BETWEEN ACC DEAMINASE AND BIOTIC STRESS TOLERANCE MECHANISMS

A gaseous plant growth hormone called ET is created endogenously by plants. Any stressful situation causes plants to produce more ET, which is why it is also referred to as the stress hormone. A small initial low-magnitude peak of ET is seen for several hours at the beginning of stress, followed by a second, much larger high-magnitude peak for one to three days [17]. The severity of the second peak of ET is responsible for the onset of a number of processes, including senescence, chlorosis, and abscission, the culmination of which frequently renders a plant incapable of surviving. By disrupting its precursor, such as ACC or S-adenosylmethionine, which effectively lowers ET levels, the herb can release inhibitory levels of ET. A lot of research is being done on the enzymes ACC deaminase, SAM decarboxylase, and SAM hydrolase among them. A normal amount of photosynthetically fixed carbon in root exudates is thought to be between 5% and 30%. The root exudates of plants are where tryptophan is discovered. Rhizospheric bacteria utilize it as a precursor in the synthesis of indole-3-acetic acid (IAA), which is partially absorbed by the plant. Alongside endogenously produced plant IAA, this bacterial IAA stimulates cell growth and elongation. ET production in higher plants can also be enhanced by promoting the transcription of ACC synthase, a precursor to ET production. Rhizospheric bacteria can ingest small amounts of ACC, which is excreted from seeds or roots, leading to the production of ammonia and beta-ketobutyrate during the hydrolysis of the ET precursor ACC by rhizobacterial ACC deaminase [18]. The activity of rhizobacteria results in a decrease in ET and ACC levels in the spermosphere and rhizome, thereby eliminating plant growth inhibition. Plant roots that possess PGPR with ACC deaminase activity are better equipped to withstand the effects of biotic stress [19]. Reduced root and shoot growth occurs when plants are under stress because the gaseous ethylene (ET) hormone endogenously regulates plant homeostasis. The presence of high ET levels in the plant causes feedback inhibition of the indole-3-acetic acid (IAA) signal transduction pathway. Thus, it limits the transcription of 1-aminocyclopropane-1-carboxylate (ACC) synthase by IAA [20]. Bacteria that can synthesize IAA and have ACC deaminase activity play an important role in promoting plant growth by preventing the formation of high ET levels in plants. As a result, auxin-induced plant growth resumes without being negatively impacted by ET's feedback inhibition of IAA signal transduction.

10.3.2.1 INDIRECT MECHANISMS

Bacteria that promote plant growth have the ability to act as biocontrol agents against phytopathogens. These bacteria also indirectly promote plant growth through various mechanisms, such as induced systemic resistance (ISR), antibiotic production, plant growth promotion, and iron removal from the rhizosphere. Some biocontrol bacteria produce low molecular mass siderophores that bind iron in the rhizosphere with high avidity, preventing the growth of fungal pathogens caused by iron deficiency near the plant roots [21]. Other biocontrol bacteria that support plant growth produce enzymes like lipase, 1,3-glucanase, chitinase, and protease, which aid in the degradation of fungal cells [22]. Interestingly, pathogen-induced stress often leads to the production of the same protein sequence (protease, 1,3-glucanase, lipase, and chitinase) in stressed plants [17]. Competing biocontrol bacteria on niches and root surfaces promote plant growth by competing for nutrients. By effectively preventing pathogens from adhering to and infecting plants, they safeguard plants from phytopathogens. Various microbial metabolites and specialized biocontrol techniques can result in broad-spectrum and long-term systemic resistance against different pathogens or bacteria that promote plant growth [23, 24].

10.3.2.2 Direct Mechanisms

In order to achieve this, bacteria produce siderophores that can extract iron from the soil, deliver it to plant cells, and then take up the bacterial siderophore-iron complex. This directly stimulates the growth of plants. Additionally, microbes produce phytohormones that aid in different stages of plant growth, including auxins, gibberellins, and cytokinins. By fixing atmospheric nitrogen and supplying it to the plants, bacteria help the plant. Additionally, they dissolve minerals like phosphorus so that plants can more easily access them for growth. Moreover, they generate the enzyme ACC deaminase, which can reduce the amount of ET produced by plants [25]. Any one or more of these mechanisms may be used by a bacterium to directly influence plant development and growth. Because many of the bacteria that are beneficial to plant growth possess more than one of these characteristics, a bacterium may make use of a variety of characteristics during the course of the life cycle of the plant (Figure 10.1).

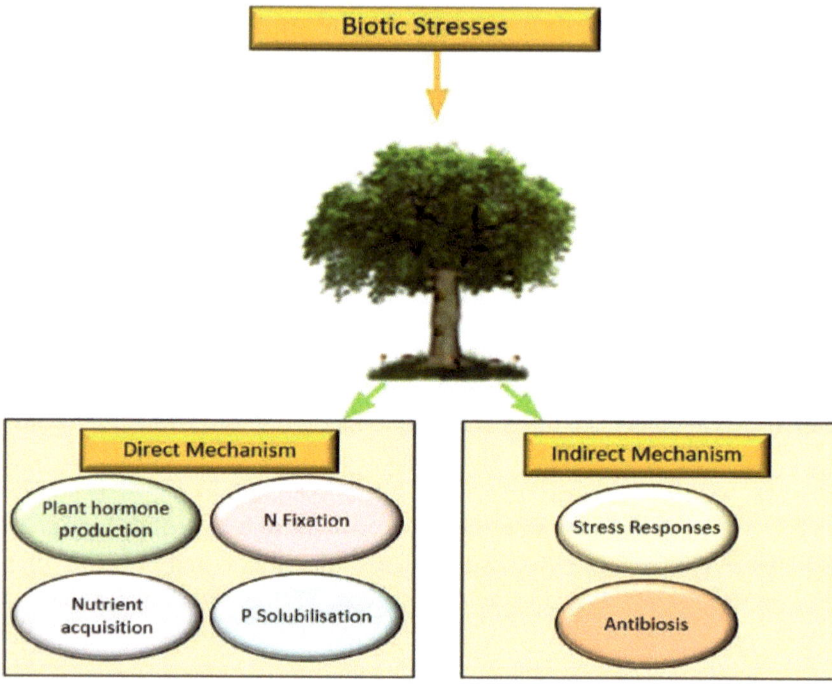

FIGURE 10.1 Diagram of mechanisms resulting from biotic stress in plants.

In addition, when plants are grown under ideal conditions without stress, the impact of bacteria that promote plant growth is typically minimal or non-existent compared to the overall growth of the plant. The study investigated the mechanism that produces the most phytohormone and the function of the phytohormone auxin in relation to the direct effects of plant growth-promoting bacteria. It has also been discovered that several bacteria beneficial to plant growth contain the enzyme ACC deaminase [26, 27]. This enzyme can break down the phytohormone ACC, which reduces the ethylene (ET) level in the plant during its growth and stress.

10.3.3 MOLECULAR CHARACTERIZATION OF ACC DEAMINASE ENZYME

Eukarya, Bacteria, and Archaea have been found to produce ACC deaminase. The yeast *Hansenula saturnus* and the bacteria *Pseudomonas sp* strain ACP were the first organisms to contain the ACC deaminase enzyme (now reclassified as *Cyberlindnera saturnus*). Additionally, gram-negative and gram-positive bacteria, as well as human hosts, rhizospheric soil, plants, water, etc., have been discovered to possess ACC deaminase activity. Various fungi inhabiting different ecological habitats have also been found to exhibit ACC deaminase activity. For instance, ACC deaminase activity is frequently observed in the rhizosphere of soil, particularly in rhizobacteria, and it aids plants in coping with stress by reducing the harmful ET concentration induced under stressful conditions [28, 29]. The bacterial ACC deaminase activity is molecularly characterized by amplifying the corresponding 1,000 bp acds gene sequence found in chromosomal DNA using PCR (polymerase chain reaction) and a pair of universal or degenerate primers. The concentration of the substrate, the availability of oxygen, and the end products all influence the expression of the Acds gene [29]. Different species have different regulatory mechanisms for the acids gene. In the presence of ACC, the regulatory protein LRP is activated and forms the LRP-octamer. This activation leads to the activation of AcdB, which in turn causes the formation of glycerophosphoryl diester phosphodiesterase that binds to ACC. The LRP-ACC-AcdB triple complex activates the expression of the acds gene, resulting in the breakdown of the ACC molecule into ketobutyrate and ammonia. Leucine, a branched chain amino acid produced from ketobutyrate, binds to the LRP octamer and breaks it down into inactive dimers, thereby inhibiting the transcription of the acds gene. Without ACC, the acdR gene's transcription results in an accumulation of LRP protein, which binds to the LRP Box promoter region and inhibits the production of LRP Box negatively. This suggests that ACC deaminase is an enzyme that is subject to regulation [30]. In some species, such as *Mesorhizobium loti,* the expression of the acds gene is controlled by the nitrogen-fixing gene known as nifA2 as well as the sigma factor. When legume plants are stressed, ACC deaminase, which generates symbiotic bacteria like *Mesorhizobium loti,* has been shown to lessen the negative effects of ET on root nodules, promoting nodulation and subsequently nitrogen fixation [31].

10.3.4 BIOCHEMISTRY OF ACC DEAMINASE

One of the enzymes that relies on the cofactor pyridoxal 5-phosphate (PLP) for its enzymatic activity is ACC deaminase. These enzymes are categorized based on their three-dimensional structures, such as tryptophan synthesis, alanine racemase, D-amino acid aminotransferase, and aspartate aminotransferase. ACC deaminase falls into the tryptophan synthesis category. ACC deaminase is responsible for converting aminocyclopropane carboxylate into ammonium (NH3) and ketobutyrate. It was initially discovered during research on *Pseudomonas* spp. bacteria that the polymeric ACC deaminase, connected to PLP, reduces the cyclopropanoid amino acid to ketobutyrate and NH_3. Thus, studies on ACC deaminase were initiated, particularly in stressful situations, to promote plant growth and mitigate the effects of stress [26, 32]. Numerous bacteria possess the ACC deaminase enzyme, which supports plant growth. These bacteria are found on the surface of the roots and utilize ACC as a source of carbon and nitrogen by converting it to NH_3 and ketobutyrate. In terms of colonizing the root rhizosphere, bacteria capable of utilizing ACC as a nitrogen and carbon source may have a competitive advantage over other bacteria (Figure 10.2).

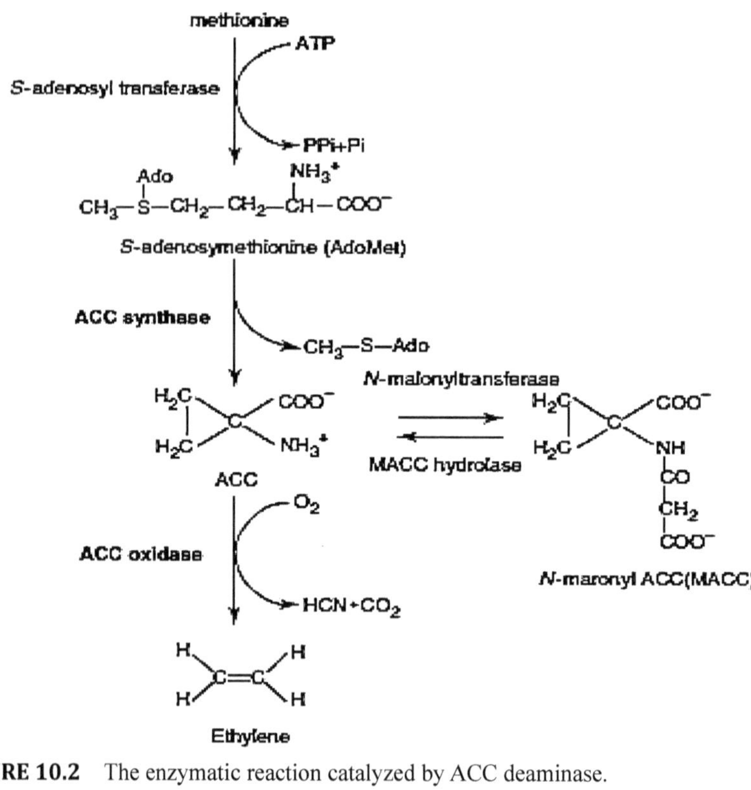

FIGURE 10.2 The enzymatic reaction catalyzed by ACC deaminase.

Source: Modified from Ref. [8].

ACC deaminase is a substance possessed by many soil microorganisms. In a study, strains of numerous species from the genera *Ralstonia, Enterobacter, Azospirillum, Burkholderia, Rhizobium, Achromobacter, Agrobacterium,* and *Pseudomonas* were found to contain ACC deaminase activity genes [26]. It was discovered that ACC deaminase was present in the majority of *Pseudomonas spp.* bacteria isolated from various regions of the world that exhibited biological control activity.

10.4 PLANT GROWTH ENHANCING EFFECTS OF ENDOPHYTIC BACTERIA (EB)

Similar to the mechanisms described for rhizobacteria that promote plant growth, PGPR (Plant Growth-Promoting Bacteria) have positive effects on their hosts. PGPRs have a direct effect on promoting plant growth by facilitating nutrient absorption or through the synthesis of beneficial compounds [33]. Additionally, PGPRs indirectly enhance plant growth by mitigating the negative effects caused by phytopathogenic organisms [34]. Furthermore, PGPRs directly support plant growth and development by acting as biocontrol agents against phytopathogens in the root zone. They can stimulate plant growth either by producing phytohormones like Indole acetic acid or by influencing the synthesis of phytohormones in the plant. Through processes such as siderophore production and phosphate solubility activity, it can assist the plant in absorbing minerals. In one study involving potatoes and alfalfa in crop rotation, it was discovered that 21% of the isolated EB supported plant growth, resulting in a 63% increase in shoot length, a 66% increase in shoot fresh weight, and a 55% increase in root wet weight [35].

Nitrogen fixation is the part of EB that is most studied for promoting plant growth. *Endophytic diazotrophs* that directly transform their biologically fixed nitrogen (Biological Nitrogen Fixation-BNF-) into a form usable by their hosts have been the subject of recent research. These symbiotic relationships have benefits such as lowering competition between EB and rhizosphere microorganisms on the surface of the roots, easy access to substrates for metabolism, creation of suitable environmental conditions, and defense against extremely high oxygen concentrations. Numerous host plants in the Graminea family have been found to contain diazotrophic bacteria that take part in biological N_2 fixation [33]. In one experiment, it was found that *in vitro*-grown bacterized potato plants were greener and had higher cytokinin levels [36]. Bacterized seedlings had higher levels of phenylalanine ammonia lyase (PAL) and free phenolics. There is a strong connection between the evolution of a plant's resistance to abiotic stress and its resistance to the attack of phytopathogens [33].

ACC and ET are synthesized as a result of biotic factors such as pathogen attack, injury, and flood. Stress causes the plant to produce large amounts of ET, which inhibits root growth and elongation. Additionally, ET can directly impact auxin signaling and transport [33, 37]. Another way that bacteria associated with plants can influence their hosts is through the ACC deaminase enzyme, which is present in the majority of bacteria that promote plant growth. According to research, ACC-deaminase controls ET levels during plant development [38]. Some bacteria found in plants can degrade the precursor to ET, ACC, using the enzyme they encode, ACC-deaminase, and then utilize the resulting products as a source of nutrients such as carbon and nitrogen. As a result, these bacteria colonize the plant tissues. Subsequently, these bacteria efficiently degrade ACC while simultaneously promoting plant growth under stressful conditions and low ET levels. Endophytic bacterial species, such as those that produce ACC-deaminase, benefit *B. cepacia* and *Burkholderia phytofirmans* isolates, *Arthrobacter* spp., *Methylobacterium fujisawaense, Pseudomonas spp.,* and *Bacillus* spp. by promoting plant growth, including an increase in root length and plant biomass [39]. It is well known that plants inoculated with ACC-deaminase-producing EB exhibit improved growth-supporting qualities and stress tolerance. Plant damage from environmental stress may occur as a direct result of the plant's response to the increase in ET levels during stress, similar to the infection caused by fungal pathogens. As a result, during the development of the plant under stress, ACC-deaminase can support the growth of the plant by reducing the rising ethylene (ET) levels. When ACC-deaminase-producing bacteria (*Achromobacter piechaudii AVR8, Klebsiella oxytoca Rs-5, Enterobacter cloacae CAL2, Serratia proteamaculans M35,* and *Pseudomonas spp.*) were inoculated on canola, cotton, peanut, tomato, wheat, and maize plants, cell content and K+ increased, while the Na+ ratio decreased, and the host plant biomass increased compared to uninoculated control plants [39].

10.5 CONCLUSIONS AND PERSPECTIVES

From an agricultural and ecological perspective, the primary objectives for addressing the future population explosion will be to enhance plant productivity, improve food quality, and preserve environmental quality for sustainable agriculture. These three areas will be the main areas of focus. However, in order to achieve these goals, it is essential to conduct mechanistic research to enhance our understanding of the interactions that occur between microbes and the soil in the rhizosphere. In this context, the application of PGPR (Plant Growth-Promoting Rhizobacteria) in agricultural settings is an intriguing technology to explore. Not only does PGPR, which produces the enzyme ACC deaminase, protect plants from the harmful effects of high ethylene (ET) concentrations during stress, but it also facilitates nutrient uptake, phytohormone production, and nutrient solubilization, among other functions. This is achieved through various additional mechanisms, including enhancing the plant's capacity. To increase crop production, it is crucial to comprehend the role of bacterial strains with ACC deaminase activity and how to utilize them to mitigate stresses caused by atmospheric pollutants. This is because these bacterial strains can be used to reduce the effects of stress caused by atmospheric pollutants. The use of microorganisms containing ACC deaminase genes is highly recommended for early root development from cuttings or seeds. To achieve this, these microorganisms protect plants against biotic stresses. In conclusion, co-administration of PGPR containing ACC deaminase with natural rhizospheric microflora can reduce the effect of stress on crop plants. This may also encourage crop yield and plant growth. Even though the bacteria in the bacteria-root association don't offer the plant many advantages, they do give the plants many advantages by lowering the level of ET. One method for reducing the negative effects of biotic stress may be to increase the use of bacteria that are friendly to plants and promote endophytic growth. In the future, in addition to bacteria that show effective ACC deaminase activity, there should be extensive research carried out on the possibilities of increasing plant antioxidant enzyme activity with plant growth-promoting bacteria and increasing their resistance to biotic stress. Additionally, appropriate and effective bacteria-plant combinations should be revealed.

KEYWORDS

- ACC deaminase enzyme
- bacterial endophytes
- biotic stress
- biotic stress tolerance
- plant physiology
- rhizobial infection
- stress reactions

REFERENCES

1. Ghosh, P. K., De, T. K., & Maiti, T. K., (2018). Role of ACC deaminase as a stress ameliorating enzyme of plant growth-promoting rhizobacteria useful in stress agriculture: A review. *Role of Rhizospheric Microbes in Soil,* 57–106.
2. Gupta, S., & Pandey, S., (2019). Unravelling the biochemistry and genetics of ACC deaminase-an enzyme alleviating the biotic and abiotic stress in plants. *Plant Gene, 18*, 100175.
3. Hameed, A., Yeh, M. W., Hsieh, Y. T., Chung, W. C., Lo, C. T., & Young, L. S., (2015). Diversity and functional characterization of bacterial endophytes dwelling in various rice (*Oryza sativa* L.) tissues, and their seed-borne dissemination into rhizosphere under gnotobiotic P-stress. *Plant Soil., 394*(1), 177e197.
4. Glick, B. R., Todorovic, B., Czarny, J., Cheng, Z., Duan, J., & McConkey, B., (2007a). Promotion of plant growth by bacterial ACC deaminase. *Crit. Rev. Plant Sci., 26*, 227–242.
5. Glick, B. R., (2014). Bacteria with ACC deaminase can promote plant growth and help to feed the world. *Microbiol. Res., 169*, 30–39.
6. Kang, S. M., Waqas, M., Khan, A., & Lee, I. J., (2013). Plant-growth-promoting rhizobacteria: Potential candidates for gibberellins production and crop growth promotion. In: Miransari, M., (ed.), *Use Microbes Alleviation Soil Stress* (pp. 1–19). Springer, New York.
7. Compant, S., Duffy, B., Nowak, J., Clement, C., & Barka, E. A., (2005). Use of plant growth-promoting bacteria for biocontrol of plant diseases: Principles, mechanisms of action, and future prospects. *Appl.Environ Microbiol., 71*, 4951–4959.
8. Saraf, M., Jha, C. K., & Patel, D., (2010). The role of ACC deaminase producing PGPR in sustainable agriculture. In: *Plant Growth and Health Promoting Bacteria* (pp. 365–385). Springer, Berlin, Heidelberg.
9. Domenech, J., Reddy, M. S., Kloepper, J. W., Ramos, B., & Gutierrez-Man˜ero, J., (2006). Combined application of the biological product LS213 with *Bacillus*, *Pseudomonas* or *Chryseobacterium* for growth promotion and biological control of soil-borne diseases in pepper and tomato. *Biocontrol, 51*, 245–258.
10. Rasche, F., Velvis, H., Zachow, C., Berg, G., Van, E. J. D., & Sessitsch, A., (2006). Impact of transgenic potatoes expressing anti-bacterial agents on bacterial endophytes is comparable with the effects of plant genotype, soil type and pathogen infection, *J. Appl Ecol., 43*, 555–566.
11. Sapre, S., Gontia-Mishra, I., & Tiwari, S., (2019). ACC deaminase-producing bacteria: A key player in alleviating abiotic stresses in plants. In: Kumar, A., & Meena, V. S., (eds.), *Plant Growth Promoting Rhizobacteria for Agricultural Sustainability*, Springer Nature Singapore Pte Ltd.https://doi.org/10.1007/978-981-13-7553-8_14.
12. Guinel, F. C., (2015). Ethylene, a hormone at the center-stage of nodulation. *Front Plant Sci., 6*, 1121.
13. Shahzad, S. M., Khalid, A., Arshad, M., Tahir, J., & Mahmood, T., (2010). Improving nodulation, growth, and yield of *Cicer arietinum* L. through bacterial ACC-deaminase induced changes in root architecture. *European Journal of Soil Biology, 46*(5), 342–347.
14. Conforte, V. P., Echeverria, M., Sánchez, C., Ugalde, R. A., Menéndez, A. B., & Lepek, V. C., (2010). Engineered ACC deaminase-expressing free-living cells of

Mesorhizobium loti show increased nodulation efficiency and competitiveness on *Lotus spp. The Journal of General and AppliedMicrobiology, 56*(4), 331–338.
15. Johnson, P. R., & Ecker, J. R., (1998). The ethylene gas signal transduction pathway: A molecular perspective. *Annu. Rev. Genet., 32*, 227–254.
16. Gontia-Mishra, I., Sasidharan, S., & Tiwari, S., (2014). Recent developments in use of 1-amino cyclopropane-1-carboxylate (ACC) deaminase for conferring tolerance to biotic and abiotic stress. *Biotechnol. Lett., 36*, 889–898.
17. Van, L. L. C., Geraats, B. P. J., & Linthorst, H. J. M., (2006). Ethylene as a modulator of disease resistance in plants. *Trends Plant Sci., 11*, 184–191.
18. Shaharoona, B., Arshad, M., & Zahir, Z. A., (2006). Effect of plant growth promoting rhizobacteria containing ACC-deaminase on maize (*Zea mays* L.) growth under axenic conditions and on nodulation in mung bean (*Vigna radiata* L.), *Lett. Appl. Microbiol., 42*, 155–159.
19. Naveed, M., Mitter, B., Reichenauer, T. G., Wieczorek, K., & Sessitsch, A., (2014). Increased drought stress resilience of maize through endophytic colonization by *Burkholderia phytofirmans* PsJN and *Enterobacter sp.* FD17. *Environ. Exp. Bot., 97*, 30–39.
20. Glick, B. R., Cheng, Z., Czarny, J., & Duan, J., (2007b). Promotion of plant growth ACC deaminase containing soil bacteria. *Eur. J. Plant Pathol., 119*, 329–339.
21. Dowling, D. N., Sexton, R., Fenton, A., Delany, I., Fedi, S., McHugh, B., Callanan, M., et al., (1996). Iron regulation in plant-associated *Pseudomonas fluorescens* M114: Implications for biological control. In: *Molecular Biology of Pseudomonads* (pp. 502–511).
22. Chet, I., & Inbar, J., (1994). Biological control of fungal pathogens. *Appl. Biochem. Biotechnol., 48*, 37–43.
23. Vallad, G. E., & Goodman, R. M., (2004). Systemic acquired resistance and induced systemic resistance in conventional agriculture. *Crop Sci., 44*, 1920–1934.
24. Verhagen, B. W. M., Glazebrook, J., Zhu, T., Chang, H. S., Van, L. L. C., & Pieterse, C. M. J., (2004). The transcriptome of rhizobacteria-induced systemic resistance in *Arabidopsis*. *Molec. Plant Microbe.Interact., 17*, 895–908.
25. Glick, B. R., (1995). The enhancement of plant growth by free-living bacteria. *Can. J. Microbiol., 41*, 109–117.
26. Blaha, D., Prigent-Combaret, C., Mirza, M. S., & Mo¨enne-Loccoz, Y., (2006). Phylogeny of the 1- aminocyclopropane-1-carboxylic acid deaminase encoding gene acdS in phytobeneficial and pathogenic proteobacteria and relation with strain biogeography. *FEMS Microbiol. Ecol., 56*, 455–470.
27. Saravanakumar, D., & Samiyappan, R., (2006). ACC deaminase from *Pseudomonas fluorescens* mediated saline resistance in groundnut (*Arachis hypogea*) plants. *J. Appl. Microbiol., 102*, 1283–1292.
28. Gamalero, E., & Glick, B. R., (2015). Bacterial modulation of plant ethylene levels. *Plant Physiol., 169*, 13–22.
29. Singh, R., Shelke, G., Kumar, A., & Jha, P., (2015). Biochemistry and genetics of ACC deaminase: A weapon to "stress ethylene" produced in plants. *Front. Microbiol., 6*, 937.
30. Gontia, I., Sasidharan, S., & Tiwari, S., (2014). Recent developments in use of 1-aminocyclopropane-1- carboxylate (ACC) deaminase for conferring tolerance to biotic and abiotic stress. *Biotechnol. Lett., 36*, 889–898.

31. Nukui, N., Minamisawa, K., Ayabe, S. I., & Aoki, T., (2006). Expression of the 1-aminocyclopropane-1- carboxylic acid deaminase gene requires symbiotic nitrogen-fixing regulator gene nifA2 in *Mesorhizobium* loti MAFF303099. *Appl. Environ. Microbiol., 72*, 4964–4969.
32. Madhaiyan, M., Poonguzhali, S., Ryu, J.,& Sa, T., (2006). Regulation of ethylene levels in canola (*Brassica campestris*) by 1-aminocycloprpane-1-carboxylate deaminase-containing *Methylobacterium fjisawaense, Planta, 224*, 268–278.
33. Akbaba, M., (2014). *Possibilities of Using Bacterial Endophytes that Increase Plant Growth in the Prevention of Cucumber Bacterial Angular Leaf Spot Disease (Pseudomonas Syringae pv. Lachrymans).* Master's thesis, Faculty of Agriculture/plant protection department/department of phytopathology, Ege University.
34. Lodewyckx, C., Vangronsveld, J., Porteous, F., Moore, E. R. B., Taghavi, S., Mezgeay, M., & Van, D. L. D., (2002). Endophytic bacteria and their potential applications. *Crit. Rev. Plant Sci., 21*, 583–606.
35. Sturz, A. V., Christie, B. R., & Matheson, B. G., (1998). Associations of bacterial endophyte populations from red clover and potato crops with potential for beneficial allelopathy. *Can. J. Microbiol., 44*, 162–167.
36. Lazarovits, G., & Nowak, J., (1997). Rhizobacteria for improvement of plant growth and establishment. *Hortscience, 32*, 188–192.
37. Strader, L. C., Chen, G. L., & Bartel, B., (2010). Ethylene directs auxin to control root cell expansion. *Plant Journal, 64*, 874–884.
38. Glick, B. R., Penrose, D. M., &Li, J., (1998). A model for the lowering of plant ethylene concentrations by plant growth-promoting bacteria. *J. Theor. Biol., 190*, 63–68.
39. Nadeem, S. M., Zahir, Z. A., Naveed, M., & Ashraf, M., (2010). Microbial ACC-deaminase: Prospects and applications for inducing salt tolerance in plants. *Critical Reviews in Plant Sciences, 29*, 360–393.

CHAPTER 11

Rhizobacteria-Induced Resistance in Plants Toward Cyst Nematodes

FREDERICK KANKAM,[1] ISAAC BOATEY AKPATSU,[1] and STEPHEN LARBI-KORANTENG[2]

[1]Department of Crop Science, Faculty of Agriculture, Food and Consumer Sciences, University for Development Studies, Tamale, Ghana

[2]College of Agricultural Education, Akenten Appiah Menka University of Skills Training and Entrepreneurial Development, Asante-Mampong, Ghana

ABSTRACT

The two main cyst nematode species that cause the most harm in crop production are the potato cyst nematodes (*Globodera* spp.) and the soybean cyst nematode (*Heterodera* spp.). They are plant parasitic nematodes, obligatory biotrophs that usually extract plant nutrients and water from their feeding sites. These sites are typically located in the rhizosphere of the plant or within the roots themselves. Their impact results in significant yield loss in crops, particularly potatoes and soybeans. If left unmanaged, yield losses associated with cyst nematode infestation can reach up to 80%. The management approaches for this disease vary across regions and farming systems, encompassing cultural methods, chemical methods, host-plant resistant approaches, and biological methods. One of the most effective biological means of managing the disease is the utilization of beneficial rhizobacteria, such as *Agrobacterium* spp. These agrobacteria have been extensively studied for their potential as crop inoculants, enhancing plant growth and development by influencing other microbial communities in the

Microbial Biostimulants: Biorational Pesticides for Management of Plant Pathogens.
Faheem Ahmad and Rakesh Pandey (Eds.)

rhizosphere, thereby promoting disease resistance in crops. Apart from being environmentally friendly, the use of carefully and specifically engineered species can also be economically friendly since they have the capability of self-replicating, hence circumventing routine application. This chapter seeks to explore the various potentials of Rhizobacteria that can be harnessed in the management of cyst nematode infection in plants.

11.1 INTRODUCTION

Plant nematode infection is one of the important economic challenges faced by most farmers and researchers in the sustainable agriculture value chain. In an ideal situation, plant parasitic organisms such as nematodes feed on water and other essential nutrients from their host. This is possible due to their ability to pierce through the vascular tissues of the host plant [1]. Managing these organisms is one of the most challenging tasks for farmers and researchers in nematode hotspot areas [2]. Nematodes are known to attack all underground parts of the plant, leading to a reduction in the plant's ability to absorb both water and essential nutrients. Depending on the severity and the susceptibility of the crop cultivar, nematode infection can cause up to an 80% yield loss if not managed [3]. Among the nematode species, two species, namely root-knot nematodes (RKN) (*Meloidogyne* sp.) and cyst nematodes (*Heterodera and Globodera* sp.), are of the most economic importance in most arable and vegetable crops. These species have a wide host range and exhibit very high antagonistic interactions with various plants.

In the past, nematode management was primarily carried out using synthetic chemical nematicides such as carbamates, fumigants, and organophosphates. The use of these nematicides has been reported to significantly increase the production of various arable and vegetable crops, as well as fiber production. However, it has also been reported that the use of these chemicals has a detrimental impact on different environments and their ecosystems [4]. As a result, many advanced countries have prohibited the use of a large portion of these chemicals in nematode management [3]. In the search for an alternative approach, one of the most reliable methods to manage nematodes with minimal or no negative impact on the environment is the utilization of biological techniques. One of the most viable biological methods in this approach is the use of plant growth-promoting agents (PGPR). Plant growth-promoting agents, such as rhizobacterial strains, have been reported to be effective nematode biocontrol agents, as well as growth-promoting agents in plants, leading to a significant boost in plant development and yield. In a symbiotic relationship, rhizobacterial strains utilize organic compounds, sugars, and amino acids that are produced and released by the plants through their roots. In exchange, the rhizobacterial strains produce highly effective plant growth-promoting chemicals with biocontrol potentials to enhance the host plant's ability to fight against various forms of pathogens [4]. Based on these facts, this chapter explores the various potentials of Rhizobacteria that can be harnessed in the management of cyst nematode infection in plants.

11.2 MANAGEMENT OF NEMATODES

Due to the significant losses caused by the cyst nematode, various approaches have been adopted to mitigate this problem. These approaches include cultural practices such as deep and timely plowing, intercropping, proper crop rotation, chemical (nematicide) application, using resistant cultivars, and the use of biocontrol agents. However, the cyst nematode is able to survive for a long time in the soil or underground plant parts, even in the absence of its host, due to survival mechanisms such as the formation of a gelatinous matrix and protective cyst. This renders most cultural methods, such as deep plowing, intercropping, and proper crop rotation, mostly ineffective in managing the nematode. It is also important to consider that while synthetic nematicides (chemical method) have been reported to be effective, many smallholder farmers in developing countries with limited purchasing power are unable to afford these chemicals. Also, the environmental and health concerns posed by these chemicals make it unsafe to continue their application. Although resistant cultivars have proven to be effective over time, continuous use of these cultivars in the same area or piece of land can lead to a breakdown in their resistance ability. This is because most phytopathogens are capable of developing highly virulent pathotypes over time. According to the literature, biological control is the most effective, economical, and environmentally friendly approach to combat the problems caused by cyst nematodes [5]. Biological control involves the application or employment of antagonistic organisms to reduce the population of the targeted organism. This is achieved without the use of synthetic nematicides, nematode resistant cultivars, or manipulation of cultural practices. The reduction in population is solely achieved through the activities of the introduced biocontrol agent(s) [6]. Specifically, the biological control method aims to control the population of harmful nematodes by increasing the populations of antagonistic organisms in the plant rhizosphere. The soil serves as a reservoir for various micro-floras, which vary in composition and activities. In plant defense mechanisms, the microflora present in the rhizosphere serves as a leading defense mechanism that protects plants against various phytopathogens. This can be exploited as commercial biocontrol agents [7]. The adaptation of biocontrol agents in the management of nematodes is a major step in maintaining ecological balance through minimal or negligible environmental pollution. Biocontrol agents, once established, maintain their efficacy in the rhizosphere for extended periods, leading to the concept of "suppressive soils." At this point, plant-parasitic nematodes (PPN) such as cyst nematodes are naturally suppressed by active soil-dwelling microorganisms [8]. In most cases, biocontrol of PPN species has been successful through the use of bacterial and fungal antagonists. Bacterial biocontrol agents can be categorized into three main groups: endoparasitic, epiphytic, and endophytic bacteria (EB). Their mode of antagonism is mainly through antibiosis, parasitism, and competition. On the other hand, the fungal group of antagonists consists of various microorganisms, including endoparasitic fungi, vesicular-*Arbuscular mycorrhiza* (VAM), nematode-trapping fungi, and toxin-producing fungi [11, 12].

11.3 IMPACTS OF RHIZOBACTERIA AS A BIOCONTROL AGENT

11.3.1 INDUCED SYSTEMATIC RESISTANCE

Plants in nature have various mechanisms that help them fight against different forms of diseases. After successfully stimulating these mechanisms to combat a wide range of diseases and most plant pests, they acquire an enhanced defense ability known as induced systemic resistance (ISR) [11]. This ISR can be induced by carefully and specifically engineered rhizobacteria. The induced resistance provides broad-spectrum protection against most phytopathogens, including nematodes, bacteria, viruses, fungi, and to a larger extent, insect pests [14]. Enzymes such as polyphenol oxidase (PPO), lipoxygenase (LOX), β-1,3-glucanase, superoxide dismutase (SOD), phenylalanine ammonia lyase (PAL), catalase (CAT), proteinase inhibitors, ascorbate peroxidase (APX), and chitinase [15, 16] play a crucial role in this process. Figure 11.1 illustrates a web of phytopathogen-induced ISR and SAR [14].

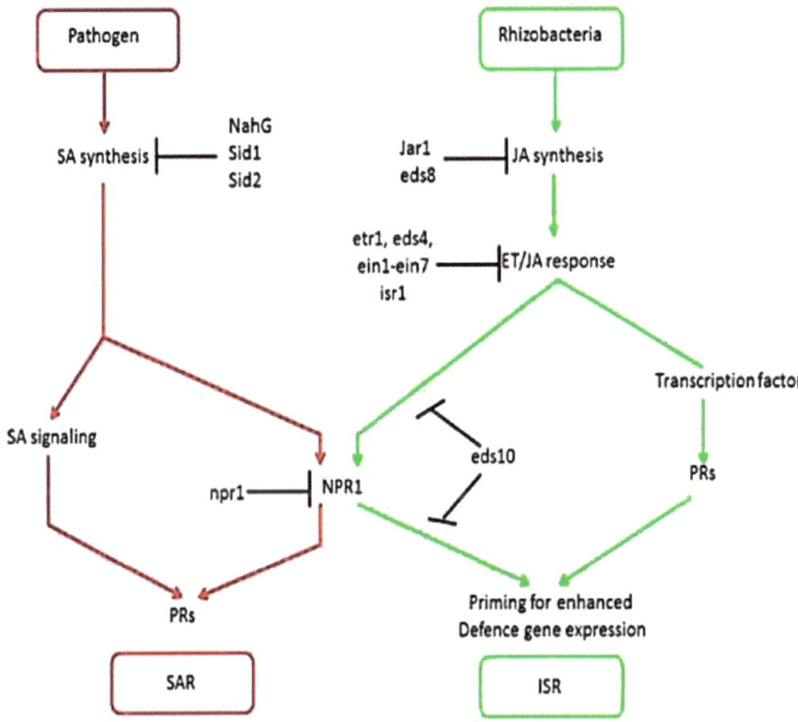

FIGURE 11.1 Chart flow illustration of pathogen ISR and SAR [14].

Furthermore, studies have postulated that rhizobacteria, by inducing ISR, make it very effective in reducing the severity of nematode infection [17, 18]. This is achieved through both physical and mechanical accentuation of the cell wall *via* the accumulation of phenolic compounds, wall thickening, and callose deposition. Other ways include up-regulating defense reactions through the synthesis of multiple biochemical compounds. Examples of these compounds are salicylic acid (SA), PR proteins, siderophores, phytoalexin, jasmonic acid (JA), lipopolysaccharides (LPS's), chitinase, and several others which may be present as secondary metabolites [15, 16]. Nevertheless, other studies have also reported findings that are in contrast with this assertion. An example of this was observed when ISR by *P. fluorescence* in tomatoes did not have any bearing on SA accumulation [17].

In other literature, cyst and root knot nematodes in potatoes were effectively controlled when *R. etli* was introduced as a catalytic agent of systemic resistance [18]. It was also revealed that *R. etli* was strongly associated with suppressing both the recognition and penetration of *M. incognita* juveniles and *G. pallida*. Similar studies reported a negative correlation between enzymatic activity and cyst nematode population in tomatoes when Bbv 57, *P. fluorescens*, and Pf128+ *B. subtilis* were applied. Thus, an increase in enzymatic activity will result in a drastic reduction of nematode population [18–20]. This highlights the potential role that carefully engineered and target-specific rhizobacteria could play in the management of cyst nematode infections in various crops through the activation and deployment of defense enzymes such as PAL, PO, and PPOs.

11.3.2 PRODUCTION OF PHYTO-HORMONE

Most rhizobacterial strains are capable of producing plant growth promoting substances (PGPR). These substances are responsible for regulating the growth and development of plants. Examples of these substances include gibberellic acid, ethylene (ET), jasmonates, strigolactones (SLs), polyamines, abscisic acid (ABA), brassinosteroids, cytokines (such as isopentenyl adenine riboside, zeatin, isopentenyl adenosine, and transzeatin ribose), and auxins (such as phenylacetic acid, indole acetic acid, and indole butyric acid) [21]. These phytohormones produced by PGPRs have various effects on the growth, development, and metabolism processes of plants. Among them, indole acidic acid is the most abundant. While these phytohormones serve many functions in plants, one of their prominent roles is to enhance the plant-bacteria relationship by promoting plant growth and development through stimulating cell division, tissue expansion, elongation, and activating other beneficial enzyme effects that contribute to both vegetative growth and yield [22, 23]. Indole acetic acids are also known to increase nutrient and mineral uptake [24]. This effect may be attributed to their ability to induce the formation and development of both adventitious and lateral roots in most green plants. Additionally, it has been suggested that PGPRs could serve as highly effective agents in preventing detrimental effects caused by various forms of environmental stress [25]. For instance, studies have shown that inoculating seeds with phytohormones that produce PGPRs has been crucial in the biological control of nematodes, as well as in enhancing plant growth and development [26]. In other studies, a significant increase was observed when eggplants were inoculated with phytohormones that produce various *Streptomyces* strains. The nematicidal effect of *Streptomyces* can significantly reduce the egg mass of nematodes [27, 28]. Moreover, the utilization of viable and well-engineered bacteria to produce phytohormones could be a promising approach in the management of cyst nematodes.

11.3.3 NUTRIENT SOLUBILIZATION

Plant nutrients such as nitrogen, phosphorus, and potassium are two of the most essential nutrients in most plants. They are very influential in plant growth and developmental processes such as protein synthesis, development of new tissues, nucleic acid composition, cell division, and other associated processes. Phosphorus increases the metabolism rate of plants, leading to a transformation of complex energy within the plant. In most agricultural fields, phosphorus is usually available in an insoluble form. To transform this into a soluble form, PGPRs facilitate synthesis processes such as acidification, chelation, and production of organic acids. Through this process, they enhance the availability of nutrients for plant uptake [21, 29, 30]. In line with this fact, several studies have reported that plants have a higher tendency to naturally fight against diseases when supplied with adequate nutrients. Some PGPRs that have been reported to be very efficient in the solubilization of phosphate are *Serratia, Enterobacter, Microbacterium, Arthrobacter, Rhodococcus, Burkholderia, Pseudomonas, Bacillus, Erwinia, Flavobacterium, Rhizobium,* and *Beijerinckia* [4, 22, 31]. In other accounts, bacterial strains such as *P. syringae, B. megaterium, P. cichorii, B. cereus,* and *B. caryophylli* have been reported to have a higher efficacy in both solubilization and mineralization of phosphates [32]. This could potentially lead to increased levels of bioactivation of present phosphate compounds when these bacteria are present in an agricultural field. Similarly, increased rates of plant growth, development, and resistance have been reported when strains of *Mesorhizobium mediterraneum* were inoculated into barley and chickpea [33]. When inoculated to plants, these phosphate solubilizing microorganisms not only have a nutrient facilitation ability but may also possess nematicidal components in their activities. For instance, *B. megaterium*, a phosphate solubilizing bacteria (PSB), has been reported to enhance growth parameters such as shoot and root biomass, shoot length, and chemical components like N, P, and K content in tomatoes [34]. The same source further reveals that inoculating *B. megaterium* in the same plant leads to a significant reduction in the populations of *M. incognita* in the plant rhizosphere.

On the other hand, potassium is one of the essential nutrients in both physiological and biochemical processes in plants. However, potassium minerals present in most agricultural lands are in a fixed form, which plants are unable to utilize. Fortunately, useful bacteria present in the plant rhizosphere act as solubilization agents, releasing potassium to plants in a form that can be efficiently utilized. Many potassium solubilizing PGPRs play critical roles in plant growth mechanisms, including chelation, acidolysis, excretion of organic acid, exchange, reduction, and complexolysis [20, 35]. *Burkholderia, Pseudomonas sp, Acidothiobacillus ferrooxidans, Paenibacillus sp, B. mucilaginosus,* and *B. edaphicus* are some of the microbial genera frequently associated with higher potassium solubilization. Further studies have shown that potassium solubilizing bacteria have a positive correlation with both plant growth and nematicidal activity when they are inoculated into tomato seeds before planting [34, 36].

11.4 CONCLUSION AND FUTURE PROSPECTS

Achieving global food security has been the main objective of many agricultural interventions over the years. In this regard, crop protection has been a consistently important topic. While protecting plants is a priority, it is crucial to ensure that efforts towards this objective are both cost-effective and environmentally friendly. Various synthetic approaches have been utilized for better crop protection, but they often come with significant negative impacts on the environment. Nematode infestation is a major factor limiting crop yields in many agricultural crops, with cyst nematodes being particularly detrimental. Effectively managing these pests is therefore crucial in meeting the food security needs of the growing population. Biological control has been identified as an economical, effective, and environmentally friendly approach for nematode management. This concept can be applied to the management of cyst nematodes. To this, the use of PGPR technology could be the breakthrough to managing cyst nematodes. These agrobacteria have widely been reviewed for use as crop inoculants, improving plant growth and development through interference with other microbial communities in the rhizosphere, thus simulating disease resistance in crops. In addition to being environmentally friendly, the use of carefully targeted and specifically engineered species can also be economically friendly, as they have the capability of self-replication, thereby circumventing routine application.

KEYWORDS

- biocontrol agent
- cyst nematodes
- nematodes
- plant growth-promoting agents
- rhizobacteria
- rhizobacterial strains

REFERENCES

1. Press, M. C., & Phoenix, G. K., (2005). Impacts of parasitic plants on natural communities. *New Phytol., 166*, 737–751.
2. Gillet, F., Bournaud, C., Junior, J., & Grossi-De-Sa, M., (2017). Plant-parasitic nematodes: Towards understanding molecular players in stress responses. *Ann. Bot, 119*, 775–778. [Online]. Available: https://academic.oup.com/aob/article-abstract/119/5/775/2901799 (accessed on 9 January 2023).
3. Schneider, S. M., Rosskopf, E. N., Leesch, J., Chellemi, D. O., Bull, C. T., & Mazzola, M., (2003). Research on alternatives to methyl bromide: Pre-plant and post-harvest. *Pest Manag. Sci., 26*, 814–826.
4. Karthik, C., et al., (2017). Characterization of multifarious plant growth promoting traits of rhizobacterial strain AR6 under chromium (VI) stress. *Microbiol Res., 204*, 65–71.
5. Timper, P., (2014). Conserving and enhancing biological control of nematodes. *J. Nematol., 46*, 75–89.
6. Tian, B. Y., Yang, J. K., Lian, L. H., Wang, C. Y., & Zhang, K. Q., (2007). Role of neutral protease from *Brevibacillus laterosporus* in pathogenesis of nematode. *App. Microbiol. Biotechnol, 74*, 372–380.
7. Mendes, R., Garbeva, P., & Raaijm, J., (2013). The rhizosphere microbiome: Significance of plant beneficial, plant pathogenic, and human pathogenic microorganisms. *FEMS Microbiol. Rev., 37*, 634–663.
8. Trudgill, D. L., et al., (2000). The importance of tropical root-knot nematodes (*Meloidogyne* spp.) and factors affecting the utility of *Pasteuria penetrans* as a biocontrol agent. *Nematol., 2*, 823–845.
9. Abd-Elgawad, M. M., & Kabeil, S. A., (2012). Biological control of *Meloidogyne incognita* by *Trichoderma harzianum* and *Serratia marcescens* and their related enzymatic changes in tomato roots. *Afr. J. Biotechnol., 11*, 16247–16252.
10. Akinbode, O. A., (2010). Evaluation of antifungal efficacy of some plant extracts on *Curvularia lunata*, the causal organism of maize leaf spot. Institute of Agricultural Research and Training, IAR&T, Ibadan, Nigeria. *African J. Environ. Sci. Technol., 4*(11), 797–800.
11. Van-Loon, L. C., (2000). Systemic induced resistance. In: Slusarenko, A. J., Fraser, R. S. S., & Van, L. L. C., (eds.), *Mechanisms of Resistance to Plant Diseases* (pp. 521–574). Kluwer. Dordrechet.
12. Pokhare, S., Pankaj, Shakil, N. A., Kumar, J., & Singh, K., (2012). Foliar application of chemical elicitors induces biochemical changes in wheat against the cereal cyst nematode, *Heterodera avenae*. *Nematol. Medit., 40*, 181–187.
13. Mhatre, P., Pankaj, S. K., Malik, Kaur, S., Singh, A. K., Mohan, A., & Sirohi, S., (2015). Histopathological changes and evaluation of resistance in Asian rice (*Oryza sativa*) against rice root knot nematode, *Meloidogyne graminicola*. *Indian J. Genet. Plant Breed, 75*, 41–48.
14. Pieterse, C. M. J., VanWees, S. C. M., Ton, J., VanPelt, J. A., & VanLoon, L., (2002). Signaling in rhizobacteria-induced systemic resistance in *Arabidopsis thaliana*. *Plant Biol., 4*, 535–544.
15. Ramamoorthy, V., Viswanathan, R., Raguchander, T., Prakasam, V., & Samiyappan, R., (2001). Induction by systemic resistance by plant growth promoting rhizobacteria in crop plants against pests and diseases. *Crop Prot., 20*, 1–11.

16. Siddiqui, Z. A., & Mahmood, I., (1999). Role of bacteria in the management of plant parasitic nematodes: A review. *Bioresour. Technol., 69*, 167–179.
17. Siddiqui, I. A., & Shaukat, S., (2004). Systemic resistance in tomato induced by biocontrol bacteria against the root-knot nematode, *Meloidogyne javanica* is independent of salicylic acid production. *J. Phytopathol., 152*, 48–54.
18. Reitz, M., Rudolph, K., Schroder, I., Hoffmann-Hergarten, S., Hallmann, J., & Sikora, R. A., (2000). Lipopolysaccharides of *Rhizobium etli* strain G12 act in potato roots as an inducing agent of systemic resistance to infection by the cyst nematode *Globodera pallid. App. Env. Microbiol., 66*, 3515–3518.
19. Mahdy, M., Hallmann, J., & Sikora, R. A., (2001). Influence of plant species on the biological control activity of the antagonistic rhizobacterium *Rhizobium etli* strain G12 toward the root-knot nematode *Meloidogyne incognita. Meded Rijksuniv Gent Fak Landbouwkd Toegep Biol. Wet., 66*, 655–662.
20. Meena, K. S., Jonathan, E. I., Devrajan, K., & Raguchander, T., (2012). *Pseudomonas fluorescens* induced systemic resistance in tomato against *Meloidogyne incognita. Indian J. Nematol., 42*, 5–10.
21. Gopalakrishnan, S., Sathya, A., Vijayabharathi, R., Varshney, R. K., Gowda, C. L. L., & Krishnamurthy, L., (2015). Plant growth promoting rhizobia: Challenges and opportunities. *Biotechnology, 5*, 355–377.
22. Ahemad, M., Zaidi, M., Khan, M. S.,& Oves, M., (2009). Biological importance of phosphorus and phosphate solubilizing microorganisms—An overview. *Phosphate Solubilizing Microbes for CropImprovement* (pp. 1–4.). Nova, New York.
23. Karthik, C., Oves, M., Thangabalu, R., Sharma, R., Santhosh, S. B., & Arulselvi, P. I., (2016). *Cellulosimicrobium funkei*-like enhances the growth of *Phaseolus vulgaris* by modulating oxidative damage under chromium (VI) toxicity. *J. Adv. Res., 7*, 839–850.
24. Arora, N., Tewari, S., & Singh, R., (2016). *Multifaceted Plant-Associated Microbes and Their Mechanisms Diminish the Concept of Direct and indirect PGPRs* (pp. 411–449). Springer. doi: 10.1007/978-81-322-1287-4_16.
25. Glick, B., (2010). Using soil bacteria to facilitate phytoremediation. *Biotech. Adv., 28*, 367–374.[Online]. Available: https://www.sciencedirect.com/science/article/pii/S0734975010000212 (accessed on 9 January 2023).
26. Khan, M. R., Khan, M. M., Anwer, M. A., & Haque, Z., (2012). Laboratory and field performance of some soil bacteria used as seed treatments on *Meloidogyne incognita* in chickpea. *Nematol. Mediterr., 40*, 143–151.
27. Ruanpanun, P., Tangchitsomkid, N., Hyde, K. D., & Lumyong, S., (2010). Actinomycetes and fungi isolated from plant-parasitic nematode infested soils: Screening of the effective biocontrol potential, indole-3-acetic acid and siderophore production. *World J. Microbiol. Biotechnol., 26*, 1569–1578.
28. Rashad, F. M., Fathy, H. M., El-Zayat, A. S., & Elghonaimy, A., (2015). Isolation and characterization of multifunctional *Streptomyces* species with antimicrobial, nematicidal and phytohormone activities from marine environments in Egypt. *Microbiol. Res., 175*, 34–47.
29. Gulati, A., et al., (2010). Organic acid production and plant growth promotion as a function of phosphate solubilization by *Acinetobacter rhizosphaerae* strain BIHB 723 isolated from the cold deserts of the trans-Himalayas. *Arc Microbiol., 192*, 975–983.

30. Oves, M., Khan, M. S., & Zaidi, A., (2013). Chromium reducing and plant growth promoting novel strain *Pseudomonas aeruginosa* OSG41 enhance chickpea growth in chromium amended soils. *Eur. J. Soil Biol., 56*, 72–83.
31. Kumar, J., Kashyap, N. P., & Sharma, S. D., (2001). Pests of maize and their management in Himachal Pradesh – A review. *Agricultural Reviews, 22*(1), 47–51.
32. Guang-Can, T., Shu-Jun, T., Miao-Ying, C., & Guang-Hui, X., (2008). Phosphate-solubilizing and -mineralizing abilities of bacteria isolated from soils. *Pedosphere, 18*, 515–523.
33. Peix, A., Rivas-Boyero, A. A., Mateos, P. F., Rodriguez-Barrueco, C., Martınez-Molina, E., & Velazquez, E., (2001). Growth promotion of chickpea and barley by a phosphate solubilizing strain of *Mesorhizobium* Mediterraneum under growth chamber conditions. *Soil Biol Biochem, 33*, 103–110.
34. El-Hadad, M., et al., (2011). The nematicidal effect of some bacterial biofertilizers on *Meloidogyne incognita* in sandy soil. *Braz J Microbiol, 42*, 105–113. [Online]. Available: https://www.scielo.br/j/bjm/a/NbKqyTk7XDSdJtktVNvLqPd/?lang=en&format=html (accessed on 9 January 2023).
35. Zhang, C., & Kong, F., (2014). Isolation and identification of potassium-solubilizing bacteria from tobacco rhizospheric soil and their effect on tobacco plants. *App. Soil Ecol., 82*, 18–25.
36. Han, H. S., & Lee, K. D., (2006). Effect of co-inoculation with phosphate and potassium solubilizing bacteria on mineral uptake and growth of pepper and cucumber. *Plant Soil Env., 52*, 130–136.

CHAPTER 12

Microbial and Algal Biostimulants: Modern Approach Toward the Next Generation of Sustainable Agriculture

GHOLAMREZA ABDI,[1] AB WAHEED WANI,[2] and
MUKUL MACHHINDRA BARWANT[3]

[1]*Department of Biotechnology, Persian Gulf Research Institute, Persian Gulf University, Bushehr, Iran*

[2]*Department of Horticulture, Lovely Professional University, Punjab, India*

[3]*Department of Botany, Sanjivani Arts, Commerce, and Science College, Kopargaon, Ahmednagar, Maharashtra, India*

ABSTRACT

Biostimulants are organic substances that promote the overall well-being, vitality, and development of plants, as well as their resistance to disease. They are effective in both horticulture and agricultural crops. The principal active ingredients used in these formulations include humic and fulvic acids, protein hydrolysates, nitrogen-containing chemicals, seaweed extracts, helpful fungi, and bacteria. Bio-stimulant formulations can be single- or multi-component, and there is evidence of synergistic activity among many separate components. Various categories of biostimulants have been identified based on their use (soil, foliar), source (plant, animal), or generation technique (hydrolysis, fermentation, extraction). Natural soil stimulants can encourage the establishment of beneficial soil organisms that support plant growth. Given the ongoing processes of soil degradation and air pollution, it is crucial to use natural preparations that do not harm the environment. This chapter provides an overview of the significance and impact of several

Microbial Biostimulants: Biorational Pesticides for Management of Plant Pathogens.
Faheem Ahmad and Rakesh Pandey (Eds.)

natural plant biostimulants on crop production and quality, formulation, mechanism, and various aspects of these substances.

12.1 INTRODUCTION

Biostimulants are becoming increasingly important in strategies for sustainable agriculture. Inappropriate and excessive application of chemical-based products on crops has resulted in instances of environmental dangers and soil pollution, which has been a major problem for the sector in recent years. The size of the world market for biostimulants was evaluated at USD 1.74 billion in 2016 and is anticipated to grow at a CAGR of 10.2% from 2017 to 2025. Over the projected period, the market is likely to be driven by a growing emphasis on increased productivity as well as rapid soil degradation. The size of the worldwide biostimulants market was predicted to be USD 2.30 billion in 2019 and USD 2.53 billion in 2020 [1]. Plant biostimulants (PBs) have seen increased interest over the past 10 years as a result of scientists, extension specialists, the private sector, and growers becoming more interested in incorporating these products into their arsenal of environmentally friendly tools to ensure improved crop performance and yield stability. The United Nations predicts that by 2050, there will be more than 9.7 billion people on the planet, up from the current estimated population of 7.5 billion. As a result, contemporary agriculture will need to become more effective at providing more food in a sustainable and environmentally benign manner. The creation of novel plant biostimulants (PBs) and efficient applications for them is an innovative technology tackling these significant difficulties [2].

Agricultural systems face numerous novel and challenging problems due to global climate change, making it increasingly difficult to sustainably feed a growing world population. This global phenomenon has led to more frequent occurrences of droughts, floods, salinization, and extreme temperatures, disrupting normal climatic patterns and water budgets. Consequently, weed and phytopathogen populations change, beneficial microbe populations decline, and plants become more vulnerable to biotic stress. On the other hand, chemical fertilizers and pesticides are being increasingly utilized in agriculture to ensure adequate yields and pest control.

In recent years, there has been a growing interest in PGPR plant biostimulants (PBs) as potential environmentally friendly alternatives to these agrochemicals. The PB market is expected to experience significant growth in the near future. By applying PGPR-based PBs to crops cultivated in agricultural or horticultural fields, it is possible to reduce abiotic and biotic stress, as well as enhance crop quality attributes (such as fruits, vegetables, ornamental plants, and medicinal plants). PGPRs regulate phytohormones, siderophores, organic acids, and enzyme production and release, thereby improving the availability of essential nutrients like iron, phosphorus, potassium, and zinc to plants.

In addition to optimizing human inputs in agro-ecosystems, the use of microbial biostimulants may offer a long-term and cost-effective solution to mitigate plant productivity losses caused by shifting climatic factors. The preliminary findings from experiments using microbial biostimulants should be shared with relevant policymakers and stakeholders, including extension services and growers, to ensure the broad application of this approach across different crops, regions, and environmental conditions.

12.2 CONCEPT OF BIO-STIMULATION, COMPOSITION, AND MICROBIAL METABOLITES

Biologically active substances derived from microbial biomass and their metabolites are referred to as bioformulations. Bioformulation is a valuable tool for plant growth, disease control, and nutrient acquisition. In essence, it involves combining an active substance, such as an enzyme, with an inert substance, like water, to create a *bio*-formulated product. By using these products instead of traditional chemical fertilizers and pesticides, we can promote plant growth, suppress phytopathogens, maintain soil fertility, and prevent disease [3]. The term "bioformulation" does not have a standardized definition and is used by scientists in various contexts. These products are designed to effectively deliver their contained mixture to the intended target while preserving their efficacy the microbes' potential and increasing their activities, according to Burges [76]. Pesticide and fertilizer alternatives can be created using bioformulations derived from microorganisms or their active gradients. An active ingredient and microbes or their products are necessary for the product to be effective [4]. Carboxymethylcellulose, peat, vermiculite, talc, and polymers such as diatomaceous earth and xanthan gum are the most common inert active substances. There is evidence that bioformulations based on inert carriers can introduce antagonistic microbes into the plant's system for extended periods of time. Additionally, these carrier-based bioformulations have a stable adduct that includes phytohormones, siderophores, antibiotics, and volatile metabolites hydrolysing enzymes when in contact with the plant [5, 6]. Additives such as silica gel, gum, starch, and methyl cellulose can be used to protect bioformulated products from harsh environmental conditions [7]. Bioformulated derivatives are necessary in modern agriculture to counteract this effect. Many of the microbial species used in bioformulated products contain beneficial rhizospheric microorganisms. These microbes directly and indirectly affect plant growth and development [8].

The bacteria that inhabit the digestive tract are referred to as intestinal microbiota. The intestinal microbiota contains a wide variety of bacterial species, each of which has a distinct capacity for interacting with both human cells and other bacteria. It is believed that the gut microbiota is responsible for producing more than half of the metabolites found in urine and feces [9]. A modified metabolite classification has been used to categorize three different types of metabolites: Metabolites produced by the intestinal microbiota (such as short-chain fatty acids), those secreted by the host cell and modified by the intestinal microbiota (such as secondary bile acids), and those produced entirely by the intestinal microbiota (SCFAs). This section covers SCFA biotransformation, the unique intestinal microbiota associated with each SCFA, and how each SCFA affects the intestinal microbiome. The specific intestinal microbiota related to SCFAs, as well as SCFA biotransformation, are discussed in this section [10]. Gut bacteria influence polyphenol bioavailability by hydrolyzing hepatic O-glucuronides and O-glycosides to release aglycones. Plants and foods primarily contain polyphenols as O-glycosides. O-glycosides are the main form of flavonoids like quercetin and kaempferol, two of the most common and abundant flavonoids. Flavonoids derived from glycosides are abundant in the red, blue, and purple pigments found in berries. The stomach's acidity, the mucosa of the small intestines, and enzymes produced by gut bacteria hydrolyze a significant portion of these glycosides into their corresponding aglycones [11]. There have also been attempts to classify microbial metabolites based on their elemental composition (e.g., proteins, polysaccharides, and organic acids, etc.), as well as their bioactivity (e.g., peptidoglycan, immunomodulatory, anti-inflammatory, antibacterial, antioxidant, antiproliferative, etc.). In the human intestine, the most common short-chain fatty acids are acetate, propionate, and butyrate. Small polycationic compounds made by mammalian cells or the intestinal microbiome (using amino acid decarboxylase enzymes as a limiting enzyme) or from food are also produced by gut bacteria [64]. Cholesterol in the liver is converted to cholic acid and chenodeoxycholic acid, the two primary bile acids (PBAs), which are then secreted into the digestive tract by bile. Commensal bacteria produce polysaccharides that are beneficial to the intestinal microbiome. The only way humans can make vitamins is through food or microorganisms [12]. Vitamin K and several B-group vitamins, including B12, B2, B6, and B9, can be produced by gut bacteria. De novo synthesis is the name given to this process. These vitamins are absorbed or converted into their derivatives by the intestinal epithelium before being released into the bloodstream [13].

12.3 PRODUCTION AND MARKETING STRATEGIES

It is challenging to simultaneously create microbial bioformulations and provide high-quality products. Maintaining the microbial load and vigor of bioformulated products, which utilize living microbial species, requires meticulous care. Furthermore, advanced manufacturing technologies necessitate the use of high-tech instruments throughout the production process. Inoculants, which are present in bioformulated products, play a crucial role in the nutrient supply chain for plants. However, in developing countries, microbial products often lose their viability, efficacy, and effectiveness due to technical difficulties, lack of information, inexperienced personnel, inadequate instruments and technologies, discrepancies in distribution, and importation laws for live inoculants [3]. The development of bioformulations may also face obstacles such as high production costs, short shelf lives, and inconsistencies in performance. While the bioformulations are produced using inoculants, it is essential to ensure that the final product remains uncontaminated. The production of microbial bioformulations is hindered by the high cost of advanced instruments required for a completely sterile environment. Ineffective handling practices and inexperienced staff may contribute to the poor performance of microbial products."Living microbial cells" in the bioformulations are extremely sensitive to temperature and other environmental factors, necessitating extreme caution during all stages of production, cultivation, transport, and application [3]. Carriers, packaging, and storage must be purchased at a significant cost for all of these processes to work. To promote the expansion of agricultural markets, it is essential for the growth of bioformulations. Companies with larger production facilities tend to invest more in networks to gain market share. Additionally, registering a product, which is costly and time-consuming, poses a significant obstacle [14]. The inconsistent performance of microbial products and the lack of demand from bioformulation producers and investors necessitate effective storage for a longer shelf life. Microbial shelf life can be affected by various factors, including production methods, the type of packaging materials used, the mode of transportation, and the distance traveled. It is true that the vast majority of businesses fail to provide an adequate and effective storage system for bioformulations due to inappropriate handling experiences, a dearth of technical expertise, and a lack of subject matter expertise. It is also challenging to successfully market microbial products in the field due to the inconsistent microbial performance that occurs when they are transferred to soil." The ability of the microbial population to thrive in a new environment is impacted by the rapid decrease in the number of active cells. Each release should be carefully considered due to the complex nature of the soil environment, which consists of a wide variety of organisms and fluctuating local conditions. Soil microbiostasis plays a crucial role in the inoculant's ability to inhibit growth and survival. Nutrient availability, soil microbial hostility, and crop growth and productivity are all influenced by the growth and survival of the inoculant, which in turn is affected by various abiotic and biotic factors. Consequently, different species will exhibit varying performance when considering their activity and survival.

12.4 BIOFORMULATION AS BIOCONTROL AGENTS

When pathogens and parasites are either active or dormant, one or more organisms are used to reduce the number of inocula and diseases they can cause. One way to accomplish this is by introducing an antagonist in large numbers or changing the host or antagonist in some natural way. In order to mass produce biopesticides, bioformulations based on biological control agents (BCA) have been used. Bioformulations such as Bioveer, Biofor PF-2 (Jaiva Kiran), Biozin-PTB, Biozin-PTB (Jaiva Kiran-2), Biozium, and the talc-based biopesticide containing three aggressive antagonist strains such as *Trichoderma* spp., *Pseudomonas aeruginosa,* have been developed by the Department of Plant Pathology, Assam Agricultural University, Assam, India. Biological control of diseases like chili wilt, tomato wilt, and ginger wilt as well as rhizome and foot root and root rot damping-off, and wilt diseases such as those that affect pepper and cabbage, has been found to be effective with the biocontrol formulation. *P. fluorescence* and *T. harzianum* are beneficial microbes in BioPf-2, which boost soil nutrients, revitalize soils, help in the growth of plants, and fight against soil-borne diseases like BioPf-2 [15, 16]. The spores of *Metarhizium anisopliae,* a soil fungus, are found in the bioinsecticide Met52. In the fight against foliar and soil-borne diseases, Taegro is a biofungicide/bactericide that is based on bacteria [17].

Additionally, one of the most widely used biopesticides is *TrichoderMax,* which contains *T. asperellum* spores. Similar to BoveMax, which utilizes *Beauveria bassiana* to control Broca (*Hedypates betulinus*) infestations in Erva-mate plantations [17]. There are several biopesticides made from microbial consortia, such as Biosona, Biometa, and Biollium, which include entomopathogens like *Bacillus thuringiensis* (Bt) and *Verticillium lecanii,* as well as microbes that promote plant growth. There are many factors that go into making an effective biostimulant or bioformulated product, such as its potency and the manner in which it is prepared. According to a recent study, it was found that combining PGPRs like Pf1, *Bacillus subtilis* Bs10, and the biocontrol fungal *T. viridae* (Tv1) bioformulations [18] was effective in reducing the incidence of peduncle blight in greenhouse conditions.

Nanoformulations of the antagonist fungus *Talaromyces flavus* were used by Naraghi et al. [19] to biologically control fungal pathogens like *Verticillium dahliae*, Fusarium *oxysporum*, and *Rhizoctonia solani* [19]. Fungal bioinoculants used in the majority of approved bioformulated products include *T. asperellum, T. harzianum, Aspergillus flavus, Chondrostereum purpureum, T. gamsii,* and *Coniothyrium minitans*.

12.5 FORMULATION AND APPLICATION METHODS

12.5.1 SOLID BIOFORMULATION

Bioformulations are commonly used in liquid and solid forms, in addition to liquid and solid forms [4]. Both types have undergone modifications, but they remain popular today despite this. The most common methods for producing solid bioformulations include the use of various grains, microgranules, wet powders, and water-dispersible granules [20–23]. Granular bioformulations consist of particles containing an active ingredient, a carrier, and a binding agent. Particle sizes in the MG (100–600 μm) and coarse (100–1000 μm) ranges are the most commonly used for bioformulated products. The active ingredient concentration in GR ranges from 5% to 20% [24]. Cornmeal, wheat GR, gluten, cornstarch, gelatinized flour, cottonseed flour, and sugar GR are some of the most common substrates used for the formation of gelatinized rice roe (GR) [23, 25]. In 2008 [26], there were reports of the harmful rhizobacteria *Pseudomonas trivialis* and the BioShield granular formulation being found in the soils of wheat and soybean crops, respectively [27]. When it comes to WP composition, the typical range is 50-80% technical powder, 15% filler, 1% dispersant, and 3-5% surfactant (except for GR-based formulations) [24]. Several commercial *Trichoderma* bioformulations have visible

WPs [28]. Post-harvest disease was controlled using a WP containing 60% freeze-dried *B. cereus* powder, 4% diatomite carrier, 6% alkyl naphthalene sulfonate, 4% sodium lignin sulfonate, and 1% K_2HPO_4. Particle sizes range from 50 to 100 micrometers in dust-type bioformulated products (typically at a 10% rate). Pests in tomato, basil, carnation, and cyclamen gardens are controlled by a bioformulated dust called Biofox C. It does not contain any potentially harmful *F. oxysporum* [29]. WP variants such as WG/WDG, which are less dusty and easier to dissolve in water than WP but contain a higher concentration of dispersing agent, are another option. WDGs are frequently used because of their durability and effectiveness in controlling nematodes. Antifungal *Ampelomyces quisqualis* is frequently used as a biocontrol agent for powdery mildew disease, according to [30].

12.5.2 LIQUID FORMULATIONS

Water, oil, or emulsion are all examples of liquid formulations [7]. The following elements are included in this: 35–65% carrier liquid (oil or water), 10–40% microorganisms, 5–15% dispersion agent, 3–8% surfactant, and 1–3% suspender are included in this formula. An ultra-low volume (ULV) suspension (SU) can be obtained, as can a suspension concentrate (SC) or an oil miscible flowable concentrate (OF) optical density (OD). Active ingredients with low water solubility and stability are added to hydrolysis before hydrolysis [31]. An emulsion of the active ingredient(s) that can be used after diluting in an organic liquid is known as an "Operational Fluid." Spraying formulations in a fine mist is accomplished using ultra-low-voltage equipment [32]. There is an oil or solvent-based medium in which the active ingredients are suspended. Soybean oil was used as a carrier for *T. asperellum* spores in the fight against cacao black pod disease. Liquid *Trichoderma* products include Trichorich-L, Trichojet, and Enpro-Derma [28]. Oil-based bioformulations used as foliar sprays can enhance entomopathogen activity [33].

12.6 MICROBIAL BIOFORMULATION-BASED PLANT BIOSTIMULANTS (PBS)

12.6.1 BACTERIAL PLANT BIOSTIMULANTS (PBS), BENEFICIAL EFFECTS, AND MODE OF ACTION

Bacteria and plants have been observed to interact in a variety of ways. Symbiosis can take on a variety of forms, from short-term partnerships to long-term partnerships, and from bacterial niches that extend from a substrate into a cell's interior to functions that impact a wide range of variables, such as the ability to participate in biogeochemical cycles and the availability of nutrition. Much recent research has focused on the potential benefits of bacterial association with plants, particularly as tools for promoting plant development and maintaining the health of the soil and crops [34]. Bacteria that promote plant growth can be found in a wide range of habitats and are commonly associated with many agricultural plant species, if not all. Plant growth-promoting rhizobacteria (PGPR) are the most well-studied type of PGPB, and they typically colonize the root surfaces and are closely attached to the soil-to-soil interface (the rhizosphere) [35]. According to recent studies, numerous PGPR can penetrate the root and establish endophytic relationships. Endophytes (those that live inside stems, tubers, leaves, and other organs) can even cross the endodermis barrier, allowing them to move from the vascular system to the root cortex [36].

The extent of endophytic connections in host plant tissues (and/or organs) can indicate how well these bacteria can adapt to various ecological niches. Despite the fact that soil bacteria have been shown to play a role in stimulating and creating plant development, it is still unclear how the bacteria accomplish this. Because of the recently developed "omics" techniques, the molecular basis for the association mechanisms between agricultural plants and bacteria that induce/stimulate physiological changes is beginning to be understood. Enhancing plant root surfaces, phosphorus solubility, nitrogen fixation, HCN production, and siderophores are just a few of the methods used to help plants absorb nutrients. Detailed information on each of these mechanisms can be found in the following sections. Promoting the growth and acquisition of nutrients for plants could benefit greatly from the ability to control the behavior of bacteria. Root development, mineral availability, and efficient nutrient use in the rhizosphere of crops are encouraged by PGPR formulations, which are an important class of biostimulants. Several PGPR have been found to stimulate phytohormone synthesis in various ways. PGPR releases hormones that induce long roots, increase root biomass, decrease stomata density, provide protection, and wall-related transcription modifications, in addition to activating auxin response genes that promote plant development [37] (Figure 12.1). Plants can't get their hands on the most important macronutrient, nitrogen (N), because it's inert. The PGPR nitrogen fixes atmospheric nitrogen (N_2) into ammonia, which agricultural plants can use to boost productivity.

Stimulants with PGPR, which release organic acids, increase soil fertility and productivity by making insoluble phosphate more accessible to crop plants. *Pseudomonas, Rhizobium, Bacillus,* and *Enterobacter* are some of the most potent P-solubilizers found in bacteria [38].

The rhizospheric deficiency of crops has always been a major obstacle to crop production because of potassium's fundamental importance to crop growth and improvement [39]. It is common for bacteria to produce siderophores or organic acids as a strategy for increasing iron (Fe) availability. Agriculture could benefit from environmentally friendly modernization by using plant biostimulants (PBs), such as AgriLife's commercial preparation of the Acidithiobacillus ferrooxidans genus. This preparation promotes flowering, fruit formation, and crop production in plants. Around 60-70% of the yield deficit caused by climate change is due to abiotic stressors, such as inadequate nutrient availability, salinity, drought, hypoxia, and heat stress. A good agronomic technique for increasing crop tolerance to difficult environmental conditions and deficient soil conditions, and reducing the negative effects of unfavorable conditions, is the use of PBs in these situations [40]. There is a biological solution to the water shortage in crop production in the form of plant growth-promoting rhizobacteria (PGPR). Drought-stressed plants can respond by increasing root changes, producing phytohormones, volatile chemicals, ACCD, exopolysaccharides, and antioxidants (AOX), controlling osmolytes and stress-responsive genes, and more [41].

12.6.2 FUNGAL-BASED BIOSTIMULANTS

Microbial endophytes parasitism has also been reported to be a mutualistic relationship between beneficial fungi and plant roots [42]. To improve ground and surface water contamination, plants and fungi have coevolved to establish the parasitism-mutualism continuum, which in turn enhances crop growth and yield [43]. The use of fungal biostimulants, such as mycorrhizal fungi (MF) and *Trichoderma* fungi, can increase crop yields in an environmentally friendly and sustainable manner, thereby reducing greenhouse gas emissions [44]. Due to its versatility as a fungal biostimulant used to mitigate the negative effects of stresses (both natural and man-made) on vegetable crops, *Trichoderma* species have become increasingly important as a biostimulant for plants beyond their traditional role in encouraging growth [45] (Table 12.1).

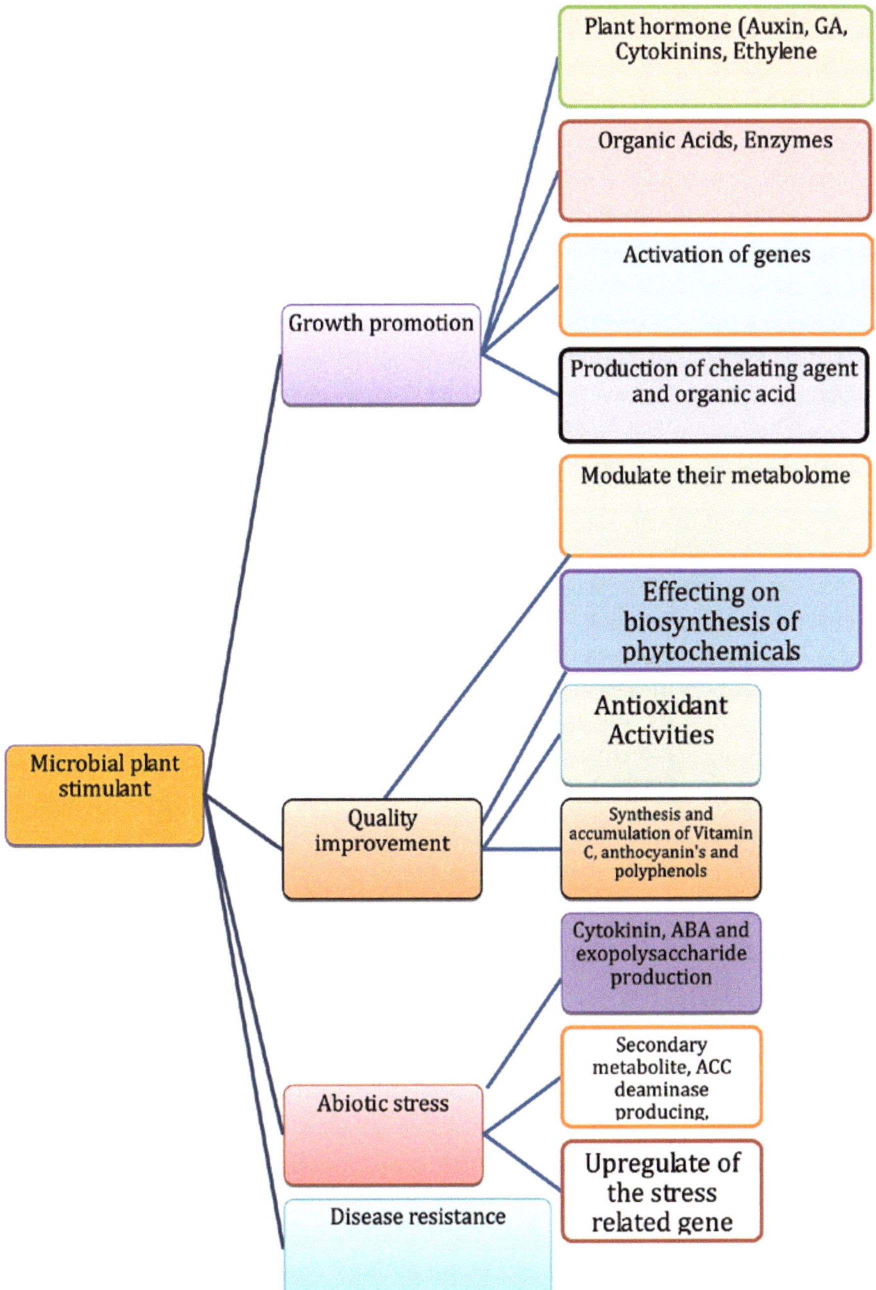

FIGURE 12.1 Mode of action of microbial biostimulant on the growth of crop plants.

TABLE 12.1 Mode of Action of Microbial Biostimulant on the Growth of Crop Plants

Report Attribute	Details
Market size value in 2022	USD 3.08 billion
Revenue forecast in 2030	USD 6.79 billion
Growth rate	CAGR of 10.4% from 2022 to 2030
Base year for estimation	2021
Historical data	2018–2020
Forecast period	2022–2030
Quantitative units	Revenue in USD million/billion, volume in million hectares, and CAGR from 2022 to 2030.
Report coverage	Revenue forecast, volume forecast, company ranking, competitive landscape, growth factors, and trends.
Segments covered	Active ingredients, crop type, application, region.
Regional scope	North America; Europe; Central and South America; Middle East and Africa; Asia Pacific.
Country scope	U.S.; U.K.; Germany; Australia; Brazil; Italy; China; India; Japan.
Key companies profiled	BASF SE; Italpollina SAP; Novozymes A/S; Sapec Agro S.A.; Biolchim SpA; Platform Specialty Products Corp.; Koppert B.V.; Biostadt India Ltd.; Isagro Group; Valagro SpA.
Customization scope	Free report customization (equivalent up to 8 analysts working days) with purchase. Addition or alteration to country, regional and segment scope.
Pricing and purchase options	Avail customized purchase options to meet your exact research needs.

Strains of *Trichoderma,* which have demonstrated their biostimulatory function through developmental processes and promoting plant growth in stressful environments, have shown the ability to counter a wide range of plant pathogens, including certain bacterial diseases and phytopathogenic fungi [46, 47]. The biostimulatory function of *Trichoderma* spp. has been shown to improve root-to-shoot communication in a study. Other volatile compounds, such as indole-3-acetaldehyde and indole-3-carboxaldehyde, have also been found to have a positive effect on plant growth. The presence of these phytochemicals has resulted in improved root, shoot, and branch architecture, as well as increased agricultural productivity, attributed to their auxin-like activity. These phytochemicals have demonstrated a wide range of benefits [48]. Additionally, mycorrhizal fungi (MF) have been found to form symbiotic relationships with over 90% of the world's plant species [49]. The most common type of mycorrhiza associated with economically significant agricultural crops is Arbuscular Mycorrhiza (AMF). AMF forms a mutual association with plants by penetrating the root cortical cells, resulting in the formation of highly branched structures called arbuscules [47]. AMF has been shown to have a significant impact on sustainable agriculture by increasing nutrient uptake, improving plant growth and development, and maintaining ion homeostasis under various stress conditions (both biotic and abiotic) [50]. The biostimulatory properties of AMF have been widely accepted in recent years, and the hyphal network of AMF has been credited with facilitating communication between individual plants within a plant community [42]. Many growers utilize beneficial fungi and their products to enhance nutrient uptake and increase plant resistance to various stresses, leading to improved plant growth and productivity. Biotrophicity and a lack of suitable Microbial bioformulation-based PBs 213 make large-scale exploitation of these organisms more difficult. However, the biotrophic nature of these fungi and the lack of suitable microbial bioformulations make large-scale exploitation of these organisms challenging. These factors also influence host specificity and population dynamics in different agro-ecosystems [48]. While the ability of AMF to colonize plant roots is well-known, other fungal biostimulants, such as *Trichoderma* spp. (Ascomycota) and Sebacinales (Basidiomycota), have also been shown to perform this function without direct contact with plants, enhancing plant nutrient uptake and modulating their antioxidant [51] Plant growth and productivity are increased by fungus-based formulations like Sebacinales and *Trichoderma* spp. *Trichoderma* spp. are well known for their biocontrol and biopesticide properties that increase plants' innate immunity. Many biotechnological applications have found use for the enzymes found in *Trichoderma* spp. [52]. More crop plants should be treated with microorganisms, and better crop management practices should be used to maximize the potential of beneficial fungi. The use of metagenomics to develop new sustainable agricultural strategies that increase crop growth and productivity while decreasing the need for chemical fertilizers is an exciting new tool in the metagenomics toolbox.

12.7 MECHANISMS IMPLICATED IN PLANT BIOSTIMULATORY EFFECTS ON CROP PRODUCTIVITY

Even though modern biotechnological tools have made it easier to decipher plant physiological changes, the majority of these advancements are only applicable to model plants and artificial environments. In the year 2014

[53], we now face a new challenge: applying these insights to the precise structural and functional annotation of biostimulatory effects on various agriculturally important plants (Figure 12.2) using meta and functional genomics techniques (Figure 12.2). The development of high-throughput phenomics techniques for identifying mutants generated in plants as a result of this interaction has improved our comprehension of how bacterial or fungal biostimulants affect different plant genotypes that thrive in less-than-ideal environmental conditions [54].

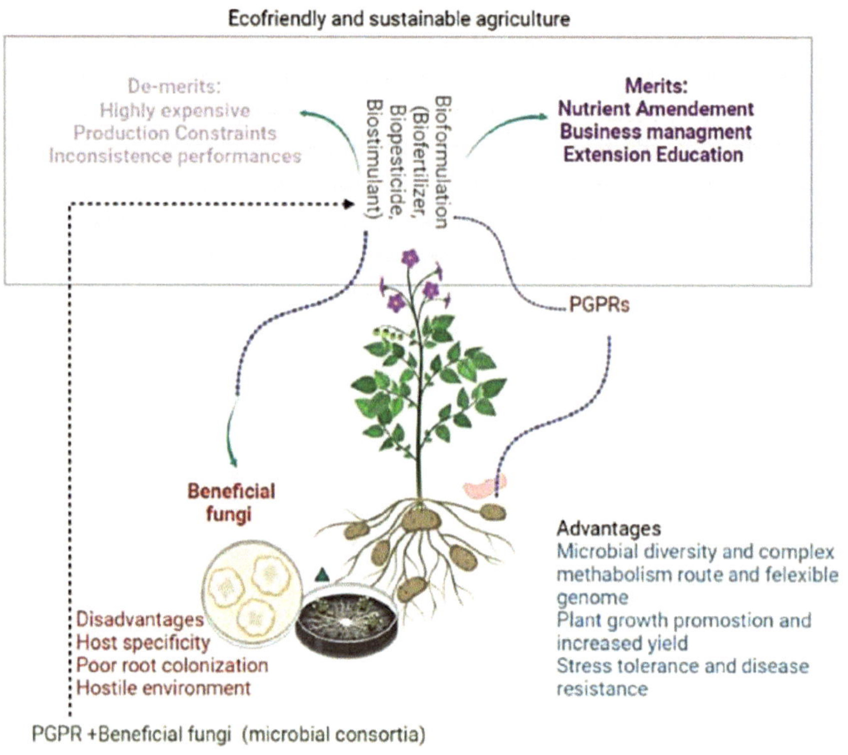

FIGURE 12.2 Strategies for research and development of bioformulation technology.

To bridge the gap between the biostimulatory effect of beneficial microorganisms and their interaction with plants, it is crucial to utilize data from both laboratory and field studies (Figure 12.3). In the past, traditional pharmacological methods have been employed to develop biostimulants, involving the evaluation of potential microorganisms in both controlled laboratory settings and real-world field conditions [55].

Although biostimulant production has become more efficient due to careful selection processes, the high cost associated with it poses challenges in terms of marketability [56]. By observing and characterizing beneficial microorganisms and their biostimulatory impact on various plants, field data can be collected to aid in the development of biostimulants. Notably, scientists have documented detailed field observations of different beneficial microorganisms and their respective interactions, such as their influence on the growth and development of maize plants or bacterial and fungal populations [57]. These discoveries contribute to our understanding of how plant variation, whether inter- or intra-specific, affects the interactions with plant growth-promoting rhizobacteria (PGPRs). The standardization and commercialization of local microbiota, rather than synthetic microbial products, are being pursued. In cases where microbial biostimulants alone are insufficient to maintain the activity of beneficial microorganisms in the rhizosphere, this mechanism serves as a "bridge" [15, 16]. To ensure the widespread use of biostimulants in the modern era, it is essential to monitor their effectiveness in terms of interactions, long-term effects, environmental impact, and adaptation to local and temporal biological cycles. Field and landscape users should have access to these monitoring tools. Additionally, factors such as the timely availability of goods influence the effective adoption of this technology in developing countries like India. Agricultural extension programs should also demonstrate the positive effects of biostimulants, as promising high returns and increased crop productivity can persuade many small-scale farmers.

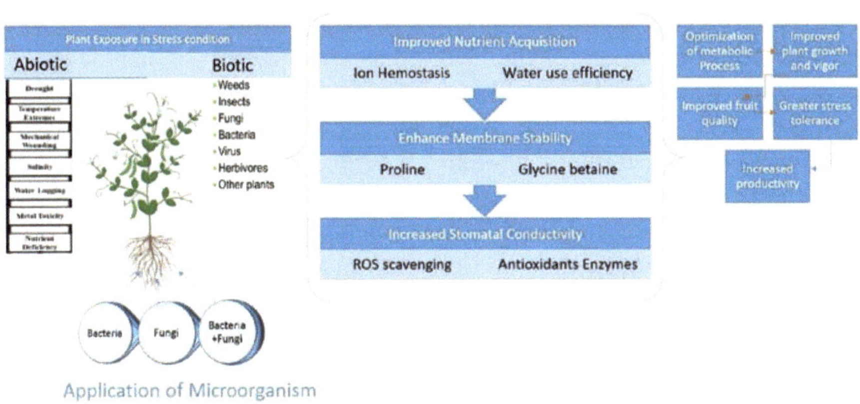

FIGURE 12.3 Key biostimulatory mechanisms targeted by beneficial bacteria and fungi upon interaction with plants.

12.8 CURRENT SCENARIO/MARKET TRENDS

The type of bioformulated product must be taken into account in order to successfully market bioformulated products that result in commercialization because the market demands products that are both effective and easy to use. The worldwide use of microbe-based formulations for the promotion of plant growth and suppression of disease has to be encouraged [58]. However, consumers are discouraged from using bioformulated products because of the confusing and inconsistent names given to them. In both "biofertilizers" and "bioinoculants," living organisms (microbes) are used to promote plant growth, increase nutrient uptake, and boost crop productivity. Most developed countries, on the other hand, use biofertilizer and biopesticides to boost crop yields. During the global economic recovery of 2021, the Microbial Biostimulants market is expected to undergo major changes. For the period from 2022 to 2030, the global biostimulants market is expected to grow at a compound annual growth rate (CAGR) of 10.4%. Environmental regulations on the use of chemicals in agriculture and the need for sustainable and eco-friendly alternatives to boost agricultural growth and productivity are cited as the main causes of this trend. Substances or microorganisms can be added to plants to enhance their characteristics and increase yield. As people become more aware of the potentially harmful long-term effects of synthetic chemicals used in plant growth, the demand for this product is rising in tandem with it. The market is expanding due to the increasing demand for organic and natural products. In 2021, Europe accounted for more than 37.5% of the global market's total revenue. This can be attributed to consumers' growing preference for organic foods and manufacturers' increasing preference for environmentally friendly production methods. Asia Pacific, with its numerous manufacturing companies and the rapid growth of agriculture-based economies like India and China, was the second-largest regional market in 2021. The growing population, soil erosion, increasing awareness among farmers, as well as strict government regulations and encouragement to use biostimulants, have led to a growing demand for biostimulants. According to a study, India has 151 biofertilizer manufacturing facilities run by both public and private organizations. Mahajan & Gupta [59] found that 95% of the microorganisms used in biopesticides were based on Bt. The United States is the world's largest market for biopesticides, with an annual average growth rate of 15.0%. Europe, on the other hand, is the fastest-growing regional market, with an average annual growth rate of 15.0%[32].

12.9 REGULATORY FRAMEWORK

The regulatory environment for biostimulants is currently complex. Regulatory agencies might be reluctant to recognize biostimulants as a valid class of drugs in the absence of a clear definition. A formal legal framework for biostimulants is lacking in many EU and non-EU nations because there is no agreed-upon definition. The two main marketing channels for biostimulants in Europe are EU biostimulant regulations (2018) and European pesticide laws, which are governed by both European and national laws, according to Caradonia et al. [60].

The status of substances as biostimulants differs under REACH (EC No. 1907/2006), which regulates the registration, evaluation, authorization, and restriction of chemicals. The regulatory status of biostimulants is influenced by various factors, including their registration under the REACH regulation, their classification as fertilizers or pesticides under national law, their authorization or lack thereof in organic production, and other variables. However, REACH primarily focuses on chemicals and disregards microorganisms, a significant class of biostimulants. In contrast, EU Regulation No. 1107/2009 on Plant Protection Products defines biostimulants more broadly ("PPPs"). *Trichoderma* spp. is an example of a biocontrol agent that utilizes PPPs and has obtained federal approval and registration [28]. In addition to controlling plant diseases, *Trichoderma* fungi inoculation offers advantages such as enhancing systemic resistance, stimulating root and shoot growth, and facilitating nutrient absorption and release. Various strains of this beneficial fungus have been employed as biocontrol agents in commercial bioformulations [61].

Thanks to biostimulants like arbuscular mycorrhizal fungi (AMF), where plants can develop systemic resistance to disease and pest attacks [62]. This is based on Ref. [63]. Similarly to Europe, there is no uniform definition of biostimulants across the 50 states, and some biostimulants are regulated by state-specific fertilizer regulations. According to the American Association of Plant Food Control Officials, one of the organization's primary responsibilities is to define and recognize various types of fertilizers. In the future, we will learn more about the regulation of biostimulants. Currently, the Environmental Protection Agency may regulate some biostimulants while registering others as fertilizers or soil amendments at the state level. A common legal framework has been established for plant biostimulants (PBs), which are currently dispersed across EU member states. Recent amendments to fertilizer regulations have been made to facilitate internal marketing operations for fertilizer products [2, 60, 64–69]. According to the recently established regulation, PBs will be CE marked as fertilizing products that stimulate the nutritional aspects and processes of the plant independently of the products' nutrient content, with the sole aim of improving one or more of the following characteristics of the plant and the plant rhizosphere or phyllosphere: improved nutrient use efficiency, crop quality and yield, as well as increased accessibility to cons. Since the group includes a variety of inorganic and organic substances or microorganisms, such as protein hydrolysates, humic acids, and seaweed extracts, as well as PGPRs and endophytic fungi, N-fixing bacteria, and MF, it is best to define PBs by their proposed effects, or more specifically, "by the plant response they elicit rather than by their makeup" [70–72].

12.10 ALGA-BASED BIOSTIMULANT

Microalgae are categorized mostly based on their cell structure, life cycle, and pigmentation. It is believed that there are 800,000 different species of microalgae, of which only 50,000 have been identified. This vast diversity of species offers a wide range of potential uses. In practice, it is possible to select specific strains that have different biochemical contributions and can thrive in various conditions. Scientists classify algae into four general types based on their morphological and size properties: unicellular, colonial, filamentous, and multicellular. All of these types fall within the algae category. Following this classification, algae are further divided into macroalgae and microalgae. Macroalgae are multicellular, macroscopic organisms with a maximum length of 65 m. On the other hand, microalgae are tiny, unicellular, or filamentous organisms with sizes ranging from 1 to 900 m. The use of microalgae extracts as biostimulants (MBS) and biofertilizers (MBF) in agricultural production has been increasing. A growing number of agrochemical companies are incorporating microalgal extracts into their novel MBS and MBF formulations, which are now commercially available in various forms. Microalgal extracts, in particular, offer several advantageous benefits, including improved crop production and quality, enhanced nutrient absorption, and increased resistance to abiotic stressors [73–75]. However, to better understand how and when microalgal extracts can be used in crop production to enhance agricultural sustainability, several issues still need to be addressed. Further research is needed on microalgae productivity, as well as the preparation and use of macro- or microalgal extracts:

- There are several macro- or microalgae strains; hence, various MBS and MBF can be suggested.
- Several abiotic variables can affect the biochemical composition of microalgae during their development.
- Even after processing the same microalga, numerous MBS and MBF may be produced, each with a distinct biochemical composition and set of properties.
- Even the same MBS and MBF might have distinct effects on various crop species.
- The amount that should be applied, as well as how frequently, as MBS and MBF.
- When to apply MBS and MBF to crops may depend on their phenological stage.
- The bioactive substances in the MBS and MBF may, at various doses, activate many physiological processes.
- To fully understand the impact of MBS and MBF on crop yields, it is necessary to conduct agronomic, physiological, chemical, biochemical, and molecular investigations. These investigations will provide detailed information on the optimal preparation and utilization of MBS and MBF. By utilizing this information, specific procedures can be developed to assist farmers and agrochemical industries in the effective development and application of MBS and MBF, respectively..

12.11 CONCLUSION

Feeding a growing population remains a significant challenge despite agriculture dealing with a variety of complex issues brought on by environmental changes as a result of global climate change. Due to the shifting of typical climatic patterns, such as water budgets, this global phenomenon contributes to and exacerbates already existing abiotic stresses, including droughts, floods, salinization, and temperature extremes. Due to weeds and phytopathogens shifting their distribution patterns, there are fewer beneficial microbial organisms associated with plants and influencing plant health, and as a result, plants are more susceptible to biotic stress. In order to ensure a sufficient yield and to manage pests, farming increasingly relies on chemical fertilizers and pesticides, which have negative environmental effects. As a result, the creation of environmentally sound substitutes for PGPR PBs has attracted attention on a global scale in recent years. The PB market is anticipated to grow quickly in the near future, with an expected exponential growth rate. When used on horticultural and agricultural crop plants, PGPR-based BPBs have demonstrated efficacy in the use of nutrition, the mitigation of abiotic/biotic stress, and/or crop quality characteristics (fruits, vegetables, ornamental plants, and medicinal plants). To increase the availability of soil components like iron, phosphorus, potassium, and zinc, PGPR controls the release of these substances as well as the production of siderophores, organic acids, and enzymes. Microbial biostimulants can be used in agricultural systems to increase human inputs while offering a long-term, affordable solution to plant productivity losses brought on by changing climates and other environmental factors. To make sure that this methodology can be broadly applied to a variety of crops, regions, and environmental conditions, experiments using microbial biostimulants should be widely disseminated to pertinent policymakers and stakeholders, such as extension services and growers.

KEYWORDS

- **agriculture**
- **algal biostimulants**
- **bioformulation**
- **crop productivity**
- **microbial bioformulation**
- **plant biostimulation effects**
- **sustainableagricultural strategies**

REFERENCES

1. Hamid, B., Zaman, M., Farooq, S., Fatima, S., Sayyed, R. Z., Baba, Z. A., Sheikh, T. A., et al., (2021). Bacterial plant biostimulants: A sustainable way towards improving growth, productivity, and health of crops. *Sustainability, 13*(5), 2856.
2. Rouphael, Y., & Colla, G., (2018). Synergistic biostimulatory action: Designing the next generation of plant biostimulants for sustainable agriculture. *Frontiers in Plant Science,9*, 1655.
3. Rouphael, Y., & Colla, G., (2020). Toward a sustainable agriculture through plant biostimulants: From experimental data to practical applications. *Agronomy, 10*(10), 1461.

4. Rani, U., & Kumar, V., (2019). Microbial bioformulations: Present and future aspects. In: *Nanobiotechnology in Bioformulations* (pp. 243–258). Springer, Cham.
5. Ardakani, S. S., Heydari, A., Khorasani, N., & Arjmandi, R., (2010). Development of new bioformulations of *Pseudomonas fluorescens* and evaluation of these products against damping-off of cotton seedlings. *J. Plant Pathol.*, 83–88.
6. Jorjani, M., Heydari, A., Zamanizadeh, H. R., Rezaee, S., & Naraghi, L., (2011). Development of *Pseudomonas fluorescens* and *Bacillus coagulans* based bioformulations using organic and inorganic carriers and evaluation of their influence on growth parameters of sugar beet. *J. Biopesticides, 4*(2), 180.
7. Schisler, D. A., Slininger, P. J., Behle, R. W., & Jackson, M. A., (2004). Formulation of *Bacillus* spp. for biological control of plant diseases. *Phytopathology*, *94*(11), 1267–1271.
8. Lugtenberg, B., & Kamilova, F., (2009). Plant-growth-promoting rhizobacteria. *Annu. Rev. Microbiol.*, *63*, 541–556.
9. Lin, Y. H., Chen, Y., Smith, T. C., Karna, S. R., & Seshu, J., (2018). Short-chain fatty acids alter metabolic and virulence attributes of *Borrelia burgdorferi*. *Infection and Immunity*, *86*(9), e00217–18.
10. Postler, T. S., & Ghosh, S., (2017). Understanding the holobiont: How microbial metabolites affect human health and shape the immune system. *Cell Metabolism*, *26*(1), 110–130.
11. Stevens, J. F., & Maier, C. S., (2016). The chemistry of gut microbial metabolism of polyphenols. *Phytochemistry Reviews*, *15*(3), 425–444.
12. Wen, L., Ley, R. E., Volchkov, P. Y., Stranges, P. B., Avanesyan, L., Stonebraker, A. C., & Chervonsky, A. V., (2008). Innate immunity and intestinal microbiota in the development of type 1 diabetes. *Nature*, *455*(7216), 1109–1113.
13. Mayorgas, A., Dotti, I., & Salas, A., (2021). Microbial metabolites, postbiotics, and intestinal epithelial function. *Molecular Nutrition & Food Research*, *65*(5), 2000188.
14. Ehlers, R. U., (2006). Einsatz der biotechnologie im biologischen pflanzenschutz. *Schriftenreihe der Deutschen Phytomedizinischen Gesellschaft,* eV. 8, 17–31.
15. Kumar, A., Verma, H., Singh, V. K., Singh, P. P., Singh, S. K., Ansari, W. A., & Pandey, K. D., (2017a). Role of *Pseudomonas* sp. in sustainable agriculture and disease management. In: *Agriculturally Important Microbes for Sustainable Agriculture* (pp. 195–215). Springer, Singapore.
16. Kumar, R., Singh, S. K., Kumari, A., Kumar, A., & Yadav, S., (2017b). *In vitro* evaluation of various bio-control agents against *Fusarium solani* isolated from papaya (*Carica papaya* L.). *Int. J. Curr. Microbiol. App. Sci.*, *6*(4), 40–50.
17. Mehnaz, S., (2016). An overview of globally available bioformulations. *Bioformulations: For Sustainable Agriculture*, 267–281.
18. Durgadevi, D., Srivignesh, S., & Sankaralingam, A., (2018). Effect of consortia bioformulation of rhizobacteria on induction of systemic resistance in tuberose against peduncle blight disease. *Int. J. Bio-Res. Stress Manag.*, *9*(4), 510–517.
19. Naraghi, L., Negahban, M., Heydari, A., Razavi, M., & Afshari-Azad, H., (2018). Growth inhibition of *Fusarium oxysporum* f. sp. lycopercisi, the causal agent of tomato *Fusarium* WILT disease by nanoformulations containing talaromyces flavus. *Ekoloji*, *27*(106), 103–112.

20. Larena, I., Melgarejo, P., & De Cal, A., (2003). Drying of conidia of *Penicillium oxalicum*, a biological control agent against *Fusarium* wilt of tomato. *J. Phytopathol.*, *151*(11, 12), 600–606.
21. Abadias, M., Teixidó, N., Usall, J., Solsona, C., & Viñas, I., (2005). Survival of the postharvest biocontrol yeast candida sake CPA-1 after dehydration by spray-drying. *Biocontrol Scienceand Technology*, *15*(8), 835–846.
22. Guijarro, B., Melgarejo, P., & De Cal, A., (2007a). Effect of stabilizers on the shelf-life of *Penicillium* frequentans conidia and their efficacy as a biological agent against peach brown rot. *Int. J. Food Microbiol.*, *113*(2), 117–124.
23. Guijarro, B., Melgarejo, P., Torres, R., Lamarca, N., Usall, J., & De Cal, A., (2007b). Effects of different biological formulations of*Penicillium* frequentans on brown rot of peaches. *Biol. Control*, *42*(1), 86–96.
24. Brar, S. K., Verma, M., Tyagi, R. D., & Valéro, J. R., (2006). Recent advances in downstream processing and formulations of *Bacillus thuringiensis* based biopesticides. *Process Biochem.*, *41*(2), 323–342.
25. Navon, A., (2000). *Bacillus thuringiensis* application in agriculture. In: *Entomopathogenic Bacteria: From Laboratory to Field Application* (pp. 355–369). Springer, Dordrecht.
26. Batta, Y. A., (2008). Control of main stored-grain insects with new formulations of entomopathogenic fungi in diatomaceous earth dusts. *International Journal of Food Engineering*, *4*(1).
27. Young, S. D., Townsend, R. J., Swaminathan, J., & O'Callaghan, M., (2010). Serratia entomophilacoated seed to improve ryegrass establishment in the presence of grass grubs. *New Zealand Plant Protection*, *63*, 229–234.
28. Woo, S. L., Ruocco, M., Vinale, F., Nigro, M., Marra, R., Lombardi, N., & Lorito, M., (2014). *Trichoderma*-based products and their widespread use in agriculture. *The Open Mycology Journal*, *8*(1).
29. Kaur, R., Kaur, J., & Singh, R. S., (2011). Nonpathogenic *Fusarium* as a biological control agent. *Plant Pathol. J.*, *9*(3), 79–91.
30. Falk, S. P., Gadoury, D. M., Pearson, R. C., & Seem, R. C., (1995). Partial control of grape powdery mildew by the mycoparasite *Ampelomyces quisqualis*. *Plant Dis.*, *79*(5), 483–490.
31. Tadros, T., (2013). *Encyclopedia of Colloid and Interface Science*. Springer.
32. Singh, K. N., & Merchant, K., (2012). The agrochemical industry. In: Kent, J. A., (ed.), *Handbook of Industrial Chemistry and Biotechnology* (pp. 643–699). Springer Science Business Media, New York.
33. Feng, M. G., Pu, X. Y., Ying, S. H., & Wang, Y. G., (2004). Field trials of an oil-based emulsifiable formulation of *Beauveria bassiana* conidia and low application rates of imidacloprid for control of false-eye leafhopper *Empoasca vitis* on tea in Southern China. *Crop Prot.*, *23*(6), 489–496.
34. Mishra, J., & Arora, N. K., (2016). Bioformulations for plant growth promotion and combating phytopathogens: A sustainable approach. In: *Bioformulations: For Sustainable Agriculture* (pp. 3–33). Springer, New Delhi.
35. Papik, J., Folkmanova, M., Polivkova-Majorova, M., Suman, J., & Uhlik, O., (2020). The invisible life inside plants: Deciphering the riddles of endophytic bacterial diversity. *Biotechnology Advances*, *44*, 107614.

36. Nguyen, M. L., Spaepen, S., Du Jardin, P., & Delaplace, P., (2019). Biostimulant effects of rhizobacteria on wheat growth and nutrient uptake depend on nitrogen application and plant development. *Archives of Agronomy and Soil Science*, *65*(1), 58–73.
37. Spaepen, S., Bossuyt, S., Engelen, K., Marchal, K., & Vanderleyden, J., (2014). Phenotypical and molecular responses of *Arabidopsisthaliana* roots as a result of inoculation with the auxin-producing bacterium *Azospirillum brasilense*. *New Phytologist*, *201*(3), 850–861.
38. Ji, J., Yuan, D., Jin, C., Wang, G., Li, X., & Guan, C., (2020). Enhancement of growth and salt tolerance of rice seedlings (*Oryza sativa* L.) by regulating ethylene production with a novel halotolerant PGPR strain *Glutamicibacter* sp. YD01 containing ACC deaminase activity. *Acta Physiologiae Plantarum*, *42*(4), 1–17.
39. Suleman, M., Yasmin, S., Rasul, M., Yahya, M., Atta, B. M., & Mirza, M. S., (2018). Phosphate solubilizing bacteria with glucose dehydrogenase gene for phosphorus uptake and beneficial effects on wheat. *PloS One*, *13*(9), e0204408.
40. Abbas, R., Rasul, S., Aslam, K., Baber, M., Shahid, M., Mubeen, F., & Naqqash, T., (2019). Halotolerant PGPR: A hope for cultivation of saline soils. *Journal of King Saud University-Science*, *31*(4), 1195–1201.
41. Wen, W., Timmermans, J., Chen, Q., & Van, B. P. M., (2020). A review of remote sensing challenges for food security with respect to salinity and drought threats. *Remote Sensing*, *13*(1), 6.
42. Behie, S. W., & Bidochka, M. J., (2014). Nutrient transfer in plant-fungal symbioses. *Trends Plant Sci.*,*19*, 734–740.
43. Johnson, N. C., & Graham, J. H., (2013). The continuum concept remains a useful framework for studying mycorrhizal functioning. *Plant Soil, 363*, 411–419.
44. Johnson, D., & Gilbert, L., (2015). Interplant signaling through hyphal networks. *New Phytol.*,*205*, 1448–1453.
45. Simard, S. W., Beiler, K. J., Bingham, M. A., Deslippe, J. R., Philip, L. J., & Teste, F. P., (2012). Mycorrhizal networks: Mechanisms, ecology, and modelling. *Fungal Biol. Rev.*, *26*, 39–60.
46. Candido, V., Campanelli, G., Addabbo, T. D., Castronuovo, D., Renco, M., & Camele, I., (2013). Growth and yield promoting effect of artificial mycorrhization combined with different fertilizer rates on field-grown tomato. *Ital. J. Agron.*,*8*, 168–174.
47. Candido, V., Campanelli, G., D'Addabbo, T., Castronuovo, D., Perniola, M., & Camele, I., (2015). Growth and yield promoting effect of artificial mycorrhization on field tomato at different irrigation regimes. *Sci. Hortic.*,*187*, 35–43.
48. Colla, G., Rouphael, Y., Di Mattia, E., El-Nakhel, C., & Cardarelli, M., (2015). Co-inoculation of glomus intraradices and *Trichoderma atroviride* acts as a biostimulant to promote growth, yield, and nutrient uptake of vegetable crops. *J. Sci. Food Agric.*, *95*, 1706–1715.
49. Sarkar, A., Asaeda, T., Wang, Q., & Rashid, M. H., (2015). Arbuscular mycorrhizal influences on growth, nutrient uptake, and use efficiency of *Miscanthus sacchariflorus* growing on nutrient-deficient river bank soil. *Flora Morphol. Distrib. Funct. Ecol. Plants, 212*, 46–54.
50. Colla, G., Rouphael, Y., Canaguier, R., Svecova, E., & Cardarelli, M., (2014). Biostimulant action of a plant-derived protein hydrolysate produced through enzymatic hydrolysis. *Front. Plant Sci.*,*5*, 1–6.

51. Mukherjee, P. K., Horwitz, B. A., Herrera-estrella, A., Schmoll, M., & Kenerley, C. M., (2012). *Trichoderma* research in the genome era. *Annu. Rev. Phytopathol., 51*, 105–129.
52. Nicolás, C., Hermosa, R., Rubio, B., Mukherjee, P. K., & Monte, E., (2014). *Trichoderma* genes in plants for stress tolerance-status and prospects. *Plant Sci., 228*, 71–78.
53. Przybysz, A., Gawronska, H., & Gajc-wolska, J., (2014). Biological mode of action of a nitrophenolates-based biostimulant: Case study. *Front. Plant Sci., 5*, 1–15.
54. Bahadur, I., Maurya, B. R., Meena, V. S., Saha, M., Kumar, A., & Aeron, A., (2017). Mineral release dynamics of tricalcium phosphate and waste muscovite by mineral-solubilizing rhizobacteria isolated from indo-gangetic plain of India. *Geomicrobiol J.,34*(5), 454–466.
55. Dotaniya, M. L., Meena, V. D., Basak, B. B., & Meena, R. S., (2016). Potassium uptake by crops as well as microorganisms. *Potassium Solubilizing Microorganisms for Sustainable Agriculture* (pp. 267–280). Springer, New Delhi.
56. Ghosh, P. K., Kumar De, T., & Maiti, T. K., (2015). Production and metabolism of indole acetic acid in root nodules and symbiont (*Rhizobium undicola*) isolated from root nodule of aquatic medicinal legume *Neptunia oleracea* Lour. *Journal of Botany, 2015.*
57. Jaiswal, D. K., Verma, J. P., Prakash, S., Meena, V. S., & Meena, R. S., (2016). Potassium as an important plant nutrient in sustainable agriculture, a state of the art. In: *Potassium Solubilizing Microorganisms for Sustainable Agriculture* (pp. 21–29).Springer, New Delhi.
58. Naderifar, M., & Daneshian, J., (2012). Effect of different nitrogen and biofertilizers effect on growth and yield of *Brassica napus* L. *International Journal of Agriculture and Crop Sciences (IJACS*), *4*(8), 478–482.
59. Mahajan, A., & Gupta, R. D., (2009). *Integrated Nutrient Management (INM) in a Sustainable Rice—Wheat Cropping System.* Dordrecht: Springer Netherlands.
60. Caradonia, F., Valerio, B., Lorenzo, R., Giacomo, P., & Anna La, T., (2019). Plant biostimulant regulatory framework: Prospects in Europe and current situation at international level. *Journal of Plant Growth Regulation, 38*(2), 438–448.
61. Lorito, M., & Woo, S. L., (2015). *Trichoderma*: A multi-purpose tool for integrated pest management. In: *Principles of Plant-Microbe Interactions* (pp. 345–353). Springer, Cham.
62. Rouphael, Y., Franken, P., Schneider, C., Schwarz, D., Giovannetti, M., Agnolucci, M., De Pascale, S., et al., (2015). Arbuscular mycorrhizal fungi act as biostimulants in horticultural crops. *Scientia Horticulturae, 196,* 91–108.
63. Cameron, D. D., Andrew, L. N., Saskia, C. M. V. W., & Jurriaan, T., (2013). Mycorrhiza-induced resistance: More than the sum of its parts? *Trends in Plant Science,18*(10), 539–545.
63. Rouphael, Y., Spíchal, L., Panzarová, K., Casa, R., & Colla, G., (2018). High-throughput plant phenotyping for developing novel biostimulants: From lab to field or from field to lab? *Frontiers in Plant Science*, *9*, 1197.
64. Nakajima, T., Katsumata, K., Kuwabara, H., Soya, R., Enomoto, M., Ishizaki, T., & Sugimoto, M., (2018). Urinary polyamine biomarker panels with machine-learning differentiated colorectal cancers, benign disease, and healthy controls. *International Journal of Molecular Sciences*, *19*(3), 756.
65. Gašić, S., & Tanović, B., (2013). Biopesticide formulations, possibility of application and future trends. *Pesticidi i Fitomedicina*, *28*(2), 97–102.

65. Rouphael, Y., Lucini, L., Miras-Moreno, B., Colla, G., Bonini, P., & Cardarelli, M., (2020). Metabolomic responses of maize shoots and roots elicited by combinatorial seed treatments with microbial and non-microbial biostimulants. *Frontiers in Microbiology, 11*, 664.
66. Aamir, M., Krishna, K. R., Andleeb, Z., Manish, K. D., Sunil, K., Vaishali, S., & Ram, S. U., (2020). "Microbial bioformulation-based plant biostimulants: A plausible approach toward next generation of sustainable agriculture. In: *Microbial Endophytes* (pp. 195–225). Woodhead Publishing.
69. EU, (2019). Regulation of the European parliament and of the council laying down rules on the making available on the market of EU fertilizing products and amending regulations (EC) No 1069/2009 and (EC) No 1107/2009 and repealing Regulation (EC) No 2003/2003. *Off. J. Eur. Union, 170*, 1–114.
70. Fiorentino, N., Ventorino, V., Woo, S. L., Pepe, O., De Rosa, A., Gioia, L., & Rouphael, Y., (2018). *Trichoderma*-based biostimulants modulate rhizosphere microbial populations and improve N uptake efficiency, yield, and nutritional quality of leafy vegetables. *Frontiers in Plant Science, 9*, 743.
71. González-Pérez, B. K., Rivas-Castillo, A. M., Valdez-Calderón, A., & Gayosso-Morales, M. A., (2022). Microalgae as biostimulants: A new approach in agriculture. *World Journal of Microbiology and Biotechnology, 38*(1), 1–12.
72. Jägermeyr, J., (2020). Agriculture's historic twin-challenge toward sustainable water use and food supply for all. *Frontiers in Sustainable Food Systems, 4*, 35.
73. Igor, N. C., Pedro, A. M., Pedro, H. R. A., Leonardo, V. T., Rodrigo, F. C., Jeiveison, G. S. S. M., João, B. V. B., & Mauro, S. G. P., (2022). Continuous design and economic analysis of a *Sargassum muticum* biorefinery process. *Bioresource Technology, 343*, 74.
74. Danilo, L., Giuseppe, C., & Barbara, D., (2022). Improving aerial and root quality traits of two landscaping shrubs stem cuttings by applying a commercial brown seaweed extract. *Horticulturae, 9*, 806.
75. Pushp, S. S., Mantin, E. G., Adil, M., Bajpai, S., Alan, T. C., & Balakrishnan, P., (2019). *Ascophyllum nodosum*-based biostimulants: Sustainable applications in agriculture for the stimulation of plant growth, stress tolerance, and disease management. *Frontiers in Plant Science, 10*, 35.
76. Burges, H. D., & Jones, K. A., (1998). Trends in formulation of microorganisms and future research requirements. *Formulation of Microbial Biopesticides* (pp. 311–332). Springer, Dordrecht.

CHAPTER 13

Microbiota in Edible Mushroom Industry: Disease Management, Yield, and Quality Improvement

IVANA POTOČNIK,[1] SVETLANA MILIJAŠEVIĆ-MARČIĆ,[1]
GABRIELLA KANIŽAI ŠARIĆ,[2] and IVANA MAJIĆ[2]

[1]*Institute of Pesticides and Environmental Protection, Belgrade-Zemun, Serbia*

[2]*Faculty of Agrobiotechnical Sciences Osijek, Josip Juraj Strossmayer University of Osijek, Osijek, Croatia*

ABSTRACT

A significant problem in the mushroom industry worldwide is a lack of effective chemicals in disease control. Pathogen resistance to fungicides and the harmful impact of the chemicals imply disease management based on biopesticides. Nevertheless, declining substrate quality due to the intensive production of its components, plants, and poultry manure indicate disturbed genuine microbiome. Microbiota from mushroom substrate has already exhibited significant effects both in disease control and yield improvement: bacteria (*Bacillus* spp., *Pseudomonas* spp.), fungi (*Scytalidium* spp.), and actinobacteria (*Streptomyces* spp., etc.). Many of them are registered as biopesticides or biostimulators as they suppress the pathogen and improve the yield by colonizing the host environment. The introduction of well-timed and appropriate microbial combinations would create new possibilities for biorational edible mushroom protection with improved yield and quality.

Microbial Biostimulants: Biorational Pesticides for Management of Plant Pathogens.
Faheem Ahmad and Rakesh Pandey (Eds.)

13.1 INTRODUCTION

Button mushroom [*Agaricus bisporus* (Lange) Imbach], oyster mushroom (*Pleurotus* spp.) and shiitake (*Lentinula edodes* (Berk) Pegler) are the most commonly cultivated edible fungi throughout the world. Mushrooms are produced on composted raw plant materials (hay, straw, corn stover, etc.), animal manure (horse, poultry), gypsum, and other ingredients. Making of appropriate substrate and integrated pest and disease control are major cost inputs in the mushroom industry [1]. The commercial cultivation of edible mushrooms is one of the most technologically demanding among all growing vegetables [2, 3]. Yield production of consistent quantity and quality is quite difficult [4]. The nature of the growing substrates indicates delicate mushroom cultivation which is seriously affected by various parasites.

The most important diseases of cultivated mushrooms are caused by viruses (La France and Mushroom Virus X) [5], pseudomonads bacteria (bacterial brown and ginger blotch, drippy gills and mummy disease) [6] and several mycopathogenic soil-inhabiting cosmopolitan fungi (dry bubble disease - *Lecanicillium (Verticillium) fungicola* (Preuss) Zare & W. Gams; wet bubble disease - *Mycogone perniciosa* (Magnus) Delacroix, cobweb disease - *Hypomyces* spp. and green mold disease - *Trichoderma* spp.) [7–10]. The most important pests are flies [dipteran families: Cecidomyiidae – *Mycophila* spp., *Heteropeza* spp.), Phoridae – *Megaselia halterata* (Wood), and Sciaridae – *Lycoriella* spp., *Bradysia* spp.], mites (gamasids, tarsonemids, tyroglyphids, and pyemotid or red pepper mites), and nematodes (genus Caenorhabditis, family Cephalobidae) [11]. The dominant problematic species have regional, annual, and seasonal variations [12]. The greatest impact of pests is their role as disease vectors, especially for sticky spores of bubble diseases (*L. fungicola* and *M. perniciosia*) or green mold disease *Trichoderma* spp. [12–14]. Sciarid larvae have a very low economic threshold and therefore pest control must enable their complete absence [15].

Disease and pest control involves prophylactic steps in mushroom farms: strict sanitation measures, and usage of disinfectants and pesticides. Chemical pesticides are harmful to non-target organisms, people, and the environment. According to the European Commission Regulation 1107/2009, the application of bio-rational compounds gives other options than chemical pesticides [16]. Ongoing pesticide reviews have resulted in the withdrawal of many chemicals from the market and only bioinsecticides are approved for use in mushroom cultivation in many countries. Major considerations for mushroom growers are lack of effective pesticides and resistance development of pathogens to chemicals [17]. A promising alternative to chemical control in mushroom production is to make their environment in accordance with the natural one – by application of antagonistic microorganisms, bacteria, actinobacteria, fungi, and nematodes.

13.2 MUSHROOM DISEASES

13.2.1 FUNGAL DISEASES OF MUSHROOMS

Fungal pathogens of edible mushrooms have been known since the early days of commercial mushroom production: *L. fungicola* (dry bubble), *M. perniciosa* (wet bubble), *Cladobotryum* (*Dactylium*) spp. (cobweb disease) and *Trichoderma* spp. (green mold disease) [7–10]. They cause serious fungal diseases on button and oyster mushrooms. In the past two decades green mold, caused by *Trichoderma* spp., is the most serious fungal disease out of all three main edible mushroom species, reducing yield between 60% and 100%. Button mushroom is affected by Trichoderma *aggressivum* Samuels & W. Gams (Hypocreales: Hypocreaceae) [10], oyster mushroom by *Trichoderma pleurotum* S.H. Yu & M.S. Park and *Trichoderma pleuroticola* S.H. Yu & M.S. Park, and shiitake by *Trichoderma harzianum* Rifai species complex (THSC) [18]. There are two main varieties of dry bubble disease agents of button and oyster mushrooms: *L. fungicola* var. fungicola in Europe and *Lecanicillium fungicola* var. *aleophilum* W. Gams & Zaayen 1982 in North America [19]. Australian *L. islae* is resistant to the most commonly used fungicide prochloraz and is highly aggressive to the host [17]. Symptoms and disease incidence depend on the mushroom growth at the moment of the pathogen penetration [7]. The primary source of dry bubble infection is casing soil [20]. Fungal spores require 7–14 days for symptoms to appear from the moment of infection. Besides *A. bisporus*, only two oyster mushroom species, *Pleurotus eryngii* (D.C.) Quél. 1886 and *Pleurotus nebrodensis* (Inzenga) Quél. 1886, are susceptible to *M. perniciosa*, a causal agent of wet bubble disease [21]. The pathogen grows superficially but causes immense cytological changes inside the fruiting bodies of mushrooms. When the mushroom pins are infested, large and tumorous lumps are formed. Tear drops are also formed on the mushroom surface, as a result of extracellular exudation [22]. Causal agents of cobweb disease are *Cladobotryum* spp., and symptoms on white button mushrooms are: cottony fluffy white, yellowish to pink colored colonies on mushroom casing, rapid colonization of casing surface, eventually causing decay of the host basidiomata. The major epidemic cause was the clonal fungicide-resistant strain of *Cladobotryum mycophilum* (Oudemans) W. Gams & Hoozemans [23]. It has a faster growth rate, simple phialide tips without extensions (rachises), more profuse and earlier sporulation, without distinct odor, and two, three-, and four-celled spores rather than the two-celled conidia characteristic for the species *Cladobotryum dendroides* (Bulliard: Fries) W. Gams & Hoozemans [24]. Strains identified as *T. harzianum* species complex (THSC) induce destructive symptoms in mushroom facilities of *A. bisporus*, *Pleurotus ostreatus* (Jacq. ex Fr.) P. Kumm. (1871), and *L. edodes* [25]. Afterwards, in the mid-1980s, *T. aggressivum* emerged as an aggressive substrate mold on button mushrooms that decimated the production. When encountered in cultivation facilities, *T. aggressivum* found a niche ideally suited to its features: cellulolytic activities, fast growth rate, and temperature optimum close to that of cultivated mushrooms. Infected mushroom substrate showed a dark green color after *T. aggressivum* colonization and sporulation. Red-pepper mites (*Pygmephorus* spp.) commonly occur in infested substrate, and seriously diseased areas produce no mushrooms [26]. Green mold disease has been spread worldwide in button mushroom farms, and is mainly caused by *T. aggressivum* f. *europaeum* (Ta2) Samuels & W. Gams in Europe, or *T. aggressivum* f. *aggressivum* (Ta4) Samuels & W. Gams in North America [27]. Severe green mold infections caused by *T. pleurotum* and *T. pleuroticola* of cultivated oyster mushrooms have been reported in Asia [28, 29] and Europe [30–32], which also indicate a worldwide threat. The species complex of *T. harzianum* (THSC) causes green mold in substrate and fruit body infection of *L. edodes* [33]. Furthermore, THSC, *Trichoderma ghanense* Yoshim. Doi. Y. Abe & Sugiyama, 1987, *Trichoderma longibrachiatum* Rifai, *Trichoderma asperellum* Samuels, Lieckf & Nirenberg, and *Trichoderma atroviride* P. Karsten were found in substrates of button and oyster mushrooms, but aggressive colonization has not been proven for these species [32]. Bacteria and the sticky fungal spores of *L. fungicola*, *M. perniciosa* and *Trichoderma* spp. can be spread through pests, yet the dry, large, and aerodynamic conidia of *Cladobotryum* spp. are transmitted by ventilation system [34]. The conidial dispersal by splashing water, workers, and equipment are also means of spread of all fungal pathogens. Significant source of the disease inocula could be both wild and cultivated infested mushrooms [20].

13.2.2 BACTERIAL DISEASES OF MUSHROOMS

Brown blotch, a bacterial disease responsible for severe outbreaks throughout the world, is caused by both *Pseudomonas tolaasii* Paine 1919 and *Pseudomonas reactans* Migula 1894 [35]. When bacterial brown blotch is found in mushroom facilities, it is almost impossible to eradicate. The disease affects not only the button mushroom but also *Pleurotus* and *Lentinula* mushrooms [6, 35]. Bacterial occurrence on mushroom pilei is characterized by the appearance of spot symptoms, which render them unmarketable. Furthermore, after severe infections, spots become dark and sunken. Lesions may coalesce to cover the entire mushroom surface [36]. Species *Pseudomonas agarici* Young 1970 is responsible for the drippy gills and the cap discoloration symptoms; *Pseudomonas gingeri* Preece and Wong 1982 causes ginger blotch; both *Pseudomonas aeruginosa* (Schröter, 1872). Migula 1900 and *Pseudomonas fluorescens* (Flügge, 1886). Migula 1895 are known to initiate the mummy disease [37]. The method including irrigation with 0.3% calcium chloride and the electrolyzed oxidizing acidic water was recorded as the most effective in symptom reduction and enhancement of the initial and postharvest mushroom whiteness (during 7-day storage, at 12°C) [38]. The other successful method is addition of 0.75% hydrogen peroxide (H_2O_2) and/or 0.3% calcium chloride to irrigation water starting 11 days after casing. This also significantly enhances mushroom whiteness and shelf life during 6-day postharvest storage at 12°C.

13.2.3 MUSHROOM VIRUSES

During the 1960s, the La France virus disease was probably the first major pathogen that affected white button mushroom cultivation immensely. Disease-suppressed vegetative mycelia and induced abnormal carpophores. It was also detected in mycelial cultures of *L. edodes* [39]. La France virus was predominantly spread by both spores and mycelium from infected mushrooms, proceeding from one crop to the other [40]. Spread of spores could be prevented by placing filtration at openings and cooking out spent mushroom substrate at the end of the growing cycle. Mushroom virus X (MVX) is characterized by a variable number of viral dsRNAs, first noticed in the late 1990s [41]. MVX of *A. bisporus* has different symptoms: cropping areas without mushrooms, disruption of mushroom pins, crop delay, premature veil opening, discoloration, and malformations of the mushroom caps.

Viruses are the most problematic when infected spores induce disease at the time of spawning and are less serious if infestation occurs later. On the other hand, mushroom mycelium bearing MVX induces heavy effects if minute quantities of infected mycelium are present at both spawning or casing time. Early infection of a crop produces no brown mushroom symptoms, whereas infection of a crop at casing time produces brown symptomatic mushrooms. MVX was controlled and prosperous through improving farm hygiene [5].

13.3 MUSHROOM PESTS

13.3.1 PATHOGENIC NEMATODES

The primary source of nematodes is raw compost or improper preparation of the casing soil, and they can survive during the whole mushroom growing cycle [42]. During composting of mushroom substrate, nematodes can survive fermentation in the outer parts of the compost pile. After surviving pasteurization in a single infested batch, nematodes reach the spawning machine and contaminate mushroom beds after spawning [42]. Parasitic and saprobic nematodes are significant for mushroom cultivation. Parasitic nematodes consume mushroom mycelium, while saprobic nematodes ingest different microorganisms present in the substrate. Parasitic fungal-feeding nematodes are rare in mushroom facilities. The number of nematodes can increase 30 to 100 times in two weeks [42]. After destroying hyphal structures by nematodes, entire barren bed areas could be found and the mushroom crop could be totally reduced. When the nematode population exhausts the nutrients from the compost, they swarm to the surface to be readily picked up by vectors. Dried nematodes could be distributed by air movements. Saprobic nematodes are more commonly present in mushroom facilities than the parasitic species. They could cause severe indirect disintegration of the mycelium [42]. Mostly present species are: *Acrobeloides apliticus, Acrobeloides buetschii* de Man. 1884 (de Man. 1921), *Caenorhabditis elegans* (Maupas, 1899) Dougherty 1953, *Cruzenema lambdiensis* (Maupas, 1900), *Panagrolaimus rigidus* (Schneider, 1866), *Pelodera strongyloides*, *Rhabditis (Cephaloboides) oxycerca* (de Man, 1896) Dougherty, 1955, *Rhabditis (Choriorhabditis) longicaudatus* (Bastian, 1865; Andrássy, 2005), and *Rhabditis terricola* (Rhatte) [42]. Saprobic nematodes are parthenogenetic, and they multiply even faster than parasitic ones, 100-fold in three days. The crop loss could be the result of the nematodes' toxin production when bacteria, its primary food source, are very abundant in excessively wet compost [42]. When numbers of nematodes decrease, 100 of them join together and move towards the surface of the substrate, where they can adhere to vectors. Woodwork and insulation in the growing rooms contaminated with nematodes could inoculate successively. Nematicides are not accessible in the mushroom industry, and measures to control them are restricted to climate conditions regulation and sanitation measures. A nematode-trapping fungus (*Arhtrobotrys irregularis*) may improve prevention of their occurrence [42]. Nematodes inhibit the growth of the mushroom mycelium in the compost, spread bacterial blotch, cause distortion of the mushroom fruiting bodies, and consequently reduce yields [43]. The nematode species belonging to the genera *Ditylenchus, Pratylenchus, Psilenchus, Acrobeloides, Caenorhabditis,* and *Rhabditis* were recovered from the peat samples collected after disinfection, and prior casing time, in a total population of 9.5 nematodes per 100 mL sample [43]. From the peat samples collected 32 days after casing time, on average 630 nematodes were detected. *Caenorhabditis* was the most dominant genus found in all samples ranging from 100 to 510 nematodes per 100 mL peat sample. Nematodes of the family Cephalobidae comprise the second most dominant group in the peat samples collected 32 days after casing, ranging from 50 to 150 nematodes per sample. Species *Helicotylenchus* sp. *and Ditylenchus* sp. were recorded in small populations, an average of 20 and 15 nematodes per sample, respectively. Other isolated nematode species were identified as *Acrobeloides, Heterocephalobus, Panagrolaimus, Plectus,* and *Diplogaterids*. Diplogateridae were found in an average of 37.5 nematodes per sample [43]. From the peat collected 46 days after casing time, an average of 880 *Caenorhabditis* sp. nematodes per 100 mL sample were detected. The *Caenorhabditis* sp. was also found in the samples of distorted mushrooms, an average of 10 nematodes per 100 g mushrooms. Some studies showed that rhabditid nematodes were the most dominant group and causal agent of distorted mushrooms. *Caenorhabditis* sp. should be considered as major threat in mushroom cultivation, since these nematodes survive pasteurization of compost, disinfection of the casing material, sanitation measures, while chemical control during cultivation process is not allowed [43]. Several *Aphelenchiodes* and *Ditylenchus* species have been identified as particularly problematic nematodes in oyster mushroom [44]. Though the details of the feeding mechanisms in the two genera differ, both of them cause severe mycelium destruction [44, 45].

13.4 ARTHROPOD PESTS

13.4.1 PATHOGENIC FLIES

Mushroom flies are a regular problem in mushroom cultivation. Three groups appear most often in mushroom facilities: the sciarid flies, the phorid flies, and the cecid flies [47]. The sciarid larvae feed on the substrate, mushroom mycelia, and the mushroom caps. Phorid larvae ingest mushroom mycelia and reduce yield. Larvae of cecid flies consume the mushroom stipes or lamellae, which also decrease the crop yield. The most important influence of sciarid and phorid adults is the spread of disease agents inside the mushroom growing chambers [47].

13.4.1.1 SCIARID FLIES

Sciarid flies, *Lycoriella mali* (Fitch) and *Lycoriella ingenua* (Dufour), are small, 3–5 mm long black insects, with long antennae and gray wings [47]. Females are attracted to lights. Fly traps are used to monitor the number of female flies present in the mushroom growing chambers throughout the cropping cycle [47]. The sciarid larvae have large and dark heads [47]. Females invade the compost when it cools down and the mesophilic fungi begin to grow. Females of sciarid flies are able to find small cracks in the walls of the cultivation facilities. Adults prefer outdoor temperatures between 10°C to 24°C [47]. They are able to detect and avoid residues of insecticide diflubenzuron. A female may lay up to 150 small, white, and oval eggs, preferentially in areas contaminated with fungus *Trichoderma* (green mold disease). Bacterial and fungal pathogens quickly infect the tunnels made by sciarid larvae in the mushroom fruiting bodies. Fully spawn-run compost is covered with calcium oxalate crystals, and as a water repellent, prevents larvae from feeding [47]. The first generation of fly adults emerges synchronously before the first break, as oviposition takes place over a short period of time from cooldown of the mushroom substrate to the spawning. Flies reach their greatest numbers during the second generation. The life cycle of flies depends on different strains of *A. bisporus*, different species of *Agaricus* (e.g., *Agaricus blazei* Murill, also known as *Agaricus brasiliensis* Peck 1983, or *Agaricus bitorquis* var. *valid*a Moll), different species of edible mushrooms (such as button mushroom, oyster mushroom, or shiitake), or resistance to pesticides. The first generation larvae cause minimal damage to the mushroom yield. However, when *Trichoderma green* mold is present, flies significantly worsen the severity of the disease by spreading fungal conidia. On the other hand, the *Trichoderma* spores inside the flies promote their fertility [47]. Large numbers of larvae ingesting the mushroom mycelium could inhibit the mushroom fruit body production. The making tunnels into the mushroom stipe of the second-generation larvae result in "black stem," which makes the mushrooms unmarketable [47].

13.4.1.2 PHORID FLIES

Phorid fly, *M. halterata*, has secondary importance in many countries [47]. Flies are 2 to 3 mm small, with very small antennae. Phorid flies enter the mushroom facilities after the second break, later in the cultivation cycle than sciarid ones. They are more susceptible to temperature variations than sciarid flies, adapted to warmer and drier conditions in the mushroom substrate. Larvae are up to 6 mm long and creamy white. The larvae's head is pointed and the end is blunt [47]. Phorid larvae ingest the mushroom mycelium and are more susceptible to temperature changes than sciarids. The larvae feed selectively and can be tolerated in mushroom beds 50 to 100 times more than sciarids. They are attracted by volatiles from the mushroom mycelium and are a major vector of *Lecanicillium* disease [47].

13.4.1.3 CECID FLIES

The most potentially damaging mushroom flies are cecids (*Mycophila speyeri* [46], *Heteropeza* Winnertz 1846) [47]. Under most conditions, larvae are present instead of adult flies. When larvae reach a length of 2 millimeters, they develop into mother larvae and give birth to up to 30 daughter larvae through pedogenic parthenogenesis. Cecid larvae have bluntly pointed ends [47]. White larvae belong to the species *Heteropeza pygmaea*, while orange larvae belong to *M. speyeri* or *Mycophila barnesi* Edwards, 1938. *H. pygmaea* is commonly found in *Agaricus, Pleurotus,* and other edible mushroom species, and is an important vector for browning bacteria [12, 48]. Larvae can significantly reduce yield by feeding on the mushroom mycelium, stipes, and gills. When populations grow, the larvae gather together and spread in large groups. Among cecid larvae, the most common genus is *Heteropeza*. However, orange *Mycophila* species cause the most damage [47]. Cecid larvae can survive high temperatures during spent mushroom cookouts within wooden structures inside the growing room [47].

13.4.2 PATHOGENIC MITES (ACARIDS)

Red pepper mites or pygmy mites (*Pygmephorus* spp., *Siteroptes mesebrinae* Canestrini) do not harm the mushroom mycelium, but rather feed on fungi such as the pathogen *Trichoderma* spp. [49, 50]. Their presence indicates poor compost conditions and they can completely ruin a harvest by spreading the spores of the pathogenic fungi they feed on [51, 52]. Mites *Tyrophagus* spp. feed on the crop mycelium [52, 53]. They cause secondary damage to the sporocarps by spreading bacteria, and *Tarsonemus* spp. are capable of transmitting mushroom viral diseases [54]. Predatory species (*Arctoseius* spp., *Parasitus* spp.) feed on nematodes and mushroom flies [51, 55].

13.5 CHEMICAL DISEASE CONTROL

Since mushrooms are considered a small crop, disease control includes only a few fungicides authorized for use: chlorothalonil in a few European countries (France, Poland) and in the USA and Canada; thiabendazole in Australia, Republic of South Africa, and the USA and Canada; imazalil in Australia; prochloraz in Europe, South Africa, and Australia; and metraphenone in several European countries (France, Spain), Australia, New Zealand, Republic of South Africa, and the USA [17, 24, 56–58]. The control of fungal diseases of edible mushrooms is especially demanding as the host and the pathogens are both fungi. Selective fungicides should suppress the spreading of the pathogen without harming the growth of the cultivated mushroom [59]. After several years of intensive usage of fungicides from the methyl benzimidazole carbamates (MBC) group, resistant strains of *L. fungicola* [7, 60–62], *Cladobotryum* spp. [63–66], and *T. aggressivum* [67, 68], especially to benomyl, carbendazim, thiophanate-methyl, and thiabendazole, were reported. After reports of benzimidazole mutagenicity, this class of fungicides has been withdrawn, and prochloraz remains the only recommended chemical pesticide for use in the EU [17]. Fungicide chlorothalonil was ineffective against *Trichoderma* on *A. bisporus* either in toxicity tests in laboratory conditions or efficacy tests in the mushroom growing rooms [69]. Reduced sensitivity of *L. fungicola* and *Cladobotryum* spp. to fungicide prochloraz was also noted [62, 64, 70].

Lately, the fungicide metrafenone has been used against fungal pathogens *Cladobotryum* spp. and *L. fungicola* in Spain, France, Belgium, the UK, and the USA [58, 71]. However, there is a lack of data on *Trichoderma* spp.'s susceptibility to fungicides, which raises concerns about controlling outbreaks of green mold disease [17]. Recent findings have shown that *Trichoderma* spp. strains from North Macedonia, Croatia, Serbia, and Hungary, originating from three edible mushroom species, were susceptible to prochloraz [18, 72]. *In vitro* tests have revealed that *Trichoderma* strains isolated from *Agaricus, Pleurotus,* and *Lentinula* from North Macedonia, Serbia, Croatia, and Hungary were considerably susceptible to the fungicide metrafenone [18]. The most sensitive taxon to both prochloraz and metrafenone was the *T. harzianum* species complex (THSC) isolated from oyster mushrooms. Prochloraz suppressed pathogenic strains of *Trichoderma* without inhibiting the growth of oyster mushrooms [73]. Furthermore, fungicides are generally more toxic to soil organisms than herbicides and insecticides [74], and they also have the potential to affect beneficial fungi [75]. It is considered that introduced microbial inocula are short-lived as they become a food source for the natural population of microorganisms, which then amplify and exhibit antagonistic effects in the substrate [76, 77]. The fungicide chlorothalonil suppressed respiration and stimulated dehydrogenase activities of soil microorganisms [78]. The mycopathogenic bacterium *P. tolaasii* is a causal agent of bacterial blotch epidemics in edible mushrooms, which is difficult to control. The development and implementation of biological control for pests and diseases is very important due to the decreasing efficacy of chemicals and the increasing resistance of pathogenic organisms to pesticides [79, 80]. Antagonistic fluorescent pseudomonads have higher biological effects than other bacterial species, as they belong to the same taxonomic group and require the same ecological conditions as the pathogenic strains, such as *P. tolaasii* [81, 82]. A sufficient amount of antagonistic pseudomonads in the casing soil could successfully compete for nutrients and space [83]. Bacterial blotch disease (*P. tolaasii*) is commonly controlled with hypochlorite. Treatment with chlorine compounds can partially control the disease, but it has negative effects on yield [84], browning symptoms [20], and human health. Research on biological disease control is being intensified [85, 86]. Calcium hydroxide is applied to the affected area to prevent the disease from spreading [73]. Unsatisfactory results have been reported with chlorine, while calcium chloride and chlorinated compounds are commonly used against agents of brown blotch disease [87]. Low pH values and anaerobic conditions are unfavorable for fungal pathogens of edible mushrooms, such as the agents of dry bubble disease. Peat soil taken from lower ground layers is free of the dry bubble agent, *L. fungicola*. Danish mushroom growers use a casing soil mixture made from sphagnum peat, sand, and carbonate instead of clay, loam, and humus to prevent soil-borne fungal pathogens [88]. Optimal growth of oyster mushrooms is achieved at an alkaline pH (pH = 8–9), while fungal pathogens including *Trichoderma* spp. prefer an acidic to neutral environment (pH 5–7). Therefore, increasing the pH of the casing soil to 8–9 could enable the spread of green mold infection [30, 72]. A synergistic effect between peracetic acid and the fungicide prochloraz was observed in the control efficacy of cobweb disease (93%) on *A. bisporus* [89]. Conversely, an antagonistic reaction between the fungicide prochloraz and colloidal silver was observed against *Cladobotryum* spp. in the same trial. Peracetic acid showed better control of cobweb disease than colloidal silver. Therefore, peracetic acid could be used as an eco-friendly disinfectant of the casing layer against the fungal pathogen *Cladobotryum* spp. in button mushrooms [89]. Colloidal silver and H_2O_2 are also environmentally friendly disinfectants of the casing layer, and exhibit remarkable efficacy against the causal agent of bacterial brown blotch, *P. tolaasii* [90]. Treatment of the mushroom spawn (1.15 g of fungicide per ton of compost) showed higher efficacy than treatment of the mushroom compost (70 g of product per ton of compost). Initially, benzimidazole fungicides were applied to the spawn, but resistance problems have emerged. Fungicide imazalil sulfate, from the group of imidazoles, has proven to be highly effective as a spawn and supplement treatment in suppressing aggressive pathogenic organisms that are resistant to benzimidazole fungicides [67]. Another issue is the degradation of fungicides by microbes present in the casing soil. For example, prochloraz significantly dropped, up to 15% by day 45, by the end of the second flush [91, 92]. The composition of the casing soil for mushroom growing is very similar to mineral soils, although it contains a high level of organic matter from the microorganisms present [93]. Fungicide prochloraz degrades rapidly in the presence of residual liquid from a sprayer that contains prochloraz-degrading microbes, so it should be emptied after use [91].

13.6 MICRO-BIOLOGICAL DISEASE CONTROL

Diseases are commonly controlled by chemical pesticides and cultural practices. There has been significant public concern regarding the widespread use of chemical pesticides in agriculture and their harmful effects on non-target organisms, human health, and the environment since 1962 [94, 95]. Additionally, the problem of pathogen resistance to pesticides, the gradual elimination of previously available chemicals, and difficulties with the registration process for new pesticides have been observed in large-scale plant cultivation [96]. As a result, awareness has led to the development and implementation of biocontrol methods, including the use of antagonistic microorganisms and natural enemies that are compatible with non-target organisms [96–99]. The introduction of selected thermophilic microorganisms during the fermentation phase of composting provides a certain level of control over disease-causing organisms [100–102]. Indigenous bacterial species present in mushroom substrate, such as *Pseudomonas* spp. and *Bacillus* spp., exhibit antifungal potential. Beneficial bacteria produce various antibiotics, with dipeptides and cyclic peptides being the most prevalent [103]. Some beneficial bacteria also produce extracellular metabolites that exhibit fungicidal effects [103, 104]. For example, *Bacillus subtilis* (Ehrenberg) Cohn is known to produce hydrolytic enzymes (cellulase, chitinase, and β-1,3-glucanase) that are responsible for degrading fungal cell walls and membranes [105]. Several biopesticides containing *P. fluorescens* as an active ingredient have been discovered [106]. The application of fluorescent pseudomonads in the control of bacterial blotch in button mushrooms resulted in a significant reduction in disease severity (70–100%) [107]. Suspensions (SU) of fluorescent pseudomonad strains (10^{11} cfu/ml) were applied at three different growth stages of the mushrooms: five days after casing, at the button stage, and before harvest. The best results were obtained when Pseudomonad strains M-4/2 and M-5/3 were applied at the button stage, reducing disease severity by 88.1% and 85.3%, respectively. Considering the lack of inhibitory effects of fluorescent pseudomonad strain M-5/3 in *in-vivo* experiments against *P. tolaasii*, it seems that competition was the main control mechanism. On the other hand, the siderophore-producing strain M-4/2 competes for Fe^{3+} ions Significant yield improvement was noted as a result of decrease of symptoms by 50% using biological control [108]. Two applications of fluorescent pseudomonads were sufficient for disease control of *P. tolaasii*, first on day one (casing time) and secondly during formation of pins [107]. The combined use of solarization and biofumigation is recommended for effective inhibition of the pathogens and significant reduction of soil microflora [109].

The addition of grain spawn and nitrogen supplements, the presence of nematodes and insects in all their metamorphic stages, and the mushroom mycelium itself create an environment with conditions conducive to the formation of short food chains [110]. A typical example of a food chain in this environment can be found in the wheat straw that ferments in areas of the columns where the moisture content is too high: flies living on mushroom mycelium and laying eggs in the fermented substrate; predacious beetles and nematodes feeding on the flies' larvae; predacious mites and fungi feeding on the nematodes. Insects and nematodes present in the mushroom substrate have many predator species that are being investigated for their potential for non-disruptive pest control [111]. The use of *Bacillus thuringiensis* (Bt) Berliner 2015 var. *israelensis* and predatory nematodes in the control of mushroom flies and predatory mites in the control of cecid larvae is promising [54]. The effectiveness of predaceous fungi (*Arthrobotrys* spp. and *Dactylaria candidula* (Hohn.) Bhatt and Kendrick) is also being investigated [112–115]. Control of harmful mycophagous nematodes is another ongoing investigation in mushroom cultivation [44, 116, 117]. Environmental conditions, such as the pH value of the soil, can also affect fungal behavior. These fungi are predaceous in alkaline or neutral soils, whereas saprophytic in acid surroundings [118]. Reducing of fungal predaceous effects were recorded, perhaps, not only due to the environmental factors, but also because of unsuitable nutritional requirements in the mushroom substrate [119]. Ecological and nutritional requirements of *Arthrobotrysoligospora* (Fresen.) 1850 (Helotiales: Orbiliaceae) are similar to *A. bisporus* [114]. The fungus *Dactylella oviparasitica* Stirling & Mankau parasitizes nematode eggs [120] and produces networks and knobs around nematode *Panagrolaimus* sp. on agar media [119]. The joint use of different fungal strains with various modes of action is recommended for the control of pathogenic nematodes, due to fluctuations in both nematode populations and fungal predation [119, 121]. Among the five predaceous fungal strains, *Dactylella dasguptae* showed the most remarkable efficacy against the *Panagrolaimus* sp., a saprophagous nematode population, by 35%, during the cultivation of button mushrooms [119]. Oyster mushrooms (*Pleurotus* spp.) themselves have been indicated as being able to attack nematodes in the substrate [122]. Specialized secretory cells arising from aerial hyphae of *P. ostreatus* produce droplets with the toxin ostratin, which quickly inactivates nematodes and allows the fungus to colonize them [123]. After penetrating the host, the fungus secretes an antibiotic called pleurotolysin, which inhibits bacterial growth and colonizes and digests nematodes [124]. Oyster mushroom, *P. ostreatus*, produces the secondary metabolite nematoxin, which is neither located in the sporocarps of the mushroom nor produced in liquid culture [123]. The nematocidal metabolite, muscarine, was found in nature and isolated from *P. ostreatus*, *Pleurotus strigosus* Berk. & M.A. Curtis, *Pleurotus subareolatus* Peck 1887, *Pleurotus cornucopiae* (Paulet) Rolland 81910), *Pleurotus sajor-caju* (Fr.) Fr. (1838), *Pleurotus cystidiosus* O.K. Mill 1969, *Pleurotus levis* (Berk. & M.A. Curtis) Singer, *Pleurotus populinus* O. Hilber & O.K. Mill (1993),

Pleurotus pulmonarius (Fr.) Quél. (1972), Pleurotus drynus (Pers.) P. Kumm (1871), *Pleurotus euosmus* (Berk.) Sacc., *Pleurotus tuber-regium* Fr. 1882, and *Pleurotus eryngii* [125]. Under very low nutrient conditions, oyster mushroom (*P. ostreatus*) could also decompose and feed on bacterial micro-colonies [123]. Nematocidal effect of extracts and fractions of *L. edodes* ESC-401 was also confirmed [126]. Remains of bacterium *B. subtilis* could also serve as a nutrient source for button mushroom [127]. It is possible that the ostreatin toxin from *P. ostreatus* is not only restricted to bacteria and nematodes, but to other microorganisms that may serve as nutrient sources as well [123].

Synchronous development of sciarid flies enables the application of biological control measures, including insect growth regulators and predatory mites such as *Hypoaspis miles* Berlese, just before casing time or at the beginning of spawn run (830 mites m^{-2}). Insect growth regulators and predatory mites are effective against both immature and mature sciarids towards the end of the cultivation cycle. Insect-predaceous nematodes are useful parasites of mushroom phorids [44, 111]. Two species of rhabditid nematodes (*Steinernema* spp. and *Heterorhabditis* spp.) and *Howardula husseyi* were tested against the fly *M. halterata* [128, 129]. The rhabditid nematodes eventually achieved better biological control and are currently being used in commercial formulations against sciarid flies. The application of predatory nematodes *Steinernema feltiae* (Filipjev) (1.5×10^6 nematodes m^{-2}) at casing time suppresses the number of sciarids in the casing layer. After the application of *S. feltiae* (0.5×10^6 nematodes m^{-2}) and *Steinernema carpocapsae* (0.5×10^6 nematodes m^{-2}), the species *Lycoriella auripila* was suppressed by both treatments, while the population of the fly *M. halterata* was not affected [130]. Correct use of nematodes could provide reasonable control, for instance if mixed in cold water, continual disturbance of the substrate throughout application, and application at the recommended rate. Beneficial nematodes should be applied during pasteurization, in two split doses, firstly at casing time and subsequently one week after casing; in the phase III compost, using 5–15 days after casing time or 18 days before the expected fly outbreak. Entomopathogenic nematodes are most effective against the pest larvae 18 days before maturation. Thus, it is important to apply nematodes at the proper time during the cultivation cycle, for example, 18 days before the multiplication of flies. The recommended dose is 3×10^6 nematodes m^{-2} (or two split doses of $1–1.5 \times 10^6$ nematodes m^{-2}). If the timing is uncertain, it is recommended to use the highest application rate. If there are no insect larvae in the casing, 80% of entomopathogenic nematodes will die. In each insect larva infected by entomopathogenic nematodes, 4,000 new nematodes will multiply [130–132]. The application of *H. miles* requires a 12-fold increase in workforce compared to the application of *S. feltiae*, and its life cycle lasts 28 days [132]. Phorid flies are frequently parasitized by *Howardula husseyi*.

13.6.1 COMPOSTING MICROFLORA

Mushrooms are saprophytic organisms and thus obtain nutrients from the products of organic decay, decomposing dead material in nature [133]. The most commonly used plant material for mushroom compost in Europe is wheat straw (*Triticum aestivum*) mixed with chicken manure. In improperly prepared substrate, competitive organisms are able to multiply more quickly than mushroom mycelium and can control their territories by producing antibiotics. Similar to nature, a succession of organisms decompose organic matter in the mushroom substrate. The process of fermentation begins when the compost is moistened. Primary organisms are fast-growing microbes that can extract energy from the plant material [133]. Microorganisms produce ammonia and convert simple carbohydrates into larger molecules. These complex carbohydrates provide nourishment for mushroom mycelium to grow. During thermophilic fermentation, low microbial activity is observed due to the high temperature inhibiting the activity of heterotrophic bacterial microorganisms [134, 135]. Excessive temperature can hinder the composting process [136, 137]. Compost microflora convert ammonia to microbial protein during the pasteurization stage at high temperatures. Unfavorable conditions in the composting material, such as density and moisture, create anaerobic conditions and lower the temperature, which leads to the growth of anaerobic organisms [137]. An excess of carbohydrates promotes the growth of competing fungi such as *Aspergillus*, which can outgrow mushroom mycelium. During pasteurization, different organisms are present. They consume ammonia, obtaining nitrogen molecules, and convert them into proteins. These organisms then serve as a protein source for mushroom mycelium. If the temperature is not optimal during pasteurization, the microbes that convert ammonia will be inhibited. If the compost temperature is too high, fermentation will start again, resulting in excessive ammonia production that can disrupt mushroom growth. Microbes that convert ammonia cannot survive if the temperature is too low. Pasteurization also helps control pests [137]. Various harmful organisms, such as nematodes and molds, can be present in the compost. During composting, high temperatures help eliminate some pathogens while allowing beneficial organisms to survive. Beneficial microflora protect mushrooms from potential newly introduced pathogens. Microflora exclude pathogenic organisms through competition or by producing antibiotics [137]. The optimal temperature range for the composting process is between 35°C and 55°C, and the released ammonia is efficiently removed by the activity of microorganisms [138]. Mesophilic. The original amount of microorganisms recorded in the mix of bagasse and coast-cross straw was 3.2×10^8 CFU/g, and bacterial species (94%) were predominant. The consistency of the population duplicates from day 2 to day 10 of the composting onset and later slightly decreases to 5.7×10^8 CFU/g by day 14. Pasteurization decreased the total population of microorganisms additionally by heating (> 60°C) [141]. The mix of bagasse and coast-cross straw is rich in bacteria, particularly in *Bacillus* species during the whole composting process, with an average density of population of 3×10^8 CFU/g. Bacteria are the most prevailing group during all stages of the composting process, succeeded by the actinobacteria and the micromycetes. Comparable findings were noted during the composting process of wheat [142] and rice straw [140], where prokaryotes along with several fungi were the most represented organisms in the beginning of the composting process, when the actinobacteria predominated during the last stage. Both gram-negative [143] and gram-positive bacteria [144] were detected during and after the composting process in the substrate. The obtained bacteria were quite ubiquitous in both soil and the agricultural waste. From the family Enterobacteriaceae, 18 gram-negative and six *Pseudomonas* species were detected. Regarding the gram-positive species, eight were from the genus *Bacillus*. It was found that ammonia was assimilated and transformed by bacterial species of *Bacillus* and *Pseudomonas*, beside members of the actinobacteria [134, 135], while *Pseudomonas putida* (Trevisan) Miguala promoted the forming of fruiting bodies of *A. bisporus* [145]. Throughout fermentation process, pathogenic species *Pantoea agglomerans* [146], *Escherichia coli* (Migula, 1895) [147] and *Klebsiella pneumoniae* (Schroeter, 1886) Trevisan 1887were detected from coast-cross straw and bagasse. Species *E. coli* was even found 30 h after pasteurization. Optimal temperature for the activity and diversity of microorganisms in the mushroom substrate were in the range of 54–70C [144]. Mesophilic bacteria such as pathogenic *E. coli, Salmonella*, and beneficial *B. subtilis* and *Bacillus pumilus* [148], were obtained from both composting phases, I and II, and were able to survive temperatures >55°C [149]. Bacteria *Bacillus circulans* Jordan and *Bacillus licheniformis* (Weigmann, 1898) Chester 1901 predominated before and from the start of the composting process until the second day in the coast-cross straw and bagasse. The bacterial population changed considerably throughout the composting process. From the onset of compost mixing, the following Bacillus species have been recorded: *B. subtilis, Bacillus amyloliquefaciens* (Fucumoto, 1943), *B. pumilus, Bacillus sphaericus* [150], and *B. anthracis* [150]. Species of the genus *Bacillus* arouse amplification of the thermophilic cellulolytic actinobacteria. Actinobacteria were detected two days after the beginning of the composting process, in which case the variety of *Bacillus* was the highest. In the wheat straw compost for *P. ostreatus* cultivation, the most predominant Gram-positive bacteria were *Paenibacillus* and *Bacillus* species [143, 151]. They stimulate the growth of mushrooms by producing enzyme laccases and suppressing the spread of the green mold pathogen *T. harzianum*. Beneficial bacteria *Bacillus* and *Paenibacillus* restrain the growth of yeasts, Gram-positive and Gram-negative bacteria, by synthesis of the toxic peptides bacteriocins [152]. The peak of Gram-negative bacteria was found from the beginning of composting until day 8 [141], and the species detected were comparable to those obtained from another sort of compost [153]. Species *Pseudomonas* and *Actinetobacter* (Gram-negative bacteria) own high oxidative activities to promote the mycelial growth and fructification in *P. ostreatus* [154].

Few bacteria were observed on the wheat leaf surface prior to ensiling, which would later degrade the wheat straw [155]. Stable litter, the compost precursor, contains at least eight types of dense bacterial populations [142]. Microorganisms multiply significantly seven days after the start of the composting process, unlike uncolonized straw. Maximal colonization coincides with the beginning of phase II – pasteurization, and the number of microorganisms diminishes later [127]. The amount of bacteria declines during mushroom production and increases again on day 112, at the end of the cultivation cycle. Degradation patterns were found in the wheat straw, and the bacterial colonies are fixed to the cell walls by the polysaccharide fibers of a glycocalyx, contributing to straw erosion [156]. The glycocalyx restrains the separation of the bacterial cells throughout the process of mixing the composting pile. The dark brown color of plant material in the mature compost is due to a combination of polysaccharides combined with humic and fulvic acids, which is important for the spread of the mushroom mycelium [157]. Devastation of the secondary thickening of straw has been limited to the parenchyma cells, even though detriment of vascular tissue was found later in the process. Bacteria propagate very fast in the vascular bundles of the straw, and the highest density was recorded in the phloem, particularly at the beginning of the composting process. Actinobacteria multiply in the substrate from the start of the second phase of composting (e.g., day 14) to the finish of the cropping cycle, and sporulate profusely on the straw surface. An enormous number of conidia of the thermophilic actinobacteria *Thermopolyspora flexuosa* Meyer 1989, *Thermomonospora alba, Thermomonospora curvata* Henssen 1957, and *Thermomonospora fusca* Henssen 1957 were isolated from the air throughout the spawning process of the mushroom substrate, with over 10^9 CFU m^{-3} of air [158]. Actinobacterial hyphae were widespread inside the straw cells, and thermophilic actinobacteria *T. fusca* was detected in lignocellulosic pulps [159].

Actinobacteria accounted for 18–86% of all bacterial types up to the end of the composting process [160]. A variety of total actinobacterial and fungal species expanded during the composting process [161]. Dominance of actinobacteria was recorded exclusively at the end of the compost maturation [162]. Although aerobic disintegration of organic matter mostly appear during composting, some anaerobic degradation also exists [163]. Species *Clostridium thermocellum* [164] is cellulose-degrading strain and found to have a significant role in the biocomposting process under high temperature conditions [165]. Actinobacteria were more abundant during the composting process than clostridia and fungi. A significant number of actinobacteria were previously recorded in the raw straw, including *Corynebacterium* (only present in the thermophilic phase), *Saccharopolyspora*, and *Prauserella*. Both groups of actinobacteria and clostridia continue to multiply during the thermophilic phase, while they diminish throughout pasteurization and the growth and fructification of mushroom mycelium [166]. During pasteurization, more actinobacteria and fungi were obtained from the subsurface of the substrate compared to the interior of the pile [166]. Present suborders included *Pseuodonocardineae, Micromonosporineae, Micrococcineae, Streptomycineae* (after initial growth of mushrooms – Streptomyces), *Streptosporangineae* (Thermopolyspora possessing hemicellulase activity, *Microbispora* in mature compost with enzymatic activities of cellulase, xylanase, and mannase), *Propionibacterineae, Corynebacterineae,* and *Acidimicrobineae* [166].

The variety of fungal populations diminished towards the end of the composting process [166]. The fungal species *Acremonium, Thermomyces*, and *Humicola* were the most abundant during the composting process. The most prevalent thermophilic fungal genera at the end of the composting process were *Thermopolyspora, Microbiospora, and Humicola.* During pasteurization, the most prevalent genus was *Humicola* (75%), followed by

Chaetomium (15%) and *Myceliophthora* (10%). The species *Acremonium* prevailed at the beginning and during fermentation but almost completely disappeared after pasteurization [166]. The species *Thermomyces* predominated in the thermophilic phase but was absent in the others. The species *Humicola* grisea Traaen is a growth-promoting fungus and possibly has a positive impact on mushroom yield [167, 168]. The species *Chaetomium thermophilum* La Touche (1950) and *Myceliophthora thermophila* (A.E. Apinis) C.A. van Oorschot 1997 produce the enzyme cellulase [169, 170]. Species from *Sordariomycetes* and *Eurotiomycetes* were recorded throughout the entire composting process, including all phases of succession, and were prevalent towards the end of the pasteurization phase. Species from the ascomycetous class *Dothideomycetes* were only present at the beginning of the composting process but disappeared completely later on.

The natural microbiota from agricultural residues interact throughout the composting process and have a significant impact on the subsequent colonization by the mushroom [141]. The techniques used in the cultivation of *A. brasiliensis* are similar to *A. bisporus*, with a few changes in the compost making process [171]. The goal is to find more suitable, affordable, and inexpensive materials for the mushroom substrate, instead of the commonly used wheat straw, chicken/horse manure, and gypsum, which should have a C:N ratio ranging from 25:1 to 50:1 [172]. Potential substrates for *A. brasiliensis* cultivation are made from sugar cane bagasse and coast-cross straw. The process lasts for 14 days, with the substances being mixed every two days, followed by a two-step steam pasteurization (55-65°C, 15 hours each step). During the pasteurization process, the substrate is heated by the activities of natural microorganisms, which eliminate intolerant pathogenic, pest, and competing organisms. Subsequently, actinobacteria and other thermophilic microorganisms multiply and synthesize microbial proteins by converting toxic released ammonia, enabling the nutrition of the cultivated fungi [138, 167, 168, 173]. The predominant microorganisms during the composting process were bacteria, mainly *Bacillus* and *Paenibacillus* spp., and members of the Enterobacteriaceae, with an average count of 3 x 10^8 CFU/g. Actinobacteria, particularly members of the genus *Streptomyces*, were also abundant, with a count of 2-3 x 10^8 CFU/g. Filamentous fungi were the least present and less diverse compared to other microorganisms. The fungal species *Aspergillus fumigatus* Fresenius 1893 was detected throughout the entire composting process and after pasteurization.

13.6.2 SUBSTRATE MICROFLORA

The sources of compost, specifically made for the commercial growth of button mushrooms, are plant-derived materials supplemented with animal manure [174]. Throughout the composting process, microorganisms naturally inhabiting the materials decompose the straw as well as the cellulose and hemicellulose, preparing the nutrients for *A. bisporus* [127, 175–178]. Button mushroom (*A. bisporus*) inoculation shows a significant impact on the microbial activity in the substrate environment. Optimal temperature is essential for mycelial spawn run, as low temperature decreases mycelium growth, while high temperature weakens mushroom mycelium. Mycelium debris is a food source for the competitor organisms. Wet mushrooms are an ideal habitat for pathogenic bacteria such as *P. tolassi*, the causal agent of mushroom blotch [133]. Mushroom *A. bisporus* is able to access lignin, not readily available to other microorganisms [179–182]. Cellulose degradation is consistent with mushroom production [183]. For example, *A. bisporus* is able to use actinobacteria as a nutrient source [184]. The mycelium of *A. bisporus* grows over straw surfaces; however, only 30% of the straw surface is penetrated by mycelium. Many extracellular hydrolytic enzymes are exuded by mushroom mycelium [174, 185]. The mycelium on the straw surfaces also produces an abundance of crystals of calcium oxalate, completely covering the hyphal tips [186]. The microbial biomass in composted straw can provide up to 10% of the nutritional requirement for button mushrooms. Dead bacterial cells are also used as a supply of nutrients for *A. bisporus* [184, 187]. Considerable degradation of straw is produced by the synergistic interactions of the compost-inhabiting microorganisms [188].

Growing techniques such as quick spawning, adding compost at casing soil, and reducing the number of breaks increase pest control [133]. Sporophore formation in button mushrooms starts with the accumulation of mycelium into thick strands, followed by the formation of primordia [189, 190]. The casing layer possesses chemical and microbiological characteristics that provoke pin formation [189, 191]. The appearance of some bacteria in the casing, such as *P. putida*, stimulates the formation of mushroom primordium [192, 193]. Button mushroom mycelium produces volatile organic compounds (VOCs) [194–196]. These compounds (2-ethyl-1-hexanol and 1-octen-3-ol) are metabolized by increasing the dominance of significant pseudomonad populations in the casing soil, thus enabling primordium formation [145, 197, 198]. It is noteworthy that mixed bacterial strains have a more stimulatory effect on primordium formation than individual pseudomonads [145]. Bacteria remove self-inhibitory substances produced by the vegetative mushroom mycelium [145, 191]. A strain of *Pseudomonas poae* [199] stimulated primordia formation more than *P. putida* strains, while *Pseudomonas veronii* had little or no stimulatory effect on *A. bisporus* primordium formation [196]. Other pseudomonads (*P. aeruginosa, P. agarici, P. fluorescens, P. reactans*,and *P. tolaasii*), did not stimulate *A. bisporus* [192, 193, 200]. Different *P. putida* strains showed variations in their stimulatory effect on *A. bisporus* mycelium growth [193, 200]. The aerobic bacterial microflora in the casing soil, with fluorescent *Pseudomonads* comprising 50% of the population [201], induced the initiation of mushroom basidiomes [145, 198]. The species with the greatest significance in fruiting body initiation is *P. putida* [192, 193]. Casing soil is primarily composed of sphagnum peat, although mushroom growers sometimes add up to 40% black peat in the casing layer. Its role is to regulate pinning during mushroom production and also contribute to the quality of harvested mushrooms in terms of shape, size, maturity, and solids content [38]. Aerobic bacterial populations registered in the casing soil ranged between 8.2 and 8.5 log CFU/g [201]. *Pseudomonads* represent up to 50% of the total bacteria in the casing soil [202]. A succession of casing bacteria was observed during the cultivation of *A. bisporus* [203]. The amount of fluorescent pseudomonads can vary between 14% and 41% of the total bacteria present in the casing, and increases during mushroom fruiting. Regarding aerobic bacterial population densities, the major bacterial genera found in the casing soil were from the *Pantoea* and *Pseudomonas* genus [38]. The genus *Pseudomonas* is widely utilized for disease control and yield improvement. As for the genus *Pantoea*, only one species, *Pantoea* agglomerans, was detected. In terms of the *Pseudomonas* genus, two species were predominantly recorded: *P. aeruginosa* and *P. fluorescens*.

Along with bacteria, casing soil also involves a high level of other beneficial organisms such as yeasts, molds, and actinobacteria. Sphagnum peat, as a relatively new constituent of casing soil, frequently contains fungi *Trichoderma* and *Streptomyces*. Being the predominant genera of molds, a number of *Penicillium* species are recorded in the casing soil: *Penicillium decumbens* Thom C. 1910, *Penicillium chrysogenum* Thom (1910), *Penicillium glabrum* (Wehmer) Westling (2011), *Penicillium citreonigrum* Dierckx 1901, and *Penicillium digitatum* (Pers.) Sacc. As for other molds, *Aspergillus* niger is rarely found in the casing soil [38]. Actinobacteria and fungi are usually situated in the upper level of the peat pile [204, 205]. Subsequent addition of calcium carbonate favors the development of actinobacteria. The sphagnum peat fungal species belonging to the genera *Trichoderma* and *Streptomyces* are known to be effective against pathogenic fungi: *Fusarium* spp., *Rhizoctonia solani, Phythium* spp., and Alternaria brassicicola Schwein [206]. Animal manure, a constituent of compost, is a source of several foodborne pathogens: *Salmonella* spp., *E. coli* O157:H7, *Listeria monocytogenes* Pirie 1940, and *Campylobacter jejuni* Veron & Chatelain 1973 [207–209]. These pathogenic bacteria may commonly be present in the casing soil or can be introduced by contaminated supplements or irrigation water. During the second phase of composting, the population of L. *monocytogenes* and *Salmonella* could be reduced by seven logs [38]. Steam treatment for disinfection of casing soil is not a common practice due to increased costs and reduced mushroom yields resulting from the elimination of beneficial microorganisms. In sterile soil, the survival of *Salmonella enterica* Kauffmann and Edwards 1952 var. *dusseldorf* was reduced in the presence of *Streptomyces bikiniensis* Johnstone and Waksman 1947, a known streptomycin producer [203]. Additionally, the addition of clay - 2% (w/w) bentonite - to improve soil properties influences the growth of *P. fluorescens* and enhances its antagonism against the pathogen *Fusarium culmorum* (Wm. G. Sm.) Sacc. 1892 [210]. Well-aggregated soils with high organic content and soil microbial activity have shown low persistence of introduced organisms [211]. Species of *Streptomyces* are producers of antibiotics and secondary metabolites with anti-cancer and anti-inflammatory activities. The reintroduction of *Streptomyces* spp. in sterile casing soil inhibited the growth of *L. monocytogenes* and *Salmonella* spp. [38]. Furthermore, actinobacteria isolated from the casing soil: *Streptomyces* spp., *Micromonospora* sp., *Nocadia* sp., were able to inhibit the growth of human pathogens [*Staphylococcus aureus* Rosenbach 1884, *Proteus vulgaris* Hauser 1885, *Proteus aeruginosa* (Schröter, 1872) Migula 1900, *E. coli, B. subtilis, Bacillus megaterium* de Bary 1884, *Klebsiella pneumonia* (Schroeter 1886) Trevisan 1887, *Candida albicans* (C.P. Robin) Berkhout 1923, *Aspergillus niger* van Tieghem 1867

Addition of olive mill solid waste (10.6% w/w) to mushroom compost enhances the total number of heterotrophic microbes and actinobacteria. This results in higher biological efficiency against pathogens, improved quality, and increased yield of the mushrooms compared to non-amended compost [214]. After pasteurization, the biomass ratio of fungi to prokaryotes in the compost was 1.8:1 [187]. Ratios of 0.9:1 and 2.3:1 were found for conventional and experimental compost, respectively [215].

13.7 BENEFICIAL MICROORGANISMS

13.7.1 BENEFICIAL FUNGI

Species *Talaromyces thermophilus* Stolk and *Thermomyces lanuginosus* Tsiklinsky (syn. *Humicola lanuginosa*) were found in the mature compost. Additionally, *Scytalidium thermophilum* (Cooney & Emerson) Austwick (syn. *Torula thermophila, Humicola grisea* var. *thermoidea, Humicola insolens*), and *Chaetomium thermophilum* grow rapidly and exhibit strong cellulose degradation. The number of CFU of *S. thermophilum* in the fresh matter during pasteurization was $10^6 g^{-1}$[176].

13.7.1.1 SCYTALIDIUM THERMOPHILUM

Among the succession of microorganisms during the composting process, thermophilic fungi appear to play a significant role in creating a compost structure suitable for mushroom growth. The thermophilic fungi isolated from the mushroom compost include *S. thermophilum, C. thermophilum, Malbranchea sulfurea* (Miehe) Sigler & J.W. Carmich., and *T. lanuginosus* [216, 217]. *In vitro*, a mixed inoculum of *M. sulfurea* and *S. thermophilum* showed the best promotion of *A. bisporus* mycelia growth, resulting in a yield twice as high as the pasteurized control. Additionally, thermophilic fungi provide a protective shield for bacteria present in the compost and support the mycelial growth of *A. bisporus*. Among the thermophilic fungi, *S. thermophilum* appears to be the most abundant in mushroom compost [216, 218–220]. It is worth noting that fungicides can inhibit the thermophilic fungi present in the substrate during the composting process [221]. The presence of fungicide residues (prochloraz, fluzilazol, etc.) in straw during the composting process can restrict colonization by useful thermophilic fungi [184, 218–220, 222]. Although the sensitivity of fungi to residues is generally higher in *in-vitro* experiments than *in situ*, fungicide residues represent a risk in wheat straw and have to be considered in mushroom compost production [66, 216]. Nine species out of 34 thermophilic fungi promoted mycelial growth of *A. bisporus* in compost: *C. thermophilum, Chetomium* sp., *M. sulfurea, Myriococcum thermophilum* (Fergus) (Cooney and Emerson 1964; Van der Aa 1973), *S. thermophilum, Stibella thermophila, Thielavia terrestris* Zopf 1876, and two unidentified basidiomycetes [167].

Better growth of *A. bisporus* was recorded after a pre-incubation period of straw with the thermophilic fungus *S. thermophilum*, although no effect was found when *H. grisea* Traaen var. *thermoidea* Cooney & Emerson or *T. lanuginosus* were used [223]. Additionally, better growth of *A. bisporus* on compost treated with either *S. thermophilum* or *H. grisea* var. *thermoidea* was noted in comparison to non-amended substrate [224]. The composting process was faster when a combination of straw and organic nitrogen was joined with the addition of *S. thermophilum* [212, 218, 219]. This inoculation effect was not recorded after finished pasteurization as sufficient indigenous inoculum was present. Mycelial growth of *A. bisporus* is strongly stimulated by pre-incubating the sterilized compost with the thermophilic fungus *S. thermophilum* [220]. Mushroom compost is almost completely colonized with *S. thermophilum*. A positive relationship was found between mushroom yield and the density of *S. thermophilum*. The dominant fungal species, *S. thermophilum*, after the composting process was completed, were also present in the compost ingredients (fresh straw, horse manure, and drainage from compost) [167]. Mushroom yield gained from the compost inoculated with *S. thermophilum* was almost doubled in comparison to the control substrate lacking the fungus [168]. Inoculation of compost with the afore-mentioned fungus may be of great significance if the compost is exposed to high temperatures in tunnels.

13.7.1.2 PANDORA GLOESPORA

Some research indicated *Pandora gloespora* is a promising sciarid control agent [225], by invading the bodies of pest flies.

13.7.1.3 NEMATOPHAGOUS FUNGI

Nematophagous fungi were effective in eliminating the soil nematode, *Butlerius* sp., and the saprophagous nematode, *Panagrolaimus* sp. [119]. Nematophagous fungi include: *A. oligospora, Dactylaria scaphoides* Peach, *Dactylaria dasguptae* Shome et Shome, and *D. oviparasitica* (egg parasite). Species *D. dasguptae* reduced the nematode population by 35%.

13.7.2 BENEFICIAL BACTERIA

Bacterial genera *Bacillus* sp., *Alcaligenes* sp., *Bradirhizobium japonicum* Kirchner 1896, *Azotobacter vinelandii* Limpman 1903, *Rhodopseudomonas palustris* (Molisch 1907) van Niel 1944, and the yeast *Lipomyces starkeqi* Lodder et Kreger-van Rij are known to stimulate mushroom sporophore formation [226–228].

13.7.2.1 RHODOPSEUDOMONAS SP.

Spraying suspensions of photosynthetic bacteria *R. palustris* (3.3×10^9 CFU/mL per 0.54 m^2) onto the casing soil during the period between the emergence of new mushrooms has been shown to increase mushroom yield (39.5%) [228]. These photosynthetic bacteria have the ability to accumulate 5-aminolevulinic acid [229, 230], which has been proven to enhance the yield of cultivated plants [231]. Spraying *R. palustris* onto the casing soil offers an advantage over supplementing the casing with nutrient release. Bacterial suspension can be easily produced using agricultural wastes or their by-products [232].

13.7.2.2 BACILLUS SPP.

Species of the genus *Bacillus* are thermostable and spore-forming organisms, which survive the composting temperature and subsequently compete with remaining pathogens. Species of the genus *Bacillus*, obtained from mushroom compost, showed strong antifungal activity against the mushroom pathogen *Chaetomium olivaceum* Cooke and Ellis, the causal agent of green olive mold [4]. Furthermore, Bacillus sp. introduced to *C. olivaceum* infested compost enhanced mycelial development of *A. bisporus* and significantly increased mushroom yield. The antibiotic chaetomacin, extracted from the *Bacillus* sp., is only produced at mesophilic temperatures [4]. This antibiotic was effective against *Yarrowia (syn. Candida) lipolytica* Harison and some other microorganisms. The addition of *Bacillus* sp. to the compost increased the rate of mushroom mycelial growth by 86%, and the biological efficiency against olive green mold was significantly better than in the inoculated control. Moreover, mushroom plots containing only *C. olivaceum* produced the mushroom fruiting bodies a week later compared to the plots treated with both *Bacillus* AOG and *C. olivaceum* [4]. It is noteworthy that *T. aggressivum* was reduced by some strains belonging to the *Bacillus species,* although it was resistant to other bacteria isolated from the compost [233]. Furthermore, *B. pumilus* and *B. subtilis* group are considered safe for environment and harmless to human health which makes them generally recognized as safe (GRAS) organisms [234]. Species *B. subtilis* and *B. licheniformis* exhibited significant inhibition of the growth of *T. harzianum, T. viride,* and *Trichoderma pseudokoningii* Oudem (1902) on shiitake mushrooms *P. sajor-caju* and *L. edodes* [235]. The biofungicide based on *Bacillus velezensis* (Ehrenberg) Cohn, strain QST713, has been registered against many plant pathogens [236, 237]. It is also registered for the treatment of mushroom compost against *T. aggressivum* f. *europaeum* in France [238]. Noticeable antagonistic activity of *Bacillus* spp. isolated from livestock manure and cotton-waste composts was registered against *T. harzianum, Trichoderma koningii* (Oudem.) 1902, and *T. viridescens* (A.S. Horne & H.S. Will.) on *Flamulina velutipes* (Curtis) Singer (straw mushroom*), L. edodes*, and *P. ostreatus* in Korea [239]. Also, *B. subtilis* showed a positive impact on the yield in oyster mushroom production [240]. The strain *Bacillus velezensis* QST713 (10^8 CFU) did not inhibit the growth of mycelia of *A. bisporus* when coated on mushroom grain spawn [241]. No significant differences were found between the treatments with the fungicide prochloraz (applied on casing soil) and *B. velezensis* QST713 (coated on mushroom grain spawn) against *T. aggressivum* f. *europaeum* T77. Strain *B. velezensis* QST713 demonstrated considerable effectiveness in preventing green mold disease (*T. harzianum*) compared to a biofungicide based on tea tree oil, but lower than the fungicide prochloraz [10]. On the contrary, the *B. velezensis* QST 713 strain decreased button mushroom yield compared to both controls and prochloraz fungicide treatment. Furthermore, biofungicide should be applied alone since antagonistic reaction between fungicide prochloraz and *B. velezensis* QST713 was registered. In *in vivo* experiments against *T. pleurotum*, a strain of *B. amyloliquefaciens* was found to be effective in both disease suppression and oyster mushroom yield improvement (10%) [242]. Numerous bacterial strains were isolated from mushroom substrate materials such as straw, chicken manure, mushroom compost during different stages of the composting process, and casing soil [104]. In these studies, 23 bacterial strains belonging to the genus *Bacillus* showed high inhibition of the growth of *T. aggressivum* f. *europaeum, T. harzianum,* and *T. koningii,* while 13 strains exhibited antifungal activity against *T. atroviride.* The isolated strains belonged to *B. subtilis, B. amyloliquefaciens, B. licheniformis,* and *B. pumilus.* However, only two bacterial strains, identified as *B. amyloliquefaciens,* successfully inhibited the growth of all tested *Trichoderma* spp. The most sensitive pathogen to antagonistic bacteria was *T. aggressivum* f. *europaeum.* At least one *Bacillus* strain was obtained from each sample, considering their application in every stage during composting and mushroom growing. Around 10 native *B. subtilis* inhibited the growth of *T. pleuroti* (54.4–62.2%) and *T. pleuroticola* (55.56–69.62%), with the highest activity shown by *B. subtilis* B-358 [104]. Another biofungicide based on *B. subtilis* strain Ch-13 exhibited higher effectiveness compared to *B. velezensis* QST713 against the compost pathogen *T. aggressivum* in small-scale trials [243]. Moreover, *B. subtilis* Ch-13 increased yield by 12% compared to the control and significantly more than *B. velezensis* QST713 when applied at higher concentrations. The better yield promotion and effectiveness in the suppression of the pathogen by *B. subtilis* Ch-13 at lower concentrations compared to *B. velezensis* QST 713 could be due to different formulations of the products or the strain features [243, 244]. These findings suggest that biofungicides based on *Bacillus* spp. could serve as an alternative to fungicide application in mushroom farms, particularly during severe outbreaks of compost green mold caused by *T. aggressivum* [243]. The strain of *B. subtilis,* Ch-13, resulted in a higher mushroom yield when applied three times as split applications (30 mL on the second day after casing + 15 mL two weeks after casing + 15 mL after the first flush) rather than being used two times (30 mL on the second day after casing + 30 mL after the first flush), despite using an equal final amount in both cases. This biofungicide increased the total mass of harvested champignons by 8.41% compared to the untreated control, and by 10.53% compared to the fungicide prochloraz, when applied in three split doses [245].

13.7.2.3 BACILLUS THURINGIENSIS (BT) SUBSPECIES ISRAELENSIS

Bacterium *Bacillus thuringiensis* subspecies *israelensis* (Bti) is widely used in pest biocontrol [48]. The crystal produced by the bacteria degrades in the larva's gut, paralyzing the insect's gut and killing the larvae within 48 hours. Bti has proven to be effective in controlling phorids and sciarids when applied to compost or casing soil. Small-scale trials have achieved an 85% efficacy in sciarid control. However, casing application of Bti may result in excessive mycelial growth or decreased yield. Younger sciarid larvae are more susceptible to Bti than older ones [48].

13.7.2.4 PSEUDOMONAS FLUORESCENS BIOVAR V

A strain of *P. fluorescens* biovar V showed some promising effects against mycopathogenic bacteria, *P. tolaasii* or *P. gingeri,* causal agents of bacterial blotch [48]. This antagonistic bacterium has a preventive effect, as it acts as a competitor for space and nutrients with bacterial blotch causal agents. It competitively excludes the colonization and development of the bacterial population [48].

13.7.2.5 PSEUDOMONAS PUTIDA

Plant growth promoting rhizobacteria (PGPR) have been found to have a positive effect on the yield and quality of button mushrooms when inoculated in both the casing soil and substrate [246]. The tested species included *B. japonicum, Rhizobium leguminosarum* (Frank 1879) Frank 1889, and *P. putida.* Among these species, *P. putida* was the most successful when inoculated in both the substrate and casing soil, resulting in an increase in mushroom yield (12.1–13.43%) and biological efficiency (20.64–21.08%) [246].

13.7.2.6 LACTIC ACID BACTERIA (LAB)

Lactic acid bacteria (LAB) are commonly used as inoculants by food fermentation industries to produce a variety of fermented foodstuffs [247]. Furthermore, the broad antifungal activity of LAB has also been confirmed [248]. LAB *Lactobacillus plantarum* Orla-Jensen 1919 was isolated from spent mushroom substrate for silage production [248]. Fresh spent mushroom substrates can also be used as animal feed [249]. Therefore, an effective storage technique using LAB is a desirable method among various storage methods.

13.7.3 BENEFICIAL ACTINOBACTERIA

Thermophilic actinobacteria play a significant role in environments where organic matter decomposes under elevated temperatures and aerobic conditions (e.g., hay, cereal grains, manure, straw, etc.) [250–252]. They are considered important biological regulators of soil and contribute to organic matter decomposition. Actinobacteria are a group of branching single-cell microorganisms capable of forming mycelium, which can be substrate or vegetative, or both substrate and aerial [38]. In most types of soil, spores are more abundant than hyphae, except in nutrient-rich compost [253]. Spores can remain viable for long periods and only germinate when exogenous nutrients are available. Thermophilic actinobacteria strains have been identified in various mushroom composts, belonging to five genera: *Pseudonocardia, Saccharomonospora, Saccharopolyspora, Streptomyces,* and *Thermobifida,* within the order *Actinomycetales.* Additionally, 16 strains were classified into the genus *Thermoactinomyces* within the family *Bacillaceae* [254]. The majority of strains belong to two genera: *Streptomyces* and *Thermoactinomyces,* isolated mainly from composts made from waste cotton and hay, respectively. Among them, two strains were placed in distinct taxonomic positions, although they formed phylogenetic lineages related to the genus *Streptomyces* and to the family *Streptosporangiaceae,* respectively. The evaluated generation time for *Streptomyces* in soil (1.7 days) differs from those calculated in nutrient media under laboratory conditions (1.14–1.88 h) [255]. The genus *Streptomyces* is aerobic, mesophilic, and Gram-positive actinobacteria. When Streptomyces produce antibiotics in soil, they are detected sporadically in tiny secluded sites [256]. Regardless of their unstableness in soil, they could be significant to the mushroom producers. Thermophilic actinobacteria (*Streptomyces thermovulgaris* Henssen 1957, *Thermoactinomyces vulgaris* Tsilinsky 1899, *Saccharomonospora viridis* Nanomura and Ohara 1971) are components of mushroom composts and play an important role in the aerobic decay of organic matter. They serve as a significant source of carbon and nitrogen for *A. bisporus* [184, 257]. Microbial supplements can act as slowly released nutrients that aim to prolong the production of mushroom fruiting bodies during the subsequent stages of cropping [184, 258]. The addition of Actinobacteria to mushroom substrate enhanced the mushroom yield by 10-32% [259], while also increasing the resistance of *A. bisporus* to the pathogen *T. aggressivum*. The strain *Streptomyces flavovirens* Waksman and Henrici 1848 A06, isolated from mushroom substrate on the 3rd day of fermentation, inhibited 42.4% of the *in vitro* growth of *T. harzianum* T54 and 27.6% of *T. aggressivum* f. *europaeum* T77 [260]. Strain *S. flavovirens* A06 significantly enhanced mushroom yield by 31.5% compared to the uninoculated control. The presence of the compost mold *T. aggressivum* further improved yield by 26.1%, while casing green mold *T. harzianum* only improved yield by 8%. There were no significant differences in efficacy against *T. aggressivum* between *S. flavovirens* A06 and the fungicide prochloraz. However, the efficacy of actinobacteria against *T. harzianum* was lower [260].

13.7.4 BENEFICIAL BACTERIOPHAGES

Bacteriophages that infect specific bacteria by killing them also have been used against causal agents of bacterial blotch. In combination with other beneficial organisms, phages could be highly effective [48].

13.7.5 BENEFICIAL NEMATODES

13.7.5.1 HOWARDULA HUSSEYI

Endoparasitic nematode, *Howardula husseyi,* occasionally appears in the phorid population [48]. It can be found both in compost and inside flies. Female phorids have the ability to lay eggs and deposit nematode larvae on spawned mushroom compost. Fertile females enter the phorid larvae or pupae and continue their development within the phorid body cavity. This process affects the fly's food reserves and egg-laying capabilities. Once the nematodes hatch, they enter the ovaries and oviducts of the flies. The oviposition of female phorids leads to the release of nematodes. The presence of developing nematodes reduces phorid egg production by 50-100%. However, there are currently no commercially available products based on this nematode species [48].

13.7.5.2 STEINERNEMA FELTIAE

Entomopathogenic nematode, *S. feltiae,* has been introduced as a commercial bioinsecticide against sciarids [48]. Strains of *S. feltiae* from Croatia, isolated from agricultural loamy soils with acidic pH, along with those from Slovenia (B30) and England (A2), form a monophyletic group in evolutionary analysis. The nematode *S. feltiae* carries bacteria that are lethal to the sciarid fly [261]. The bacteria from the nematode's gut are released into the insect. The nematode, at the third juvenile stage, enters the sciarid larvae at the third or fourth stage, either directly through the body wall or through natural body openings. The bacteria from the nematodes are released inside the body cavity of the sciarid larvae. Within 48 hours, the bacteria kill the sciarid larvae through blood poisoning. The nematodes develop fully by consuming the bacteria and insect tissues, and then they begin the reproduction process. Once the food supply has been depleted, the young nematodes exit the cadaver. The infective nematodes should be applied to the mushroom substrate during casing time and/or throughout the mushroom cultivation. The mushroom substrate does not serve as a food source for the nematodes. In on-farm trials, high levels of mortality were observed among *L. mali* larvae when *S. feltiae* was applied at rates of 81–300 million per 100 m^2 [48].

13.7.6 BENEFICIAL MITES

The mite species *Hypoaspis miles* Berlese 1892 is a known predator of cecid, phorid, and sciarid larvae in the mushroom crop [48]. A commercial formulation of this predator is available in the market for application in greenhouses. In the field, lower efficacy in sciarid control was observed compared to small-scale experiments. The advantages of their application are numerous: they are mobile, the growing conditions are suitable for their reproduction, and they can live for several weeks without food. Additionally, mites do not have negative effects on yield. Despite the low control rate, predatory mites are a safe alternative for chemical control and could be included in integrated pest control in mushroom production [48].

13.8 CONCLUSION

Making a proper substrate and integrated disease control are major challenges for mushroom growers. The lack of chemicals due to pathogen resistance and the harmful impact of pesticides implies pest and disease management based on biopesticides. The decrease in substrate quality after intensive production of substrate components, cereals, and animal manure leads to an unsuitable microbiome of both compost and casing soil. Microbiota from mushroom substrate have already exhibited positive effects in pest and disease control and yield improvement. These beneficial organisms include bacteria (*Bacillus* spp., *Pseudomonas* spp.), fungi (*Scytalidium* spp.), actinobacteria (*Streptomyces* spp., etc.), and nematodes (*Steinernema* spp.). Many of them are registered as biopesticides or biostimulators as they suppress pathogen growth and improve yield by colonizing the host environment. The introduction of well-timed and appropriate microbial combinations, similar to their natural environment, would create new possibilities for biorational edible mushroom cultivation with improved yield and quality.

KEYWORDS

- **beneficial microorganisms**
- **beneficial nematodes**
- **edible mushroom**
- **lactic acid bacteria**
- **microbial biopesticides**
- **micro-biological disease control**
- ***Pseudomonas putida***

REFERENCES

1. Royse, D. J., (2010). Effects of fragmentation, supplementation, and the addition of phase II compost to 2nd break compost on mushroom (*Agaricus bisporus*) yield. *Bioresour. Technol., 101*, 188–192. doi: 10.1016/j.biotech.2009.07.073.
2. Smith, J., (1969). Commercial mushroom production. *Process Biochem., 4*(43–46), 52.
3. Hayes, W. A., & Wright, S. H., (1979). Edible mushrooms. *Economic Microbiology* (pp. 141–176). New York, Academic Press, Inc.
4. Tautorus, T. E., & Townsley, P. M., (1983). Biological control of olive green mold in *Agaricus bisporus*cultivation. *Applied and Environmental Microbiology, 45*(2), 511–515. doi: 10.128/aem.45.2.511-515.1983.
5. Grogan, H. M., Adie, B. A. T., Gaze, R. H., Chalen, M. P., & Mills, P. R., (2003). Double-stranded RNA elements associated with the MVX disease of *Agaricus bisporus. Mycol. Res., 107*(2), 147–154. doi: 10.1017/s0953756203007202.
6. Milijašević-Marčić, S., Todorović, B., Potočnik, I., Stepanović, M., & Rekanović, E., (2012). First report of *Pseudomonas tolaasii* on *Agaricus bisporus* in Serbia. *Phytoparasitica, 40*(3), 299–303. doi: 10.1007/s12600-001-0215-z.
7. Potočnik, I., Vukojević, J., Stajić, M., Tanović, B., & Todorović, B., (2008). Fungicide sensitivity of selected *Verticillium fungicola* isolates from *Agaricus bisporus*farms. *Arch. Biol. Sci., 60*(1), 151–158. doi: 10.2298/ABS0801151P.
8. Potočnik, I., Vukojević, J., Stajić, M., Rekanović, E., Stepanović, M., Milijašević, S., & Todorović, B., (2010). Toxicity of biofungicide timorex 66 EC to *Cladobotryum dendroides* and *Agaricus bisporus. Crop Prot., 29*, 290–294. https://doi.org/10.1016/j.cropro.2009.07.016.
9. Potočnik, I., Vukojević, J., Stajić, M., Tanović, B., & Rekanović, E., (2010). Fungicide and essential oils sensitivity of *Mycogone perniciosa*originated from *Agaricus bisporus* farms. *Int. J. Med. Mushrooms 12*(1), 91–98.
10. Kosanović, D., Potočnik, I., Duduk, B., Vukojević, J., Stajić, M., Rekanović, E., & Milijašević-Marčić, S., (2013). *Trichoderma*species on *Agaricus bisporus*farms in

Serbia and their biocontrol. *Ann. Appl. Biol.*, *163*, 218–230. https://doi.org/10.1111/aab.12048.

11. Snetsinger, R., (1972). Laboratory studies of mushroom-infesting arthropods. *Mushroom Sci.*, *8*, 199–208.
12. Rinker, D. L., (2017). Insect, mite, and nematode pest of commercial mushroom production. *Edible and Medicinal Mushrooms: Technology and Applications* (1st edn., pp. 221–237). John Wiley, New Jersey, NJ.
13. Clift, A. D., & Terras, M. A., (1995). Mites as indicators of compost conditioning. *Mushroom Sci.*, *XIV*, 491–496.
14. Shamshad, A., (2010). The development of integrated pest management for the control of mushroom sciarid flies, *Lycoriella ingenua* (Dufour) and *Bradysia ocellaris* (Comstock), in cultivated mushrooms. *Pest Manag. Sci.*, *66*(10), 1063–1074. https://doi.org/10.1002/ps.1987.
15. White, P. F., (1986). The effect of sciarid larvae (*Lycoriella auripila*) on cropping of the cultivated mushroom (*Agaricus biospus*). *Ann. Appl. Biol.*, *109*(1), 11–17. Doi.org/10.1111/j.1744-7386.tb03179.x.
16. Villaverde, J. J., Sevilla-Moran, B., Sandin-Espana, P., Lopez-Goti, C., & Alonso-Prados, J. L., (2014). Biopesticides in the framework of the European Pesticide Regulation (EC) No. 1107/2009. *Pest Manag. Sci.*, *70*(1), 2–5. Doi.org/10.1002/ps.3663.
17. Grogan, H. M., (2008). Challenges facing mushroom disease control in the 21st century. *Proceeding of the Sixth International Conference on Mushroom Biology and Mushroom Products* (pp. 120–127). WSMBMP, Bonn, Germany.
18. Luković, J., Milijašević-Marčić, S., Hatvani, L., Kredics, L., Szücs, A., Vàgvölgyi, C., Duduk, N., et al., (2021). Sensitivity of *Trichoderma* strains from edible mushrooms to the fungicides prochloraz and metrafenone. *J. Environ. Sci. Health B*, *56*, 54–63. doi: 10.1080/03601234.2020.1838821.
19. Berendsen, R. L., Baars, J. J. P., Kalkhove, S. I. C., Lugones, L. G., Wösten, H. A. B., & Bakker, P. A. H. M., (2010). *Lecanicillium fungicola*: Causal agent of dry bubble disease in white-button mushroom. *Mol. Plant Pathol.*, *11*(5), 585–595. doi: 10.1111/j.1364-3703.2010.00627.x.
20. Wong, W. C., & Preece, T. F., (1987). Sources of *Verticillium fungicola* on a commercial mushroom farm in England. *Plant Pathol.*, *36*, 577–582. doi: 10.1111/j.1365-3059.1987.tb02276.x.
21. Sisto, D., Faggiano, S., & Rana, G. L., (1997). *Mycogone perniciosa*, a potential threat for cultivated mushrooms in southern Italy. *Petria*, *7*(3), 159–164.
22. Umar, M. H., Van, G. L. J. L. D., (1999). Studies on the morphogenisis of *Agaricus bisporus*: Dilema of normal versus abnormal fruit body development. *Mycol. Res.*, *103*, 1235–1244. https://doi.org/10.1017/S0953756299008473.
23. McKay, G. J., Egan, D., Morris, E., Scott, C., & Brown, A. E., (1999). Genetic and morphological characterization of *Cladobotryum* species causing cobweb disease of mushrooms. *App. Environ. Microbiol.*, *65*(2), 606–610. doi: 10.1128/AEM.65.2.606–610.1999.
24. Grogan, H. M., & Gaze, R. H., (2000). Fungicide resistance among *Cladobotryum* spp. – causal agents of cobweb disease of the edible mushroom *Agaricus bisporus*. *Mycol. Res.*, *104*(3), 357–364. https://doi.org/10.1017/S0953756299001197.

25. Seaby, D. A., (1996). Investigation of the epidemiology of green mould of mushroom (*Agaricus bisporus*) compost caused by *Trichoderma harzianum*. *Plant Pathol., 45*, 913–923. https://doi.org/10.111/j.1365-3059.1996.tb02902.x.
26. Krupke, O. A., Castle, A. J., & Rinker, D. L., (2003). The North American mushroom competitor, *Trichoderma aggressivum* f. *aggressivum*, produces antifungal compounds in mushroom compost that inhibit mycelial growth of the commercial mushroom *Agaricus bisporus*. *Mycol. Res., 107*, 1467–1475. https://doi.org/10.1017/S0953756203008621.
27. Ospina-Giraldo, M. D., Royse, D. J., Thon, M. R., Chen, X., & Romaine, C. P., (1998). Phylogenetic relationships of *Trichoderma harzianum* casing mushroom green mold in Europe and North America to other species of *Trichoderma* from world-wide sources. *Mycologia, 90*, 76–81. https://doi.org/10.2307/3761014.
28. Park, M. S., Bae, K. S., & Yu, S. H., (2004). Molecular and morphological analysis of *Trichoderma*isolates associated with green mold epidemic of oyster mushroom in Korea. *J. Huazhong Agricultural Univ., 23*, 157–164.
29. Park, M. S., Bae, K. S., & Yu, S. H., (2006). Two new species of *Trichoderma* associated with green mold of oyster mushroom cultivation in Korea. *Mycobiology, 34*, 11–113. doi: 10.4489/MYCO.2006.34.3.111.
30. Woo, S. L., Di Benedetto, P., Senatore, M., Abadi, K., Gigante, S., Soriente, I., Ferraioli, S., et al., (2004). Identification and characterization of *Trichoderma*species aggressive to *Pleurotus* in Italy. *J. Zhejiang Univ. (Agric. Life Sci.), 30*, 469–470.
31. Kredics, L., Hatvani, L., Antal, Z., Manczinger, L., Druzhina, I. S., Kubicek, C. P., Szekeres, A., et al., (2006). Green mold disease of oyster mushroom in Hungary and Transylvania. *Acta Microbiol. Immunol. Hung., 53*, 306–307.
32. Hatvani, L., Antal, Z., Manczinger, L., Szekeres, A., Druzhina, I. S., Kubicek, C. P., Nagy, A., et al., (2007). Green mold diseases of *Agaricus* and *Pleruotus* are caused by related but phylogenetically different *Trichoderma*species. *Phytopathology, 97*, 532–537. doi: 10.1094/PHYTO-97-4-0532.
33. Tokimoto, K., & Komatsu, M., (1979). Effect of carbon and nitrogen sources in media on the hyphal interference between *Lenditunus edodes* and some species of *Trichoderma*. *Ann. Phytopathol. Soc. Jpn., 45*, 261–264.
34. Adie, B., Grogan, H., Archer, S., & Mills, P., (2006). Temporal and spatial dispersal of *Cladobotryum* conidia in the controlled environment of a mushroom growing room. *Appl. Environ. Microbiol., 72*, 7212–7217. doi: 10.1128/AEM.01369-06.
35. Paine, S. G., (1919). Studies in bacteriosis II: A brown blotch disease of cultivated mushrooms. *Ann. Appl. Biol., 5*, 206–219. doi: 10.1111/j.1744-7348.1919.tb05291.x.
36. Rainey, P. B., Brodey, C. L., & Johnston, K., (1992). Biology of *Pseudomonas tolaasii*, causal of brown blotch disease of the cultivated mushrooms. *Advanced Plant Pathology* (pp. 95–117). Academic Press, New York, NY.
37. Geels, F. P., Hessen, L. P. W., & Van, G. L. J. L. D., (2008). Brown discoloration of mushrooms caused by *Pseudomonas agarici*. *J. Phytopathol., 140*, 249–259. doi: 10.1111/j.1439-0434.1994.tb.04814.x.
38. Chikthimmah, N., (2006). *Microbial Ecology of Mushroom Casing Soils and Preharvest Strategies to Enhance Safety and Quality of Fresh Mushrooms*. PhD Thesis, Pennsylvania State University, Pennsylvania, PL.
39. Rytter, J. L., Royse, D. J., & Romaine, C. P., (1991). Incidence and diversity of double-stranded RNA in *Lentinula edodes*. *Mycologia, 83*(4), 506–510. https://doi.org/10.2307/3760362.

40. Schisler, L. C., Sinden, J. W., & Sigel, E. M., (1967). Etiology, symptomology, and epidemiology of a virus disease of cultivated mushrooms. *Phytopathology, 57*, 519–526.
41. Gaze, R. H., Calvo-Bado, L., Challen, M. P., Adie, B., & Romaine, C. P., (2000). A new virus disease of *Agaricus bisporus*? *Mushroom Sci., 15*, 701–705.
42. Coles, P. S., (2002). Pest species biology and control: 5. Nematodes.*Mushroom Integrated Pest Management Handbook* (pp. 78–84). AMI, Pennsylvania State University, University Park, PA. http://www.cas.psu.edu(accessed on 9 January 2023).
43. Majić, I., Sarajlić, A., Kanižai, Š. G., Milijašević-Marčić, S., & Potočnik, I., (2019). Diversity of nematodes associated with white button mushroom. *Book of Abstracts of the VIII Congress on Plant Protection: Integrated Plant Protection for Sustainable Crop Production and Forestry* (pp. 190, 191). DZBS, Zlatibor, Serbia.
44. Hesling, J. J., (1972). Nematode pests of mushrooms. *Economic Nematology* (pp. 435–468). Academic Press, London, UK.
45. Bloom, J. R., Kaufman, T. D., & Lukezic, F. L., (1984). Reduced mushroom yields caused by nematode toxins. *Sci. Agric., 31*, 8, 9.
46. Barnes, H. F., (1926). An undescribed mushroom-feeding gall-midge. *Entomol. Mon. Mag., 62*, 89–92.
47. Keli, C., (2002). Arthropod pests. In: *Mushroom Integrated Pest Management. Handbook* (pp. 47–50). Pennsylvania State University, University Park, PA, American Mushroom Institute. http://www.cas.psu.edu(accessed on 9 January 2023).
48. Rinker, D. R., (2002). Biological control. In*: Mushroom Integrated Pest Management. Handbook* (pp. 33–36). AMI, Pennsylvania State University, University Park, PA. http://www.cas.psu.edu(accessed on 9 January 2023).
49. Terras, M. A., & Hales, D. F., (1995). Red pepper mites are vector of *Trichoderma. Sci. Cultiv. Edible Fungi, 2*, 485–490.
50. Wetzel, H. A., Wuest, P. J., Snetsinger, R., Royse, D. J., & Tetrault, R. C., (1982). Integrated pest management for mushroom farming. *Handbook for Commercial Mushroom Growers* (pp. 61–67). Pennsylvania State University, Pennsylvania, PA.
51. Fletcher, J. T., White, P. F., & Gaze, R. H., (1989). *Mushroom Pest and Disease Control* (2nd edn., p. 174). Intercept Ltd, Andover, Hants, UK.
52. Van Griensven, L. J. L. D., (1988). History and development. *The Cultivation of Mushrooms* (pp. 11–28). Darlington Mushroom Laboratories, Rustington, Sasex, England.
53. Stamets, P., & Chilton, J. C., (1983). *The Mushroom Cultivator: A Practical Guide to Growing Mushrooms at Home.* Agaricon Press, Olympia, Washington D.C.
54. Khanna, A. S., & Sharma, N. K., (1997). Nematodes in mushrooms: Management and control. *Advances in Mushroom Biology and Production – Proceedings of the Indian Mushroom Conference* (p. 431). MSI, NRCM, Solan, India.
55. Dmoch, J., (1995). *Arctoseius semiscissus* (Berlese, 1892) phoretic on mushroom sciarid flies in Poland – a possible agent for biological control of sciarids in mushroom houses. *Science and Cultivation of Edible Fungi* (pp. 533–537). A.A. Balkema, Rotterdam, the Netherlands.
56. Beyer, D. M., & Kremser, J. J., (2004). Evaluation of fungicide tolerance and control for three fungal diseases of mushrooms. Science and cultivation of edible and medicinal fungi. *Mushroom Sci.* (Vol. 16, pp. 521–529). Pennsylvania, PA, Penn State University, University Park.

57. Chakwiya, A., Van Der Linde, E. J., & Korsten, L., (2015). *In vitro* sensitivity testing of *Cladobotryum mycophilum* to carbendazim and prochloraz manganese. *S. Afr. J. Sci., 111*, 1–7. doi: 10.17159/sajs.2015/20140408.
58. Carrasco, J., Navarro, M. J., Santos, M., & Gea, F. J., (2017). Effect of five fungicides with different modes of action on mushroom cobweb disease (*Cladobotryum mycophilum*) and mushroom yield. *Ann. Appl. Biol., 171*(1), 62–69. https://doi.org/10.1111/aab.12352.
59. Chrysay-Tokousbalides, M., Kastanias, M. A., Philippoussis, A., & Diamantopoulou, P., (2007). Selective fungitoxicity of famaxadone, tebuconazole and trifloxystrobin between *Verticillium fungicola* and *Agaricus bisporus*. *Crop Prot., 26*, 469–475.
60. Bollen, G. J., & Van Zaayen, A., (1975). Resistance to benzimidazole fungicides in pathogenic strains of *Verticillium fungicola*. *Netherlands. J. Plant Pathol., 81,* 157–167.
61. Bonnen, A. M., & Hopkins, C., (1997). Fungicide resistance and population variation in *Verticillium fungicola*, a pathogen of the button mushroom, *Agaricus bisporus*. *Mycoll. Res., 101*, 89–96. https://doi.org/10.1017/S0953756296002237.
62. Grogan, H. M., Keeling, C., & Jukes, A. A., (2000). *In vivo* response of the mushroom pathogen *Verticillium fungicola* (dry bubble) to prochloraz-manganese. *Proceeding of Brighton Crop Protection Conference: Pests & Diseases* (Vol. 1, pp. 273–278). BCPC, Farnham, Surrey, UK.
63. McKay, G. L., Egan, D., Morris, E., & Brown, A. E., (1998). Identification of benzimidazole resistance in *Cladobotryum dendroides* using a PCR-based method. *Mycoll. Res., 102*(6), 671–676.
64. Grogan, H. M., (2006). Fungicide control of mushroom cobweb disease caused by *Cladobotryum* strains with different benzimidazole resistance profiles. *Pest Manag. Sci., 62*(2), 153–161. https://doi.org/0.1002/ps.1133.
65. Potočnik, I., Vukojević, J., Stajić, M., Rekanović, E., Milijašević, S., Todorović, B., & Stepanović, M., (2009). *In vitro* toxicity of selected fungicides from the groups of benzimidazoles and demethylation inhibitors to *Cladobotryumdendroides* and *Agaricus bisporus*. *J. Environ. Sci. Health B, 44*(4), 365–370. https://doi.org/10.1080/03601230902801059.
66. Potočnik, I., Vukojević, J., Stajić, M., Rekanović, E., Milijašević, S., Stepanović, M., & Todorović, B., (2009). Toxicity of fungicides with different modes of action to *Cladobotryum dendroides* and *Agaricus bisporus*. *J. Environ. Sci. Health B, 44*(8), 823–827. https://doi.org/10.1080/03601230903238699.
67. Romaine, C. P., Royse, D. J., & Schlagnhaufer, C., (2005). Superpathogenic *Trichoderma*resistant to Topsin M found in Pennsylvania and Delaware. *Mushroom News, 53*, 6–9.
68. Romaine, C. P., Royse, D. J., & Schlagnhaufer, C., (2008). Emergence of benzimidazole-resistant green mould *Trichoderma aggressivum*, on cultivated *Agaricus bisporus*in North America. *Science and Cultivation of Edible and Medicinal Fungi: Mushroom Science XVII, Proceeding of the 17th Congress of the International Society for Mushroom Science* (pp. 510–523). ISMS, Cape Town, South Africa.
69. Rinker, D. L., & Alm, G., (2008). Management of casing *Trichoderma*using fungicides. *Science and Cultivation of Edible and Medicinal Fungi: Mushroom Science XVII, Proceeding of the 17th Congress of the International Society for Mushroom Science* (pp. 496–509). ISMS, Cape Town, South Africa.

70. Gea, F. J., Navarro, M. J., & Tello, J. C., (2005). Reduced sensitivity of the mushroom pathogen *Verticillium fungicola* to prochloraz-manganese *in Vitro. Mycol. Res., 109*, 741–745. https://doi.org/10.1017/S095375620500242X.
71. Anonymous, (2019). *Grower Resources, Integrated Pest Management, Pesticides, Fungicides*. American Mushroom Institute. https://www.americanmushroom.org/integrated-pest-management/fungicides/ (accessed on 9 January 2023).
72. Kosanović, D., Potočnik, I., Vukojević, J., Stajić, M., Rekanović, E., Stepanović, M., & Todorović, B., (2015). Fungicide sensitivity of *Trichoderma*spp. from *Agaricus bisporus*farm in Serbia. *J. Environ.Sci. Health B, 50*, 607–6013. doi: 10.1080/03601234.2015.1028849.
73. Hatvani, L., Kocsubé, S., Mnczinger, L., Antal, Z., Szekeres, A., Druzhina, I. S., Komoń-Zelazowska, M., et al., (2008). The green mould disease global threat to the cultivation of oyster mushroom (*Pleurotus ostreatus*): A review. *Science and Cultivation of Edible and Medicinal Fungi: Mushroom Science XVII, Proceeding of the 17th Congress of the International Society for Mushroom Science* (pp. 485–495). ISMS, Cape Town, South Africa.
74. Bünemann, E. K., Schwenke, G. D., & Zwieten, L., (2006). Impact of agricultural inputs on soil organisms: A review. *Aust. J. Soil Res., 44*, 379–406. doi: 10.1071/SR05125.
75. Fai, P. B., & Grant, S., (2009). A rapid resazurin bioassay for assessing the toxicity of fungicides. *Chemosphere, 74*(9), 1165–1170. doi: 10.1016/j.chemosphere.2008.11.078.
76. Kim, K. Y., Jordan, D., & McDonald, G. A., (1997). Effect of phosphate-solubilizing bacteria and vesicular-arbuscular mycorrhizae on tomato growth and soil microbial activity. *Biol. Fertil. Soils, 26*, 79–87. doi: 10.1007/s003740050347.
77. Bashan, Y., (1999). Interactions of *Azospirillum* spp. in soil: A review. *Biol. Fertil. Soils, 29*, 246–256. doi: 10.1007/s003740050549.
78. Chen, S. K., Edwards, C. A., & Subler, S., (2001). Effects of the fungicides benomyl, captan and chlorothalonil on soil microbial activity and nitrogen dynamics in laboratory incubations. *Soil Biol. Biochem., 33*, 197–1980. doi: 10.1016/S0038-0717(01)00131-6.
79. Geels, F. P., Van Griensven, L. J. L. D., & Rutjens, A. J., (1991). Chlorine dioxide and the control of bacterial blotch. *Mushroom Sci., XIII*, 437–442.
80. Royse, D. J., & Wuest, P. J., (1980). Mushroom brown blotch: Effect of chlorinated water on disease intensity and bacterial populations in the casing soil and on the plate. *Phytopathology, 70*, 902–905.
81. Baker, K. F., & Scher, F. M., (1987). Enhancing the activity of biological control agents. *Innovative Approaches to Plant Disease Control* (pp. 1–19). John Willey and Sons, New York, NY.
82. Miller, F. C., & Spear, M., (1995). Very large scale commercial trial of a biological control for mushroom blotch disease. *Mushroom Sci., XIV*, 635–642.
83. Compeau, G., Al-Achi, B. J., Platsouka, E., & Levy, S. B., (1988). Survival of rifampicin resistant mutants of *Pseuodomonas fluorescens* and *P. putida* in soil systems. *Appl. Environ. Microbiol., 54*, 2432–2438. doi: 10.1128/aem.54.10.2432-2438.1988.
84. Olivier, J. M., & Guillaumes, J., (1981). Essais de lutte biologique contre la tache bacterienne. *Mushroom Sci., XI*, 353–368.
85. Munjal, A., Khanna, P. K., & Garcha, H. S., (1989). *In vitro* chemical and biological control of bacterial blotch of *Agaricus bisporus*. *Mushroom Sci., XII*, 667–677.

86. Khanna, P. K., Olivier, J. M., Samson, R. S., & Guillaumes, J., (1990). Isolation, characterization, and screening of antagonistis of *P. tolaasii* for biological control. *Indian Phytopahtol.*, *43*, 352–360.
87. Solomon, J. M., Beelman, R. B., & Bartley, C. E., (1991). Addition of calcium chloride and stabilized chlorine dioxide to irrigation water to improve quality and shelf life of *Agaricus bisporus*. *Mushroom Sci.*, *13*, 695–701.
88. Bech, K., & Riber-Rasmussen, C., (1967). Experiments with soil disinfectants for casing material and their effect on yield. *Mushroom Sci.*, *6*, 515–521.
89. Potočnik, I., Rekanović, E., Stepanović, M., Milijašević-Marčić, S., Todorović, B., Nikolić-Bujanović, L., & Čekerevac, M., (2014). Possibility of environmentally-safe casing soil disinfection on the cobweb disease control of button mushroom. *Pesticides and Phytomedicine*, *29*(4), 283–289. doi: 10.2298/PIF1404283P.
90. Todorović, B., Milijasević-Marčić, S., Potočnik, I., Stepanović, M., Rekanović, E., Nikolić-Bujanović, L., & Čekerevac, M., (2012). *In vitro* activity of antimicrobial agents against *Pseudomonas tolaasii*, pathogen of cultivated button mushroom. *J. Environ. Sci. Health B*, *47*(3), 175–179. https://doi.org/10.080/03601234.2012.632282.
91. Grogan, H. M., & Jukes, A. A., (2003). Persistence of the fungicides thiabendazole, carbendazim and prochloraz-Mn in mushroom casing soil. *Pest Manag. Sci.*, *59*, 1225–1231. doi: 10.1002/ps.759.
92. Papadopoulos, G., (2006). *The Fate of Prochloraz in Mushroom Casing*. Ph.D. Thesis University of Reading, UK.
93. Yarden, O., Salomon, R., Katan, J., & Aharonson, N., (1990). Involvement of fungi and bacteria in enhanced and nonenhanced biodegradation of carbendazim and other benzimidazole compounds in soil. *Can. J. Microbiol.*, *36*, 15–23. https://doi.org/10.1139/m90-004.
94. Cook, R., & Baker, K. F., (1983). *The Nature and Practice of Biological Control of Plant Pathogens* (p. 539). The APS. St. Paul, MN.
95. Carson, R., (1992). *Silent Spring*. Houghton Mifflin Harcourt: Boston, USA.
96. Ardakani, S. S., Heydari, A., Khorasani, N. A., Arjmandi, R., & Ehteshami, M., (2009). Preparation of new biofungicides using antagonistic bacteria and mineral compounds for controlling cotton seedling damping-off disease. *J. Plant Prot. Res.*, *49*(1), 49–55. doi: https://doi.org/10.2478/v10045-009-0007-3.
97. Hussey, N. W., Read, W. H., & Hesling, J. J., (1974). *The Pests of Protected Cultivation: The Biology and Control of Glasshouse and Mushroom Pests* (2nd edn.). Arnold, London, UK.
98. Georghiou, G. P., (1986). The magnitude of the resistance problem. *Pesticide Resistance: Strategies & Tactics for Management* (pp. 14–43). National Academy of Science, Washington D.C.
99. Carlile, M. J., & Watkinson, S. C., (1994). *The Fungi*. Academic Press, London, UK.
100. Huhnke, W., (1970). Modern mushroom farming. *Sci. J.*, *6*, 62–66.
101. Nair, N. G., & Fahy, P. C., (1972). Bacteria antagonistic to *Pseudomonas tolassii* and their control of brown blotch of the cultivated mushroom *Agaricus bisporus*. *J. Appl. Bacteriol.*, *35*, 439–442. https://doi.org/10.1111/j.1365-2672.1972.tb03720.x.
102. Townsley, P. M., (1974). Pure cultural industrial fermentation in the production of mushrooms. *Canadian Inst. Food Sci. Technol. J.*, *7*, 254–255. https://doi.org/10.1016/S0315-5463(74)3922-0.

103. Loeffler, W., Katzer, W., Kremer, S., Kugler, M., Petersen, F., Jung, G., Rapp, C., & Tschen, J. S. M., (1990). Gegen pilze wirksame antibiotika der *Bacillus subtilis*-GB 03-gruppe. *Forum Microbiol.*, *3*, 156–163.
104. Stanojević, O., Milijašević-Marčić, S., Potočnik, I., Stepanović, M., Dimkić, I., Stanković, S., & Berić, T., (2016). Isolation and identification of *Bacillus* spp. from compost material, compost, and mushroom casing soil active against *Trichoderma*spp. *Arch. Biol. Sci.*, *68*(4), 845–852. doi: 10.2298/ABS151104073S.
105. Maarten, P., Small, K., & Berg, G., (2000). Genotypic and phenotypic differentiation of an antifungal biocontrol strain belonging to *Bacillus subtilis*. *J. Appl. Microbiol.*, *89*, 463–471. https://doi.org/10.1045/j.1365-2672.2000.01136.x.
106. Fermor, T. R., Henry, M. B., Fenlon, J. S., Glensister, M. J., Lincoln, S. P., & Lynch, J. M., (1991). Development and application of a biocontrol system for bacterial blotch of the cultivated mushroom. *Crop Prot.*, *10*, 271–278. doi: 10.1016/0261-2194(91)90005-C.
107. Őzaktan, H., Bora, T., Alsan, E., & Uslu, A., (2005). Biological control potentials of fluorescent pseudomonad strains against bacterial blotch disease (*Pseudomonas tolaasii*) of cultivated mushroom. *Proceeding of the 1st International Symposium on Biological Control of Bacterial Plant Diseases* (pp. 144, 145). MBBLF, Seeheim / Darmstadt, Germany.
108. Fermor, T. R., (1986). Bacterial diseases of edible mushrooms and their control. *Proceeding of International 1st Symposium. Scientific and technical Aspects of Cultivating Edible Fungi* (pp. 361–370). The Pennsylvania State University, PA.
109. Martinez, M. A., Guerrero, M. M., Martinez, M. C., Barceló, N., Ros, C., Lacasa, A., & Tello, J., (2004). Efecto de la biofumigación con solarización sobre la microflora del suelo en cultivos de pimiento en inemadero. *Actas del XII Congreso de la Socedad Española de Fitopatologia* (p. 89). Lloret de Mar, Girona, Spain.
110. Cunningham, W. P., & Siago, W. B., (1992). *Environmental Science: A Global Concern* (p. 622). Wm. C. Brown Publishers. Dubuque, IA.
111. Grewal, P. S., Ehlers, R. U., & Shapiro-Illan, D. I., (1999). Critical issues and research needs for expanding the use of nematodes in biocontrol. *Nematodes as Biocontrol Agents* (pp. 479–489). CABI Publishing, Walingford, UK.
112. Galper, S., Eden, L. M., Stirling, G. R., & Smith, L. J., (1995). Simple screening methods for assessing the predacious activity of nematode-trapping fungi. *Nematologica* (Vol. 41, No. 1–4, pp. 130–140). Leiden, Netherlands, Academic Publishers.
113. Glockling, S. L., & Dick, M. W., (1994). *Dactylella megalobrocha,* a new species of nematophagous fungus with constricting ring. *Mycol. Res.*, *98*, 845–853. https://doi.org/10.1016/S0953-7562(09)80252-9.
114. Van, G. M., & Eicker, A., (1992). The morphology and physiology of *Arthrobotrys oligospora,* a weed mould on mushroom beds of *Agaricus bisporus*. *Mushroom Res.*, *1*, 33–38.
115. Gray, N. F., (1985). Ecology of nematophagous fungi: Effect of soil moisture, organic matter, pH and nematode density on distribution. *Soil Biol. Bichem.*, *17*, 499–507. https://doi.org/10.1016/0038-0717(85)900017-3.
116. Cayrol, J. C., & Frankowski, J. P., (1979). Perfecting a new method of biological control against mycophagous nematodes in mushroom farms. *Mushroom Scie.*, *X (Part 2)*, 407–413.

117. Dowsett, J. A., & Feid, J., (1979). Observations on the trapping of nematodes by *Dactylatia scaphoides* using optical, transmission and scanning-electron-microscopic techniques. *Mycologia, 71*, 379–391. https:/doi.org/10.1080/00275514.1979.12021016.
118. Wallace, H. R., (1963). *The Biology of Plant Parasitic Nematodes*. Edward Arnold Ltd., London, UK.
119. Koning, G., Hamman, B., & Eicker, A., (1996). The efficacy of nematophagous fungi on predaceous nematodes in soil compared with saprophagous nematodes in mushroom compost. *S. Afr. J. Bot.*, *62*(1), 49–53. https://doi.org/10.1016/S0254-6299(15)30578-0.
120. Stirling, G. R., & Mankau, R., (1978). *Dactylella oviparasitica,* a new fungal parasite of *Melodogyne* eggs. *Mycologia, 70*, 774–783. https://doi.org/10.2307/3759357.
121. Cooke, R. C., (1963). The predaceous activity of nematode-trapping fungi added to the soil. *Ann. Appl. Biol.*, *51*, 259–299. https://doi.org/10.1111/j.1744-7348.1963.tb03696.x.
122. Thorn, R. G., & Barron, G. L., (1984). Carnivorous mushrooms. *Science, 224*, 76–78. Washington, D.C.
123. Barron, G. L., & Thorn, R. G., (1986). Destruction of nematodes by species of *Pleurotus. Can. J. Bot.*, *65*, 774–778. doi: 10.1139/b87-103.
124. Bernheimer, A. W., & Avigad, L., (1979). A cytoliytic protein from the edible mushroom, *Pleurotus ostreatus. Biochim. Biophys. Acta*, *585*, 451–461. doi: 10.1016/0304-4165(79)90090-4.
125. Rodríguez-Barrera, T. M., Téllez-Téllez, M., Sánchez, J. E., Castañeda-Ramirez, G. S., De Acosta-Urdapilleta, M. L., Bautista-Garifias, C. R., & Aguilar-Marcelino, L., (2021). Edible mushrooms of the genus *Pleurotus* as biocontrol agents of parasites of importance for livestock. *Scientia Fungorum, 52,* e1375. doi: 10.33885/sf.2021.52.1375125.
126. Pineda-Alegría, J. A., Sánchez, J. E., Ventura-Zapata, E., González-Cortazar, M., & Aguilar-Marcelino, L., (2021). Nematicidal effect of shiitake (*Lentinula edodes*) extracts against *Haemonchus contortus. J. Med. Food,24*(9), 953–959. doi: 10.1089/jmf.2020.0168.
127. Fermor, T. R., & Wood, D. A., (1979). The microbiology and enzymology of wheat straw mushroom compost production. *Straw Decay and its Effect on Disposal and Utilization* (pp. 105–112). John Wiley, New York, NY.
128. Riding, I. L., & Hague, N. G. M., (1974). Some observations on a tylenchid nematode *Howardula* sp. parasitizing the mushroom phorid *Megasalia halterata* (*Phoridae, Diptera*). *Ann. Appl. Biol.*, *78*(3), 205–211. https://doi.org/10.1111/h.1744-7348.1974.tb01499.x.
129. Richardson, P. N., (1987). Susceptibility of mushroom pests to the insect-parasitic nematodes *Steinernema feltiae* and *Heterorhabditis heliothidis. Ann. Appl. Biol.*, *111*, 433–438. https://doi.org/10.1111/j.1744-7348.1987.tb01471.x.
130. Navaro, J. M., & Gea, F. J., (2014). Entomopathogenic nematodes for the control of phorid and sciarid flies in mushroom crops. *Pesqui. Agropecu. Bras.*, *49*(1), 11–17. doi: 10.1590/S0100-204X2014000100002.
131. Walsh, G., Kellegher, T., & Grogan, H., (2011). Disease control workshops. *Mushrooms – A Teagasc Advisory Newsletter*, *36*, 1–3.
132. Jess, S., & Schweiter, H., (2009). Biological control of *Lycoriella ingenua* (*Diptera*: *Sciaridae*) in commercial mushroom (*Agaricus bisporus*) cultivation: A comparison between *Hypoaspis miles* and *Steinernema feltia. Pest Manag. Sci.*, *65*(11), 1195–1200. https://doi.org/10.1002/ps.1809.

133. Coles, P. S., & Berber, W., (2002). Cultural control. In: *Mushroom Integrated Pest Management. Handbook* (pp. 27–32). AMI, Pennsylvania State University, University Park, PA. http://www.cas.psu.edu(accessed on 9 January 2023).
134. Ryckeboer, J., Mergaert, J., Coosemans, J., Deprins, K., & Swings, J., (2003). Microbiological aspects of biowaste during composting in a monitored compost bin. *J. Appl. Microbiol.*, *94*, 127–137. doi: 10.1046/j.1365-2672.2003.01800.x.
135. Ryckeboer, J., Mergaert, J., Vaes, K., Klammer, S., De Clerco, D., Coosemans, J., Insam, H., & Swings, J., (2003). A survey of bacteria and fungi occurring during composting and self-heating process. *Ann. Microbiol.*, *53*, 349–410.
136. MacGregor, S. T., Miller, F. C., Psarianos, K. M., & Finstein, M. S., (1981). Composting process-control based on interaction between microbial heat output and temperature. *Appl. Environ. Microbiol.*, *41*, 1321–1330. doi: 10.1128/em.41.6.1321-1330.1981.
137. Sharma, H. S. S., Lyons, G., & Chambers, J., (2000). Comparison of the changes in mushroom (*Agaricus bisporus*) compost during windrow and bunker stages of phase I and II. *Ann. Appl. Biol.*, *136*, 59–68. https://doi.org/10.1111/j.1744-7348.2000.tb00009.x.
138. Colak, M., (2004). Temperature profiles of *Agaricus bisporus*in composting stages and effects of different composting stages and effects of different composts formulas and casing materials on yield. *Afr. J. Biotechnol.*, *3*, 456–462.
139. Smith, J. F., Wood, D. A., & Thurston, C. F., (1995). Growth measurement of *Agaricus* mycelium in composed substrates as an indicator of compost selectivity and mushroom productivity. *Science and Cultivation of Edible Fungi* (pp. 293–301). Balkema, Rotterdam, the Netherland.
140. Cahyani, V. R., Matsuya, K., Asakawa, S., & Kimura, M., (2004). Succession and phylogenetic profile of eukaryotic communities in the composting process of rice straw estimated by PCR-DGGE analysis. *Biol. Fertil. Soils*, *40*, 334–344. https://doi.org/10.1111/j.1747-0765.2007.00187.x.
141. Ferreira, S. C., Santos, A. R., Braga, C., Da Silva, R., Souza, D. E., & Freitas, S. R., (2009). Microbial diversity in a bagasse-based compost prepared for the production of *Agaricus brasiliensis*. *Braz. J. Microbiol.*, *40*, 590–600. https://doi.org/10.1590/S1517-83822009000300023.
142. Atkey, P. T., & Wood, D. A., (1983). An electron microscope study of wheat straw composted as a substrate for the cultivation of the edible mushroom (*Agariucs bisporus*). *J. Appl. Bacteriol.*, *55*, 293–304. doi: 10.1111/j.1365-2672.1983.tb01326.x.
143. Gbolagade, J. S., (2006). Bacteria associated with compost used for the cultivation of Nigerian edible mushroom *Pleurotus tuber-regium* (Fr.) singer and *Lentinus squarrosulus* (Berk.). *Afr. J. Biotechnol.*, *5*, 338–342.
144. Miyatake, F., & Iwabuchi, K., (2005). Effect of high compost temperature on enzymatic activity and species diversity of culturable bacteria in cattle manure compost. *Biores. Technol.*, *96*, 1821–1825. doi: 10.1016/j.biortech.2005.01.005.
145. Eger, G., (1972). Experiments and comments on the action of bacteria on sporophore initiation in *Agaricus bisporus. Mushroom Sci.*, *8*, 719–725.
146. Gavini, F., Mergaert, J., Beji, A., Mielcarek, C., Izard, D., Kersters, K., & De Ley, J., (1989). Transfer of *Enterobacter agglomerans* (Beijerinck 1888) Ewing and Fife 1972 to *Pantoea* gen. nov. as *Pantoea agglomerans* comb. nov. and description of *Pantoea dispersa* sp. nov. *Int. J. Syst. Bact.*, *39*, 337–345.

147. Castellani, A., & Chalmers, A. J., (1919). *Manual of Tropical Medicine* (3rd edn.). Williams Wood and Co., New York, USA.
148. Gottheil, O., (1901). Botanische beschreibung einiger bodenbakterien. *Zent. fuer Bakt., Paras., Infek., und Hyg.*, *2*(7), 680–691.
149. Droffner, M. L., Brinton, Jr. W. F., & Evans, E., (1995). Evidence for the prominence of well characterized mesophilic bacteria in thermophilic (50–70°C) composting environments. *Biomass Bioenergy*, *8*, 1991–1995.
150. Meyer, A., & Neide, E., (1904). Botanische Beschreibung einiger sporenbildenden Bakterien. *Zent. fuer Bakt., Paras., Infek., und Hyg.*, *2*(12), 337–352.
151. Velázquez-Cedeño, M., Farnet, A. M., Mata, G., & Savoie, J. M., (2008). Role of *Bacillus* spp. in antagonism between *Pleurotus ostreatus* and *Trichoderma harzianum* in heat-treated wheat-straw substrates. *Biores. Technol.*, *99*(15), 6966–6973.
152. Watabe, M., Rao, J. R., Murphy, A. R., & Moore, J. E., (2003). Inhibition of *Listeria ivanovii* by *Paenibacillus lentimorbus* isolated from phase II mushroom compost. *World J. Microbiol. Biotechnol.*, *19*, 875–877.
153. Subbarao, N. S., (1992). *Biofertilizers in Agriculture* (2nd edn.). Oxford and IBH Publishing, New Delhi, India.
154. Cho, Y. S., Kim, J. S., Crowley, D. E., & Cho, B. G., (2003). Growth promotion of the edible fungus *Pleurotus ostreatus* by fluorescent pseudomonads. *FEMS Microbiol. Lett.*, *218*, 271–276. https://doi.org/10.1016/S0378-1097(02)01144-8.
155. Moon, N. J., & Henk, W. G., (1980). Progression of epiphytic microflora in wheat and alfalfa silages as observed by scanning electron microscopy. *Appl. Environ. Microbiol.*, *40*, 1122–1129.
156. Costerton, J. W., Geesey, G. G., & Cheng, K. J., (1978). How bacteria stick. *Sci. Am.*, *238*, 86–95.
157. Eddy, B. P., & Jacobs, L., (1976). Mushroom compost as a source of food for *Agaricus bisporus. Mushroom J.*, *38*, 56–59. doi: 10.1128/aem.40.6.1122-1129.1980.
158. Van Den Bogart, H. G., Van Den Ende, G., Van Loon, P., C. C., & Van Griensven, L. J. L. D., (1993). Mushroom worker's lung: Serologic reactions to thermophilic actinomycetes present in the air of compost tunnels. *Mycopathologia*, *122*, 21–28.
159. Crawford, D. L., (1974). Growth of *Thermomonospora fusca* on lignocellulosic pulps of varying lignin content. *Can. J. Microbiol.*, *20*, 1069–1072. https://doi.org/10.1139/m74-167.
160. Wang, C., Guo, X., Deng, H., Dong, D., Tu, Q., & Wu, W., (2014). New insights into the structure and dynamics of actinomycetal community during manure composting. *Appl. Microbiol. Biotechnol.*, *98*(7), 3327–3337. doi: 10.1007/s00253-013-5424-6.
161. Peters, S., Koschinsky, S., Schwieger, F., & Tebbe, C. C., (2000). Succession of microbial communities during hot composting as detected by PCR-single-strand-conformation polymorphism-based genetic profiles of small-subunit rRNA genes. *Appl. Environ. Microbiol.*, *66*(3), 930–936. doi: 10.1128/AEM.66.3.930-936.2000.
162. Vajna, B., Nagy, A., Sajben, E., Manczinger, L., Szijarto, N., Kadar, Z., Bordas, D., & Marialigeti, K., (2010). Microbial community structure changes during oyster mushroom substrate preparation. *Appl. Microbiol. Biotechnol.*, *86*(1), 367–375. doi: 10.1007/s00253-009-237-3.
163. Mehta, C. M., Palni, U., Franke-Whittle, I. H., & Sharma, A. K., (2014). Compost: Its role, mechanism, and impact on reducing soil-borne plant diseases. *Waste Manag.*, *34*(3), 607–622. doi: 10.1016/j.wasman.2013.11.012.

164. Viljoen, J., Fred, E., & Peterson, W., (1926). The fermentation of cellulose by thermophilic bacteria. *J. Agric. Sci., 16*, 1–17.
165. Izquierdo, J. A., Sizova, M. V., & Lynd, L. R., (2010). Diversity of bacteria and glycosyl hydrolase family 48 genes in cellulolytic consortia enriched from thermophilic biocompost. *Appl. Environ. Microbiol., 76*(11), 3545–3553. doi: 10.128/AEM.02689-09.
166. Zhang, X., Zhong, Y., Yang, S., Zhang, W., Xu, M., Ma, A., Zhuang, G., et al., (2014). Diversity and dynamics of the microbial community on decomposing wheat straw during mushroom compost production. *Biores. Technol., 170*, 183–195. doi: 10.1016/jbiortech.2014.07.093.
167. Straatsma, G., Banson, R. A., & Olijnsma, T. W., (1994). Ecology of thermophilic fungi in mushroom compost, with emphasis on *Scytalidium thermophilum* and growth stimulation of *Agaricus bisporus*mycelium. *Appl. Environ. Microbiol., 60*, 454–458. doi: 10.128/aem.60.2.454–458.1994.
168. Straatsma, G., Olijnsma, T. W., Gerrits, J. P. G., Amsing, J. G. M., Op Den Camp, H. J. M., & Van Griensven, L. J. L. D., (1994b). Inoclataion of *Scytalidium thermophilum* in button mushroom compost and its effect on yield. *Appl. Environ. Microbiol., 60*(9), 3049–3054. doi: 10.1128/aem.60.9.3049-3054.1994.
169. Maheshwari, R., Bharadwaj, G., & Bhat, M. K., (2000). Thermophilic fungi: Their physiology and enzymes. *Mycrobiol. Mol. Biol. Rev., 64*(3), 461–488. doi: 10.1128/MMBR.64.6.461-4888.2000.
170. El-naghy, M. A., El-Katatny, M. S., & Attia, A. A., (1991). Degradation of cellulosic materials by *Sporotrichum thermophile* culture filtrate for sugar production. *Int. Biodeterior., 27*(1), 75–86. https://doi.org/10.1016/0265-3036(91)90025-M.
171. Firenzuoli, F., Gori, L., & Lombardo, G., (2007). The medical mushroom *Agaricus blazeil* Murril. Literature and pharmaco-toxicological problems. *E-CAM, 5*, 3–15. doi: 10.1093/ecam/nem007.
172. Have, R. T., Wijngaard, H., Aries-Kronenburg, N. A. E., Straatsma, G., & Schaap, P. J., (2003). Lignin degradation by *Agaricus bisporus*accounts for a 30% increase in bioavailable holocellulose during cultivation on compost. *J. Agric. Food Chem., 51*, 2242–2245. doi: 10.1021/jf021131h.
173. Chang, S. T., & Miles, P. G., (2004). *Mushrooms: Cultivation, Nutritional Value, Medicinal Effect, and Environmental Impact*. CRC Press, Boca Raton, FL.
174. Wood, D. A., & Fermor, T. R., (1981). Nutrition of *Agaricus bisporus*in compost. *Mushroom Sci., 11*, 63–73.
175. Chang, Y., (1967). The fungi of wheat straw compost. II biochemical and physiological studies. *Trans. Br. Mycol. Soc., 50*, 667–667.
176. Chang, Y., & Hudson, H. J., (1967). The fungi of wheat straw compost. I ecological studies. *Trans. Br. Mycol. Soc., 50*, 649–666.
177. Gerrits, J. P. G., (1969). Organic compost constituents and water utilized by the cultivated mushroom during spawn run and cropping. *Mushroom Sci., 7*, 111–126.
178. Hedger, J. N., & Hudson, H. J., (1974). Nutritional studies of *Thermomycs lanuginosus* from wheat straw compost. *Trans. Br. Mycol. Soc., 62*, 129–143.
179. Waksman, S. A., & Nissen, W., (1932). On the nutrition of the cultivated mushroom *Agaricus campestris,* and the chemical changes brought about by this organism in manure compost. *Am. J. Bot., 19*, 514–537. https://doi.org/10.1002/j.1537-2197.1932.tb09668.x.

180. Durrant, A. J., Wood, D. A., & Cain, R. B., (1991). Lignocellulose biodegradation by *Agaricus bisporus*during solid substrate fermentation. *J. Gen. Microbiol.*, *137*, 751–755. https://doi.org/10.1099/00221287-137-4-751.
181. Bonnen, A. M., Anton, L. H., & Orth, A. B., (1994). Lignin-degrading enzymes of the commercial button mushroom, *Agaricus bisporus*. *Appl.Environ. Microbiol.*, *60*, 960–965. doi: 10.1128/aem.60.3.960-965.1994.
182. Ten, H. R., Wijngaard, H., Aries-Kronenburg, N. A. E., Straatsma, G., & Schaap, P. J., (2003). Lignin degradation by *Agaricus bisporus*accounts for a 30% increase in bioavailable holocellulose during cultivation on compost. *J. Agric. Food Chem.*, *561*, 2242–2245. doi: 10.1021/jf021131h.
183. Adams, J. D. W., & Frostick, L. E., (2008). Investigating microbial activities in compost using mushroom (*Agaricus bisporus*) cultivation as an experimental system. *Biores. Technol.*, *99*, 1097–1102. https://doi.org/10.1016/j.biortech.2007.02.019.
184. Fermor, T. R., & Grant, W. D., (1985). Degradation of fungal and actinomycete mycelia by *Agariucs bisporus*. *J. Gen. Microbiol.*, *131*, 1729–1737.
185. Wood, D. A., & Goodenough, P. W., (1977). Fruiting of *Agaricus bisporus*. Changes in extracellular enzyme activities during growth and fruiting. *Arch. Microbiol.*, *114*, 161–165.
186. Piquemal, M., Belot, L., Viala, G., & Latine, J. C., (1977). Influence de la composition du milieu de culture sur la biosynhèse d'acide oxalique par le mycelium d'*Agariucus bisporus*. *C. R. Hebd. Sie. Acad. Sci. D*, *284*, 741–744.
187. Sparling, G. P., Fermor, T. R., & Wood, D. A., (1982). Evaluation of methods for measuring microbial biomass in composted wheat straw; the significance of biomass in nutrition of *Agaricus bisporus*. *Soil Biol. Biochem.*, *14*, 609–611.
188. Cheng, K. J., Fay, J. P., Coleman, R. N., Milligan, L. P., & Costerton, J. W., (1984). Formation of bacterial microcolonies on feed particles in the rumen. *Appl. Environ. Microbiol.*, *41*, 298–305. doi: 10.1128/aem.41.1.298-305.1981.
189. Flegg, P. B., (1979). Effect of temperature on sporophore initiation and development in *Agaricus bisporus*. *Mushroom Sci.*, *10*, 595–602.
190. Noble, R., Fermor, T. R., Lincoln, S., Dobrovin-Pennington, A., Mead, A., Li, R., & Evered, C., (2003). Primordia initiation of mushroom (*Agariucus bisporus*) strains on axenic casing materials. *Mycologia*, *95*, 620–629. doi: 10.1080/15572536.2004.11833066.
191. Wood, D. A., (1976). Primordium formation in axenic cultures of *Agaricus bisporus*(Lange) sing. *J. Gen. Mycrobiol.*, *95*, 313–323. https://doi.org/10.1099/00221287-95-2-313.
192. Hayes, W. A., Randle, P. E., & Last, F. T., (1969). The nature of the microbial stimulus affecting sporophore formation in *Agaricus bisporus*(Lange) sing. *Ann. Appl. Biol.*, *64*, 177–187. https://doi.org/10.1111/j.1744-7348.1969.tb02867.x.
193. Rainey, P. B., Cole, A. L. J., Fermor, T. R., & Wood, D. A., (1990). A model system for examining involvement of bacteria in basidiome initiation of *Agaricus bisporus*. *Mycological Res.*, *94*, 191–194. doi: 10.1016/S0953-7562(09)80612-6.
194. Grove, J. F., (1981). Volatile compounds from the mycelium of the mushroom *Agaricus bisporus*. *Phytochemistry*, *20*, 2021, 2022.
195. Combet, E., Henderson, J., Eastwood, D. C., & Burton, K. S., (2006). Eight carbon volatiles in mushrooms and fungi: Properties, analysis, and biosynthesis. *Mycoscience*, *47*, 317–326. https://doi.org/10.1007/S10267-006-0318-4.

196. Noble, R., Dobrovin-Pennington, A., Pederby, J., & Rodger, A., (2009). Volatile C8 compounds and pseudomonads influence primordium formation of *Agaricus bisporus. Mycologia, 101*(5), 583–591. doi: 10.3852/07-194.
197. Long, P. E., & Jacobs, L., (1974). Aseptic fruiting of the cultivated mushroom *Agaricus bisporus*. *Trans. Br. Mycol. Soc.*, *63*, 99–107.
198. Hayes, W. A., & Nair, N. G., (1976). Effects of volatile metabolic by-products of mushroom mycelium on the ecology of the casing layer. *Mushroom Sci.*, *9*(1), 259–268.
199. Behrendt, U., Ulrich, A., & Schumann, P., (2003). Fluorescent pseudomonads associated with the phyllosphere of grasses; *Pseudomonas trivialis* sp. nov., *Pseudomonas poae* sp. nov. and *Pseudomonas congelans* sp. nov. *Int. J. Syst. Evol. Microbiol.*, *53*, 1461–1469.
200. Fermor, T. R., Lincoln, S., Noble, R., & Dobrovin-Pennington, A., (2000). Microbiological properties of casing. *Science and Cultivation of Edible Fungi* (pp. 447–454). Balkema, Rotterdam, the Netherland.
201. Doores, S., Kremer, M., & Beelman, R., (1986). Evaluation and bacterial populations associated with fresh mushrooms (*Agaricus bisporus*). Developments in crop science (10),*Proceeding of the International Symposium on Technical Aspects of Cultivating Edible Fungi* (pp. 283–294). Penn State University, University Park, PA.
202. Samson, R., (1986). Variability of fluorescent *Pseudomonas*populations in compost and casing soils used for mushroom cultures. *Proceeding of International Symposium of Scientific and Technical Aspects of Cultivating Edible Fungi* (pp. 19–25). Pennsylvania State University; University Park, PA.
203. Miller, N., Gillespie, J. B., & Doyle, O. P. E., (1995). The involvement of microbiological components of peat based casing materials in fructification of *Agaricus bisporus, Mushroom Sci.*, *14*(1), 313–321.
204. Rao, M. G., & Subrahmanyan, V., (1929). Studies on soil actinomycetes. Part III. Standardization of a plate method of counting soil actinomycetes. *J. Indian Inst. Sci.*, *XII(A)*(57–68), 253–273.
205. Waksman, S. A., & Purvis, E. R., (1932). The microbiological population of peat. *Soil Sci.*, *34*, 95–113.
206. Tahvonen, R., (1993). The disease suppressiveness of light-colored sphagnum peat and biocontrol of plant diseases with *Streptomyces* sp. *Acta Horiticult.*, *342*, 37–42.
207. Atwill, E. R., (2005). *Microbial Pathogens Excreted by Livestock and Potentially Transmitted to Humans Through Water*. Tulare, CA, USA, Veterinary Medicine Teaching and Research Center, School of Veterinary Medicine, University of California, Davis, CA. http://nature.Berkeley.edu/forestry/rangelandwq/pdfs/AtwillArcfinal.pdf (accessed on 9 January 2023).
208. Himathongkham, S., Riemann, H., Bahari, S., Nuanualsurwan, S., Kass, P., & Cliver, D. O., (2000). Survival of *Salmonella typhimurium* and *Escherichia coli* O157:H7 in poultry manure and manure slurry at sublethal temperatures. *Avian Diseas.*, *44*, 853–860.
209. Heuvelink, A. E., Zwartkruis-Nahuis, J. T. M., Van Den Biggelaar, F. L. A. M., Van Leeuwen, W. J., & De Boer, E." Instead of "Heuvelink, A. E., Zwartkruis-Nahuis, J. T. M., Vand Den, B., F. L., Van, L., W. J., & De Boer, E., (1996). Isolation and characterization of verocytotoxin-producing *Escherichia coli* O157 from slaughter pigs and poultry. *Int. J. Food Microbiol.*, *52*, 67–75. doi: 10.1016/s0168-1605(99)00119-1.
210. Turpin, P. E., Dhir, V. K., Maycroft, K. A., Rowlands, C., & Wellington, E. M. H., (1992). The effect of *Streptomyces* species on the survival of *Salmonella* in soil. *FEMS Microbiol. Lett.*, *101*, 271–280.

211. Kurek, W., & Jaroszuk-Scisel, J., (2002). Rye (*Secale cereale*) growth promotion by *Pseudomonas fluorescens*strains and their interactions with *Fusarium culmorum* under various soil conditions. *Biol. Control, 26*, 48–56. https://doi.org/10.1016/S1049-9644(02)00115-9.
212. Rakshanya, U., Hema, S. J., & Kanchana, D. D., (2011). Antagonistic activity of actinomycetes isolates against human pathogen. *J. Microbiol. Biotechnol. Res., 1*(2), 74–79.
213. Wu, H., Lian, Y. P., Liu, B., Ren, Y. L., Yin, P. S., & Huang, F. C., (2014). *Thermotunica guangxiensis* gen. Nov., sp nov., isolated from mushroom residue compost. *Int. J. Syst. Evol. Microbiol., 64*(5), 1593–1599. doi: 10.1099/ijs.0.057562–0.
214. Parati, F., Altieri, R., Esposito, A., Lobianco, A., Pepi, M., Montesi, L., & Nair, T., (2011). Validation of thermal composting process using olive mill solid waste for industrial scale cultivation of *Agaricus bisporus. Int. Biodeterior. Biodegradation, 65*(1), 160–163.
215. Wiegant, W. M., Wery, J., Buitenhuis, E. T., & De Bont, J. A. M., (1982). Growth-promoting effect of thermophilic fungi on the mycelium of the edible mushroom *Agaricus bisporus. Appl. Envrion. Microbiol., 58*, 2654–2659. doi: 10.1128/aem.58.8.2654-2659.992.
216. Chalaux, N., Savoie, J. M., & Olivier, J. M., (1993). Growth inhibition of *Agaricus bisporus*and associated thermophilic species by fungicides used in wheat cultivation. *Agronomie, 13*, 407–412.
217. Salar, R. K., & Aneja, K. R., (2007). Significance of thermophilic fungi in mushroom compost preparation: Effect on growth and yield of *Agaricus bisporus*(Lange) sing. *J. Agric. Technol., 3*(2), 241–253.
218. Ross, R. C., & Harris, P. J., (1983). An investigation into the selective nature of mushroom compost. *Sci. Hortic., 19*, 55–64.
219. Ross, R. C., & Harris, P. J., (1983b). The significance of thermophilic fungi in mushroom compost preparation. *Sci. Hortic., 20*, 61–70.
220. Straatsma, G., Gerrits, J. P. G., Augustijn, M. P. A. M., Op Den Camp, H. J. M., Vogels, G. D., & Van Griensven, L. J. L. D., (1989). Population dynamics of *Scytalidium thermophilum* in muhsroom compost and stimulatory effects on growth rate and yield of *Agaricus bisporus. J. Gen. Microbiol., 135*, 751–759.
221. Straatsma, G., Gerrits, J. P. G., Gerrits, T. M., Op Den, C. H. J. M., Van Griensven, L. J. L. D., (1991). Growth kinetics of *Agaricus bisporus*mycelium on solid substrate (mushroom compost). *J. Gen. Microbiol., 137*, 1471–1478.
222. Gandy, D. G., (1985). Bacterial and fungal diseases. *The Biology and Technology of the Cultivated Mushroom* (pp. 261–277). John Wiley & Sons Ltd., Chichester, UK.
223. Renard, Y., & Cailleux, R., (1973). Contribution á l' étitude des microorganismes du compost destiné á la culture du champignon de couche. *Revue du Myologie, 37*, 36–47.
224. Olivier, J. M., & Guillaumes, J., (1979). Evolution microbiologique des compost pendant la croissance mycelienne du champignon de couche. *Mushroom Sci., X*, 311–334.
225. Foieri, A., Pedrini, N., & Toledo, A., (2017). Natural occurrence of the entomopathogenic genus *Pandora* on spittlebug pests of crops and pastures in Argentina. *J. Appl. Entomol.*, 1–8. doi: 10.1111/jen.12476.
226. Park, J. Y., & Agnihotri, V. P., (1969). Bacterial metabolites trigger sporophore formation in *Agaricus bisporus. Nature, 222*(5197),984. doi: 10.1038/222984a0.

227. Curto, S., & Favelli, F., (1972). Stimulative effect of certain microorganisms (bacteria, yeasts, and microalgae) upon fruit-body formation of *Agaricus bisporus*(Lange) sing. *Mushroom Sci.*, *7*, 67–74.
228. Han, J., (1999). The influence of photosynthetic bacteria treatments on the crop yield, dry matter content, and protein content of the mushroom *Agaricus bisporus*. *Sci. Hortic.*, *82*, 171–178.
229. Tanaka, T., Watanabe, K., Hotta, Y., Lin, D., & Sasaki, K., (1991). Formation of 5-aminolevulinic acid under aerobic/dark condition by a mutant of *Rhodobacter spaeroides*. *Biotechnol. Lett.*, *13*, 589–594.
230. Sasaki, K., Tanaka, T., Nishio, N., & Nagai, S., (1993). Effect of culture pH on the extracellular production of ALA by *Rhodobacter sphaeroides* from volatile fatty acids. *Biotechnol. Lett.*, *15*, 859–864.
231. Tanaka, T., Hotta, T., Takeuchi, Y., & Konnai, M., (1992). Promotive effects of 5-aminolevulinic acid on yield of several crops. *Proceedings of the 19th Annual Meeting on Plant Growth Regulator Society of America* (pp. 237–241). PGRSA, San Francisco, CA.
232. Shipman, R. H., Kao, I. C., & Fan, L. T., (1977). Single-cell protein production by photosynthetic bacteria. *Advances in Applied Microbiology*, *21*, 161–183. https://doi.org/10.1016/S0065-2164(08)70041-8.
233. Savoie, J. M., Iapicco, R., & Largeteau-Mamoun, M., (2001). Factors influencing the competitive saprophytic ability of *Trichoderma harzianum* Th2 in mushroom (*Agaricus bisporus*) compost. *Mycological Res.*, *105*, 1348–1356. https://doi.org/10.1017/S0953756201004993.
234. Food and Drug Administration (FDA), (1999). *Code of Federal Regulations, Title 21: Food and Drugs, Chapter 1: Food and Drug Administration Department of Health and Human Services, Part 184: Direct Food Substances Afirmated as Generally Recognized as Safe*. US Government Printing Office, Washington, DC, USA.
235. Chittihunsa, T., Bangeekhan, E., Wongsamitkul, N., & Subsomboon, T., (2007). Screening of *Bacillus* spp. suppressing the infection of *Trichoderma*sp. in mushroom cultivation. *KMITL Sci. Technol. J.*, *7*(S1), 19–27.
236. Lahlali, R., Peng, G., Gossen, B. D., McGregor, L., Yu, F. Q., Hynes, R. K., Hwang, S. F., et al., (2013). Evidence that the biofungicide serenade (*Bacillus subtilis*) suppresses clubroot on canola via antibiosis and induced host resistance. *Phytopathology*, *103*, 245–254. doi: 10.1094/PHYTO-06-12-0123-R.
237. Serrano, L., Manker, D., Brandi, F., & Cali, T., (2013). The use of *Bacillus subtilis* QST 713 and *Bacillus pumilus* QST 2808 as protectant fungicides in conventional application programs for black leaf streak control. *Acta Hortic.*, *986*, 149–155.
238. Védie, R., & Rousseau, T., (2008). Serenade biofungicide: Une innovation mjeure dans les champignonnières françaises pour lutter contre *Trichoderma aggressivum*, agent de la moisissure verte du compost. *La Lettre du CTC*, *21*, 1, 2.
239. Kim, W. G., Weon, H. Y., Seok, S. J., & Lee, K. H., (2008). *In vitro*antagonistic characteristics of bacilli isolates against *Trichoderma*spp. and three species of mushrooms. *Mycobiology*, *36*(4), 266–269. doi: 10.4489/MYCO.2008.36.4.266.
240. Nagy, A., Manczinger, L., Tombácz, D., Hatvani, L., Györfi, J., Antal, Z., Sajben, E., et al., (2012). Biological control of oyster mushroom green mould disease by antagonistic *Bacillus* species. *IOBC-WPRS Bull*, *78*, 289–293.

241. Potočnik, I., Milijašević-Marčić, S., Stanojević, O., Berić, T., Stanković, S., Kredics, L., & Hatvani, L., (2019). The activity of native *Bacillussubtilis* strains in control of green mould disease of oyster mushroom (*Pleurotus* spp.). *Pestides and Phytomedicine*, *34*(2), 97–102. doi: https://doi.org/10.2298/PIF1902097P.
242. Gong, A. D., Li, Q. S., Song, X. S., Yao, W., He, W. J., Zhang, J. B., & Liao, Y. C., (2015). Antagonistic mechanism of iturin a and plipastatin A from *Bacillus amyloliquefaciens* S76-3 from wheat spikes against *Fusarium graminearum*. *PloS One*, *10*(2), e0116871.
243. Potočnik, I., Todorović, B., Rekanović, E., Luković, J., Paunović, D., & Milijašević-Marčić, S., (2018). Impact of *Bacillus subtilis* QST713 mushroom grain spawn treatment on yield and green mould control. *Pestides and Phytomedicine*, *33*(3, 4), 205–211. https://doi.org/10.2298/PIF180420SP.
244. Potočnik, I., Rekanović, E., Todorović, B., Luković, J., Paunović, D., Stanojević, O., & Milijašević-Marčić, S., (2019). The effects of casing soil treatment with *Bacillus subtilis* Ch-13 biofungicide on green mould control and mushroom yield. *Pestides and Phytomedicine*, *34*(1), 53–60. doi: https://doi.org/10.2298/PIF1901053P.
245. Potočnik, I., Todorović, B., Milijašević-Marčić, S., Luković, J., Kanižai, Š. G., Majić, I., & Rekanović, E., (2021). A large-scale study on the effectiveness of a *Bacillus subtilis* Ch-13-based biofungicide against green mould disease and mushroom yield improvement. *Pesticides and Phytomedicine*, *36*(2), 83–90. doi: https://doi.org/10.2298/PIF2102083P.
246. Khalili, H. R., Olfati, J. A., & Fallah, A., (2013). Plant growth promoting rhyzobacteria affect button mushroom yield and quality. *South West. J. Hortic., Biol. Environ.*, *4*(2), 83–89. doi: 10.5455/ESEBB.20151029051037.
247. Kim, Y. I., & Kwak, W. S., (2012). Isolation and identification of lactic acid bacteria from spent mushroom substrate for silage making and determination of optimal medium conditions for growth. *J. Anim. Sci. Technol.*, *54*(6), 435–442. https://doi.org/10.5187/JAST.2012.54.6.435.
248. Kanižai, Š. G., Rapčan, I., Potočnik, I., Sarajlić, A., & Majić, I., (2021). Potential of lactic acid bacteria in biocontrol of *Aspergillus Niger, Penicillium chrysogenum* and*Fusarium graminearum* in culture media and natural substrate. *Zemdirbyste-Agric.*, *108*(2), 153–158.
249. Suzuki, Y., Okano, K., & Kato, S., (1995). Characteristic of white-rotted woody materials obtained from Shiitake mushroom (*Lentinus edodes*) and Nameko mushroom (*Pholita nameko*) cultivation with *in vitro*rumen fermentation. *Anim. Feed Sci. Technol.*, *54*, 227–236. https://doi.org/10.1016/0377-8401(95)00769-j.
250. Unaogu, I. C., Gugnani, H. C., & Lacey, J., (1994). Occurrence of thermophilic actinomycetes in natural substrates in Nigeria. *Antonie van Leeuwenhoek*, *65*, 1–5. doi: 0.1007/BF00878272.
251. Korn-Wndish, F., Rainey, F., Kroppensted, R. M., Kempf, A., Majazza, A., Jutzner, H. J., & Stakerbrandt, E., (1995). *Thermocrispum* gen. Nov., a new genus of the order *Actinomycetales,* and description of *Thermocrispum municipale* sp. Nov. and *Thermocrispum agreste* sp. Nov. *Int. J. Syst. Bacteriol.*, *45*, 67–77.
252. Lacey, J., (1997). Actinomycetes in composts. *Ann. Agric. Environ. Med.*, *4*, 113–121.
253. Mayfield, C. I., Williams, S. T., Ruddick, S. M., & Htfield, R. L., (1972). Studies on the ecology of actinomycetes in soil. IV. Observations on the form and growth of *Streptomycetes* in soil. *Soil Biol. Biochem.*, *4*, 79–91. https://doi.org/0.1016/0038-0717(72)90045-4.

254. Song, J., Weon, H. Y., Yoon, S. H., Park, D. S., Go, S. J., & Suh, J. W., (2001). *Thermoactinomyces* spp. isolated from mushroom composts in Korea based on 16S rRNA gene sequence analysis. *FEMS Microbiol. Lett.*, *202*, 87–102.
255. Flowers, T. H., & Williams, S. T., (1977). Measurment of growth rates of *Streptomycetes*: Comparison of turbidimetric and gravimetric techniques. *J. Gen. Microbiol.*, *28*, 285–289.
256. Williams, S. T., (1982). Are antibiotcs produced in soil? *Pedobiologia*, *23*, 427–435. doi: 10.1099/0221287-98-1-285.
257. Park, J. Y., (1971). A stimulative effect of *Streptomyces thermovulgaris* on the mycelium growth of *Agaricus bisporus*. *Agricola*, 254.
258. Grant, W. D., Fermor, T. R., & Wood, D. A., (1984). Production of bacteriolytic enzymes and degradation of bacterial cell walls during growth of *Agaricus bisporus*on *Bacillus subtilis*. *J. Gen. Microbiol.*, *130*, 761–769.
259. Jarak, M., Govedarica, M., Milošević, N., & Klokočar-Šmit, Z., (2000). Primena aktinomiceta u proizvodnji šampinjona. Zbornik radova naučnog instituta za ratarstvo i povrtarstvo[Application of actinomycetes in the production of button mushrooms]. *Proceedings of the Institute of Field and Vegetable Crops* (Vol. 34, pp. 77–81). Novi Sad.
260. Šantrić, L., Potočnik, I., Radivojević, L., Gajić, U. J., Rekanović, E., Duduk, B., & Milijašević-Marčić, S., (2018). Impact of a native *Streptomyces flavovirens* from mushroom compost on green mold control and yield of *Agaricus bisporus*. *J. Environ. Sci. Health B*, *53*(10), 677–684. doi.org/10.1080/03601234.2018.1474559.
261. Majić, I., Saralić, A., Lakatos, T., Tóth, T., Raspudić, E., Zebec, V., Kanižai, Š. G., Kovačić, M., & Laznik, Ž., (2018). First report of entomopathogenic nematode *Steinernema feltiae* (*Rhabditida: Steinernematidae*) from Croatia. *Helminthologia*, *55*(3), 256–260. doi: 10.2478/helm-2018-0024.

CHAPTER 14

Microbial Consortia as Plant Biostimulants: Definition, Concept, Categories, and Regulation

NAFISSA SOUDANI,[1] KHAOULA TOUMI,[2] MOUSSA EL-JARROUDI,[3] and SAMI FATTOUCH[4]

[1]*Agricultural Sciences Department, Mohamed Khider University, Biskra, Algeria*

[2]*University of Jendouba, Higher School of Agriculture of Kef (ESAK), LR14AGR04: Support for the Sustainability of Agricultural Production Systems in the North West Region, Le Kef, Tunisia.*

[3]*Environmental Sciences and Management Department, University of Liège–Arlon Campus Environment, Belgium*

[4]*National Institute of Applied Sciences and Technology (INSAT), University of Carthage, Tunis, Tunisia*

ABSTRACT

Microbial consortia include cellulolytic bacteria, fungi, yeasts, and microbial-based bio-films, which can be found in, or are constituents of, food products, soils, wastewater, etc., flourishing in nature. The characterization, design, and management of these consortia are feasible, and their application in diverse fields, such as agriculture, is under research. During the last decades, considerable attention has been paid to microbial consortia as plant biostimulants. They are an ecological alternative for sustainable agriculture and a win-win strategy to grow valuable plants without releasing fertilizers and/or pesticides into the environment. Plant biostimulants (PBs) are

Microbial Biostimulants: Biorational Pesticides for Management of Plant Pathogens.
Faheem Ahmad and Rakesh Pandey (Eds.)

natural preparations of a diverse class of substances and/or microorganisms that positively affect plant growth conditions, stimulate natural processes, improve yield and chemical composition, and increase tolerance to biotic and abiotic stress. The purpose of this chapter is to provide an up-to-date overview of microbial consortia used as PBs. Aiming to describe the emerging definitions of plant biostimulants, their concept, the functional categories, and their potential application as a significant tool for modern eco-friendly agriculture. The chapter also focuses on reporting the regulatory framework and the current market situation of plant biostimulants, as well as their advances in research.

14.1 INTRODUCTION

The world is facing several constraints and challenges, such as the reduced availability of natural resources, economic profitability, and consumer expectations to transition towards healthier and environmentally friendly products. Consequently, the increasing demand for such products and the associated challenges have strongly encouraged efforts to explore new and effective substances that offer eco-friendly options.

Plants harbor a broad and complex range of microorganisms in their phyllosphere, rhizosphere, and endosphere [1]. Microbial consortia are among the promoting microorganisms used as biological tools to enhance plant growth in modern agriculture [2]. Therefore, in recent years, biostimulants have become a key element in achieving sustainable agriculture [2], and the use of plant biostimulants (PBs) has steadily grown in agriculture and crop production [1]. As the environment is continuously harmed by the increased use and production of pesticides, PBs represent a new generation of products available on the market that are safe for both human beings and the environment. They are particularly valuable for reducing chemical usage in agriculture [2, 3]. Introducing non-pathogenic microorganisms as microbial biostimulants is particularly interesting and carries a very small chance of posing environmental risks to soil and biodiversity functioning [4, 5].

PB's are substances that can act directly on plants and have the ability to positively affect natural physiological functions within different parts of a plant, including roots, stems, leaves, flowers, and fruits [4]. They are not long-term accumulative, have low toxicity, low mobility in the soil, are less prone to resistant pests and pathogens, leave no residues in food, and require low application rates. These characteristics make them a favorable alternative to xenobiotic agrochemicals with positive implications [1, 6].

Therefore, this review addresses the need to explore microbial consortia as an innovative and trendy tool to meet the growing demands for sustainable food products for the increasing population.

14.2 MICROBIAL CONSORTIA

14.2.1 DEFINITION OF PLANT BIOSTIMULANTS (PBS)

A broad and comprehensive definition of the term 'biostimulants' is given. According to the European Regulation 2019/1009, PBs are defined as products that stimulate plant nutrition processes independently of their nutrient content, with the aim of improving characteristics such as nutrient use efficiency, tolerance to abiotic stress, quality traits, and availability of confined nutrients in soil or rhizosphere. Fertilizers, on the other hand, are not considered biostimulants as they directly deliver nutrients to plants [7]. Biostimulants are not plant nutrients and cannot be the subject of nutrient claims or guarantees, nor do they have a direct action against pests [8].

However, based on their effects on plants, biostimulants can be substances derived from natural sources and/or microorganisms. They promote plant growth, improve abiotic stress tolerance and crop quality, enhance nutrient uptake and efficiency, and reduce nutrient losses to plants and the environment. They can also act as soil amendments, while avoiding undesirable side effects [1, 7–10]. Microbial and non-microbial PBs are commonly applied to field and greenhouse crops, including fruit trees, berry crops, grapes, vegetables, ornamentals, cereals, and turf. They can be applied to seeds, soil, or other growing media [8].

14.2.2 CONCEPT

Chemical fertilizers and pesticide use have adversely affected the natural environment, including soil and water, and have raised global concern. This is particularly significant as food production is expected to increase [10]. However, microbial consortia are naturally present and have the potential to produce highly functional foods, benefiting human health while being harmless and adaptable to ecosystems and natural environments [11].

Microbial consortia offer several important benefits. They can replace chemical fertilizers with bio-fertilizers and can biodegrade xenobiotics, such as pesticides, through complex processes. These strains support cross-feeding, which enhances their metabolic processes and stability in the environment [10, 11]. When pesticides enter soil and groundwater systems, microbial consortia have a remarkable capacity to break them down. This reduces the amount of persistent pesticides released into the environment through various metabolic pathways [11].

The consortium is composed of both similar and different microbial genera, as well as a range of species. The constituents interact through a variety of mechanisms, namely neutralism, commensalism, amensalism, competition, predation, and cooperation [11].

14.2.3 CATEGORIES AND APPLICATIONS

Several compounds with bioactive properties can be applied in small doses as biostimulants to improve plant growth and increase yields, especially under normal and stressful conditions. They achieve this by reducing stress, improving soil water retention capacity, and promoting root growth [3, 12–14].

PBs have the ability to affect a number of metabolic processes, including respiration, photosynthesis, nucleic acid synthesis, and ion uptake. They achieve this by increasing the number of absorbed micro- and macronutrients in plants, thereby positively altering root morphology and plant growth [1, 3]. PBs, also known as plant growth promoters (PGRs), can include organic materials, a mixture of two or more PGRs, or a combination of these with other substances such as amino acids, nutrients, and vitamins [3].

Biostimulants are produced using a multitude of sources and technologies. Based on their functionality, use, and type of activity, there are many available classes and groups [15]. Some examples of heterogeneous categories include non-microbial biostimulants such as humic substances (HS), protein hydrolysates from vegetal sources with free amino acids or peptide chains, seaweed extracts, inorganic salts (including phosphite), silicon, chitin, and chitosan derivatives, antiperspirants, other nitrogen-containing substances, and more. The list continues to grow as new products and technologies are developed [1, 8, 9]. Additionally, an important category is emerging, which consists of microbial biostimulants derived from beneficial microorganisms such as fungi, yeast, and eubacteria. This includes endophytes, mycorrhizal fungi (MF), and plant promoting growth bacteria (PPGB) such as *Azotobacter* spp., *Azospirilium* spp., and *Rhizobium* spp. [1].

Biostimulants are effective and generally beneficial to all plants, as shown in Figure 14.1. They are available for application during the entire crop cycle to the soil, or as a foliar spray, in-furrow fertigation at root or seed transplanting [16]. Certain biostimulants have been demonstrated to activate plant hormone precursors and stimulate specific gene expression, and they are neither pesticides nor herbicides.

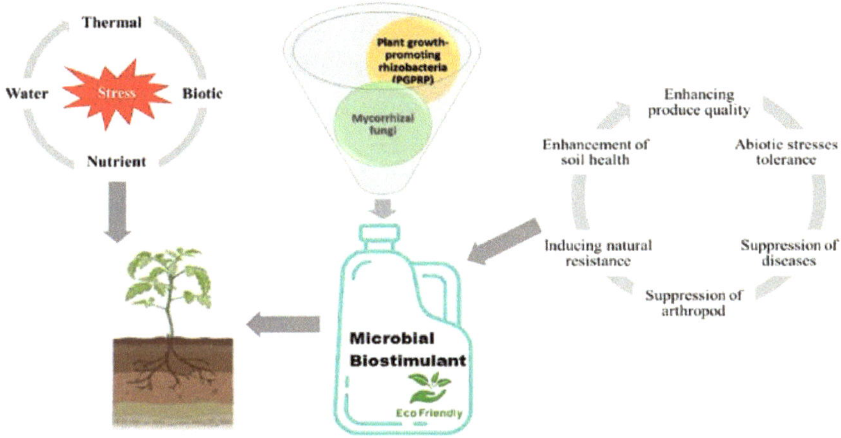

FIGURE 14.1 Microbial biostimulant types and their roles on plant growth under stress.

14.3 MICROBIAL BIOSTIMULANTS

Plant parts consist of a distinguishable association of microbes. The colonizing microbes are classified into epiphytes, which live on the surface, endophytes, which reside inside the plant tissues, and epiphytes on the leaf surface, as well as rhizomes in the soil near the roots. Among these, the rhizosphere system is considered the most dynamic, significantly influencing nutrition and growth [17].

Microorganisms generally exhibit high competition, adaptation, and functional redundancy [4]. The microbial consortia have specialized functions, with multiple species performing complementary metabolic tasks. Species such as MF, *Azotobacter* spp., *Azospirillum* spp., and *Rhizobium* spp. are part of the diverse regulatory mechanism [11, 18]. Therefore, there is great interest in using microbial biostimulants, as plants host a complex and diverse spectrum of microorganisms in their phyllosphere, rhizosphere, and endosphere [1]. Both endophytes and surface microbiota have been shown to improve plant growth, disease resistance, and stress tolerance [17]. Additionally, AMF, PGPR, and other biostimulants fix, mobilize, solubilize, or enhance the bioavailability of nutrients in soils, thereby improving soil health and fertility [16, 17, 19].

14.3.1 MYCORRHIZAL FUNGI (MF)

Arbuscular mycorrhizal fungi (AMF) have been widely used in plant production due to their biostimulatory properties. They can establish a symbiotic relationship with the roots of most agricultural crops [20]. For example, they enhance soil quality and plant health, leading to improved yield and growth of various horticultural crops under different stress conditions (both unstressed and stressed regimes) [16, 21, 22].

AMF can influence the atmospheric CO_2 fixation in host plants by increasing the movement of photo-assimilates from the aerial parts to the roots, thus acting as a "sink". Additionally, they can develop a network of external fungal hyphae that can increase the surface area by up to 40 times. This promotes soil organic matter decomposition processes and enhances the uptake and assimilation of essential elements (such as P, Ca^{2+}, N, Mg^{2+}, and K^{+}), which in turn modulate the production of phytochemicals [20, 22].

Furthermore, AMF can help host plants up-regulate tolerance mechanisms and prevent the down-regulation of key metabolic pathways (both primary and secondary). This involvement of key enzymes leads to the accumulation of secondary metabolites or phytochemicals [20, 22, 23].

14.3.2 PLANT GROWTH-PROMOTING RHIZOBACTERIA (PGPR)

The rhizosphere environment supports an abundance of microorganisms that influence and depend on each other due to the presence of organic matter, root exudates, nutrients, and moisture. They are able to exert a beneficial effect on plant growth under nutritional and abiotic stress or during the restoration of polluted soils [2]. Among these microorganisms, several plant growth-promoting rhizobacteria (PGPRs) have been revealed to be capable of enhancing plant growth and yields, regulating plant hormones, producing antibiotics and organic compounds, fixing nitrogen, solubilizing phosphate, improving soil structure, as well as increasing nutrient and water uptake, and alleviating plant stresses [2, 24].

14.3.2.1 AZOTOBACTER SPP.

Research on Azotobacters has revealed their role in acting as biofertilizers by integrating with organic matter and fixing nitrogen, thus improving plant growth and yield. Azotobacter also works as a biostimulant by boosting plant growth via growth hormone production, such as indole acetic acid, cytokinins, and gibberellins phytohormones [25]. Indirectly, Azotobacter acts as a bioprotectant due to its ability to produce siderophore, IAA, and exopolysaccharides. Additionally, it protects plants by enhancing plant growth and soil fertility [25, 26].

14.3.2.2 AZOSPIRILLUM SPP.

Azospirillum species can be associated with plants and other environments under extreme conditions of temperature or contamination. Azospirilla are considered ecologically important free-living microbes that promote plant growth (PGPB) through nitrogen fixation, phytohormone production, and the synthesis of polyamines and trehalose [27, 28]. Their mechanism and mode of action can vary depending on soil, climatic conditions, and mineral solubility [29].

14.3.2.3 RHIZOBIUM SPP.

Rhizobium species belong to the group of plant growth promoting microbes, and Rhizobium spp. is a specific genus that contains rhizobia, which function as bio-inoculants and nitrogen fixers [30–32]. They are promising plant biostimulants, as they have the ability to directly control pests through antibiosis and competition processes, or indirectly by enhancing legumes' resistance to pests [31, 33].

In a study, simultaneous exposure to graphene oxide (GO) nanosheets and *Rhizobium sp.* E20-8 demonstrated the mitigation of drought-induced stress on maize seedlings. This effect was attributed to the osmotic and antioxidant protective properties of GO. Additionally, *Rhizobium sp.* E20-8 appeared to attenuate the effects of GO on plant biochemistry [34].

14.4 IMPLICATIONS OF BIOSTIMULANTS

14.4.1 AGRONOMIC BENEFITS

14.4.1.1 ENHANCING PRODUCE QUALITY

Regarding quality, biostimulants have the ability to enhance the functional quality (physico-chemical properties) of plant production. Biostimulants can affect various physiological functions, such as promoting increased root growth and biomass, improving nutrient uptake and translocation, activating hormone signaling pathways in gene expression, protecting photosynthetic activity during abiotic stress, and enhancing the rhizosphere of plants, among others [16].

The application of biostimulants can also improve water use efficiency, enhance resistance and recovery from abiotic stress (such as drought and salinity), and increase chlorophyll concentration [19, 25].

14.4.1.2 ABIOTIC STRESSES TOLERANCE

Plant tolerance to salinity stress is generally higher during the early growth phases (seedlings, transplants) compared to later growth phases. Microbial biostimulants can enhance plant tolerance to abiotic stresses. For instance, studies have shown that applying microbial biostimulants to lettuce and tomato seedlings, which were subjected to different water salinity levels, improved both plant growth capacity and response to salinity stress [35].

Under severe abiotic stresses, microbial communities in consortia can respond through various mechanisms and pathways, including the production of enzymes, hormones, and metabolites [36]. Arbuscular mycorrhizal fungi (AMFs) can alleviate multiple stresses or combinations of them, such as drought, salinity, temperature, and nutrient deficiency [22, 36, 37]. Furthermore, microbial biostimulant consortia could be utilized to assist crops in enhancing tolerance to heavy metals [36].

14.4.1.3 SUPPRESSION OF DISEASES

Diseases are a significant burden that affects plants, leading to a decrease in quality, mass production, and nutritional value [38]. As a result, researchers have investigated this issue and have demonstrated the effectiveness of PBs in disease management.

For instance, seedling cucumbers treated with 500 ppm chitosan biostimulant exhibited 100% resistance to *Phytophthora capsici* damping-off disease [39]. *Streptomyces* strain UC1A-3 has shown potential as an *in vitro* antagonist, promoting plant growth and possessing biocontrol abilities against tested phytopathogens [40]. Additionally, Plant Growth Promoting Microbes (PGPMs) are known to sensitize and trigger plant immune defense mechanisms, such as producing antagonistic metabolites that suppress pathogens and enhance the immunity of potential crops [36].

14.4.1.4 ENHANCEMENT OF SOIL HEALTH

Constructing soil health is a major key to sustainable agriculture. Therefore, the use of biostimulants can enrich soils with organic molecules, beneficial microorganisms, or minerals, and impact soil structure. This, in turn, can improve heavy metal mobilization and leachability, as well as plant uptake. Furthermore, bioremediation of soils involves the removal of xenobiotics such as pesticides, antibiotics, pharmaceuticals, etc. This process can fix, mobilize, solubilize, or increase the bioavailability of nutrients [16].

Some biostimulants can be used as organic soil amendment products as they may affect the soil microbial population and nutrient uptake [41]. Treatment of castor oil cake and vermicompost with biostimulants (Azotobacter+PSB+Jivamrut) has been found to effectively improve soil health [42]. Additionally, the addition of biostimulants has been shown to impact soil fertility [43]. As a result, a 77% increase in calcium content in sandy loam soil and a 38% increase in potassium content in sandy soil have been observed [44].

14.4.1.5 INDUCING NATURAL RESISTANCE

Some researchers have reported that increased levels of phenolic compounds in plant tissues and the antioxidant properties greatly enhance resistance, preserve the natural microflora, and provide a combined approach to control postharvest decay [45]. Additionally, canola plants treated with PGPR alone or in combination with humic acid exhibited the highest tolerance against *B. brassicae* [46]. On the other hand, *Ascophyllum nodosum* extract (ANE) induces active defense in strawberry plants against powdery mildew by stimulating secondary metabolism and regulating enzymatic activities associated with defense [47].

14.4.1.6 SUPPRESSION OF ARTHROPOD

Microorganisms biostimulant works to boost both establishment and bloom to stimulate arthropods visitation and ensure low-input success for white mint [48]. Testing of plant defense elicitors for arthropod pest control and induction of the PR-1 gene in pear orchards has concluded that elicitors have the potential to suppress pear pests [49]. Additionally, kaolin particle film, spirodiclofen, and harpin protein have been shown to be effective as a broad-spectrum insecticide to control pear psylla infestations [50].

14.4.2 ECONOMIC BENEFITS

Biostimulants have proven to be tools for increasing the economic profitability of growers by improving their harvest yields and increasing marketable production per unit area. This can be achieved by enhancing the quality characteristics of products, creating a good opportunity to increase the selling price, as well as reducing production costs, mainly the amount of fertilizer and pesticides used and their leaching.

14.5 BIOSTIMULANTS INDUSTRY AND MARKET

The development of biomass stimulants involves scientific, technical, and regulatory challenges, which mostly consist of the categorization and pre-market assessment of biostimulants, as well as intellectual property [3]. Therefore, several major companies, including BASF SE (Germany), UPL (India), Valagro S.p.A (Italy), Gowan Group (USA), FMC Corporation (USA), and ILSA S.p.A (Italy), actively cooperate with universities, independent research institutes, and individual growers to assess the health benefits and consistency of biostimulant products, thereby regulating the global biostimulant market [6, 51].

The European market holds nearly half of the global biostimulants market, which is valued at between 1.5 and 2 billion USD by 2022 [52]. Meanwhile, the global market size for biostimulants (PBs) in 2021 was roughly USD 3.2 billion and is expected to increase at a compound annual growth rate (CAGR) of 12.1% to reach USD 5.6 billion by 2026, particularly in the European Union (EU) [51].

Regardless of the variety of sources and complexity of the resulting product, the market for biostimulants is expected to remain stable in the coming years due to consumers' awareness of the benefits, the introduction of organic food regulations, and the adoption of GLOBALGAP (global good agricultural practices) policies [51].

14.6 REGULATORY LEGISLATION

The strict EU regulations on environmental safety and support for organic farming have opened doors to an expanding biostimulants market in Europe, mainly in Germany, Italy, the UK, Spain, and France.

Therefore, the European Biostimulant Industry Council (EBIC), founded in 2011, is attempting to encourage innovation in this field by requesting operational solutions for the harmonized regulatory treatment of the EU market, including safety requirements [2]. Hence, the German Federal Office for Consumer Protection and Food Safety has drawn up a regulatory framework for biostimulants [52].

However, the lack of a clear definition of biostimulants as a distinct product group results in incomplete registration and classification, hindering trade in the market and future development. This is why many countries, states, and administrative regions have established separate categories of registration for prospective biostimulants [15].

Moreover, if a biostimulant is deemed useful with a positive effect on plant production and is suitable for use by consumers, based on its safety and proven efficacy, it can be regulated and traded between different countries [15, 52].

By 2030, several opportunities within EU countries will be available for both traditional and innovative interventions to boost crop productivity on a sustainable basis, notably the most challenging and emerging agricultural practices, the so-called PBs.

However, at present, national markets and regulatory requirements for PBs vary across countries, and non-harmonized regulation can lead to unbalanced competition between the players. For this reason, evaluation procedures for PBs are needed to build up a common marketplace for such substances and to ensure reliable competition [6].

14.7 CONCLUSION

In conclusion, supporting research on plant biostimulants (PBs), particularly regarding their chemical composition, dosage, application timing, and mode of action in different environmental conditions, has proven to be of great importance. Additionally, optimizing the nutritional interactions between plants and microorganisms can contribute to the development of more sustainable agricultural systems, ensuring both environmental and consumer safety.

Furthermore, there is a pressing need for further exploration and investment in the advancement of biostimulant technologies. This will help us gain a precise understanding of the specific functions that can be influenced and how they are affected. Equally important is providing producers with reliable and comprehensive knowledge about the types and applications of biostimulants, as it is crucial for their success.

KEYWORDS

- **abiotic stresses tolerance**
- **arthropod**
- **microbial consortia**
- **natural resistance**
- **plant biostimulants**
- **plant growth-promoting rhizobacteria**
- **soil health**

REFERENCES

1. Sangiorgio, D., Cellini, A., Donati, I., Pastore, C., Onofrietti, C., & Spinelli, F., (2020). Facing climate change: Application of microbial biostimulants to mitigate stress in horticultural crops. *Agronomy, 10*(6), 794. https://doi.org/10.3390/agronomy10060794.
2. Castiglione, A. M., Mannino, G., Contartese, V., Bertea, C. M., & Ertani, A., (2021). Microbial biostimulants as response to modern agriculture needs: Composition, role, and application of these innovative products. *Plants, 10*(8), 1533. https://doi.org/10.3390/plants10081533.
3. Rafiee, H., Naghdi, B. H., Mehrafarin, A., Qaderi, A., Zarinpanjeh, N., Sekara, A., & Zand, E., (2016). Application of plant biostimulants as new approach to improve the biological responses of medicinal plants – A critical review. *Journal of Medicinal Plants, 15*, 6–39.
4. Joner, E., Skjerve, E., Sundheim, L., Tronsmo, A., Wastson, Y., Eckner, K., Kapperud, G., et al., (2019). Microorganisms in biostimulants. *European Journal of Nutrition & Food Safety*, 310–311. https://doi.org/10.9734/ejnfs/2019/v9i330074.
5. Barros-Rodríguez, A., Rangseekaew, P., Lasudee, K., Pathom-aree, W., & Manzanera, M., (2020). Regulatory risks associated with bacteria as biostimulants and biofertilizers in the frame of the European regulation (EU) 2019/1009. *Science of The Total Environment, 740*, 140239. https://doi.org/10.1016/j.scitotenv.2020.140239.

6. Caradonia, F., Battaglia, V., Righi, L., Pascali, G., & La Torre, A., (2018). Plant biostimulant regulatory framework: Prospects in Europe and current situation at international level. *Journal of Plant Growth Regulation, 38*(2), 438–448. https://doi.org/10.1007/s00344-018-9853-4.
7. Drobek, M., Frąc, M., & Cybulska, J., (2019). Plant biostimulants: Importance of the quality and yield of horticultural crops and the improvement of plant tolerance to abiotic stress—A review. *Agronomy, 9*(6), 335. https://doi.org/10.3390/agronomy9060335.
8. Calvo, P., Nelson, L., & Kloepper, J. W., (2014). Agricultural uses of plant biostimulants. *Plantand Soil, 383*(1, 2), 3–41. https://doi.org/10.1007/s11104-014-2131-8.
9. Du Jardin, P., (2015). Plant biostimulants: Definition, concept, main categories, and regulation. *Scientia Horticulturae*, *196*, 3–14. https://doi.org/10.1016/j.scienta.2015.09.021.
10. Gavelienė, V., Šocik, B., Jankovska-Bortkevič, E., & Jurkonienė, S., (2021). Plant microbial biostimulants as a promising tool to enhance the productivity and quality of carrot root crops. *Microorganisms, 9*(9), 1850. https://doi.org/10.3390/microorganisms9091850.
11. Bhatt, P., Bhatt, K., Sharma, A., Zhang, W., Mishra, S., & Chen, S., (2021). Biotechnological basis of microbial consortia for the removal of pesticides from the environment. *Critical Reviews in Biotechnology, 41*(3), 317–338. https://doi.org/10.1080/07388551.2020.1853032.
12. Khan, W., Rayirath, U. P., Subramanian, S., Jithesh, M. N., Rayorath, P., Hodges, D. M., Critchley, A. T., et al., (2009). Seaweed extracts as biostimulants of plant growth and development. *Journal of Plant Growth Regulation, 28*(4), 386–399. https://doi.org/10.1007/s00344-009-9103-x.
13. Toscano, S., Romano, D., Massa, D., Bulgari, R., Franzoni, G., & Ferrante, A., (2018). Biostimulant applications in low input horticultural cultivation systems. *Italus Hortus*, *35*, 27–36. https://doi.org/10.26353/j.itahort/2018.1.2736.
14. Shahrajabian, M. H., Chaski, C., Polyzos, N., & Petropoulos, S. A., (2021). Biostimulants application: A low input cropping management tool for sustainable farming of vegetables. *Biomolecules, 11*(5), 698. https://doi.org/10.3390/biom11050698.
15. Yakhin, O. I., Lubyanov, A. A., Yakhin, I. A., & Brown, P. H., (2017). Biostimulants in plant science: A global perspective. *Frontiers in Plant Science*, 7. https://doi.org/10.3389/fpls.2016.02049.
16. Dara, S. K., (2021). *Advances in Biostimulants as an Integrated Pest Management Tool in Horticulture* (pp. 1–35). University of California Cooperative.
17. Vishwakarma, K., Kumar, N., Shandilya, C., Mohapatra, S., Bhayana, S., & Varma, A., (2020). Revisiting plant–microbe interactions and microbial consortia application for enhancing sustainable agriculture: A review. *Frontiers in Microbiology*, *11*. https://doi.org/10.3389/fmicb.2020.560406.
18. Qian, X., Chen, L., Sui, Y., Chen, C., Zhang, W., Zhou, J., & Ochsenreither, K., (2020). Biotechnological potential and applications of microbial consortia. *BiotechnologyAdvances*, *40*, 107500.
19. De Pascale, S., Rouphael, Y., & Colla, G., (2018). Plant biostimulants: Innovative tool for enhancing plant nutrition in organic farming. *European Journal of Horticultural Science, 82*(6), 277–285. https://doi.org/10.17660/ejhs.2017/82.6.2.
20. Ramírez-Flores, M. R., Bello-Bello, E., Rellán-Álvarez, R., Sawers, R. J. H., & Olalde-Portugal, V., (2019). Inoculation with the mycorrhizal fungus *Rhizophagus*

*irregularis*modulates the relationship between root growth and nutrient content in maize (*Zea Mays* Ssp. mays L.). *Plant Direct, 3*(12). https://doi.org/10.1002/pld3.192.
21. Rouphael, Y., Franken, P., Schneider, C., Schwarz, D., Giovannetti, M., Agnolucci, M., Pascale, S. D., et al., (2015). Arbuscular mycorrhizal fungi act as biostimulants in horticultural crops. *Scientia Horticulturae, 196*, 91–108. https://doi.org/10.1016/j.scienta.2015.09.002.
22. Begum, N., Qin, C., Ahanger, M. A., Raza, S., Khan, M. I., Ashraf, M., Ahmed, N., & Zhang, L., (2019). Role of arbuscular mycorrhizal fungi in plant growth regulation: Implications in abiotic stress tolerance. *Frontiers in Plant Science, 10*. https://doi.org/10.3389/fpls.2019.01068.
23. Bücking, H., Liepold, E., & Ambilwade, P., (2012). The role of the mycorrhizal symbiosis in nutrient uptake of plants and the regulatory mechanisms underlying these transport processes. *Plant Sci., 4*, 108–132.
24. Ruzzi, M., & Aroca, R., (2015). Plant growth-promoting rhizobacteria act as biostimulants in horticulture. *Scientia Horticulturae, 196*, 124–134. https://doi.org/10.1016/j.scienta.2015.08.042.
25. Sumbul, A., Ansari, R. A., Rizvi, R., & Mahmood, I., (2020). Azotobacter: A potential bio-fertilizer for soil and plant health management.*Saudi Journal of Biological Sciences, 27*(12), 3634–3640.
26. Hindersah, R., Kamaluddin, N. N., Samanta, S., Banerjee, S., & Sarkar, S., (2020). Role and perspective of azotobacter in crops production. *SAINS TANAH – Journal of Soil Science and Agroclimatology, 17*(2), 170. https://doi.org/10.20961/stjssa.v17i2.45130.
27. Bashan, Y., & De-Bashan, L. E., (2010). How the plant growth-promoting bacterium *Azospirillum* promotes plant growth—A critical assessment. *Advances in Agronomy, 108*, 77–136.
28. Reis, V. M., Teixeira, K. R. D. S., & Pedraza, R. O., (2011). What is expected from the genus *Azospirillum* as plant growth-promoting bacteria? In: *Bacteria in Agrobiology: Plant Growth Responses* (pp. 123–138). Berlin, Heidelberg, Springer.
29. Cruz-Hernández, M. A., Mendoza-Herrera, A., Bocanegra-García, V., & Rivera, G., (2022). *Azospirillum* spp. from plant growth-promoting bacteria to their use in bioremediation. *Microorganisms, 10*(5), 1057. https://doi.org/10.3390/microorganisms10051057.
30. Giller, K. E., & Ronner, E., (2019). The story of N2Africa: Putting nitrogen fixation to work for smallholder farmers in Africa: A flavor of the excitement and the richness of learning from N2Africa. *N2Africa*. https://doi.org/10.18174/527074.
31. Ochieno, D. M. W., Karoney, E. M., Muge, E. K., Nyaboga, E. N., Baraza, D. L., Shibairo, S. I., & Naluyange, V., (2021). Rhizobium-linked nutritional and phytochemical changes under multitrophic functional contexts in sustainable food systems. *Frontiers in Sustainable Food Systems*, 4. https://doi.org/10.3389/fsufs.2020.604396.
32. Hendriksen, N. B., (2022). Microbial biostimulants – the need for clarification in EU regulation. *Trends in Microbiology, 30*(4), 311–313. https://doi.org/10.1016/j.tim.2022.01.008.
33. Rouphael, Y., Spíchal, L., Panzarová, K., Casa, R., & Colla, G., (2018). High-throughput plant phenotyping for developing novel biostimulants: From lab to field or from field to lab? *Frontiers in Plant Science*, 9. https://doi.org/10.3389/fpls.2018.01197.
34. Lopes, T., Cruz, C., Cardoso, P., Pinto, R., Marques, P. A. A. P., & Figueira, E., (2021). A multifactorial approach to untangle graphene oxide (GO) nanosheets effects on

plants: Plant growth-promoting bacteria inoculation, bacterial survival, and drought. *Nanomaterials, 11*(3), 771. https://doi.org/10.3390/nano11030771.

35. Miceli, A., Moncada, A., & Vetrano, F., (2021). Use of microbial biostimulants to increase the salinity tolerance of vegetable transplants. *Agronomy, 11*(6), 1143. https://doi.org/10.3390/agronomy11061143.
36. Padmaperuma, G., Butler, T. O., Shuhaili, F. A. A., Almalki, W. J., & Vaidyanathan, S., (2020). Microbial consortia: Concept and application in fruit crop management. In: *Fruit Crops* (pp. 353–366). Elsevier.
37. Ebrahimi, F., Salehi, A., Movahedi, D. M., Mirshekari, A., Hamidian, M., & Hazrati, S., (2021). Biochemical response and nutrient uptake of two arbuscular mycorrhiza-inoculated chamomile varieties under different osmotic stresses. *Botanical Studies, 62*(1). https://doi.org/10.1186/s40529-021-00328-3.
38. Gebashe, F., Gupta, S., & Van, S. J., (2021). Disease management using biostimulants. In: *Biostimulants for Crops from Seed Germination to Plant Development* (pp. 411–425). Academic Press.
39. Zohara, F., Surovy, M. Z., Khatun, A., Prince, Md. F. R. K., Akanda, Md. A. M., Rahman, M., & Islam, Md. T., (2019). Chitosan biostimulant controls infection of cucumber by *Phytophthora capsici* through suppression of asexual reproduction of the pathogen. *Acta Agrobotanica, 72*(1). https://doi.org/10.5586/aa.1763.
40. Manigundan, K., Joseph, J., Ayswarya, S., Vignesh, A., Vijayalakshmi, G., Soytong, K., & Radhakrishnan, M., (2020). Identification of biostimulant and microbicide compounds from *Streptomyces* sp. UC1A-3 for plant growth promotion and disease control. *Int. J. Agric. Technol., 16*, 1125–1144.
41. Mofokeng, M. M., Araya, H. T., Araya, N. A., Makgato, M. J., Mokgehle, S. N., Masemola, M. C., & Amoo, S. O., (2021). Integrating biostimulants in agrosystem to promote soil health and plant growth. In: *Biostimulants for Crops from Seed Germination to Plant Development* (pp. 87–108). Academic Press.
42. Jat, R. S., Basak, B. B., & Gajbhiye, N. A., (2021). Organic manures and biostimulants fostered soil health and increased the harvest quality of the medicinal herb ashwagandha. *Agronomy Journal, 113*(1), 504–514. https://doi.org/10.1002/agj2.20457.
43. Kaur, I., (2020). Seaweeds: Soil health boosters for sustainable agriculture. In: *Soil Health* (pp. 163–182). Springer, Cham.
44. Yousfi, S., Marín, J., Parra, L., Lloret, J., & Mauri, P. V., (2021). A rhizogenic biostimulant effect on soil fertility and roots growth of turfgrass. *Agronomy, 11*(3), 573. https://doi.org/10.3390/agronomy11030573.
45. Romanazzi, G., Sanzani, S. M., Bi, Y., Tian, S., Gutiérrez, M. P., & Alkan, N., (2016). Induced resistance to control postharvest decay of fruit and vegetables. *Postharvest Biology and Technology, 122*, 82–94. https://doi.org/10.1016/j.postharvbio.2016.08.003.
46. Sattari, N. R., Pahlavan, Y. M., & Bozorg-Amirkalaee, M., (2018). Effects of humic acid and plant growth-promoting rhizobacteria (PGPR) on induced resistance of canola to *Brevicorynebrassicae* L. *Bulletin of Entomological Research, 109*(4), 479–489. https://doi.org/10.1017/s0007485318000779.
47. Bajpai, S., Shukla, P. S., Asiedu, S., Pruski, K., & Prithiviraj, B., (2019). A biostimulant preparation of brown seaweed *Ascophyllum Nodosum* suppresses powdery mildew of strawberry. *The Plant Pathology Journal, 35*(5), 406–416. https://doi.org/10.5423/ppj.oa.03.2019.0066.

48. Harris, B. A., Poole, E. M., & Pennisi, S. V., (2022). Impact of biostimulant and cultural factors on whorled mountainmint (*Pycnanthemum verticillatum* (Michx.) Pers. Var. *Pilosum* (Nutt.) Cooperr.): Growth performance and assessment of arthropod visitation. *Native Plants Journal, 23*(1), 97–114. https://doi.org/10.3368/npj.23.1.97.
49. Orpet, R. J., Cooper, W. R., Beers, E. H., & Nottingham, L. B., (2021). Test of plant defense elicitors for arthropod pest suppression and PR-1 gene induction in pear orchards. *Entomologia Experimentalis et Applicata, 169*(12), 1137–1146. https://doi.org/10.1111/eea.13110.
50. Saour, G., Ismail, H., & Hashem, A., (2009). Impact of kaolin particle film, spirodiclofen acaricide, harpin protein, and an organic biostimulant on pear psylla *Cacopsylla pyri* (*Hemiptera: Psyllidae*). *International Journal of Pest Management, 56*(1), 75–79. https://doi.org/10.1080/09670870903156632.
51. www.marketsandmarkets.com (accessed on 9 January 2023).
52. www.biostimulants.eu (accessed on 9 January 2023).

CHAPTER 15

Conditioning Root-Associated Microbiota Through Root-Exudate Metabolites Protects Plants Against Soil-Borne Diseases

SUNITA SINGH, RUCHI DUBE, and ARPITA GUPTE

School of Biotechnology and Bioinformatics, D. Y. Patil Deemed to be University, CBD Belapur, Navi Mumbai, Maharashtra, India

ABSTRACT

Despite major advancements in crop protection, plant diseases globally affect yields of food and income crops by 20%. To protect plant cells from the challenges they face, plants have evolved a variety of sophisticated immunological responses, one of which is the release of root exudates. Initiating and indicating interactions between plant roots and soil microorganisms, root exudates act as a signaling message. There are two recognized categories of root exudates: the first category includes high-molecular-weight substances like polysaccharides and proteins, while the second category includes low-molecular-weight substances like amino acids, organic acids, sugars, phenolic substances, and other secondary metabolites. The qualitative and quantitative composition of root exudates is controlled by environmental factors such as soil type, pH, and temperature, along with cultivars, plant species, and stage of development. These differences result in rhizosphere microbial communities that are specific to each plant species. In this chapter, we look at how root exudates can affect exchanges between bacteria in the rhizosphere, which might result in more environmentally friendly farming methods.

Microbial Biostimulants: Biorational Pesticides for Management of Plant Pathogens.
Faheem Ahmad and Rakesh Pandey (Eds.)

15.1 INTRODUCTION

Plants are sessile organisms with roots that anchor them in the earth. They are the primary food producers in terrestrial ecosystems, supporting the majority of other life. In nature, plants are constantly subjected to diseases and pests, as well as severe climatic circumstances such as drought, soil salinity, extreme temperatures, nutritional deficits, and heavy metal exposure. To protect plant cells from these dangers and to survive biotic stressors, plants have developed several complex immune responses [1].

For decades, researchers have examined plant-pathogen interactions through the lens of a single plant-microbe relationship, disregarding the intricacy of such interactions and the participation of numerous other microorganisms that impact the course of infection [2]. The impact of microbial populations linked with plants (plant microbiome) on plant development and health has received a lot of attention in recent years. Growing data suggest that plant-associated bacteria provide a plethora of metabolites that can aid the plant host's immune system. Some theories support that plants can interact with soil microbes through the production of chemicals in the rhizosphere that selectively promote bacteria beneficial to plant health [3].

Before we delve into the understanding of the process and significance of conditioning root-associated microbiota through root-exudate metabolites to enhance plant immunity against soil borne pathogens, there are some fundamental questions. For instance:

- What are root exudates?
- Which root-associated metabolite exudate conditions the rhizosphere microbiota?
- What decides the signaling between the root exudates and the microbe metabolites?
- Is this association always beneficial to plants by conditioning the root microbiota?
- How does the conditioning by root exudates help the rhizospheric microbiota?

This chapter is a small effort to compile the present state of knowledge about the makeup and activity of the microbial populations associated with the roots, as well as how various plant compounds can alter the communities' organizational framework. It is also a small attempt to enhance our understanding of these questions.

15.2 ROOT EXUDATES

Plants produce several substrates and signaling molecules to interact with the resident microbes in the rhizosphere. A crucial aspect of this communication is the release of root exudates. Root exudates are chemicals that have an impact on the development and function of soil microorganisms that reside in the rhizosphere. Different compounds released by plant roots act as regulators in microbial interactions, which can have beneficial, neutral, or detrimental effects on plants. Therefore, root exudates play a key role in shaping the microbial community in a plant's rhizosphere.

Proteins, cyclic adenosine monophosphate (cAMP), esters, aliphatic acids, amino acids, carbohydrates, and fatty acids are a few examples of these. Additional compounds include CO_2, various anions and cations, terpenoids, thiazoles/pyrazidines, and terpenoids [4]. Plants can initiate the chemotaxis process and move chemotactically with the aid of these various chemicals, attracting microbial populations for beneficial and harmful interactions at the rhizosphere interface. In addition to the substances used to promote beneficial interactions, plant roots also release numerous chemicals that inhibit the proliferation of pathogenic bacteria in the rhizosphere. These chemicals include inositols, cucurbitacin A, bithienyl, and its derivatives [5].

Plant root exudates in the rhizosphere, besides providing defense against harmful microorganisms and serving as a foundation for specific bacterial species to be attracted or repelled by chemotaxis, perform critical tasks such as maintaining soil moisture and humidity, facilitating nutrient migration, modifying the soil's chemical composition, stabilizing soil particles in the roots' immediate surroundings, and preventing competitive plant growth [6]. Compounds or substances such as secreted ions, free oxygen and water, enzymes, mucilage, and a wide variety of metabolites are continually secreted by the roots into the rhizosphere [7].

It is important to understand that while every plant produces exudates, the precise quantity and makeup differ from plant to plant. The age of the plant, the species of the plant, environmental factors such as high and/or low temperatures, the amount of precipitation, the availability of light, and agricultural practices being used in the ecosystem are other factors that influence the congregation and makeup of the root exudate. Additionally, elements such as soil type, texture, pH, moisture content, porosity, and the availability of nutrients affect the makeup of root exudates and the plant microbiome [8].

15.3 ROOT-ASSOCIATED METABOLITE EXUDATE CONDITION THE RHIZOSPHERE MICROBIOTA

15.3.1 CARBOHYDRATES AND AMINO ACIDS

The interaction between plants and PGPR is influenced by root exudates, which play a role in attracting PGPR to plant roots. Through various direct and indirect processes, PGPR can provide benefits to plants. Signals composed of amino acids and carbohydrates are believed to be responsible for attracting PGPR.

Recent research has shown that arabinogalactan proteins (AGPs), a hydroxyproline-rich glycoprotein superfamily of plant cell wall proteins, are involved in interactions between roots and rhizospheric microorganisms [9]. The colonization of *Rhizobium* sp. is influenced by AGPs released by *Arabidopsis* root cap cells and border-like cells, indicating the importance of AGPs in the recognition and adherence of rhizobia to plant roots [10].

Furthermore, the growth of bacteria on the roots of legumes and non-legumes, as well as the stimulation of polar surface attachment by *Rhizobium leguminosarum,* has been found to depend on a plant protein that resembles arabinogalactan [11]. However, it remains unknown how AGPs affect the composition of the microbial community, their role in the establishment and colonization of beneficial microorganisms on plant roots, and their involvement in other important rhizosphere activities.

15.3.2 SECONDARY METABOLITES AND HORMONES

Plants create and release a large number of secondary metabolites and hormones into the rhizosphere, in addition to carbohydrates and amino acids. Many of these substances are involved in interactions with microbes. They are used by plants to attract beneficial soil microbes and protect against diseases. For example, flavonoids act as chemoattractants by controlling the activity of the nod gene, which produces Nod factors (lipochito, oligosaccharides), attracting rhizobia to the roots. Nod factors are crucial for nodulation development [12]. Infection with foliar pathogens of *A. thaliana* leads to the generation of malic acid in root exudates, which attracts the PGPR *Bacillus subtilis* to the rhizosphere. Further research has shown that the presence of *B. subtilis* activates the signaling pathways for abscisic acid (ABA) and salicylic acid (SA) in *A. thaliana,* limiting the pathogen's access to the plants.

15.3.3 PROTEINS

In addition to primary and secondary metabolites, plants also generate proteins in the form of root exudate. However, it is still unclear how these proteins affect microbial interactions in the rhizosphere. The significance of root-released proteins in identifying harmful and unharmful bacteria has been demonstrated in a few studies. Lectins, which serve as protective and identification elements in symbiotic contacts, are among the most extensively studied proteins in this context. Furthermore, proteomic examination of *A. thaliana* root exudates revealed that as plants aged, their roots produced more defense-related proteins during flowering time, including chitinases, glucanases, and myrosinases [13].

Our understanding of plant rhizosphere-microbe interactions will be incomplete unless the role of proteins produced by the roots in these relationships is clarified. Combining a strategy of transcriptomic and proteomic techniques could elucidate the role that proteins play in rhizosphere-microbial interactions.

15.4 SIGNALING BETWEEN THE ROOT EXUDATES AND THE MICROBE METABOLITES

The first step in establishing beneficial relationships with microorganisms is root colonization. Nutrients released by plant roots attract a large number of microorganisms, known as the "rhizosphere effect." The proximity to plant roots leads to an increase in the number and activity of microorganisms. Plant roots not only provide a carbon-rich environment, but they also initiate communication with soil microbes by emitting signals that the microorganisms recognize. In response, the microorganisms generate signals that trigger necessary plant responses for colonization. Both competitive pathogenic and beneficial bacteria benefit from motility, as it allows them to colonize plant roots and participate in this interaction. Chemotaxis is the chemical process through which soil microorganisms are attracted to plant roots [14].

Electric potentials in plant roots, which are created by electrogenic ion transport at the root surface, are used in a recently discovered method to attract swimming oomycete plant pathogen zoospores to root surfaces [15]. These results also suggest that chemical signals may be masked by electrical impulses when oomycete zoospores respond to root surfaces in the short term. It is unknown whether the perception of chemotactic or electrotactic signals affects the tendency of soil microorganisms acting as pathogens or symbionts to infect an organism.

15.5 ASSOCIATION BETWEEN PLANTS AND ROOT EXUDATES AND ITS EFFECT ON CONDITIONING THE ROOT MICROBIOTA

Microbial root colonization causes a variety of interactions between bacteria and roots in the rhizosphere. When plant pathogenic bacteria are present, these interactions can be advantageous on one side and harmful on the other, ranging from neutral to negative. To further complicate matters, depending on the environment, bacteria can switch between a harmful and symbiotic form. Numerous non-pathogenic soil bacteria are referred to as rhizobacteria because they can aid in plant growth (PGPR).

PGPRs indirectly enhance plant growth and production through the reduction of phytopathogens. Some examples of competitive pathogen-control methods involve the capacity to synthesize siderophores (iron chelators), which prevent pathogens from accessing iron, the production of antifungal metabolites, enzymes that break down cell walls, or hydrogen cyanide (HCN), all of which stop the growth of fungal infections. Along with the capacity to develop systemic resistance, they also demonstrate the capacity to outcompete pathogens for nutrients or specialized root niches. Examples of direct processes include the synthesis of phytohormones, the solubilization of phosphorus, and the fixation of atmospheric nitrogen.

15.5.1 DIRECT MECHANISMS

15.5.1.1 NITROGEN-FIXATION

The process that diazotrophs use to convert atmospheric nitrogen to ammonia has received the greatest attention from researchers. Leguminous plants and rhizobia have a highly developed symbiotic relationship in which the rhizobia stimulate the plant to form root nodules, a specialized structure in which the rhizobia dwell and provide the plant with ammonia, a form of readily available nitrogen [16]. The diazotrophic bacterium *Azospirillum* not only fixes nitrogen but also secretes auxins, cytokinins, and gibberellins, which are plant hormones that directly promote plant development [17].

15.5.1.2 PLANT GROWTH HORMONES

The diazotroph Azospirillum fixes nitrogen while also secreting a variety of plant hormones, including auxins, cytokinins, and gibberellins, which directly promote plant growth. Among these hormones, auxins are secreted in the highest quantity by *Azospirillum* and play a crucial role in enhancing root growth, thereby facilitating overall plant development [18].

Another mechanism through which PGPR stimulates plant growth is the synthesis of 1-aminocyclopropane-1-carboxylate (ACC) deaminase. Ethylene (ET), a plant hormone involved in stress signaling and root elongation inhibition, is directly derived from ACC. The bacterial enzyme ACC deaminase breaks down ACC into ammonia and ketobutyrate, effectively reducing the amount of root growth inhibition caused by ET [19]. This interaction is advantageous for the bacteria as ammonia and α-ketobutyrate serve as sources of nitrogen and carbon, respectively.

15.5.1.3 SIDEROPHORES

Iron competition can suppress various soil borne diseases. Iron often acts as a limiting factor in the rhizosphere due to its extremely low solubility. As a result, most bacteria produce siderophores that bind to iron, which is then taken up by membrane receptors. For example, *WCS358* secretes the fluorescent siderophores *pseudobactin 358* when iron is scarce. *Pseudobactin-358* binds to iron, making it unavailable to species without the specialized receptor. Consequently, *WCS358r* can effectively inhibit the growth or activity of harmful organisms by monopolizing the iron in the environment [20].

15.5.2 INDIRECT MECHANISMS

15.5.2.1 LYTIC ENZYMES

Several components present in the cell walls of fungi and oomycetes can be broken down by lytic enzymes. Cellulases, glucanases, proteases, and chitinases are among the many bacterial lytic enzymes known. For example, *Pseudomonas cepacia* significantly reduces the incidence of diseases caused by *Rhizoctonia solani, Sclerotium rolfsii,* and *Pseudomonas ultimum.* However, it has been discovered that, besides the generation of lytic enzymes, other processes such as the creation of antibiotics also contribute to pathogen suppression [21].

15.5.2.2 Antibiosis Activity

Antibiosis is the term used to describe the antagonistic behavior of beneficial microorganisms, which is triggered by the production of secondary metabolites such as antibiotics or biosurfactants. Antibiotics are small molecules produced by microbes that inhibit the growth or metabolism of other microorganisms. Antibiotics like phenazines, DAPG, pyoluteorin, and pyrrolnitrin are widely available and have been extensively studied for their role in the biological control of plant diseases [22].

Antibiotic synthesis is often considered a crucial mechanism for biological control, as it is relatively easy to understand and can provide a highly effective mode of action. Biosurfactants, which are amphiphilic substances, have the ability to damage cellular membranes, leading to leakage and cytolysis. They exhibit antimicrobial properties and can combat a variety of pathogens.

15.5.3 NEGATIVE EFFECTS OF ROOT EXUDATES

Microbe interactions influence plant growth favorably through a variety of methods, including biological nitrogen fixation, enhanced biotic and abiotic stress tolerance, and direct and indirect benefits provided by rhizobacteria that promote plant growth. Aside from creating antibiotics that are employed as biocontrols against plant diseases, bacteria can also interact favorably with plants by degrading substances in the soil that would otherwise be allelopathic or even autotoxic. However, through pathogenic or parasitic infection, rhizospheric bacteria may harm or be detrimental to the well-being and life of plants.

15.5.3.1 ROOT EXUDATES AND THEIR ALLELOPATHIC EFFECTS

Plant preservation and increased plant disease resistance are the primary objectives of utilizing allelopathy in biological pest and weed control. In natural or agricultural/horticultural ecosystems, physical and chemical processes commonly alter the development and growth of plants due to adjacent plants. The significant allelopathic potential of some species against nearby plants may contribute to their dominance in the field. Allelopathic substances, when released from root exudates, generally penetrate the rhizosphere. In the rhizosphere, these substances are susceptible to sorption (physical), chemical oxidation (chemical), and microbial breakdown (biological), even though they may be released in large quantities. Microbial degradation can limit the stability of several secondary metabolites of plant origin, in addition to chemical oxidation [23].

15.6 THE EFFECT OF CONDITIONING BY ROOT EXUDATES ON THE RHIZOSPHERIC MICROBIOTA

Plant defenses target pathogenic microbes, but little is known about potential effects on the local rhizosphere microbiota. Understanding these effects will enable us to utilize beneficial rhizosphere bacteria for disease biological management in a more informed manner. While extensive research has been conducted on the interactions between plants and diseases, there is limited knowledge about how plants interact with non-pathogenic microorganisms.

According to one theory, highly vulnerable plants have a more diverse and abundant microbial community compared to more resilient ones. This phenomenon can be triggered in various ways due to the complex association between microorganisms and plants. As mentioned earlier, exudates can attract specific bacteria, leading to the development of a rhizosphere microflora that promotes plant growth.

There are numerous ways to generate perturbations due to the intricate interaction between microbes and plants. As previously mentioned, exudates from plant roots can attract specific microbes, leading to the development of a beneficial rhizosphere microflora that promotes plant growth. The chemical composition of root exudates, influenced by both biotic and abiotic factors, reflects the physiological condition of the plant.

15.6.1 ABIOTIC STRESS

Inducible plant defenses require energy to be expressed. The discovery that activating defense responses has a considerable fitness cost for Arabidopsis serves to illustrate this energy requirement. Activated plant defenses need energy to prioritize the production of defense-related chemicals, as shown by local declines in photosynthetic activity. Under stressful circumstances, it has been found that apoplastic invertases are expressed more frequently, increasing the local sink's capacity to produce hexoses that reduce stress [24]. Pathogens populating apoplasts are deprived of easily available nutrition by increased transit of carbohydrates into the cells. Similar to how alterations in the plants' energy equation may impact pathogenic and non-pathogenic microorganisms, it should be expected that secondary metabolites with antimicrobial activity generated as a result of defense response stimulation will have a more immediate effect on the unharmful microbiota.

Many secondary metabolites are secreted by SA-treated *Arabidopsis* roots. The amounts found in the exudates include butanoic acid, ferulic acid, and 3-indole propionic acid; all of them exhibit in vitro antibacterial activity against *P. syringae, Xanthomonas campestris, and Erwinia* spp. Non-pathogenic *P. fluorescens* was less susceptible to these exudates, but when field-grown *Arabidopsis* had its SA-dependent defenses activated, the range of bacteria in leaves decreased [25].

Researchers have discovered that citrus root exudates can influence the genotype and abiotic stress conditions of the rhizobacteria *P. putida KT2440* and *Novosphingobium* Sp. *HR1a.* This generally supports the notion that plants challenged by heat (40°C) or salt (60 or 90 mM NaCl) experience increased development. Additionally, both bacterial strains tend to express more *putA* and *pahA* when stressed plants produce root exudates, demonstrating their ability to distinguish proline and salicylates in these exudates [26]. This research examined the function and composition of exudates in peanut roots under water stress in the interaction between peanut and PGPR. Water shortage had an impact on the initial interactions between peanuts and PGPR, and root exudates encouraged bacterial chemotaxis, motility, and adhesion to peanut roots. Restrictive water conditions changed the pattern of molecules secreted by roots, causing them to secrete more naringenin, oleic FA, citric, and lactic acids and to stimulate the emission of terpenes with recognized anti-inflammatory and antimicrobial properties. The existence of bacterial inocula *hypogaea-Bradyrhizobium* and *A. hypogaea-Bradyrhizobium-Azospirillum* interaction, which counteracts the detrimental effects of a water-restrictive condition on peanut growth, were examined in the 7-day root exudate of peanut plants [27].

The bioavailability of heavy metals can change depending on the type of root exudate. Among other factors, low molecular weight organic acids (LMWOAs) play a crucial role in this process. Three plant species — *Poa annua, Medicago polymorpha,* and *Malva sylvestris* — have been examined for their potential application in phytoremediation, specifically in the absorption of metals and elimination of LMWOAs in contaminated areas with varying concentrations of Cadmium, Copper, and Zinc. This effective exclusion mechanism enables the growth of plants even in highly contaminated environments, as it reduces metal uptake. Generally, plants respond to increased heavy metal stress by releasing more LMWOAs [28].

15.6.2 BIOTIC STRESS

The root exudate community of bacteria can also affect the attack by bacterial, fungal, or viral pathogens. For example, *Fusarium oxysporum* causes watermelon fusarium wilt, which significantly reduces watermelon yields when continuously produced. However, intercropping aerobic rice and watermelon can effectively prevent fusarium wilt. The negative impacts of *Fusarium oxysporum,* which harm watermelon yields in continuous cultivation methods, are significantly reduced when watermelon and aerobic rice are cultivated together. Additionally, the allelopathic effects of root exudates on soil microorganisms are reportedly associated with the presence of P-coumaric acid and Ala in the root exudates of rice and watermelon, respectively [29].

Rhizobacterium Bacillus subtilis FB17 is specifically signaled and recruited in a dose-dependent manner by the tricarboxylic acid cycle intermediate L-malic acid (MA), which is released by the roots of *Arabidopsis thaliana*. Root metabolites play a controlling role in the acquisition of advantageous microorganisms, as it has been found that roots preferentially release L-MA and successfully prompt favorable *rhizobacteria.* This underscores the complexity and range of plant-microbial interactions. *Pseudomonas syringae pv* tomato (Pst DC3000), a foliar pathogen, induces root secretions of L-MA, and high levels of L-MA encourage FB17 binding and biofilm formation on Arabidopsis roots [30].

15.7 CONCLUSION

Root exudation, a crucial ecological mechanism, controls plant and microbial succession in the rhizosphere. In non-sterile soil, roots release 12-18% of fixed carbon, compared to 5-10% in sterile soil. This increase is attributed to the presence of rhizosphere bacteria. Plant-microbe interactions in the rhizosphere lead to various changes, such as carbon sequestration, ecosystem function, and nutrient cycling. The ability of plants to absorb nitrogen and other nutrients depends on the type and abundance of microorganisms in the soil. Plants can influence these changes by releasing secondary compounds that either attract or deter microorganisms in the rhizosphere. Rhizodeposition consists of small-molecular-weight metabolites, amino acids, secreted enzymes, mucilage, and cell lysate. These compounds contribute to the changes in the rhizosphere environment.

We are only beginning to comprehend the intricate relationships that exist between plant roots and their incredibly varied and constantly changing microbiota. The life of plants depends on their capacity to respond differently to harmful and beneficial microbes. While the pathogens must be swiftly and effectively stopped, the benefits must be encouraged.

KEYWORDS

- **microbes**
- **microbiome**
- **nitrogen-fixation**
- **pathogens**
- **plant growth hormones**
- **plant-microbe interactions**
- **rhizosphere**
- **root exudates**
- **siderophores**

REFERENCES

1. Hacquard, S., Spaepen, S., Garrido-Oter, R., & Schulze-Lefert, P., (2017). The interplay between innate immunity and the plant microbiota. *Annual Review of Phytopathology*, *55*, 565–589. https://doi.org/10.1146/annurev-phyto-080516-035623.
2. Nugema-Ona, E., Vicré-Gibouin, M., Cannesan, M. A., & Driouich, A., (2013). Arabinogalactan proteins in root–microbe interactions. *Trends Plant Sci.*, *18*(8), 440–449 https://doi.org/10.1016/j.tplants.2013.03.006.
3. Vicré, M., Santaella, C., Blanchet, S., Gateau, A., & Driouich, A., (2005). Root border-like cells of *Arabidopsis*. Microscopical characterization and role in the interaction with rhizobacteria. *Plant Physiology*, *138*(2), 998–1008. https://doi.org/10.1104/pp.104.051813.
4. Kumar, A. S., Lakshmanan, V., Caplan, J. L., Powell, D., Czymmek, K. J., Levia, D. F., & Bais, H. P., (2012). Rhizobacteria *Bacillus subtilis* restricts foliar pathogen entry

through stomata. *The Plant Journal: For Cell and Molecular Biology*, *72*(4), 694–706. https://doi.org/10.1111/j.1365-313X.2012.05116.x.
5. Charmont, S., Jamet, E., Pont-Lezica, R., & Canut, H., (2005). Proteomic analysis of secreted proteins from *Arabidopsis thaliana* seedlings: Improved recovery following removal of phenolic compounds. *Phytochemistry*, *66*(4), 453–461. https://doi.org/10.1016/j.phytochem.2004.12.013.
6. Walker, T. S., Bais, H. P., Grotewold, E., & Vivanco, J. M., (2003). Root exudation and rhizosphere biology. *Plant Physiology*, *132*(1), 44–51. https://doi.org/10.1104/pp.102.019661.
7. Bais, H. P., Weir, T. L., Perry, L. G., Gilroy, S., & Vivanco, J. M., (2006). The role of root exudates in rhizosphere interactions with plants and other organisms. *Annual Review of Plant Biology*, *57*, 233–266. https://doi.org/10.1146/annurev.arplant.57.032905.105159.
8. Oldroyd, G. E., & Downie, J. A., (2008). Coordinating nodule morphogenesis with rhizobial infection in legumes. *Annual Review of Plant Biology*, *59*, 519–546. https://doi.org/10.1146/annurev.arplant.59.032607.092839.
9. Steenhoudt, O., & Vanderleyden, J., (2000). *Azospirillum*, a free-living nitrogen-fixing bacterium closely associated with grasses: Genetic, biochemical, and ecological aspects. *FEMS Microbiology Reviews*, *24*, 487–506. https://doi.org/10.1111/j.1574-6976.2000.tb00552.x.
10. Cooke, T. J., Poli, D., Sztein, A. E., & Cohen, J. D., (2002). Evolutionary patterns in auxin action. *Plant Molecular Biology*, *49*(3, 4), 319–338. https://doi.org/10.1023/A:1015242627321.
11. Bernard, G., Todorovic, B., Czarny, J., & Cheng, Z., (2007). Promotion of plant growth by bacterial ACC deaminase. *Critical Reviews in Plant Sciences*, *26*, 227–242. https://doi.org/10.1080/07352680701572966.
12. Mendes, R., Kruijt, M., De Bruijn, I., Dekkers, E., Van, D. V. M., Schneider, J. H., Piceno, Y. M., et al., (2011). Deciphering the rhizosphere microbiome for disease-suppressive bacteria. *Science (New York, N.Y.)*, *332*(6033), 1097–1100. https://doi.org/10.1126/science.1203980.
13. De Weger, L. A., Van, A. J. J. C. M., Recourt, K., Lugtenberg, B., Van Der, H. G. A. J. M., & Weisbeek, P. J., (1988). Siderophores-mediated uptake of Fe^{3+} by the plant growth-stimulating *Pseudomonas putida* strain WCS358 and by other rhizosphere microorganisms. *Journal of Bacteriology*, *170*(10), 4693–4698.https://doi.org/0021-9193/88/104693-0602.00/0.
14. De Boer, M., Bom, P., Kindt, F., Keurentjes, J. J., Van, D. S. I., Van, L. L. C., & Bakker, P. A., (2003). Control of *Fusarium* wilt of radish by combining *Pseudomonas putida* strains that have different disease-suppressive mechanisms. *Phytopathology*, *93*(5), 626–632. https://doi.org/10.1094/PHYTO.2003.93.5.626.
15. Raaijmakers, J. M., Paulitz, T. C., & Steinberg, C., (2009). The rhizosphere: A playground and battlefield for soilborne pathogens and beneficial microorganisms. *Plant Soil*, *321*, 341–361s. https://doi.org/10.1007/s11104-008-9568-6.
16. Inderjit, (2001). Soil: Environmental effect on allelochemical activity. *Agron. J.,93*, 79–84.
17. Bolton, M. D., (2009). Primary metabolism and plant defense–fuel for the fire. *Molecular Plant-Microbe Interactions: MPMI*, *22*(5), 487–497. https://doi.org/10.1094/MPMI-22-5-0487.

18. Kniskern, J. M., Traw, M. B., & Bergelson, J., (2007). Salicylic acid and jasmonic acid signaling defense pathways reduce natural bacterial diversity in *Arabidopsis thaliana. Molecular Plant-Microbe Interactions: MPMI, 20*(12), 1512–1522. https://doi.org/10.1094/MPMI-20-12-1512.
19. Vives-Peris, V., Molina, L., Segura, A., Gómez-Cadenas, A., & Pérez-Clemente, R. M., (2018). Root exudates from citrus plants subjected to abiotic stress conditions have a positive effect on rhizobacteria. *Journal of Plant Physiology, 228*, 208–217. https://doi.org/10.1016/j.jplph.2018.06.003.
20. Li, X. G., Zhang, T. L., Wang, X. X., Hua, K., Zhao, L., & Han, Z. M., (2013). The composition of root exudates from two different resistant peanut cultivars and their effects on the growth of the soilborne pathogen. *International Journal of Biological Sciences, 9*(2), 164–173. https://doi.org/10.7150/ijbs.5579.
21. Montiel-Rozas, M., Madejón, E., & Madejón, P., (2016). Effect of heavy metals and organic matter on root exudates (low molecular weight organic acids) of herbaceous species: An assessment in sand and soil conditions under different levels of contamination. *Environmental Pollution, 216*, 273–281. https://doi.org/10.1016/j.envpol.2016.05.080.
22. Wen-Ya, H., Ren, L., Ran, W., & Qi-Rong, S., (2010). Allelopathic effects of root exudates from watermelon and rice plants on *Fusarium oxysporum* f. sp. Niveum. *Plant and Soil, 336*, 485–497. https://doi.org/10.1007/s11104-010-0505-0.
23. Gilbert, J. A., Quinn, R. A., Debelius, J., Xu, Z. Z., Morton, J., Garg, N., Jansson, J. K., et al., (2016). Microbiome-wide association studies link dynamic microbial consortia to disease. *Nature, 535*(7610), 94–103. https://doi.org/10.1038/nature18850.
24. Thimmaraju, R., Kirk, J. C., Paul, W. P., & Harsh, P. B., (2008). Root-secreted malic acid recruits beneficial soil bacteria. *Plant Physiology, 148*(3), 1547–1556, https://doi.org/10.1104/pp.108.127613.
25. Nardi, S., Concheri, G., Pizzeghello, D., Sturaro, A., Rella, R., & Parvoli, G., (2000). Soil organic matter mobilization by root exudates. *Chemosphere, 41*(5), 653–658. https://doi.org/10.1016/s0045-6535(99)00488-9.
26. Bais, H. P., Fall, R., & Vivanco, J. M., (2004). Biocontrol of *Bacillus subtilis* against infection of *Arabidopsis* roots by *Pseudomonas syringae* is facilitated by biofilm formation and surfactin production. *Plant Physiology, 134*(1), 307–319. https://doi.org/10.1104/pp.103.028712.
27. Mukerji, K. G., Manoharachary, C., & Chamola, B. P., (2002). *Techniques in Mycorrhizal Studies. Springer Netherlands.*
28. Tiwari, K., & Singh, S., (2011). Nature and role of root exudates: Efficacy in bioremediation. *African Journal of Biotechnology, 10*, 9717–9724. https://doi.org/10.5897/AJB10.2552.
29. Gabriele, B., & Kornelia, S., (2009). Plant species and soil type cooperatively shape the structure and function of microbial communities in the rhizosphere. *FEMS Microbiology Ecology, 68*(1), 1–13. https://doi.org/10.1111/j.1574-6941.2009.00654.x.
30. Neal, A. L., Ahmad, S., Gordon-Weeks, R., & Ton, J., (2012). Benzoxazinoids in root exudates of maize attract *Pseudomonas putida* to the rhizosphere. *PLoS One, 7*(4), e35498. https://doi.org/10.1371/journal.pone.0035498.

CHAPTER 16

Potential of Entomopathogenic Fungi in Plant Health Management

PRANJALI SINHA,[1] JITENDRA KUMAR PAL,[2] and ASHISH KUMAR[3]

[1]*Department of Plant Pathology, Indira Gandhi Krishi Vishwavidyalaya, Krishak Nagar, Raipur, Chhattisgarh, India*

[2]*Biotechnology Division, School of Science and Technology, Vanita Vishram Women's University, Athwalines, Surat, Gujarat, India*

[3]*Department of Biotechnology, Sant Gahira Guru University Sarguja, Ambikapur, Chhattisgarh, India*

ABSTRACT

Entomopathogenic fungi control the growth of insects and pathogens developed in plants. Their potential for controlling insects and arthropods is already known. In addition to disease and pest control, they act as bio-fertilizers, plant growth boosters, and rhizosphere colonizers in nature. Endophytic fungi are large, polyphyletic fungi that systematically invade their grass host crops. However, the significance of cordyceps in plant disease control, bio-fertilization, and rhizosphere colonization has received less attention. Hypocreales include *Beauveria bassiana, Lecanicillium lecanii, Acremonium, Closporum, Clonostachys, Paecilomyces,* and other naturally occurring endophytes found in bearberry, Ironwood, and coffee plant tissues, as well as artificially inoculated cotton, wheat, corn, beans, tomatoes, pumpkins, and sorghum. In terms of host specificity, entomopathogenic fungi range from a small number of obligate pathogens to many facultative pathogens. *B. bassiana* reduces lepidopteran cob and stem borer damage, as well as tomato fruit worm damage. Helicoverpa is found in tomatoes,

Microbial Biostimulants: Biorational Pesticides for Management of Plant Pathogens.
Faheem Ahmad and Rakesh Pandey (Eds.)

Cosmopolites sordidus in bananas, and so on. The *B. bassiana* strain also has an antagonistic effect, suppressing the damping-off of one of the notorious soil-borne pathogenic microbes, *Rhizoctonia solani* and *Pythium myriotylum* in tomato and cotton seedlings. Pre-treatment of their spores also reduces the intensity of bacterial blight caused by *Xanthomonas axonopodis pv. alvecearum,* showing a significant decline of Zucchi yellow mosaic virus. Antibiosis, mycoparasitism, and competition are all mediated by secondary metabolites released by diverse fungal species. The whole method combines systemic induced resistance and plant growth-promoting substances. Increasing biopesticides can minimize dependence on synthetic pesticides for a more sustainable environment. As a result, this chapter discusses the most important entomopathogenic fungi for plant health management, their varied roles, possible mechanisms, and interactions with other endophytes.

16.1 INTRODUCTION

The global dependence on agricultural products has captured the market and brought forth the requirement for improved agricultural practices, mitigating the detrimental impact of these practices on the ecosystem and its surroundings, while also ensuring the production of healthy and safe food for human consumption [76]. The global loss incurred by pests, diseases, and weeds accounts for 40% of potential production [75]. The total crop loss has remained consistent over the years in proportion to the growing population. In many affluent countries, pest and disease management tactics have shifted towards using transgenic crops that express specific features resistant to insects, fungi, herbicides, or viruses. However, this transition from chemical compounds to transgenic plants does not represent a fundamental change. It is a departure from 'like for like' as the primary tool changes, but the underlying issue remains the same [52, 79]. A significant change would involve a complete shift from chemical to sustainable techniques. Among these, the principal approach includes the use of biological control agents (BCA). Entomopathogenic fungi are an affordable tool for countries transitioning to transgenic crops to manage pests and diseases.

Fungal entomopathogenic or insect pathogens have yet to reach their full potential, particularly under field conditions [76]. The published literature hints that we have overlooked their additional roles in nature as endophytic organisms, rhizosphere colonizers, plant growth promoters (PGP), and plant disease antagonists, exclusively focusing on their development as biopesticides. These newly emerging fungal entomopathogens, effectively used in integrating pest management (IPM) strategies, will also aid in implementing these techniques in integrated disease management (IDM) [76, 49]. Numerous genera of entomopathogens have been identified as endophytic fungi in a range of crops [37], antagonists of plant pathogens [29, 38, 55], plant growth promoters [20, 41, 69], and beneficial rhizosphere colonizers [15, 35, 59, 73].

But, besides its multiple roles initially, entomopathogenic fungi were used for managing the devastating silkworm industry in early 19th century France. Agostino Bassi (1773–1856) revealed that *Beauveria bassiana* (as *Botrytis bassiana*) was the causative agent of muscardine disease in silkworms. Later, it was discovered that the fungus also affected various microorganisms [7]. However, other studies found wheat cockchafers (green muscardine) were identified as being infected by *Entomopthora anisopliae* (*Metarhizium anisopliae*). Krassilstschik [47] produced the fungus in bulk, employed in the field to combat the sugar beet weevil.

Nevertheless, the discoveries suggest that, besides its multiple roles, the initial attention was focused on biopesticides. Synthetic insecticides developed in the 1940s dominated the ability of entomopathogens, and they are still improperly appraised and employed in comparison to bio-agents, leading to the misleading belief that they are as effective as chemical pesticides. The merging of manifold functions provided by fungal entomopathogens suggests a great possibility for indirect, multidimensional, and reasonable usage in sustainable agriculture. This chapter examines the colonization of various host plants by endophytic fungal pathogens, the adverse impact of invasion and colonization on insect and plant pathogens, and the potential mechanism of defense offered by endophytic fungal pathogens.

16.2 ENTOMOPATHOGENIC FUNGUSES AS ENDOPHYTES

Endophytes are microorganisms (fungi or microbes) that invade host tissue despite causing visible signs of harm to their plant hosts. Endophytes of the genus *Neoptyphodium* (Clavicipataceae) increase and provide various benefits to their hosts [48]. However, non-clavicipaeceous endophytes dominated by Ascomycetes (Hypocreales), some of which are fungal entomopathogens, have received less attention [4]. Subsequently, the discovery of numerous classes of cordyceps as endophytes for various plant species and fungal entomopathogens as fungal endophytes has developed into an exciting new topic of study that has recently integrated a 100-year-old endophyte research background. *Beauveria bassiana* (Balsamo) Vuillemin (Ascomycota: Hypocreales), a widespread soil-borne fungus that attacks a diverse variety of invertebrates (about 700 microorganisms) [37], is among the most marketable fungal biocontrol agents and has pioneered the studies on entomophilous endophytic fungi [22].

An extensive range of hosts (both commercial and noxious species) have been shown to inhabit *B. bassiana* as fungal endophytes, including potato, cocoa, banana, tobacco, grapevine, broad bean, wheat, tomato, and others. Similarly, many other entomophagous fungus species, such as *Acremonium, Cladosporium, Clonostachys,* and *Paecilomyces,* have now been identified as pre-existing endophytes from diverse sections of the coffee plant across numerous regions. *Lecanicillium lecanii* (Zimm.) Zare & Gams and *Aspergillus parasiticus* have been experimentally established as endophytic fungi in cotton, wheat, corn, bean, tomato, and pumpkin after foliar spraying with spore suspension [33].

Invasion and colonization rates of different fungal strains of plants could be due to the interaction between the leaf surface and competitive pressure with various endophytes typically existing inside plants [41, 64, 72]. Furthermore, one of the most striking features of fungal entomopathogens is their vast host range, which varies from extremely narrow for obligatory pathogens to extremely broad for facultative pathogens [81]. Surprisingly, it has been proposed that most of these microbes have not neglected their function as plant symbiotic organisms. Instead, it is hypothesized that insect pathogenicity is a response that allows a few fungal endophytes to receive distinct nitrogen sources and other essential minerals resulting from pests, and successfully transfer such insect-driven essential minerals for significant exposure to plant monosaccharides [10]. Conversely, the "bodyguard" theory proposes that crops can influence the action of entomopathogenic fungi to boost their inhibition of plant-feeding pests and enhance plant health [25]. Entomopathogenic fungi spend a lot of their lives on the plant surface, subjecting them to plant biochemistry and surface features, ultraviolet rays, and changes in weather patterns. By harboring these fungi within crops, they may be able to manage fungal entomopathogens and adjust their potency innovatively [21]. As a result, aggressive strains competent for colonizing a large host area will be efficient as biopesticides and plant pathogenic foes of insect pests.

16.3 BIO-PESTICIDES DERIVED FROM ENTOMOPATHOGENIC FUNGI

Entomopathogenic fungi are typically recognized based on fungal development on insect remains. Despite their promising applications for preservation and conventional biocontrol strategies, extensive research on entomopathogenic fungi has been focused on improving them as effective bioagents for insects, mites, and ticks [17, 29–79]. Approximately 170 products based on at least 12 fungus species have been developed [22]. Although there are 700 evaluated species of entomophagous fungi in 90 classes [68], the most commercially available fungi are *Beauveria, Metarhizium, Lecanicillium,* and *Isaria* species, which are relatively easy to produce in large quantities. The systematic aspects of biopesticide development, such as large-scale production and formulation, as well as the selection of fast-acting isolates, have received significant attention. Affordable price, long-lasting durability, and, most importantly, consistent performance under field conditions are all essential requirements for its production. Diaspores (dissemination units) are produced by inducing aerial conidiation on a solid growth medium. Blastospores are produced by yeast-like growth in nutrient broth, and hyphal biomass is grown in liquid or solid culturing media [22]. Spray treatments of suspensions of airborne conidia comprising blastospores are used to manage insect pests in the phylloplane, such as *M. anisopliae* var. *acridum* for locust suppression in Africa [51].

Hydrophobic conidia are frequently mixed in oil or applied to spray mixes, including moistening chemicals such as adjuvants to promote dispersal. Wettability compounds can be used as adjuvants in the spray formulation of hydrophilic blastospores. However, they are usually produced as wettable powders (WPs) and water-dispersible flakes. Because solar radiation affects the propagule survival of fungi on the lamina horizon, much work has been accomplished in integrating solar blockers and sunscreens to protect these entomopathogens [37]. Rhizosphere competence of *M. anisopliae* strains is influenced by vegetation cover, not necessarily the existence of a host species [36], and *B. bassiana* races occur as endophytic fungi in various plants and have the potential to inhibit pest and plant disease [36, 56, 78]. Some of the host and colonizing pests controlled by *B. bassiana* are listed in Table 16.1.

TABLE 16.1 Plant Colonization and Biological Control by *B. bassiana*

Common Name	Species	Order	Host Plant	References
Banana weevil	*Cosmopoites sordidus*	Coleoptera: Curculionidae	Banana	[3, 4]
Stem weevil	*Apion corchori*	Coleoptera: Curculionidae	White jute	[13]
Corn borer	*Ostrinia nubilalis*	Lepidoptera: Noctuidae	Maize	[12]
Stem borer	*Sesamia calamistis*	Lepidoptera: Noctuidae	Maize	[17]
Tomato fruitworm	*Helicoverpa zea*	Lepidoptera: Noctuidae	Tomato	[64]
Poppy stem gall wasp	*Iraella luteipes*	Hymenoptera: Cynipidae	Opium poppy	[65]
Beet armyworm	*Spodoptera littoralis*	Lepidoptera: Noctuidae	Sweet potato	[66]

16.4 PLANT PATHOGENIC ANTAGONISM

A biological control refers to the use of natural or engineered fungi or bacteria that inhibit plant diseases in plant pathology. The term antagonism describes a localized process that reduces the presence or disease-causing ability of a pathogen. Various strategies of plant-pathogen antagonism have been discovered, including the synthesis of antibiotics, volatile bioactive compounds (such as NH3, HCN, alkyl pyrones, alcohols, acids, esters, ketones, and lipids), and enzymes produced as metabolites. Other methods include parasitism, hypovirulence, ISR, and enhanced plant growth sensitivity (for niche/infectious location, C, N2, or different minerals) [57]. There is evidence suggesting that specific endophytes, particularly *B. bassiana* and *Lecanicillium* spp. (previously *Verticillium lecanii*), may exhibit antagonistic behavior against plant diseases in addition to their well-known biocontrol ability against insect pests. This recommends that these entomopathogens could be produced as biopesticides for various applications in IPM methods [29, 58, 78].

In tomato [19, 59], the *Beauveria bassiana* strain was identified to prevent damping-off caused by soil-borne pathogens *Rhizoctonia solani* Kuhn (Class-Basidiomycota, Order-Cantharellales) and *Pythium myriotylum* Drechsler (Class-Oomycota, Order-Pythiales) in seedlings [31, 56]. The same *B. bassiana* strain was used to pre-treat cotton seedlings, resulting in a reduction in the incidence of bacterial blight induced by *Xanthomonas axonopodis* var. *malvacearum* (Xam) [32, 56]. Several examples of entomopathogenic fungi that induce plant diseases and mitigate severity for the host are listed in Table 16.2.

TABLE 16.2 Entomopathogenic Fungus as Plant Antagonist Reported Against Plant Pathogens

Entomopathogenic Fungus	Disease (Causal Organism)	Family–Genus	Host	References
Lecanicillium sp. DAOM 198499 and *Lecanicillium muscarium* B-2	Powdery mildew (*Sphaerotheca fuliginea*)	Leotiomycetes: Erysiphaceae	Cucumber	[34]
B. bassiana	*Zucchini yellow mosaic virus*	Potyvirus: Potyviridae	Squash	[38]
Lecanicillium spp.	Downy mildew (*Plasmopara viticola*)	Oomycota: Peronosporaceae	Grapevine	[39]
M. robertsii	Root rot of bean (*Fusarium solan*)	Ascomycota: Nectriaceae	Bean	[70]
Paecilomyces inflatus	Root-knot nematode (*Meloidogyne incognita*)	Tylenchida: Meloidogynidae	Cotton	[82]

16.5 INVADING FUNGUS IN THE RHIZOSPHERE: STIMULATING PLANT GROWTH

The soil microbiota is abundant with entomopathogenic fungi from the Hypocreales family. The genus *Beauveria, Isaria* (Cordycipitaceae), and *Metarhizium* are among the widely identified entomopathogenic species of fungi from soils (Clavicipitaceae) [55]. *Beauveria* and *Metarhizium* species attack and destroy pests in the topsoil, which results in limited somatic development in the fungus-infected carcass. Free carbon is present in the rhizosphere, and studies indicate that entomophagous fungi interacting with the root system promote development and survivability [74]. Metabolic byproducts, mucilage, shed root cells, and lysed makeup constitute 10% to 40% of the carbon ingested by a plant and transmitted to the soil [6, 9]. Various saprotrophic microbes in the rhizosphere utilize this carbon [20]. Studies on crop ectoparasites, along with their bacterial adversaries, have shown that nematode management is most effective on roots to support the rhizosphere assimilation *of Pochonia chlamydosporia*, a nematode with a facultative fungus. The extent of *P. chlamydosporia* rhizosphere colonization varies among host varieties and *P. chlamydospore* strains [15, 23]. To prevent damage caused by wireworms, conidia of *M. anisopliae* were sprayed on corn grains prior to sowing, resulting in improved stand densities and fresh and dry masses of field corn [21].

16.5.1 POSSIBLE MECHANISM INVOLVED

Possible methods range from direct plant-pathogen interactions, such as mycoparasitism, competition, and antibiosis via the release of fungal secondary metabolites, to much more sophisticated oblique contacts among endophytes, entomopathogenic fungi, and related infectious pathogens facilitated by their consensual plant. These methods include the augmentation of systemic plant resistance, the arousal of plant secondary metabolites, and the boosting of plant development. Ample scientific proof implies that endophytes and entomopathogens combat plant diseases through a unique combination of these methods, rather than relying on a single strategy [58, 77].

16.5.1.1 ANTIBIOSIS, COMPETITION, AND MYCOPARASITISM

Over mycoparasitism, competition for niches and nourishment, along with the synthesis of secondary metabolites, endophytes and entomopathogens can directly reduce plant diseases. The interaction known as mycoparasitism, which involves a fungal pathogen and a fungal recipient [11, 44], has been extensively studied, particularly in the genera *Trichoderma* [34, 75] and *Lecanicillium* [58]. The release of an antimicrobial drug into the punctured hyphae inhibits the re-synthesis of the host cell wall [7, 54], while the formation of cell lysis enzymes leads to the permeation of the cell wall, followed by the collapse of cell membranes [7, 82], and the breakdown of the host hyphal cytosol [7, 54]. In plants invaded by *B. bassiana* endophytically, competition for space and nutrients is likely to be the biological control strategy against *R. solani* [31, 59] and *P. viticola* [40]. Prior to the introduction of the disease-causing fungus (*P. viticola*) into the grapevine cultivar, the proliferation of *B. bassiana* was observed, resulting in a lower prevalence and intensity of downy mildew symptoms [40]. Endophytic fungi colonize host plants through various mechanisms, including host recognition, speculation, host invasion, and tissue infestation [62]. Once the fungus successfully colonizes parts of the plant, it occupies space and relies on the nutrients required by its host. If the endophytic fungi colonize the plant first, they likely deplete the host's resources, preventing pathogen colonization. On the other hand, if the infection occurs before the endophyte, there may be an interaction between the endophytic fungi and the pathogen, which could promote disease development [1].

Regarding the invasion and attack of the endophyte, the plant responds by producing lignin and other deposits, which serve as a physical defense shield. This shield helps prevent or limit attacks by phytopathogens. Antibiosis has been identified as a potential process through which fungal entomo-pathogens provide defense against disease phytopathogens and pest infestation by synthesizing secondary metabolites [58]. Entomopathogenic fungi are rich in secondary metabolites with antibacterial, antiparasitic, and apoptotic properties [28]. For instance, *B. bassiana* produces various secondary bioactive molecules such as beauvericin, bassianin, beauverolides, brassinolide, and more. *B. bassiana* strain 11-98 [52], an epiphytic isolate known for its ability to prevent damping-off induced by *Rhizoctonia solani* and *Pythium myriotylum* in tomato and cotton, produces beauvericin during broth culture [19, 31, 56, 59]. Nevertheless, this fungal secondary product has not yet been discovered in quantities sufficient to inhibit the corresponding fungal plant pathogen in plants infected by *B. bassiana*. oosporein, and bassianolone [58].

16.5.1.2 INDUCED SYSTEMIC RESISTANCE (ISR)

Proven to be promoted by advantageous microflora, induced systemic resistance (ISR) has been considered a powerful method for preparing the entire host for increased defense against a range of crop diseases and pests [63]. The decrease of clinical signs in plant sections remote from the active site of the inducing reagent is proof of ISR. Cotton saplings infested with *B. bassiana* isolate 11–98 via root drenching and then foliage challenged with *Xanthomonas*, a bacterial blight pathogen, two weeks after acquisition, showed this. Compared to untreated control plants, root inoculation with *B. bassiana* led to a decreased disease risk index for bacterial blight on the foliage of the host. As a consequence of pre-inoculation of cucumber roots with *Lecanicillium* spp., ISR has indeed been detected against *P. ultimum* [7] and powdery mildew *S. fuliginea* [35]. *Trichoderma* is the bioagent that has been investigated the most in terms of ISR mechanism, and the numerous strains of this organism that have demonstrated ISR are given in Table 16.3.

16.5.1.3 PROMOTING PLANT GROWTH

Plant development augmentation may help phytic entomopathogenic fungi safeguard their host plants from disease infections. Fungal endophyte colonization aids in plant health by suppressing various biotic and abiotic factors, especially plant pathogens [49]. When a *Zucchini yellow mosaic virus* (ZYMV) infested squash plant was subjected to *B. bassiana,* it showed lower disease prevalence and mortality and exhibited more vigorous and rapid development compared to the untreated control crop. Plants inoculated by *M. robertsii* and introduced to *Fusarium solani* showed

TABLE 16.3 Current Shards of Evidence of ISR Induction by *Trichoderma* Species

Species	Host	Pathogen	Effects	Potency	References
T. asperellum	*Cucumis sativus*	*Pseudomonas syringae pvlachrymans*	Increase in jasmonic acid	Reduce bacterial units by 50%.	[72]
T. harzianum strain T39	*Vitis vinifera*	*Plasmopara viticola*	Salicilic acid dependent.	Leaf treatment decreases the severity.	[60]
T. virens	*Zea mays*	*Colletotrichum graminicola*	Induction of jasmonic acid.	The decreased infection in leaves from endophytic plants.	[23]

better performance with reduced disease parameters compared to crops not infested by *M. robertsii* [71]. Plant growth enhancement observed in fungal entomopathogen-infected crops could be mainly attributed to the synthesis of phytohormones or siderophores (low molecular mass iron-chelating complexes produced by microbes). For instance, *M. anisopliae* has been found to induce the synthesis of phytohormones in infested soybean plants, while *M. robertsii* and *B. bassiana* are known to synthesize siderophores in iron-deficient environments [45, 47]. Furthermore, crop yield improvement by endophytic entomopathogens may be the result of increased nutrient availability, including iron, as evidenced by hosts infested with *B. bassiana* [69], or the ability to transfer insect-driven nitrogen (N) back to plants, as observed in *B. bassiana* and several *Metarhizium* species [12].

16.6 CONCLUSION

Fungal endophytes are a distinct and domain-specific category of pathogenic microbes with several attractive characteristics that favor their progress as biopesticides [50]. Biological insecticides are a significant potential alternative to commercial pesticides and a crucial source for ecologically sound pest control, given the growing environmental problems and issues with pest resistance [29]. Similarly, a deeper understanding of the behavior and dynamics of entomopathogenic fungi would accelerate the production and adoption of commercially viable biopesticides focused on these fungi in conventional farming techniques [29, 50, 77]. Furthermore, to promote the widespread use of microbial entomopathogenic-derived biological pesticides, solutions with efficacy against multiple insects, improved delivery mechanisms, and enhanced stability are needed to integrate these techniques into the IPM program [29]. Utilizing endophytic entomopathogenic fungi as seed treatments early in crop growth circumvents many fundamental issues typically encountered when using fungal entomopathogens as conventional biological control agents.

These problems also include sensitivity to harmful environmental situations, such as harmful ultraviolet rays, humid conditions, or severe precipitation, interaction with specific other regulatory techniques, or difficulty aligning the bioagent with pests. Besides the remarkable two-fold biocontrol capability against insects and plant pathogens, endophytic fungi protect by combating hidden insects, such as borers, that might be invisible and hard to manage via topical application [43]. These may bring other advantages, such as faster seed sprouting and enhanced plant growth. Also, studies show that endophytic fungi can subsequently work synergistically well along with several biological control organisms, such as parasitoids [5, 27]. They could be employed as a supplementary resource in IPM. The challenges that need to be overcome include precision in plant colonization and eventual endophytic defense. Also, the endophytic pathogen's efficacy needs to be targeted "on-field" as the external environmental conditions remain dynamic. However, "*in vitro*" studies give a promising result as the internal conditions remain sterile in control conditions which provide limited access to plant health management.

KEYWORDS

- **antibiosis**
- **entomopathogenic fungi**
- **hypocreales**
- **induced systemic resistance**
- **mycoparasitism**
- **plant health management**
- **secondary metabolite**
- **systemic induced resistance**

REFERENCES

1. Adame-Alvarez, R. M., Mendiola-Soto, J., & Heil, M., (2014). Order of arrival shifts endophyte-pathogen interactions in the bean from resistance induction to disease facilitation. *FEMS Microbiol. Lett., 355*, 100–107.
2. Akello, J., Dubois, T., Coyne, D., & Kyamanywa, S., (2008a). Effect of endophytic *Beauveria bassiana* on populations of the banana weevil, *Cosmopolites sordidus*, and their damage in tissue-cultured banana plants. *Entomol. Exp. Appl., 129*, 157–165.

3. Akello, J., Dubois, T., Coyne, D., & Kyamanywa, S., (2008). Endophytic *Beauveria bassiana* in banana (*Musa* spp.) reduces banana weevil (*Cosmopolites sordidus*) fitness and damage. *Crop Prot., 27*(11), 1437–1441.
4. Akutse, K. S., Fiaboe, K. K. M., Van, D. B. J., Ekesi, S., & Maniania, N. K., (2014). Effects of endophyte colonization of *Vicia faba* (*Fabaceae*) plants on the life-history of leaf miner parasitoids *Phaedrotoma scabriventris* (*Hymenoptera: Braconidae*) and *Diglyphus isaea* (*Hymenoptera: Eulophidae*). *PLoS One, 9*(10), e109965.
5. Andrews, J. H., & Harris, R. F., (2000). The ecology and biogeography of microorganisms on plant surfaces. *Annual Review of Phytopathology, 38,* 145–180.
6. Askary, H., Benhamou, N., & Brodeur, J., (1997). Ultrastructural and cytochemical investigations of the antagonistic effect of *Verticillium lecanii* on cucumber powdery mildew. *Phytopathology, 87*, 359–368.
7. Audoin, V., (1837). New experiments on the nature of the contagious disease which attacks silkworms and called muscardine. *Annals of Science Naturelles, 8*, 257–270.
8. Bardgett, R. D., (2005). *The Biology of Soil: A Community and Ecosystem Approach.* Oxford University Press,New York.
9. Barelli, L., Moonjely, S., Behie, S. W., & Bidochka, M. J., (2016). Fungi with multifunctional lifestyles: Endophytic insect pathogenic fungi. *Plant Mol. Biol., 90*, 657–664.
10. Barnett, H. L., (1963). The nature of mycoparasitism by fungi. *Annu. Rev. Microbiol., 17*, 1–44.
11. Behie, S. W., & Bidochka, M. J., (2014). An additional branch of the soil nitrogen cycle: Ubiquity of insect-derived nitrogen transfer to plants by endophytic insect pathogenic fungi. *Appl. Environ. Microbiol., 80*, 1553–1560.
12. Bing, L. A., & Lewis, L. C., (1991). Suppression of *Ostrinia nubilalis* (Hübner) (*Lepidoptera*: *Pyralidae*) by endophytic *Beauveria bassiana* (Balsamo) vuillemin. *Environ. Entomol., 20*, 1207–1211.
13. Biswas, C., Dey, P., Satpathy, S., Satya, P., & Mahapatra, B. S., (2013). Endophytic colonization of white jute (*Corchorus capsularis*) plants by different *Beauveria bassiana* strains for managing stem weevil (*Apioncorchori*). *Phytoparasitica, 41*, 17–21.
14. Bourne, J. M., Kerry, B. R., & De Leij, F. A. A. M., (1996). The importance of the host plant in the interaction between root-knot nematodes (*Meloidogyne* spp.) and the nematophagous fungus *Verticillium chlamydosporium* Goddard. *Biocontrol Science and Technology, 6*, 539–548.
15. Bruck, D. J., (2009). Fungal entomopathogens in the rhizosphere. In: *The Ecology of Fungal Entomopathogens* (pp. 103–112). Springer, Dordrecht.
16. Butt, T. M., Jackson, C., & Magan, N., (2001). *Fungi as Biocontrol Agents: Progress, Problems, and Potential.* CABI Publishing, Wallingford, United Kingdom.
17. Cherry, A. J., Banito, A., Djegui, D., & Lomer, C., (2004). Suppression of the stem-borer *Sesamia calamistis* (*Lepidoptera*; *Noctuidae*) in maize following seed dressing, topical application, and stem injection with African isolates of *Beauveria bassiana. Int. J. Pest. Manage., 50*, 67–73.
18. Clark, M. M., Gwinn, K. D., & Ownley, B. H., (2006). Biological control of *Pythium myriotylum*. *Phytopathology, 96*, S25.
19. Cooke, R. C., & Whipps, J. M., (1993). *Ecophysiology of Fungi.* Wiley-Blackwell, Boston.

20. Cory, J. S., & Ericsson, J. D., (2010). Fungal entomopathogens in a tritrophic context. *Biocontrol, 55*, 129–145.
21. De Faria, M. R., & Wraight, S. P., (2007). Mycoinsecticides and mycoacaricides: A comprehensive list with worldwide coverage and international classification of formulation types. *Biol. Control., 43*, 237–256.
22. De Leij, F. A. A. M., & Kerry, B. R., (1991). The nematophagous fungus, *Verticillium chlamydosporium* Goddard, as a potential biological control agent for *Meloidogyne arenaria* (Neal) Chitwood. *Revue de Ne'matologie., 14*, 157–164.
23. Djonovic, S., Vargas, W. A., Kolomiets, M. V., Horndeski, M., Wiest, A., & Kenerley, C. M., (2007). A proteinaceous elicitor Sm1 from the beneficial fungus *Trichoderma virens* is required for induced systemic resistance in maize. *Plant Physiol., 145*, 875–889.
24. Elliot, S. L., Sabelis, M. W., Janssen, A., Van Der, G. L. P. S., Beerling, E. A. M., & Fransen, J., (2000). Can plants use entomopathogens as bodyguards? *Ecol. Lett., 3*, 228–235.
25. Garia, J. E., Posadas, J. B., Perticari, A., & Lecuona, R. E., (2011). *Metarhizium anisopliae* (Metschnikoff) Sorokin promotes growth and has endophytic activity in tomato plants. *Adv. Biol. Res., 5*, 22–27.
26. Gathage, J. W., Lagat, Z. O., Fiaboe, K. K. M., Akutse, K. S., Ekesi, S., & Maniania, N. K., (2016). Prospects of fungal endophytes in the control of *Liriomyza* leaf miner flies in common bean *Phaseolus vulgaris* under field conditions. *Biocontrol., 61*, 741–753.
27. Gibson, D. M., Donzelli, B. G., Krasnoff, S. B., & Keyhani, N. O., (2014). Discovering the secondary metabolite potential encoded within entomopathogenic fungi. *Nat. Prod. Rep., 31*, 1287–1305.
28. Goettel, M. S., Eilenberg, J., & Glare, T. R., (2005). Entomopathogenic fungi and their role in regulation of insect populations. In: Gilbert, L., Iatrou, K., & Gill, S., (eds.), *Comprehensive Molecular Insect Science* (Vol. 6,pp. 361–406). Elsevier, Boston.
29. Goettel, M. S., Koike, M., Kim, J. J., Aiuchi, D., Shinya, R., & Brodeur, J., (2008). Potential of *Lecanicillium* spp. for management of insects, nematodes. and plant diseases. *Journal of Invertebrate Pathology, 98*(3), 256–261.
30. Griffin, M. R., & Beauveria, B. A., (2007). *Cotton Endophyte with Biocontrol Activity Against Seedling Disease*. Ph.D. Dissertation. The University of Tennessee, Knoxville, TN, USA.
31. Griffin, M. R., Ownley, B. H., Klingeman, W. E., & Pereira, R. M., (2006). Evidence of induced systemic resistance with *Beauveria bassiana* against *Xanthomonas* in cotton. *Phytopathology, 96*, S42.
32. Gurulingappa, P., Sword, G. A., Murdoch, G., & McGee, P. A., (2010). Colonization of crop plants by fungal entomopathogens and their effects on two insect pests when in planta. *Biol. Control, 55*, 34–41.
33. Harmon, G. E., Howell, C. R., Viterbo, A., Chet, I., & Lorito, M., (2004). *Trichoderma* species–opportunistic, avirulent plant symbionts. *Nat. Rev. Microbiol., 2*, 43–56.
34. Hirano, E., Koike, M., Aiuchi, D., & Tani, M., (2008). Pre-inoculation of cucumber roots with *Verticillium lecanii* (*Lecanicillium muscarium*) induces resistance to powdery mildew. *Res. Bull. Obihiro Univ., 29*, 82–94.
35. Hu, G., & St. Leger, R. J., (2002). Field studies using recombinant mycoinsecticide (*Metarhizium anisopliae*) reveal that it is rhizosphere competent. *Applied and Environmental Microbiology, 68*, 6383–6387.

36. Inglis, G. D., Goettel, M. S., Butt, T. M., & Strasser, H., (2001). Use of hyphomycetous fungi for managing insect pests. In: Butt, T. M., Jackson, C. W., & Magan, N., (eds.), *Fungi as Biocontrol Agents: Progress, Problems and Potential* (pp. 23–69). *CABI International/* AAFC, Wallingford, United Kingdom.
37. Jaber, L. R., & Ownley, B. H., (2018). Can we use entomopathogenic fungi as endophytes for dual biological control of insect pests and plant pathogens? *Biological Control, 116*, 36–45.
38. Jaber, L. R., & Salem, N. M., (2014). Endophytic colonization of squash by the fungal entomopathogen *Beauveria bassiana* (*Ascomycota: Hypocreales*) for managing zucchini yellow mosaic virus in cucurbits. *Biocontrol Science and Technology*, *24*(10), 1096–1109.
39. Jaber, L. R., (2015). Grapevine leaf tissue colonization by the fungal entomopathogen *Beauveria bassianas*. L. and its effect against downy mildew. *Biocontrol.*, *60*, 103–112.
40. Jaber, L. R., & Enkerli, J., (2016). Effect of seed treatment duration on growth and colonization of *Vicia Faba* by endophytic *Beauveria bassiana* and *Metarhizium brunneum*. *Biol. Control*, *103*, 187–195.
41. Jaber, L. R., & Enkerli, J., (2017). Fungal entomopathogens as endophytes: Can they promote plant growth? *Biocontrol Sci. Technol.*, *27*, 28–41.
42. Jaronski, S. T., (2010). Ecological factors in the inundative use of fungal entomopathogens. *Biocontrol.*, *55*, 159–185.
43. Jeffries, P., (1995). Biology and ecology of mycoparasitism. *Can. J. Bot.*, *73*, 1284–1290.
44. Jirakkakul, J., Cheevadhanarak, S., Punya, J., Chutrakul, C., Senachak, J., Buajarern, T., Tanticharoen, M., & Amnuaykanjanasin, A., (2015). Tenellin acts as an iron chelator to prevent iron-generated reactive oxygen species toxicity in the entomopathogenic fungus *Beauveria bassiana*. *FEMS Microbiol. Lett.*, *362*, 1–8.
45. Kerry, B. R., (2000). Rhizosphere interactions and the exploitation of microbial agents for the biological control of plantparasitic nematodes. *Annual Review of Phytopathology*, *38*, 423–441.
46. Krasnoff, S., Keresztes, I., Donzelli, B., & Gibson, D., (2014). Metachelins, mannosylated and N-oxidized coprogen-type metachelins, mannosylated and N-oxidized coprogen-type. *J. Nat. Prod.*, *77*, 1685–1692.
47. Krassilstschik, J., (1888). La production industrielle des parasites ve´ge´taux pour la destruction des insectesnuisibles. *Bulletin Biologique de la France et de la Belgique*, *19*, 461–472.
48. Kuldau, G., & Bacon, C., (2008). Clavicipitaceous endophytes: Their ability to enhance resistance of grasses to multiple stresses. *Biol. Control*, *46*, 57–71.
49. Lacey, L. A., Grzywacz, D., Shapiro-Ilan, D. I., Frutos, R., Brownbridge, M., & Goettel, M. S., (2015). Insect pathogens as biological control agents: Back to the future. *Journal of Invertebrate Pathology, 132*, 1–41.
50. Langewald, J., & Kooyman, C., (2007). Green muscle, a fungal biopesticide for control of grasshoppers and locusts in Africa. In: Vincent, C., Goettel, M. S., & Lazarovits, G., (eds.), *Biological Control: A Global Perspective* (pp. 311–318). *CABI International/* AAFC, Wallingford, United Kingdom.
51. Leckie, B. M., Ownley, B. H., Pereira, R. M., Klingeman, W. E., Jones, C. J., & Gwinn, K. D., (2008). Mycelia and spent fermentation broth of *Beauveria bassiana* incorporated into synthetic diets affect mortality, growth, and development of larval *Helicoverpazea* (*Lepidoptera*: *Noctuidae*). *Biocontrol Science and Technology, 18*(7), 697–710.

52. Lewis, W. J., Van, L. J. C., Phatak, S. C., & TumlinsonIii, J. H., (1997). A total system approach to sustainable pest management. *Proceedings of the National Academy of Sciences, 94*(23), 12243–12248.
53. Lorito, M., Farkas, V., Rebuffat, S., Bodo, B., & Kubicek, C. P., (1996). Cell wall synthesis is a major target of mycoparasitic antagonism by *Trichoderma harzianum. J. Bacteriol., 178*, 6382–6385.
54. Meyling, N., & Eilenberg, J., (2007). Ecology of the entomopathogenic fungi *Beauveria bassiana* and *Metarhizium anisopliae* in temperate agroecosystems: Potential for conservation biological control. *Biological Control, 43*, 145–155.
55. Ownley, B. H., Griffin, M. R., Klingeman, W. E., Gwinn, K. D., Moulton, J. K., & Pereira, R. M., (2008b). *Beauveria bassiana*: Endophytic colonization and plant disease control. *Journal of Invertebrate Pathology, 98*, 267–270.
56. Ownley, B. H., & Windham, M., (2007). Biological control of plant pathogens. In: Trigiano, R. N., Windham, M. T., & Windham, A. S., (eds.), *Plant Pathology: Concepts and LaboratoryExercises* (2nd edn., pp. 423–436). CRC Press, Boca Raton, Florida.
57. Ownley, B., Gwinn, K. D., & Vega, F. E., (2010). Endophytic fungal entomopathogens with activity against plant pathogens: Ecology and evolution. *Biocontrol., 55*, 113–128.
58. Ownley, B. H., Pereira, R. M., Klingeman, W. E., Quigley, N. B., & Leckie, B. M., (2004). *Beauveria bassiana*, a dual-purpose biocontrol organism, with activity against insect pests and plant pathogens. In: Lartey, R. T., & Caesar, A., (eds.), *Emerging Concepts in Plant Health Management* (pp. 256–269). Research Signpost, Kerala.
59. Pava-Ripoll, M., Angelini, C., Fang, W., Wang, S., Posada, F. J., & St Leger, R., (2011). The rhizosphere-competent entomopathogen *Metarhizium anisopliae* expresses a specific subset of genes in plant root exudate. *Microbiology,157*, 47–55.
60. Perazzolli, M., Dagostin, S., Ferrar, A., Elad, Y., & Pertot, I., (2008). Induction of systemic resistance against *Plasmopara viticola* in grapevine by *Trichoderma harzianum* T39 and benzothiadiazole. *Biol Control., 47*, 228–234.
61. Petrini, O., (1991). Fungal endophytes in tree leaves. In: Andrews, J. H., & Hirano, S. S.,(eds.), *Microbial Ecology of Leaves* (pp. 179–197). Springer-Verlag, New York.
62. Pieterse, C. M., Zamioudis, C., Berendsen, R. L., Weller, D. M., Van Wees, S. C., & Bakker, P. A., (2014). Induced systemic resistance by beneficial microbes. *Annu. Rev. Phytopathol., 52*, 347–375.
63. Posada, F., Aime, M. C., Peterson, S. W., Rehner, S. A., & Vega, F. E., (2007). Inoculation of coffee plants with the fungal entomopathogen *Beauveria bassiana* (*Ascomycota*: *Hypocreales*). *Mycol. Res., 111*, 749–758.
64. Powell, W. A., Klingeman, W. E., Ownley, B. H., Gwinn, K. D., Dee, M., & Flanagan, P. C., (2007). Endophytic *Beauveria bassiana* in tomatoes yields mycosis in tomato fruitworm larvae. *Hort Science, 42*, 933.
65. Quesada-Moraga, E., Muñoz-Ledesma, F., & Santiago-Alvarez, C., (2009). Systemic protection of *Papaver somniferum* L. against *Iraella luteipes* (*Hymenoptera: Cynipidae*) by an endophytic strain of *Beauveria bassiana* (*Ascomycota: Hypocreales*). *Environ. Entomol., 38*, 723–730.
66. Resquín-Romero, G., Garrido-Jurado, D. C., Ríos-Moreno, A., & Quesada-Moraga, E., (2016). Transient endophytic colonizations of plants improve the outcome of foliar applications of mycoinsecticides against chewing insects. *J. Invertebr. Pathol., 136*, 23–31.

67. Roberts, D. W., & Humber, R. A., (1981). Entomogenous fungi. In: Cole, G. T., & Kendrick, B., (eds.), *Biology of Conidial Fungi* (pp. 201–236). Academic Press, New York.
68. Sánchez-Rodríguez, A. R., Del Campillo, M. C., & Quesada-Moraga, E., (2015). *Beauveria bassiana*: An entomopathogenic fungus alleviates Fe chlorosis symptoms in plants grown on calcareous substrates. *Sci. Hortic.*, *197*, 193–202.
69. Sasan, R., & Bidochka, M. J., (2012). The insect pathogenic fungus *Metarhizium robertsii* (*Clavicipitaceae*) is also an endophyte that stimulates plant root development. *American Journal of Botany, 99*(1), 101–107.
70. Sasan, R. K., & Bidochka, M. J., (2013). Antagonism of the endophytic insect pathogenic fungus *Metarhizium robertsii* against the bean plant pathogen *Fusarium solani* f. sp. phaseoli. *Can. J. Plant Pathol.*, *35*, 288–293.
71. Schulz, B., Haas, S., Junker, C., Andree, N., & Schobert, M., (2015). Fungal endophytes are involved in multiple balanced antagonisms. *Curr. Sci.*, *109*, 39–45.
72. Segarra, G., Casanova, E., Bellido, D., Odena, M. A., Oliveira, E., & Trillas, I., (2007). Proteome, salicylic acid, and jasmonic acid changes in cucumber plants inoculated with *Trichoderma asperellum* strain T34. *Proteomics*, *7*(21), 3943–3952.
73. St. Leger, R. J., (2008). Studies on adaptations of *Metarhizium anisopliae* to life in the soil. *J. Invertebr. Pathol.*, *98*, 271–276.
74. Steyaert, J. M., Ridgway, H. J., Elad, Y., & Stewart, A., (2003). Genetic basis of mycoparasitism: A mechanism of biological control by species of *Trichoderma. NZ. J. Crop Horticult.*, *31*, 281–291.
75. Thacker, J. R., (2002). *Introduction to Arthropod Pest Control*. Cambridge University Press.
76. Vega, F. E., Goettel, M. S., Blackwell, M., Chandler, D., Jackson, M. A., Keller, S., Koike, M., et al., (2009). Fungal entomopathogens: New insights on their ecology. *Fungal Ecol.*, *2*, 149–159.
77. Vega, F. E., (2008). Insect pathology and fungal endophytes. *Journal of Invertebrate Pathology*, *98*, 277–279.
78. Vincent, C., Goettel, M. S., & Lazarovits, G., (2007). *Biological Control: A Global Perspective* (pp. 353–360). CAB International/AAFC, Wallingford, United Kingdom.
79. Welsh, R., Hubbell, B., Ervin, D. E., & Jahn, M., (2002). GM crops and the pesticide paradigm. *Nature Biotechnology, 20*(6), 548, 549.
80. Wraight, S. P., Inglis, G. D., & Goettel, M. S., (2007). Fungi. In: Lacey, L. A., & Kaya, H. K., (eds.), *Field Manual of Techniques in Invertebrate Pathology* (2nd edn., pp. 223–248). The Netherlands, Dordrecht.
81. Zeilinger, S., Galhaup, C., Payer, K., Woo, S. L., Mach, R. L., Fekete, C., Lorito, M., & Kubicek, C. P., (1999). Chitinase gene expression during mycoparasitic interaction of *Trichodermaharzianum* with its host. *Fungal Genet. Biol.*, *26*, 131–140.
82. Zhou, W., Wheeler, T. A., Starr, J. L., Valencia, C. U., & Sword, G. A., (2016). A fungal endophyte defensive symbiosis affects plant-nematode interactions in cotton. *Plant Soil*. http://dx.doi.org/10.1007/s11104-016-3147-z.

Index

B

F

N

O

P